STUDY GUIDE
&
SOLUTIONS MANUAL

Paula Yurkanis Bruice

UNIVERSITY OF CALIFORNIA - SANTA BARBARA

ESSENTIAL ORGANIC CHEMISTRY

Second Edition

Prentice Hall

New York Boston San Francisco
London Toronto Sydney Tokyo Singapore Madrid
Mexico City Munich Paris Cape Town Hong Kong Montreal

Assistant Editor: Jessica Neumann
Acquisitions Editor: Dawn Giovanniello
Editor in Chief, Chemistry and Geosciences: Nicole Folchetti
Marketing Manager: Scott Dustan
Managing Editor, Chemistry and Geosciences: Gina M. Cheselka
Project Manager, Science: Shari Toron
Full Service Vendor: Elm Street Publishing
Composition: Integra
Operations Specialist: Amanda A. Smith
Supplement Cover Manager: Paul Gourhan
Supplement Cover Designer: Tina Krivoshein
Cover Photo Credits: *Biwa Inc./Photonica/Getty Images*

Pearson Prentice Hall
Pearson Education, Inc.
Upper Saddle River, NJ 07458

Printed in the United States of America

10 9 8 7 6 5 4 3 2

ISBN-13: 978-0-321-59258-3
ISBN-10: 0-321-59258-1

Prentice Hall
is an imprint of

PEARSON

www.pearsonhighered.com

to my students, who are my teachers

CONTENTS

CHAPTER 1

Electronic Structure and Covalent Bonding

1. **The mass number** = the number of protons + the number of neutrons

   The atomic number = the number of protons.

   All isotopes of a specific element have the same atomic number; the atomic number of oxygen is 8.

   Therefore:
   The isotope of oxygen with a mass number of **16** has **8** protons and **8** neutrons.
   The isotope of oxygen with a mass number of **17** has **8** protons and **9** neutrons.
   The isotope of oxygen with a mass number of **18** has **8** protons and **10** neutrons.

2. **a.** Carbon has 4 valence electrons. **c.** Oxygen has 6 valence electrons.

   b. Nitrogen has 5 valence electrons. **d.** Fluorine has 7 valence electrons.

3. Atoms in the same column of the periodic table have the same number of valence electrons. Potassium (K) is in the same column as lithium and sodium. Therefore potassium, like lithium and sodium, has one valence electron.

4. **a.** Nitrogen has 2 core electrons and 5 valence electrons.
   Phosphorus has 10 core electrons and 5 valence electrons.

   b. Oxygen has 2 core electrons and 6 valence electrons.
   Sulfur has 10 core electrons and 6 valence electrons.

5. We have determined (in Problem 2) that fluorine has 7 valence electrons. Fluorine, chlorine, bromine, and iodine are all in the same column of the periodic table. Therefore, chlorine, bromine, and iodine each have 7 valence electrons.

6. The polarity of a bond can be determined by the difference in the electronegativities (given in the text in Table 1.2 on page 9) of the atoms sharing the bonding electrons.

   For example in part a, the difference in electronegativity between H and C is 0.4 (2.5 – 2.1 = 0.4), whereas the difference in electronegativity between Cl and C is 0.5 (3.0 – 2.5 = 0.5). Therefore, because Cl and C have the greater difference in electronegativity, the bond between them is more polar.

   a. Cl — CH_3 **c.** H — F

   b. H — OH **d.** Cl — CH_3

7.   As stated in the previous problem, the polarity of a bond can be determined by the difference in the electronegativities of the atoms sharing the bonding electrons.

   a. KCl has the most polar bond; the difference in electronegativity between the bonding atoms is 0.22 ($3.0 - 0.8 = 2.2$), whereas it is 1.8 for LiBr, 1.6 for NaI, and 0 for Cl_2.

   b. Cl_2 has the least polar bond because the two chlorine atoms share the bonding electrons equally.

8.   To answer this question, compare the electronegativities of the two atoms sharing the bonding electrons using Table 1.2. (Notice that if the atoms being compared are in the same horizontal row of the periodic chart, the atom on the right is the more electronegative; if the atoms being compared are in the same column, the one closer to the top of the column is the more electronegative.)

$$\textbf{a.} \;\; \overset{\delta-}{HO}-\overset{\delta+}{H} \qquad \textbf{b.} \;\; \overset{\delta+}{H_3C}-\overset{\delta-}{NH_2} \qquad \textbf{c.} \;\; \overset{\delta-}{HO}-\overset{\delta+}{Br} \qquad \textbf{d.} \;\; \overset{\delta+}{I}-\overset{\delta-}{Cl}$$

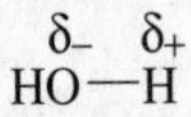

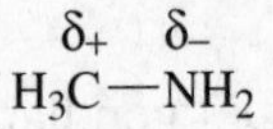

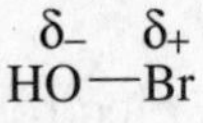

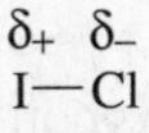

9.   **a.** LiH and HF are polar (they have a red end and a blue end).

   b. Because the hydrogen of HF is blue, we know that HF has the most positively charged hydrogen.

10.   By answering this question you will see that a formal charge is a bookkeeping device. It does *not necessarily* tell you which atom has the greatest electron density or is the most electron deficient.

   a. oxygen   **c.** oxygen

   b. oxygen (it is more red)   **d.** hydrogen (it is the most blue)

   Notice that in hydroxide ion, the atom with the formal negative charge **is** the atom with the greater electron density. In the hydronium ion, however, the atom with the formal positive charge **is not** the most electron deficient atom.

11. a. $CH_3-\overset{+}{\underset{\underset{H}{|}}{\ddot{O}}}-CH_3$ b. $H-\overset{\underset{H}{|}}{\ddot{\overset{..-}{C}}}-H$ c. $CH_3-\overset{\overset{CH_3}{+|}}{\underset{\underset{CH_3}{|}}{N}}-CH_3$ d. $H-\overset{\overset{H\;H}{+|\;\;|}}{\underset{\underset{H\;H}{|\;\;|}}{N-B}}-H$

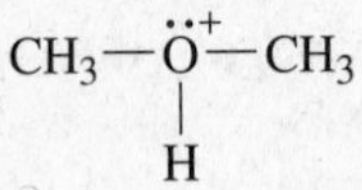

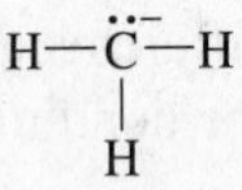

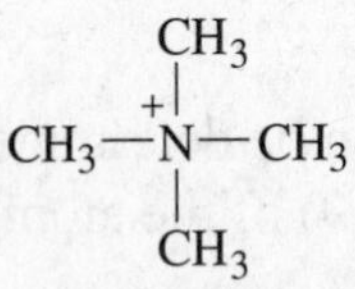

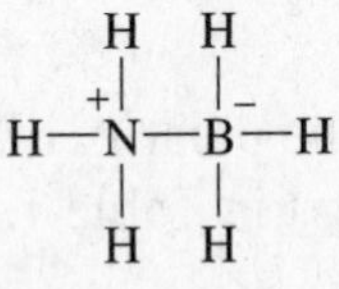

12. a. $\underset{\ddot{H}\;\ddot{H}}{\overset{H\;H+}{H:\ddot{C}:\ddot{N}:H}}$ b. $\underset{\ddot{H}\;\ddot{H}}{\overset{H\;..}{H:\ddot{C}:\ddot{C}:H}}$ c. $Na^+\; :\ddot{\ddot{O}}:H$ d. $\underset{H}{\overset{H+}{H:\ddot{N}:H}}\; :\ddot{\ddot{Cl}}:^-$ 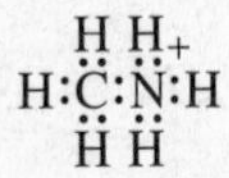

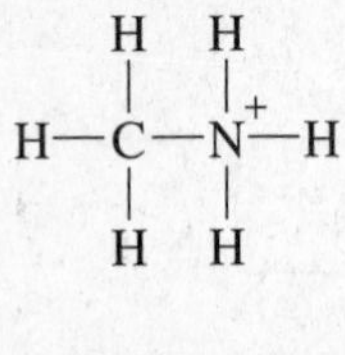

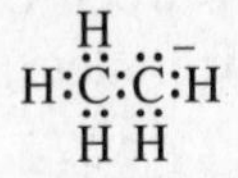

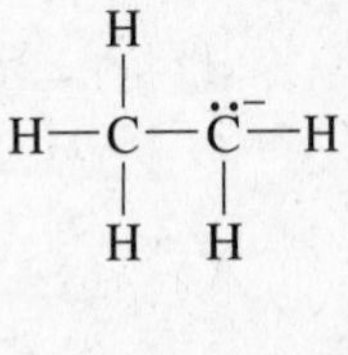

or

$H-\overset{\overset{H\;\;H}{|\;\;\;|}}{\underset{\underset{H\;\;H}{|\;\;\;|}}{C-N}}-H$ $H-\overset{\overset{H}{|}}{\underset{\underset{H\;\;H}{|\;\;\;|}}{C-\ddot{C}}}-H$ $Na^+\; :\ddot{\ddot{O}}-H$ $H-\overset{\overset{H}{|}}{\underset{\underset{H}{|}}{\overset{+}{N}}}-H\; :\ddot{\ddot{Cl}}:^-$

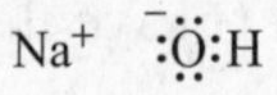

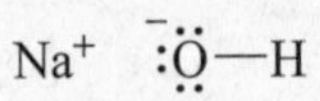

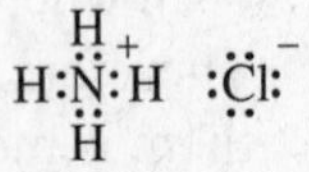

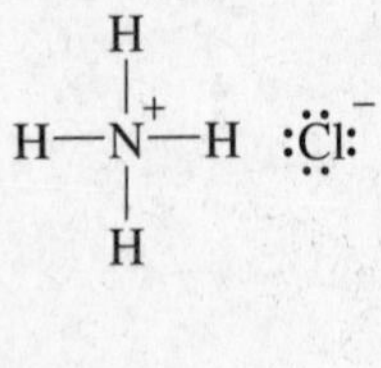

13. **a.** $CH_3CH_2\ddot{N}H_2$ **c.** $CH_3CH_2\ddot{\underset{..}{O}}H$ **e.** $CH_3CH_2\ddot{\underset{..}{Cl}}:$

b. $CH_3\ddot{N}HCH_3$ **d.** $CH_3\ddot{\underset{..}{O}}CH_3$ **f.** $H\ddot{\underset{..}{O}}\ddot{N}H_2$

14. **a.** $CH_3CH_2CH_2Cl$

c. 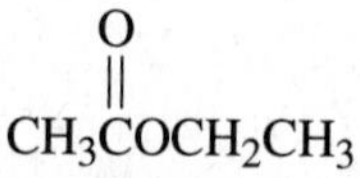

b. $CH_3CH_2C{\equiv}N$

d. 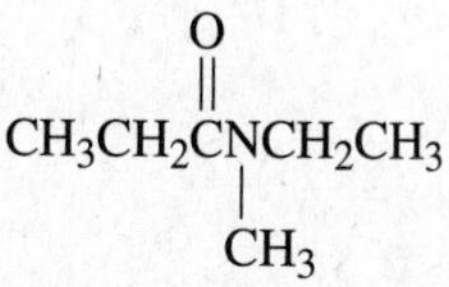

15. **a.** Cl **b.** O **c.** N **d.** C and H

16. C_2H_6 is possible: CH_3CH_3 has that molecular formula.
C_2H_7 is not possible
C_3H_9 is not possible
C_3H_8 is possible: $CH_3CH_2CH_3$ has that molecular formula.
C_4H_{10} is possible: $CH_3CH_2CH_2CH_3$ has that molecular formula.

17. **a.**

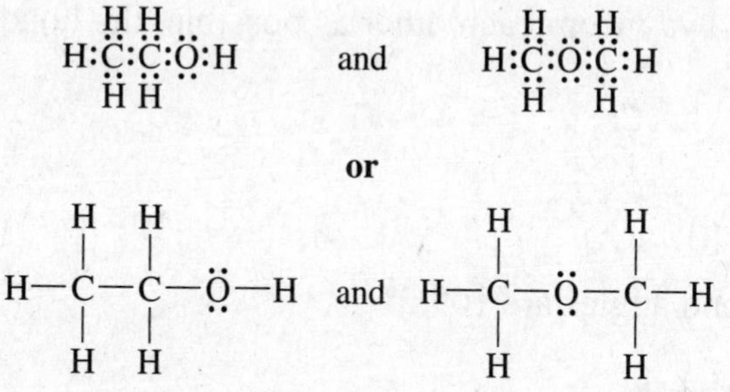

b.

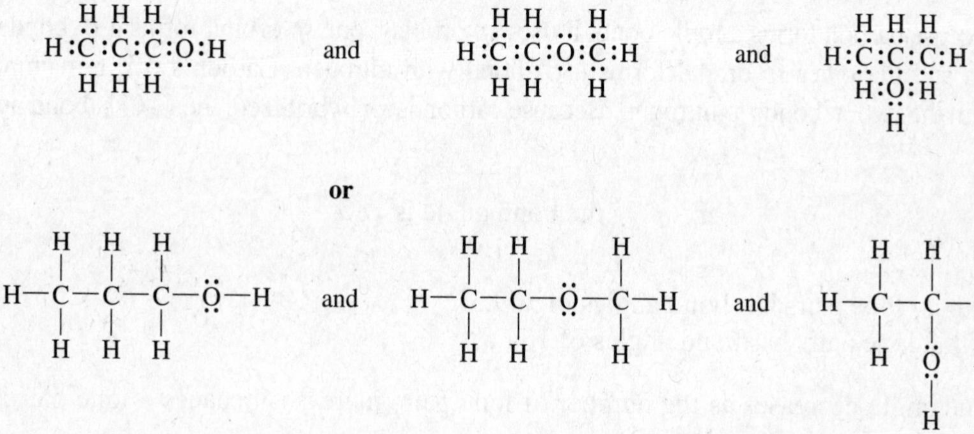

18. **a.** 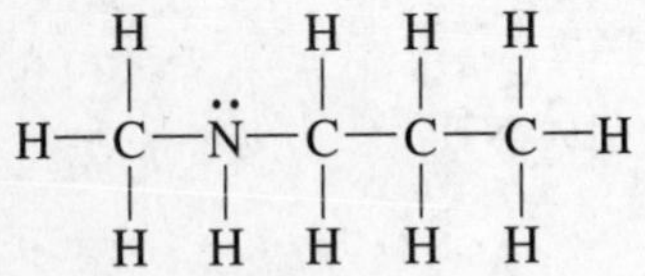 **c.** 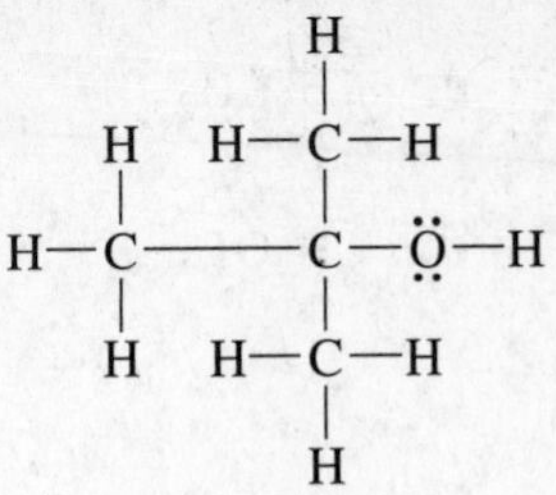

b. 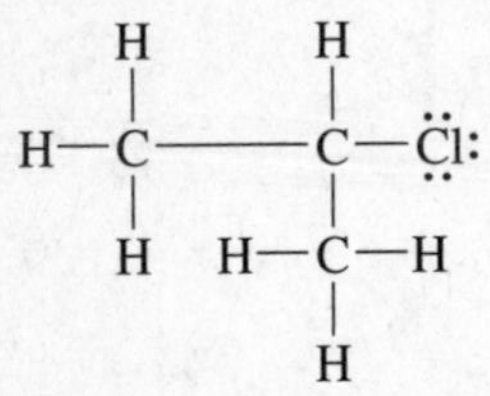 **d.** 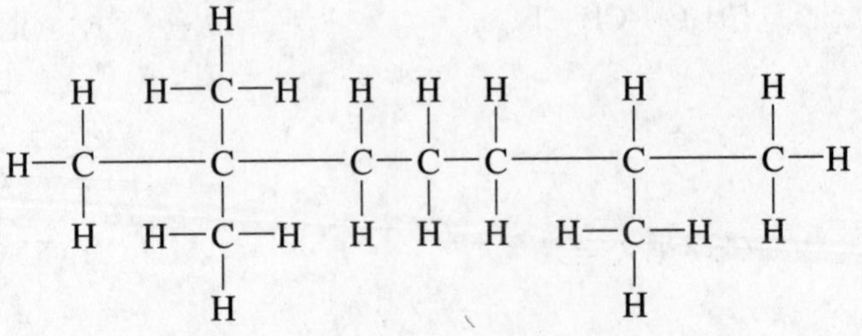

19. The two carbon-carbon bonds form as a result of sp^3— sp^3 overlap.
Each of the eight carbon-hydrogen bonds form as a result of sp^3—s overlap.

20. 1a and 1b are solved in the text.

2a.

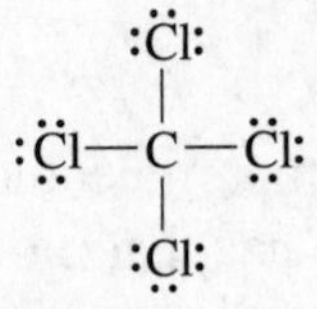

2b. Because carbon forms only single bonds, it uses sp^3 orbitals. Because the four sp^3 orbitals of carbon orient themselves to get as far away from each other as possible, the bond angles are all **109.5°**.

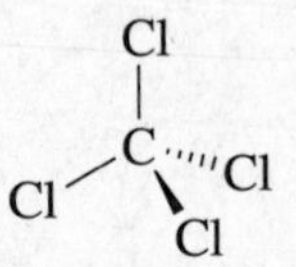

all the bond angles are 109.5°

3a. H—C≡N:

3b. Because carbon forms a triple bond, it uses *sp* orbitals; one *sp* orbital forms a σ bond with hydrogen and the other *sp* orbital forms a σ bond with nitrogen. Carbon's two remaining *p* orbitals form the two π bonds to nitrogen. Because carbon is *sp* hybridized, the H-C-N bond angle is **180°**.

H—C≡N
the bond angle is 180°

21. CH_4 with no lone pairs has bond angles of 109.5°.
H_2O with 2 lone pairs has bond angles of 104.5°.

The bond angle decreases as the number of lone pairs increases because a lone pair is more diffuse than a bonding pair.
Therefore, H_3O^+ with 1 lone pair has bond angles in between those of methane and water; its bond angles will be greater than 104.5° and less than 109.5°.

22. The hydrogens of the ammonium ion are the bluest atoms. Therefore, they have the least electron density. In other words, they have the most positive (least negative) electrostatic potential.

23. Water is the most polar—it has a deep red area and the most intense blue area.
Methane is the least polar—it is all the same color with no red or blue areas.

24. We have seen that the carbon atom in the methyl carbanion is sp^3 hybridized. Because carbon has one lone pair, its bond angle should be about 107.3°, the same as the observed bond angle in NH_3, another molecule with an sp^3 hybridized central atom and one lone pair.

25. Bonding electrons in shells farther from the nucleus form **longer** bonds; they also form **weaker** bonds due to poorer overlap of the bonding orbitals. Therefore:

a. **relative lengths** of the bonds in the halogens are: $\mathbf{Br_2 > Cl_2}$
relative strengths of the bonds are: $\mathbf{Cl_2 > Br_2}$

b. **relative lengths**: HBr > HCl > HF
relative strengths: HF > HCl > HBr

26. Notice that the shorter bond is the stronger bond

a.	1. C—Br	2. C—C	3. H—Cl
b.	1. C—Cl	2. C—H	3. H—H

27. We know the σ bond is stronger than a π bond because the σ bond in ethane has a bond dissociation energy of 90 kcal/mol, whereas the bond dissociation energy of the double bond ($\sigma + \pi$) in ethene is 174 kcal/mol, which is less than twice as strong. Because the σ bond is stronger, we know that it has more effective orbital-orbital overlap.

28. **a.**

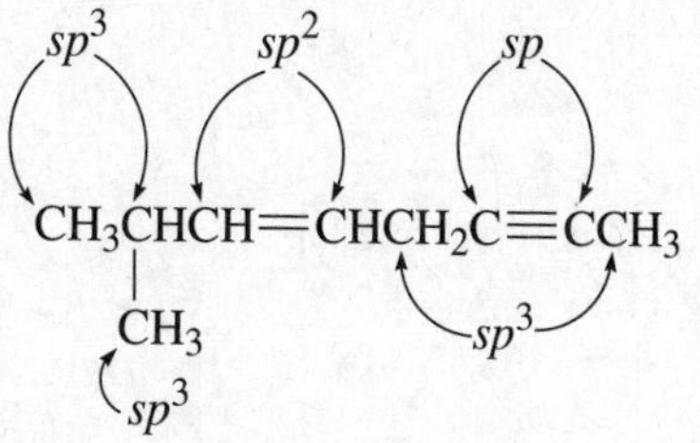
CH3CHCH=CHCH2C≡CCH3
CH3
sp3
sp2
sp

b.

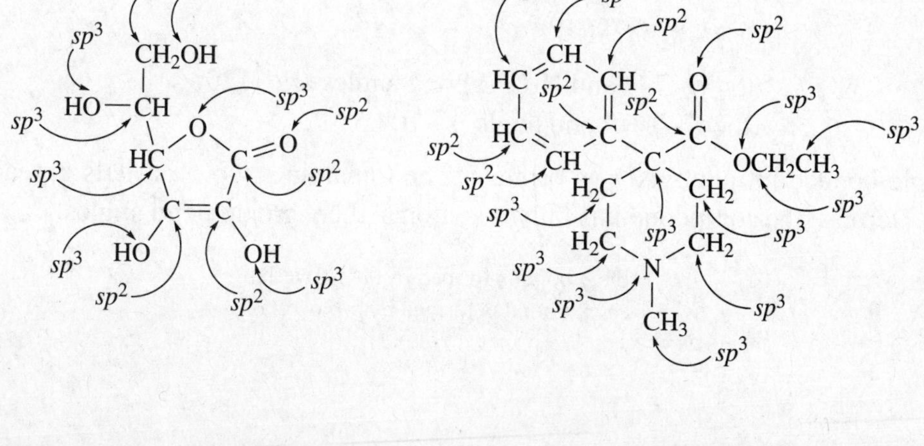
CH2OH
HO—CH
HC
O
C=O
C=C
HO
OH
CH
HC
CH
HC
C
CH
C
O
C
OCH2CH3
H2C
CH2
H2C
CH2
N
CH3
sp3
sp2

29. **a.** 109.5° **b.** 107.3° **c.** 109.5° **d.** 104.5°*

*104.5° is the correct prediction based on the bond angle in water.
However, the bond angle is actually somewhat larger (111.7°) because the bond opens up to minimize the interaction between the electron clouds of the relatively bulky CH_3 groups.

30. **a.** The single-bonded carbon and the single-bonded oxygen use sp^3 orbitals.
Each hydrogen uses an *s* orbital.

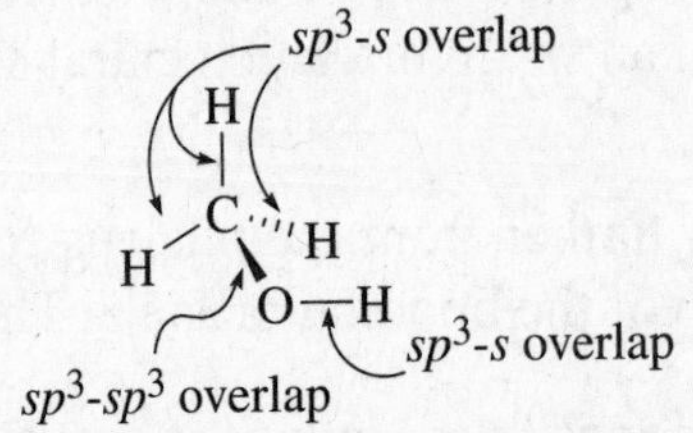
sp3-s overlap
sp3-s overlap
sp3-sp3 overlap

the H-C-H and H-C-O bond angles are ~109.5°
the C-O-H bond angle is ~104.5°

b. The single-bonded nitrogen and the single-bonded oxygen use sp^3 orbitals.
Each hydrogen uses an *s* orbital.

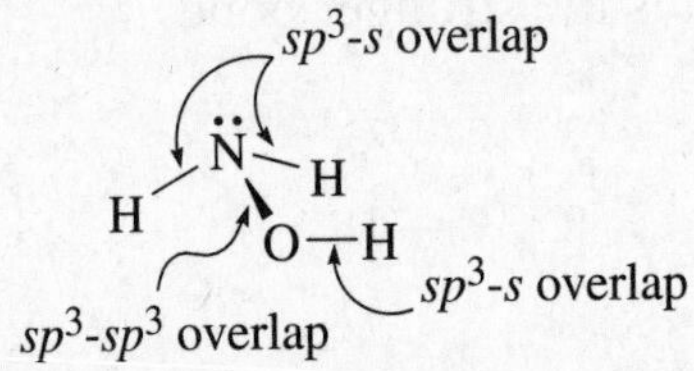
sp3-s overlap
sp3-s overlap
sp3-sp3 overlap

the H-N-H and H-N-O bond angles are ~107.3°
the N-O-H bond angle is ~104.5°

c. The double-bonded carbon and the double-bonded oxygen use sp^2 orbitals. The remaining *p* orbital on C and the remaining *p* orbital on O form a π bond.
The single-bonded oxygen uses sp^3 orbitals.
Each hydrogen uses an *s* orbital.

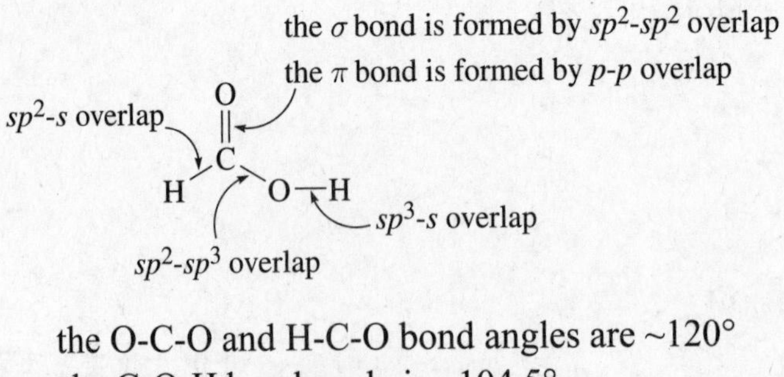
the σ bond is formed by sp2-sp2 overlap
the π bond is formed by p-p overlap
sp2-s overlap
sp3-s overlap
sp2-sp3 overlap

the O-C-O and H-C-O bond angles are ~120°
the C-O-H bond angle is ~104.5°

d. Each triple-bonded nitrogen uses *sp* orbitals. The remaining two *p* orbitals on each N form π bonds. Because the compound has only two atoms, there are no bond angles.

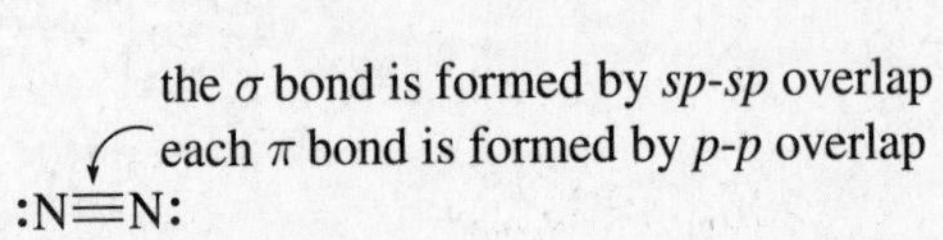
the σ bond is formed by sp-sp overlap
each π bond is formed by p-p overlap
:N≡N:

31.

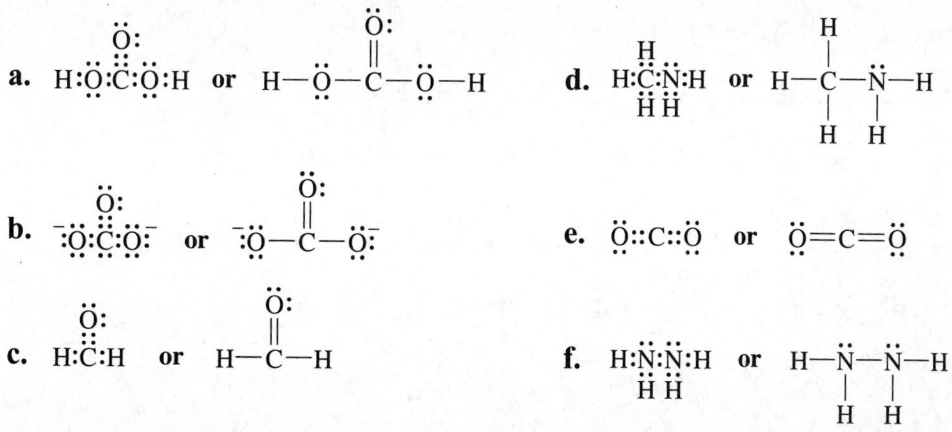

32. **a.** 4 **b.** 5 **c.** 8 **d.** 2

33.
a. sp^3, 107.3° **c.** sp^3, 107.3° **e.** sp^2, 120° **g.** sp, 180°
b. sp^3, 109.5° **d.** sp^3, 109.5° **f.** sp^2, 120° **h.** sp^3, 107.3°

34. If you need help, see the answer to Problem 8.

a. $\overset{\delta-}{F}-\overset{\delta+}{Br}$ **b.** $\overset{\delta+}{H_3C}-\overset{\delta-}{Cl}$ **c.** $\overset{\delta-}{H_3C}-\overset{\delta+}{MgBr}$ **d.** $\overset{\delta+}{H_2N}-\overset{\delta-}{OH}$

35. **a.** $CH_3CH_2CH_3$ **b.** $CH_3CH{=}CH_2$ **c.** $CH_3C{\equiv}CCH_3$ or $CH_3CH_2C{\equiv}CH$

36. **a.** 109.5° **b.** 104.5° **c.** 120° **d.** 180°

37. formal charge = the number of valence electrons – (the number of lone-pair electrons + 1/2 the number of bonding electrons)

a. formal charge = 6 – (6 + 1) = 6 – 7 = – 1
H:Ö:⁻

b. formal charge = 6 – (5 + 1) = 6 – 6 = 0
H:Ö·

c. formal charge = 5 – (4 + 2) = 5 – 6 = – 1
H—N̈⁻—H

d. formal charge = 4 – (2 + 2) = 4 – 4 = 0
H—C̈—H

38.

	$1s$	$2s$	$2p_x$	$2p_y$	$2p_z$	$3s$	$3p_x$	$3p_y$	$3p_z$	$4s$	
a. Ca	⇅	⇅	⇅	⇅	⇅	⇅	⇅	⇅	⇅	⇅	$1s^2\ 2s^2\ 2p^6\ 3s^2\ 3p^6\ 4s^2$
b. Ca^{2+}	⇅	⇅	⇅	⇅	⇅	⇅	⇅	⇅	⇅		$1s^2\ 2s^2\ 2p^6\ 3s^2\ 3p^6$
c. Ar	⇅	⇅	⇅	⇅	⇅	⇅	⇅	⇅	⇅		$1s^2\ 2s^2\ 2p^6\ 3s^2\ 3p^6$
d. Mg^{2+}	⇅	⇅	⇅	⇅	⇅						$1s^2\ 2s^2\ 2p^6$

39.   "**d**"   is the only one written correctly

   a. $CH_3CH_2CH_3$ **c.** $(CH_3)_3CCH_3$ **e.** $CH_3CH_2CH_3$

   b. CH_4 **d.** $(CH_3)_2CHCH_2CH_3$ **f.** $CH_3CH_2CH_2CH_3$

40.   **a.** C — F > C — O > C — N   **c.** H — O > H — N > H — C

   b. C — Cl > C — Br > C — I   **d.** C — N > C — H > C — C

41.   **a.**

$$\begin{array}{c} H\ \ \ O \\ |\ \ \ \| \\ H-C-C-H \\ | \\ H \end{array}$$

   b.

$$\begin{array}{c} H\ \ \ \ \ \ H \\ |\ \ \ \ \ \ | \\ H-C-O-C-H \\ |\ \ \ \ \ \ | \\ H\ \ \ \ \ \ H \end{array}$$

   c.

$$\begin{array}{c} H\ \ \ O \\ |\ \ \ \| \\ H-C-C-O-H \\ | \\ H \end{array}$$

   d.

$$\begin{array}{c} H \\ | \\ H\ \ \ \ \ \ O\ \ \ \ \ \ H \\ |\ \ \ \ \ \ |\ \ \ \ \ \ | \\ H-C\ \ \ \ \ C\ \ \ \ \ C-H \\ |\ \ \ \ H-C-H\ \ \ | \\ H\ \ \ \ \ \ |\ \ \ \ \ \ H \\ H \end{array}$$

   e.

$$\begin{array}{c} H\ \ \ H\ \ \ H \\ |\ \ \ |\ \ \ | \\ H-C-C-C-C\equiv N \\ |\ \ \ |\ \ \ | \\ H\ \ \ O\ \ \ H \\ | \\ H \end{array}$$

   f.

$$\begin{array}{c} H\ \ \ \ \ \ \ \ H\ \ H-C-H\ \ H\ \ H-C-H\ \ H \\ |\ \ \ \ \ \ \ \ |\ \ |\ \ \ \ \ \ \ \ \ \ |\ \ |\ \ \ \ \ \ \ \ \ \ | \\ H-C-C-C-C-C-C-H \\ |\ \ \ \ \ \ \ \ |\ \ \ \ \ \ \ \ \ \ \ \ \ |\ \ \ \ \ \ \ \ \ \ | \\ H\ \ H-C-H\ \ H\ \ \ \ \ \ H\ \ H-C-H\ \ H \\ |\ | \\ H\ H \end{array}$$

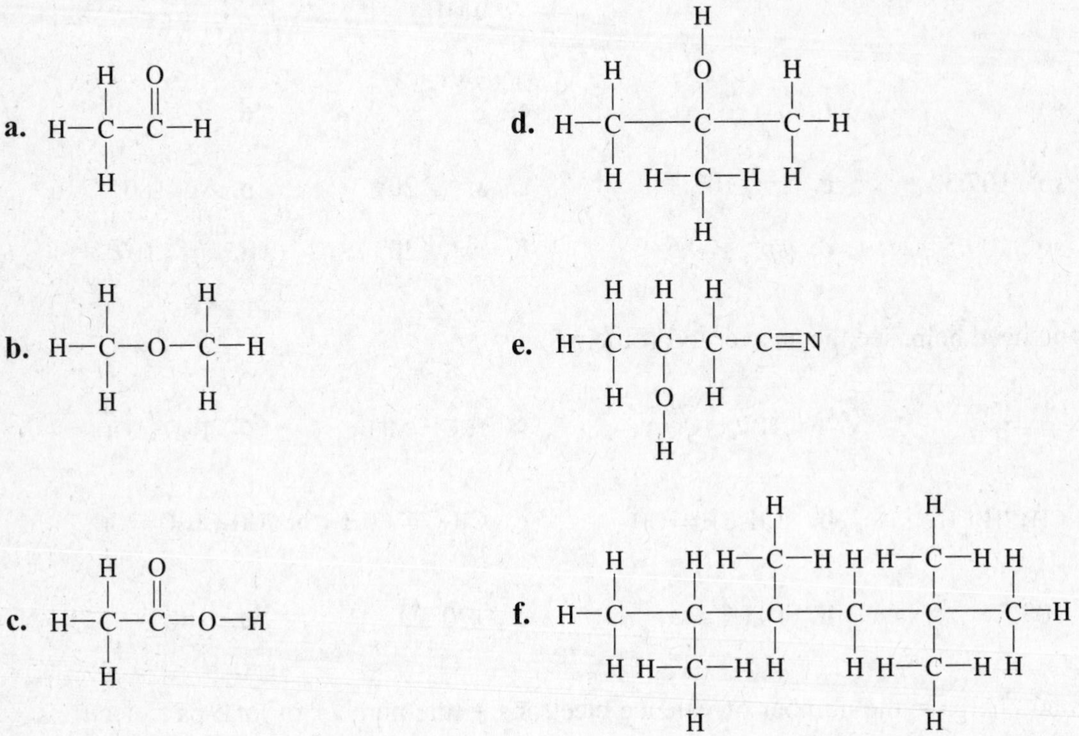

42.   Notice that none of the atoms in "**c**" has a formal charge.

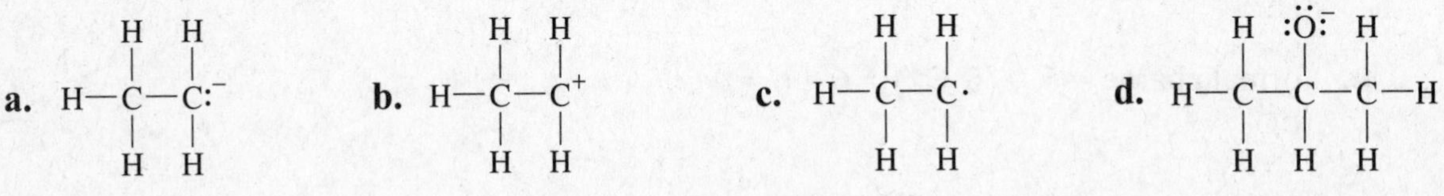

   a.

$$\begin{array}{c} H\ \ \ H \\ |\ \ \ | \\ H-C-C{:}^{-} \\ |\ \ \ | \\ H\ \ \ H \end{array}$$

   b.

$$\begin{array}{c} H\ \ \ H \\ |\ \ \ | \\ H-C-C^{+} \\ |\ \ \ | \\ H\ \ \ H \end{array}$$

   c.

$$\begin{array}{c} H\ \ \ H \\ |\ \ \ | \\ H-C-C\cdot \\ |\ \ \ | \\ H\ \ \ H \end{array}$$

   d.

$$\begin{array}{c} H\ \ \ \ddot{\overset{..}{O}}{}^{-}\ \ H \\ |\ \ \ |\ \ \ | \\ H-C-C-C-H \\ |\ \ \ |\ \ \ | \\ H\ \ \ H\ \ \ H \end{array}$$

43.   **a.**

$$\begin{array}{c} H \\ | \\ H-C-\ddot{O}-H \\ | \\ H \end{array}$$

   b.

$$\begin{array}{c} H \\ | \\ H-C-\overset{..+}{O}-H \\ | \\ H\ \ \ H \end{array}$$

   c.

$$\begin{array}{c} H \\ | \\ H-C-\ddot{\overset{..}{O}}{:}^{-} \\ | \\ H \end{array}$$

   d.

$$\begin{array}{c} H \\ | \\ H-C-\ddot{N}-H \\ |\ \ \ | \\ H\ \ \ H \end{array}$$

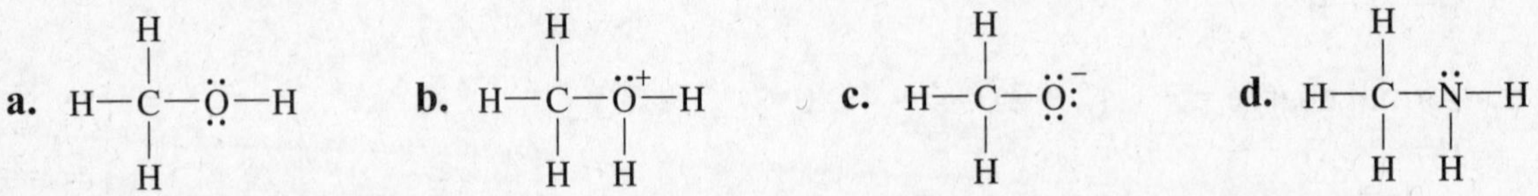

44.   The lone pair in ammonia causes the molecule to be polar; the red indicates the negative end and the blue indicates the positive end. When the lone pair is protonated to form the ammonium ion, the charge distribution in the molecule becomes symmetrical and the bond angles become larger. (The H-N-H bond angle changes from 107.3° to 109.5°.)

45.

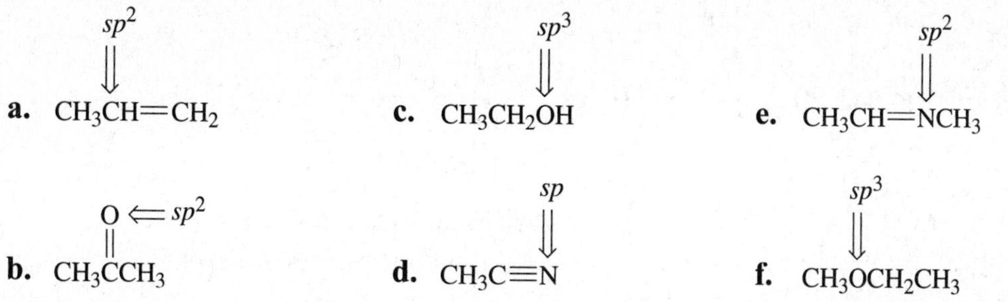

46. **a.** The open arrow points to the shorter of the two indicated bonds in each compound.

For **1**, **2**, and **3**: a triple bond is shorter than a double bond which is shorter than a single bond.

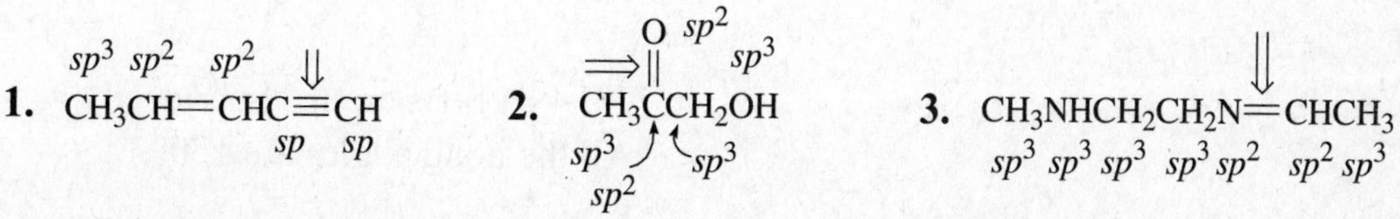

47. If the central atom is *sp*^{3} hybridized, the molecule will have tetrahedral bond angles. Therefore, the following have tetrahedral bond angles, although, because of lone pairs, some of the bond angles will be a little less than a "pure" tetrahedral bond angle of 109.5°.

H_2O H_3O^+ NH_3 $^+NH_4$ $^-CH_3$

48. In an alkene, six atoms are in the same plane: the two *sp*^{2} carbons and the two atoms that are bonded to each of the two *sp*^{2} carbons. The other atoms in the molecule will not be in the same plane with these six atoms.

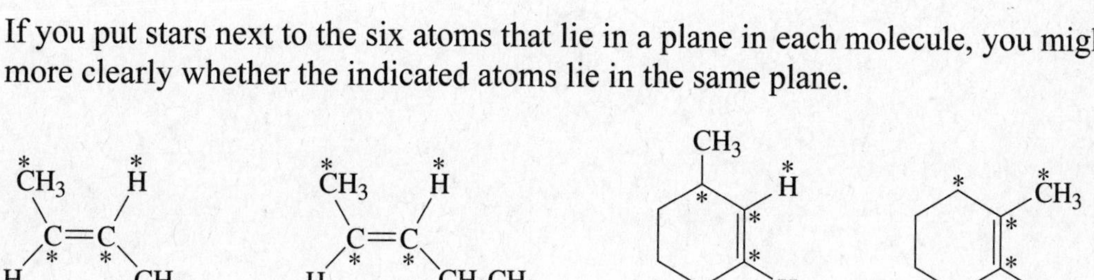

If you put stars next to the six atoms that lie in a plane in each molecule, you might be able to see more clearly whether the indicated atoms lie in the same plane.

49. **a.**

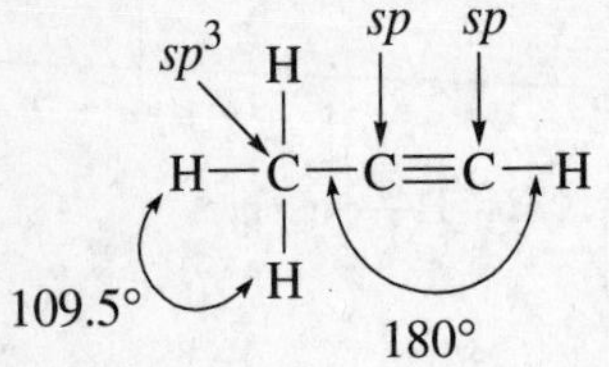

c.

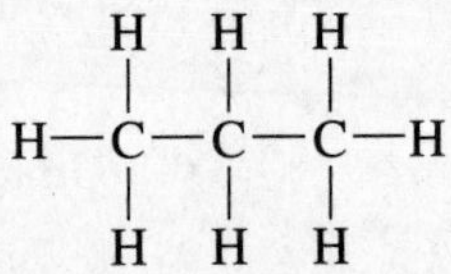

All 3 carbons are sp^3 hybridized.
All the bond angles are 109.5°.

b.

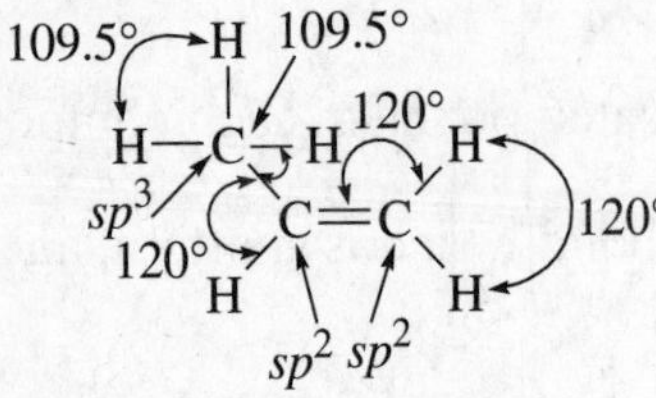

d.

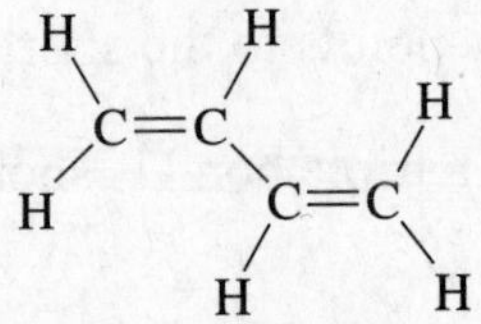

All 4 carbons are sp^2 hybridzed.
All the bond angles are 120°.

50. The bond between oxygen and sodium is ionic. All the other bonds are covalent. Thus, sodium methoxide has 4 covalent bonds.

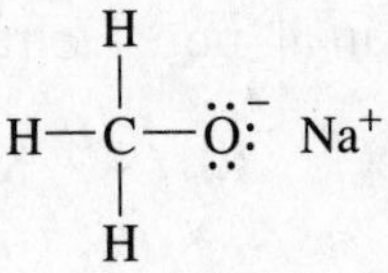

51. **a.** An H—H bond is formed by *s*—*s* overlap. A C—C bond is formed by sp^3—sp^3 overlap. Because an *s* orbital is closer to the nucleus than is an sp^3 orbital, an H—H bond is shorter than a C—C bond.

b. A C—H bond is formed by sp^3—*s* overlap. Therefore, a C—H bond will be longer than an H—H bond (O.74Å) but shorter than a C—C bond (1.54 Å).

52. Because the triple-bonded carbons are *sp* hybridized, the bond angle indicated by the arrow needs to be close to 180° if the compound is to be a stable compound; it is impossible to have a 180° bond angle in a six-membered ring. (Try to make a molecular model.)

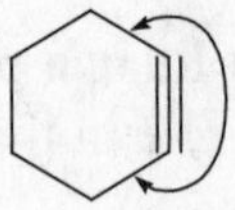

Chapter 1 Practice Test

1. What is the hybridization of the carbon atom in each of the following?

$\overset{+}{C}H_3$ $\overset{-}{C}H_3$ CH_3

2. Draw the Lewis structure for H_2CO_3.

3. Which compound has larger bond angles, H_3O^+ or $^+NH_4$?

4. Which bond is more polar?

a. C—O or C—N **b.** C—Cl or C—F

5. Give the structure of a compound that contains five carbons, two of which are sp^2 hybridized and three of which are sp^3 hybridized.

6. **a.** What orbitals do carbon's electrons occupy before promotion?

b. What orbitals do carbon's electrons occupy after promotion but prior to hybridization?

c. What orbitals do carbon's electrons occupy after hybridization?

7. Indicate whether each of the following statements is true or false:

a. A pi bond is stronger than a sigma bond. T F

b. A triple bond is shorter than a double bond. T F

c. The oxygen-hydrogen bonds in water are formed by the overlap of an sp^2 orbital of oxygen with an s orbital of hydrogen. T F

d. A double bond is stronger than a single bond. T F

e. A tetrahedral carbon has bond angles of 107.5°. T F

8. For each of the following compounds indicate the hybridization of the atom to which the arrow is pointing:

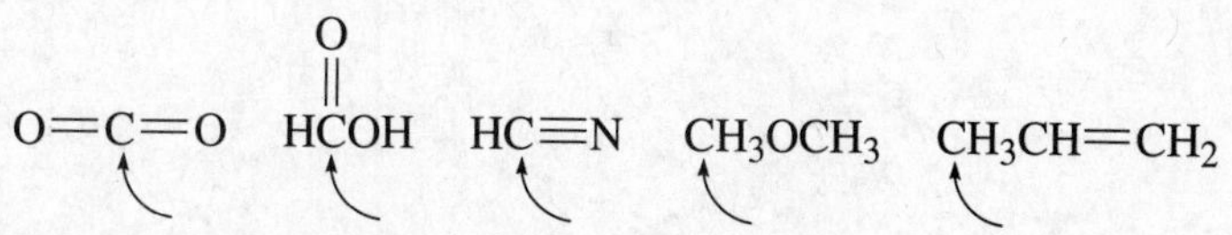
O=C=O
HCOH
HC≡N
CH3OCH3
CH3CH=CH2

ANSWERS TO ALL THE PRACTICE TESTS CAN BE FOUND AT THE END OF THE SOLUTIONS MANUAL

CHAPTER 2

Acids and Bases

1. **a.** **1.** $^{+}NH_4$ **2.** HCl **3.** H_2O **4.** H_3O^+

b. **1.** $^{-}NH_2$ **2.** Br^- **3.** NO_3^- **4.** HO^-

2. if the lone pairs are not shown:

a. CH_3OH as an acid $CH_3OH + NH_3 \rightleftharpoons CH_3O^- + \overset{+}{N}H_4$

CH_3OH as a base $CH_3OH + HCl \rightleftharpoons CH_3\overset{+}{O}H(H) + Cl^-$

b. NH_3 as an acid $NH_3 + HO^- \rightleftharpoons {}^{-}NH_2 + H_2O$

NH_3 as a base $NH_3 + HBr \rightleftharpoons \overset{+}{N}H_4 + Br^-$

if the lone pairs are shown:

a. CH_3OH as an acid $CH_3\ddot{\underset{..}{O}}H + \ddot{N}H_3 \rightleftharpoons CH_3\ddot{\underset{..}{O}}:^- + \overset{+}{N}H_4$

CH_3OH as a base $CH_3\ddot{\underset{..}{O}}H + H\ddot{\underset{..}{Cl}}: \rightleftharpoons CH_3\overset{+}{\ddot{O}}H(H) + :\ddot{\underset{..}{Cl}}:^-$

b. NH_3 as an acid $\ddot{N}H_3 + H\ddot{\underset{..}{O}}:^- \rightleftharpoons {}^{-}:\ddot{N}H_2 + H_2\ddot{O}:$

NH_3 as a base $\ddot{N}H_3 + H\ddot{\underset{..}{Br}}: \rightleftharpoons \overset{+}{N}H_4 + :\ddot{\underset{..}{Br}}:^-$

3. **a.** The lower the pK_a, the stronger the acid, so the compound with a $pK_a = 5.2$ is the stronger acid.

b. The greater the dissociation constant, the stronger the acid, so the compound with an acid dissociation constant = 3.4×10^{-3} is the stronger acid.

4. $pK_a = 4.82$; therefore, $K_a = 1.51 \times 10^{-5}$
The greater the acid dissociation constant, the stronger the acid, so butyric acid ($K_a = 1.51 \times 10^{-5}$) is a weaker acid than vitamin C ($K_a = 6.76 \times 10^{-5}$).

or

The lower the pK_a, the stronger the acid, so butyric acid ($pK_a = 4.82$) is a weaker acid than vitamin C ($pK_a = 4.17$).

5. **a.** $HO^- + HCl \longrightarrow H_2O + Cl^-$

b. $HCO_3^- + HCl \longrightarrow H_2CO_3 + Cl^- \rightleftharpoons H_2O + CO_2 + Cl^-$

c. $CO_3^{2-} + 2\,HCl \longrightarrow H_2CO_3 + 2\,Cl^- \rightleftharpoons H_2O + CO_2 + 2\,Cl^-$

6. A neutral solution has pH = 7. Solutions with pH < 7 are acidic; solutions with pH > 7 are basic.

a. basic **b.** acidic **c.** basic

7. **a.** CH_3COO^- is the stronger base.
Because HCOOH is the stronger acid, it has the weaker conjugate base.

b. $^-NH_2$ is the stronger base.
Because H_2O is the stronger acid, it has the weaker conjugate base.

c. H_2O is the stronger base.
Because $CH_3\overset{+}{O}H_2$ is the stronger acid, it has the weaker conjugate base.

8. The conjugate acids have the following relative strengths:

$$CH_3\overset{+}{O}H_2 > CH_3\overset{O}{\overset{\|}{C}}OH > CH_3\overset{+}{N}H_3 > CH_3OH > CH_3NH_2$$

The bases, therefore, have the following relative strengths:

$$CH_3\overset{-}{N}H > CH_3O^- > CH_3NH_2 > CH_3\overset{O}{\overset{\|}{C}}O^- > CH_3OH$$

9. $CH_3NH_2 + CH_3OH \rightleftharpoons CH_3\overset{+}{N}H_3 + CH_3O^-$

$pK_a = 40$ $pK_a = 15.5$

The stronger acid of the two reactants will be the acid (that is, it is the one that will donate a proton); the weaker acid will accept the proton.

10. Notice that in each case, the equilibrium goes away from the strong acid and toward the weak acid.

a. $CH_3\overset{O}{\overset{\|}{C}}OH + H_2O \rightleftharpoons CH_3\overset{O}{\overset{\|}{C}}O^- + H_3O^+$

$pK_a = 4.8$ $\qquad$ $pK_a = -1.7$

$CH_3\overset{O}{\overset{\|}{C}}OH + H_3O^+ \rightleftharpoons CH_3\overset{^+OH}{\overset{\|}{C}}OH + H_2O$

$pK_a = -1.7$ $\qquad$ $pK_a = -6.1$

$CH_3OH + HO^- \rightleftharpoons CH_3O^- + H_2O$

$pK_a = 15.5$ $\qquad$ $pK_a = 15.7$

The pK_a values are so close that there will be essentially no difference in the equilibrium arrows.

$CH_3OH + H_3O^- \rightleftharpoons CH_3\overset{+}{O}H(H) + H_2O$

$pK_a = -1.7$ $\qquad$ $pK_a = -2.5$

$CH_3NH_2 + HO^- \rightleftharpoons CH_3\overset{-}{N}H + H_2O$

$pK_a = 40$ $\qquad$ $pK_a = 15.7$

$CH_3NH_2 + H_3O^+ \rightleftharpoons CH_3\overset{+}{N}H_3 + H_2O$

$pK_a = -1.7$ $\qquad$ $pK_a = 10.7$

b. $HCl + H_2O \rightleftharpoons H_3O^+ + Cl^-$

$pK_a = -7$ $\qquad$ $pK_a = -1.7$

$NH_3 + H_2O \rightleftharpoons {}^+NH_4 + HO^-$

$pK_a = 15.7$ $\qquad$ $pK_a = 9.4$

11. **a.** HBr is the stronger acid because bromine is larger than chlorine.

b. $CH_3CH_2CH_2\overset{+}{O}H_2$ is the stronger acid because oxygen is more electronegative than nitrogen.

c. The compound on the right (an alcohol) is a stronger acid than the compound on the left (an amine) because oxygen is more electronegative than nitrogen.

12. **a.** Because HF is the weakest acid, F^- is the strongest base.

b. Because HI is the strongest acid, I^- is the weakest base.

13. **a.** oxygen **b.** H_2S **c.** CH_3SH **d.** CH_3COOH

As you saw in Problem 11, the size of an atom is more important than its electronegativity in determining stability. So even though oxygen is more electronegative than sulfur, H_2S is a stronger acid than H_2O, and CH_3SH is a stronger acid than CH_3OH. Because the sulfur atom is larger, the electrons associated with the negatively charged sulfur are spread out over a greater volume, thereby causing it to be a more stable base. The more stable the base, the stronger is its conjugate acid.

14. **a.** HO^-; if the atoms are the same, the negatively charged one is a stronger base than the neutral one.

b. NH_3; H_3O^+ is a stronger acid than $^+NH_4$ because oxygen is more electronegative than nitrogen; the stronger the acid, the weaker its conjugate base.

c. CH_3O^-; $CH_3\overset{\overset{O}{\|}}{C}OH$ is a stronger acid than CH_3OH.

d. CH_3O^-; CH_3SH is a stronger acid than CH_3OH because sulfur is larger than oxygen.

15. **a.** CH_3COO^- **c.** H_2O **e.** $^+NH_4$ **g.** NO_2^-

b. $CH_3CH_2\overset{+}{N}H_3$ **d.** Br^- **f.** $HC{\equiv}N$ **h.** NO_3^-

16. **a.**
1. neutral
2. neutral
3. charged
4. charged
5. charged
6. charged

b.
1. charged
2. charged
3. charged
4. charged
5. neutral
6. neutral

c.
1. neutral
2. neutral
3. neutral
4. neutral
5. neutral
6. neutral

17. **a.** Because the pH of the solution is greater than the pK_a value of the carboxylic acid group, the group will be in its basic form (without its proton).
Because the pH of the solution is less than the pK_a value of the ammonium group, the group will be in its acidic form (with its proton).

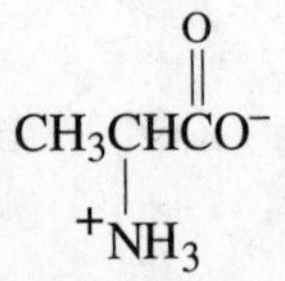

b. No, because that would require a weaker acid (the $^+NH_3$ group) to lose a proton more readily than a stronger acid (the COOH group).

18. **a.** The basic form of the buffer (CH_3COO^-) removes added H^+.

$$CH_3COO^- + H^+ \rightleftharpoons CH_3COOH$$

b. The acidic form of the buffer (CH_3COOH) removes added HO^-.

$$CH_3COOH + HO^- \rightleftharpoons CH_3COO^- + H_2O$$

Due to the rapid equilibrium, the added H^+ or HO^- readily reacts with the species (CH_3COO^- or CH_3COOH) in the solution and thereby the effect on the solution's pH is minimized.

19. **a.** 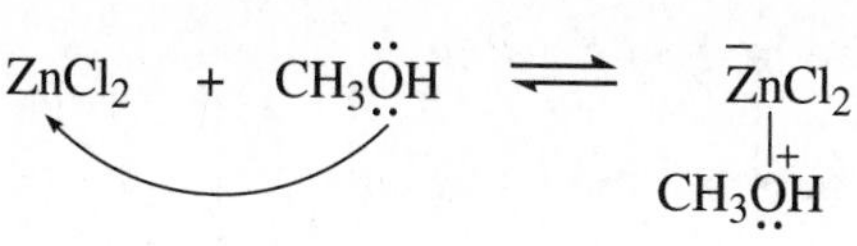

b. 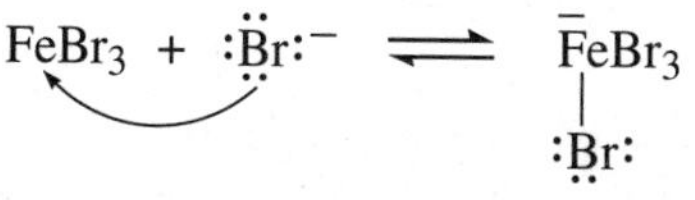

c. 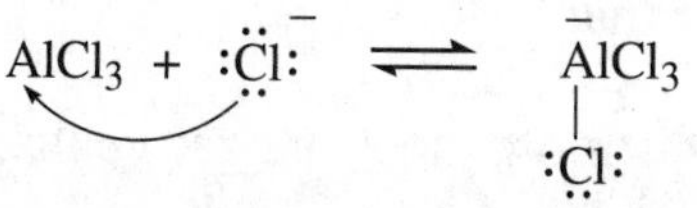

20. **a, b, c,** and **h** are Brønsted acids (protonating-donating acids). Therefore, they react with HO^- by donating a proton to it.

d, e, f, and **g** are Lewis acids. They react with HO^- by accepting a pair of electrons from it.

a. $CH_3O^- + H_2O$

b. $NH_3 + H_2O$

c. $CH_3NH_2 + H_2O$

d. $HO{-}\bar{B}F_3$

e. CH_3OH

f. $HO{-}\bar{F}eBr_3$

g. $HO{-}\bar{A}lCl_3$

h. $CH_3COO^- + H_2O$

21. If the pH of the solution is less than the pK_a of the compound, the compound will be in its acidic form (with its proton). If the pH of the solution is greater than the pK_a of the compound, the compound will be in its basic form (without its proton).

a. at pH = 3 $CH_3\overset{\overset{O}{\|}}{C}OH$

at pH = 6 $CH_3\overset{\overset{O}{\|}}{C}O^-$

at pH = 10 $CH_3\overset{\overset{O}{\|}}{C}O^-$

at pH = 14 $CH_3\overset{\overset{O}{\|}}{C}O^-$

b. at pH = 3 $CH_3CH_2\overset{+}{N}H_3$

at pH = 6 $CH_3CH_2\overset{+}{N}H_3$

at pH = 10 $CH_3CH_2\overset{+}{N}H_3$

at pH = 14 $CH_3CH_2NH_2$

c. at pH = 3 CF_3CH_2OH

at pH = 6 CF_3CH_2OH

at pH = 10 CF_3CH_2OH

at pH = 14 $CF_3CH_2O^-$

22. **a.** $CH_3\overset{\overset{O}{\|}}{C}OH + CH_3O^- \rightleftharpoons CH_3\overset{\overset{O}{\|}}{C}O^- + CH_3OH$

b. $CH_3CH_2OH + {}^-NH_2 \rightleftharpoons CH_3CH_2O^- + NH_3$

c. $CH_3\overset{\overset{O}{\|}}{C}OH + CH_3NH_2 \rightleftharpoons CH_3\overset{\overset{O}{\|}}{C}O^- + CH_3\overset{+}{N}H_3$

d. $CH_3CH_2OH + HCl \rightleftharpoons CH_3CH_2\overset{+}{O}H_2 + Cl^-$

23. **a.** Nitric acid is the strongest acid, because it has the largest K_a value.

b. Bicarbonate is the weakest acid, because it has the smallest K_a value.

c. Bicarbonate has the strongest conjugate base (CO_3^{2-}), because bicarbonate is the weakest acid.

Remember that the stronger base has the weaker conjugate acid.

24. **a.** HO^-, because H_2S is the stronger acid, since S is larger than O.

b. $CH_3\overline{N}H$, because CH_3OH is the stronger acid, since O is more electronegative than N.

c. CH_3O^-, because the oxygen is negatively charged.

d. Cl^-, because HBr is the stronger acid, since Br is larger than Cl.

25. The nitrogen in the top left-hand corner is the most basic because it has the greatest electron density (it is the most red).

26. **a.** $CH_3C{\equiv}\overset{+}{N}H$ **b.** CH_3CH_3 **c.** $F_3C\overset{\overset{O}{\|}}{C}OH$

27. As long as the pH is greater than the pK_a value of the compound, the majority of the compound will be in its basic form. Therefore, as long as the pH is greater than pH 10.4, more than 50% of the amine will be in its basic (neutral, nonprotonated) form.

28. **a.** $CH_3CH_2\underset{\underset{Cl}{|}}{C}HCOOH > CH_3\underset{\underset{Cl}{|}}{C}HCH_2COOH > ClCH_2CH_2CH_2COOH > CH_3CH_2CH_2COOH$

b. The electronegative chlorine substituent makes the carboxylic acid more acidic.

c. The closer the chlorine is to the acidic proton, the more it increases the acidity of the carboxylic acid.

d. The electronegative (electron-withdrawing) substituent makes the carboxylic acid more acidic, because it stabilizes its conjugate base by withdrawing electrons from the oxygen atom thereby decreasing its electron density.

29. Substituting an H of the CH_3 group with a more electronegative atom increases the acidity of the carboxylic acid. As the electronegativity (electron-withdrawing ability) of the subsitutent increases, the acidity of the carboxylic acid increases.

30. **a.** $CCl_3CH_2OH > CHCl_2CH_2OH > CH_2ClCH_2OH$

$5.75 \times 10^{-13} > 4.90 \times 10^{-13} > 1.29 \times 10^{-13}$

b. The greater the number of electron-withdrawing chlorine atoms, the greater will be the stability of the base, so the stronger will be its conjugate acid.

31. The equilibrium favors reaction of the stronger acid and formation of the weaker acid. Therefore, the first reaction favors formation of ethyne, and the second reaction favors formation of ammonia.

a. $HC\equiv CH + HO^- \rightleftharpoons HC\equiv C^- + H_2O$

$pK_a = 25$ $\qquad pK_a = 15.7$

b. $HC\equiv CH + {}^-NH_2 \rightleftharpoons HC\equiv C^- + NH_3$

$pK_a = 25$ $\qquad pK_a = 36$

c. $^-NH_2$ would be a better base to use to remove a proton from ethyne because it would favor formation of the desired product.

32. The reaction with the most favorable equilibrium constant is the one that has the strongest reactant acid and the weakest product acid.

a. CH_3OH is a stronger reactant acid ($pK_a = 15.5$) than CH_3CH_2OH ($pK_a = 15.9$), and both reactions form the same product acid ($^+NH_4$). Therefore, the reaction of CH_3OH with NH_3 has the more favorable equilibrium constant.

b. Both reactions have the same reactant acid (CH_3CH_2OH). The product acids are different: $^+NH_4$ is a stronger product acid ($pK_a = 9.4$) than $CH_3NH_3^+$ ($pK_a = 10.7$). Therefore, the reaction of CH_3CH_2OH with CH_3NH_2 has the more favorable equilibrium constant.

33. At a pH that is equal to the pK_a value of an acidic compound, half the compound will be in the acidic form (the form that has the proton) and half the compound will be in the basic form (the form without the proton).

At pH values less than the pK_a value, more of the compound will be in the acidic form than in the basic form.

At pH values greater than the pK_a value, more of the compound will be in the basic form than in the acidic form.

The pK_a value of carbonic acid is 6.1. Physiological pH (7.3) is greater than the pK_a value. More molecules of carbonic acid, therefore, will be present in the basic form (HCO_3^-) than in the acidic form (H_2CO_3). That means that the buffer system will be better at neutralizing excess acid.

34. From the following equilibria you can see that a carboxylic acid is neutral when it is in its acidic form (with its proton) and charged when it is in its basic form (without its proton). An amine is charged when it is in its acidic form and neutral when it is in its basic form.

$$RCOOH \rightleftharpoons RCOO^- + H^+$$

$$R\overset{+}{N}H_3 \rightleftharpoons RNH_2 + H^+$$

Charged species will dissolve in water and neutral species will dissolve in ether.

In separating compounds you want essentially all (100:1) of each compound in either its acidic form or its basic form. To obtain a 100:1 ratio of acidic form:basic form, the pH must be two pH units lower than the pK_a of the compound; and in order to obtain a 100:1 ratio of basic form:acidic form, the pH must be two pH units greater than the pK_a of the compound.

a. If both compounds are to dissolve in water, they both must be charged. Therefore, the carboxylic acid must be in its basic form, and the amine must be in its acidic form. To accomplish this, the pH will have to be at least two pH units greater than the pK_a of the carboxylic acid and at least two pH units less than the pK_a of the ammonium ion. In other words, it must be between pH 6.8 and pH 8.7.

b. For the carboxylic acid to dissolve in water, it must be charged (in its basic form), so the pH will have to be greater than 6.8. For the amine to dissolve in ether, it will have to be neutral (in its basic form), so the pH will have to be greater than 12.7 to have essentially all of it in the neutral form. Therefore, the pH of the water layer must be greater than 12.7.

c. To dissolve in ether, the carboxylic acid will have to be neutral, so the pH will have to be less than 2.8 to have essentially all the carboxylic acid in the acidic (neutral) form. To dissolve in water, the amine will have to be charged, so the pH will have to be less than 8.7 to have essentially all the amine in the acidic form. Therefore, the pH of the water layer must be less than 2.8.

35. Charged compounds will dissolve in water and uncharged compounds will dissolve in ether. The acidic forms of carboxylic acids and alcohols are neutral and the basic forms are charged. The acidic forms of amines are charged and the basic forms are neutral. Notice that one of the compounds does not have a pK_a value because it is not an acid (that is, it does not have a proton it can lose.)

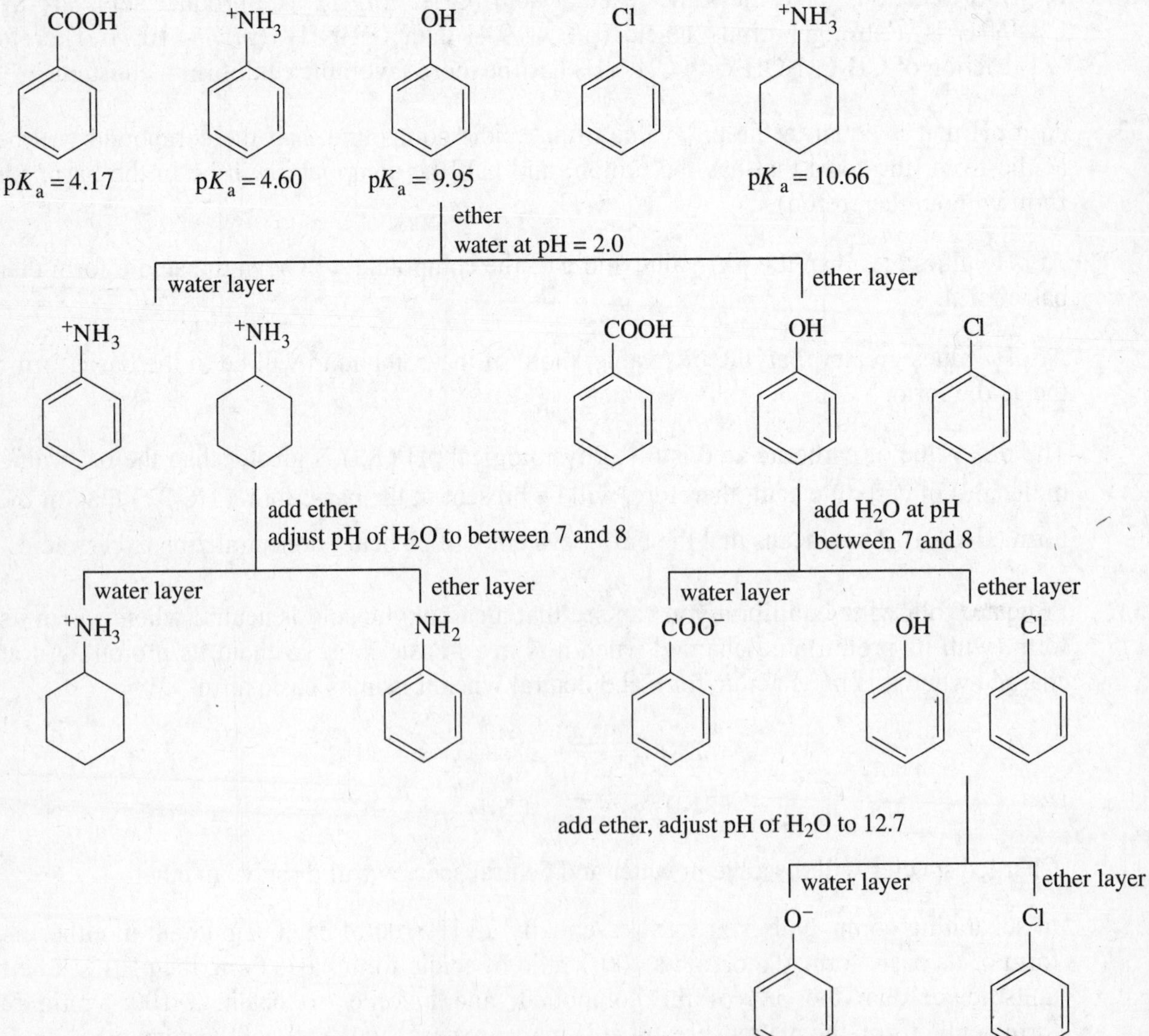
COOH
+NH3
OH
Cl
+NH3
pKa = 4.17
pKa = 4.60
pKa = 9.95
pKa = 10.66
ether
water at pH = 2.0
water layer
ether layer
+NH3
+NH3
COOH
OH
Cl
add ether
adjust pH of H2O to between 7 and 8
add H2O at pH
between 7 and 8
water layer
ether layer
water layer
ether layer
+NH3
NH2
COO−
OH
Cl
add ether, adjust pH of H2O to 12.7
water layer
ether layer
O−
Cl

Chapter 2 Practice Test

1. Answer the following:

a. Which is a stronger acid, HCl or HBr?

b. Which is a stronger acid, NH_3 or H_2O?

c. Which is a stronger base, NH_3 or H_2O?

d. Which is a stronger base, $^{-}NH_3$ or HO^-?

e. Which is a stronger base, CH_3OH or CH_3O^-?

2. The following compounds are drawn in their acidic forms, and their pK_a values are given. Draw the form in which each compound will predominate in a solution with pH = 8.

CH_3COOH	CH_3CH_2OH	$CH_3\overset{+}{O}H_2$	$CH_3CH_2\overset{+}{N}H_3$
pK_a = 4.8	pK_a = 15.9	pK_a = −2.5	pK_a = 11.2

3. **a.** What is the conjugate base of NH_3?

b. What is the conjugate base of H_2O?

c. What is the conjugate acid of H_2O?

4. **a.** What products would be formed from the following reaction?

$$CH_3OH + \overset{+}{N}H_4 \rightleftharpoons$$

b. Does the reaction favor reactants or products?

5. Which of the following is a stronger acid?

$CH_3CH(Cl)CH_2COOH$ or $CH_3CH_2CH(Cl)COOH$

6. Indicate whether each of the following statements is true or false:

a. HO^- is a stronger base than $^{-}NH_2$. T F

b. A Lewis acid is a compound that accepts a share in a pair of electrons. T F

c. $ClCH_2COOH$ is a stronger acid than CH_3COOH. T F

d. $ClCH_2COOH$ is a stronger acid than $BrCH_2COOH$. T F

CHAPTER 3

An Introduction to Organic Compounds: Nomenclature, Physical Properties, and Representation of Structure

1. **a.** *n*-propyl alcohol or propyl alcohol **b.** dimethyl ether **c.** *n*-propylamine or propylamine

2. Notice that each carbon forms four bonds and each hydrogen and bromine forms one bond.

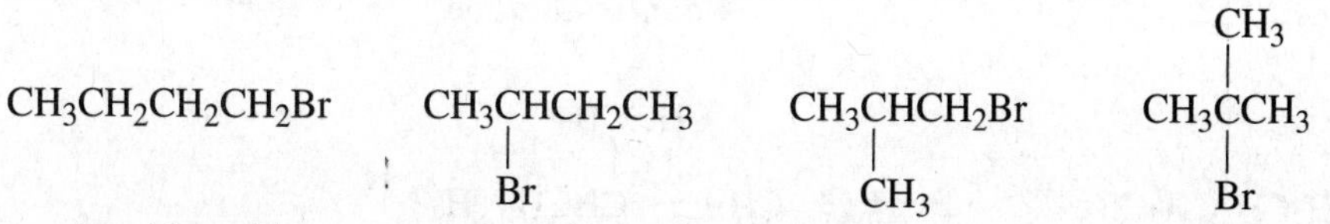

n-butyl bromide or butyl bromide; *sec*-butyl bromide; isobutyl bromide; *tert*-butyl bromide

3.

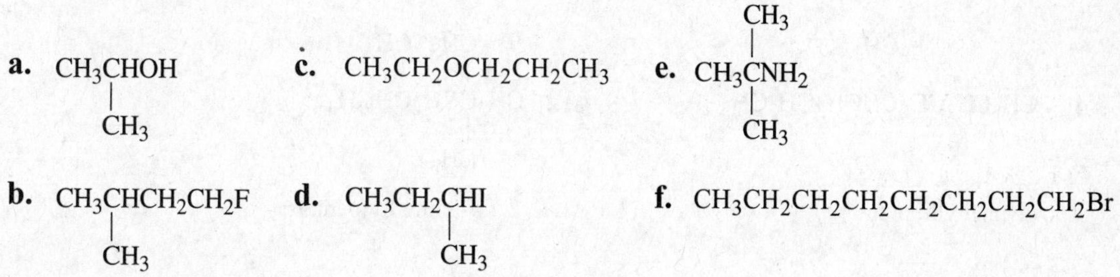

4. **a.** ethyl methyl ether **b.** methyl propyl ether **c.** *sec*-butylamine **d.** butyl alcohol or *n*-butyl alcohol **e.** isobutyl bromide **f.** *sec*-butyl chloride

5.

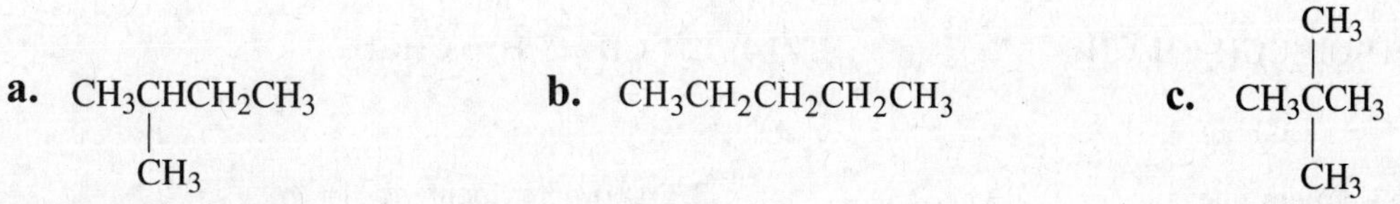

a. 2-methylbutane **b.** pentane **c.** 2,2-dimethylpropane

6.

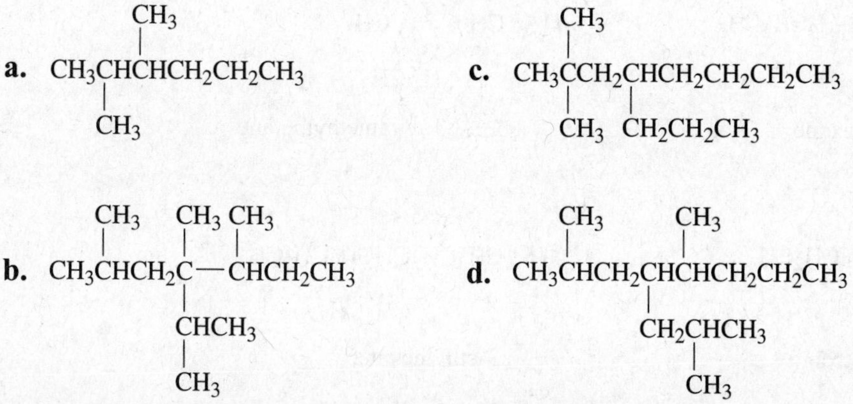

7. **a.** **#1** $CH_3CH_2CH_2CH_2CH_2CH_2CH_2CH_3$

octane

#2

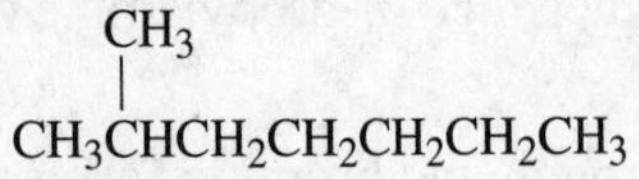

2-methylheptane

#3

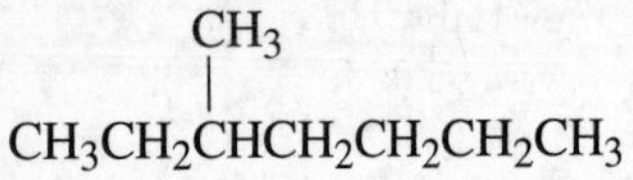

3-methylheptane

#4

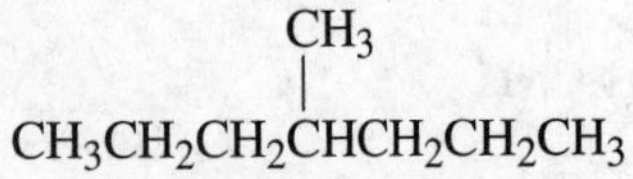

4-methylheptane

#5

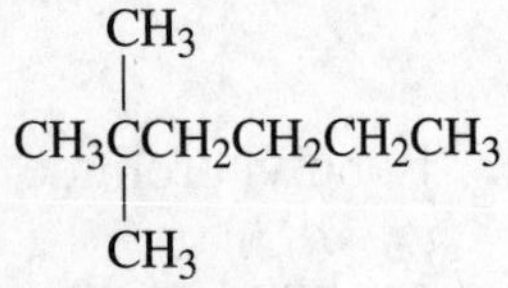

2,2-dimethylhexane

#6

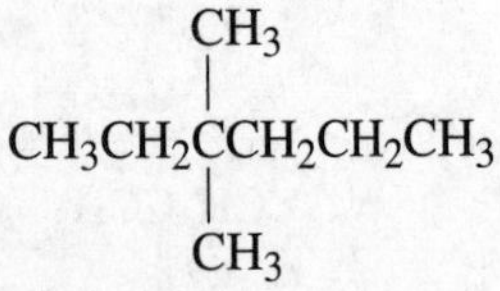

3,3-dimethylhexane

#7

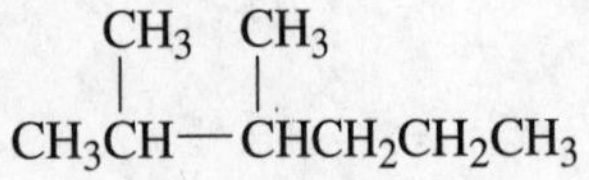

2,3-dimethylhexane

#8

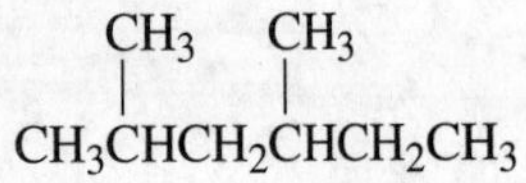

2, 4-dimethylhexane

#9

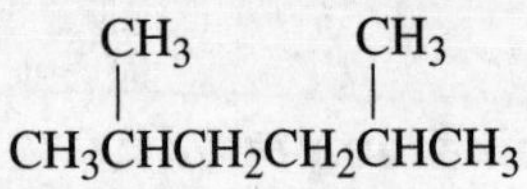

2, 5-dimethylhexane

#10

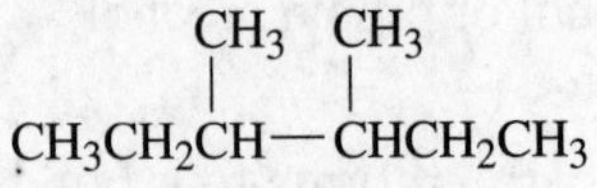

3, 4-dimethylhexane

#11

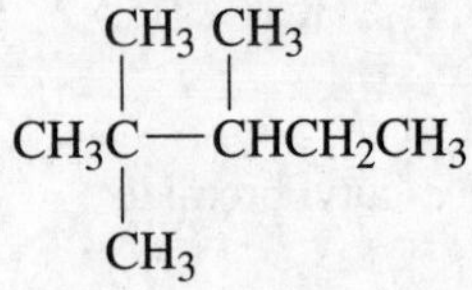

2,2,3-trimethylpentane

#12

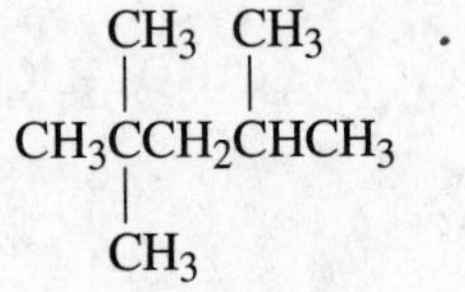

2,2,4-trimethylpentane

#13

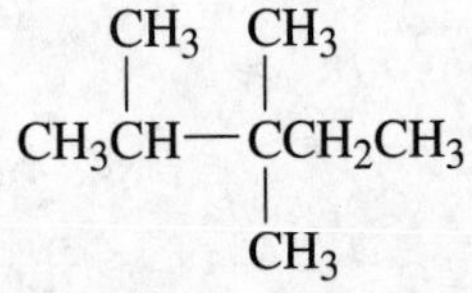

2,3,3-trimethylpentane

#14

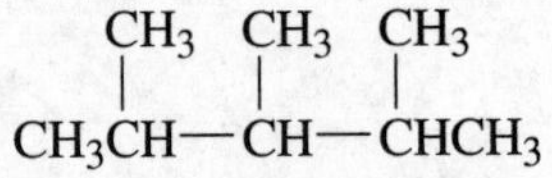

2,3,4-trimethylpentane

#15

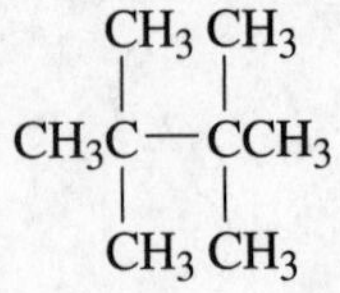

2,2,3,3-tetramethylbutane

#16

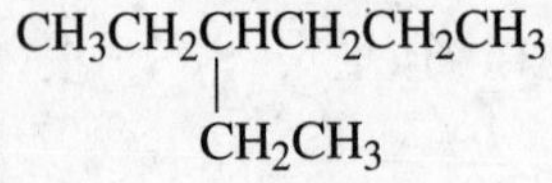

3-ethylhexane

#17 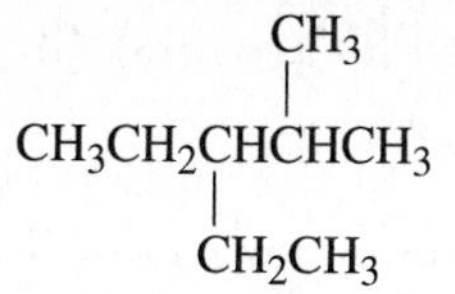

3-ethyl-2-methylpentane

#18 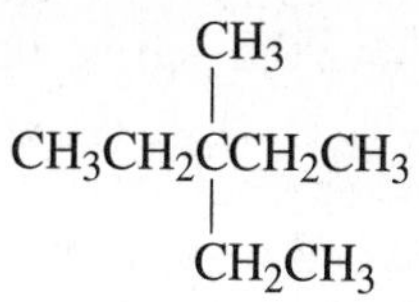

3-ethyl-3-methylpentane

b. The systematic names are under the compound.

c. #2, #7, #8, #9, #12, #13, #14, #17

d. #3, #8, #10, #11

e. #5, #11, #12, #15

8. **a.** 2,2,4-trimethylhexane

b. 2,2-dimethylbutane

c. 3,3-diethylhexane

d. 2,5-dimethylheptane

e. 4-isopropyloctane

f. 4-ethyl-2,2,3-trimethylhexane

9.

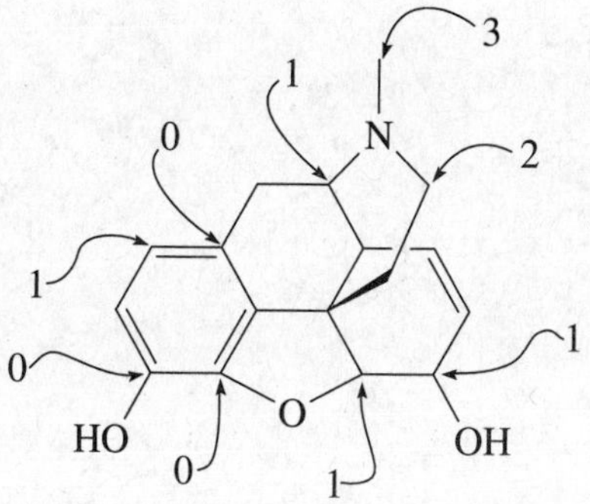

3
1
N
0
2
1
1
0
O
HO
OH
0
1

10. **a.** 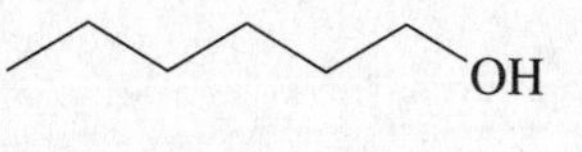
OH

b. 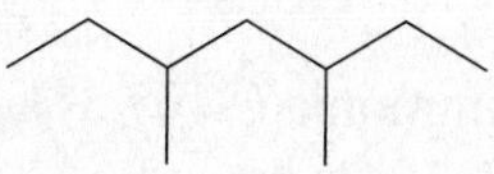

c. 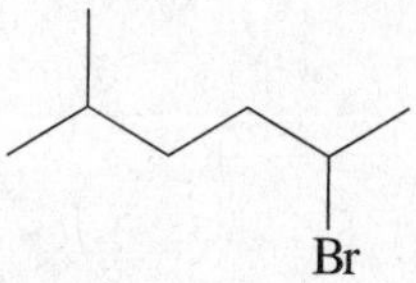
Br

d. 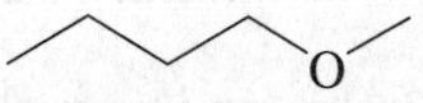
O

11. menthol = $C_{10}H_{20}O$ terpin hydrate = $C_{10}H_{20}O_2$

12. **a.** 1-ethyl-2-methylcyclopentane

b. ethylcyclobutane

c. 3,6-dimethyldecane

d. 5-isopropylnonane

13. **a.** Both compounds have the same name (1-bromo-3-methylhexane), so they are the same compound.

b. Both compounds have the same name (1-iodo-2-methylcyclohexane), so they are the same compound.

14. **a.** *sec*-butyl chloride
2-chlorobutane

b. isohexyl chloride
1-chloro-4-methylpentane

c. cyclohexyl bromide
bromocyclohexane

d. isopropyl fluoride
2-fluoropropane

15. **a.** a tertiary alkyl bromide **b.** a tertiary alcohol **c.** a primary amine

16. **a.** methylpropylamine
secondary

b. trimethylamine
tertiary

c. diethylamine
secondary

d. butyldimethylamine
tertiary

17. **a.**

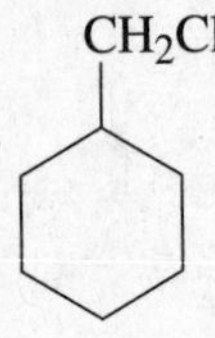
CH_2Cl

b.

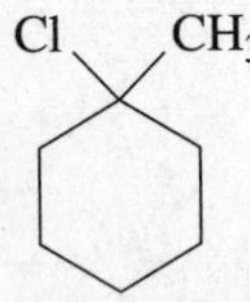
Cl
CH_3

c.

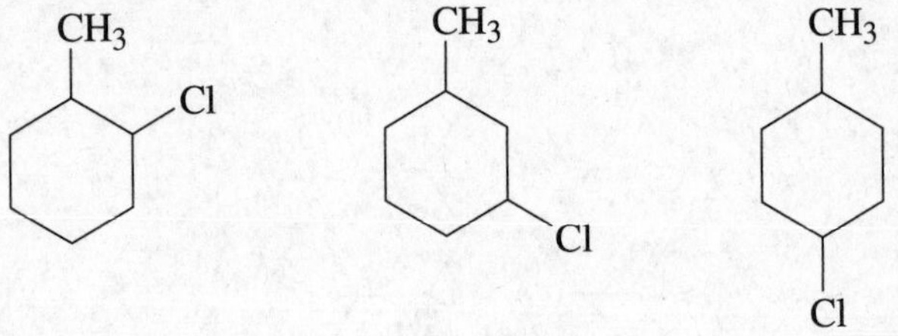
CH_3
Cl
CH_3
Cl
CH_3
Cl

18. **a.** The bond angle is predicted to be similar to the bond angle in water (~104.5°).

b. The bond angle is predicted to be similar to the bond angle in ammonia (~107.3°).

c. The bond angle is predicted to be similar to the bond angle in water (~104.5°).

19. Room temperature is about 25 °C. Therefore, pentane, with a boiling point of 36 °C is the smallest alkane that is a liquid at room temperature.

20. **a.** An O — H hydrogen bond is longer.

b. Because it is shorter, an O — H covalent bond is stronger.

21. **a.** 1, 4, and 5

b. 1, 2, 4, 5, and 6

22.

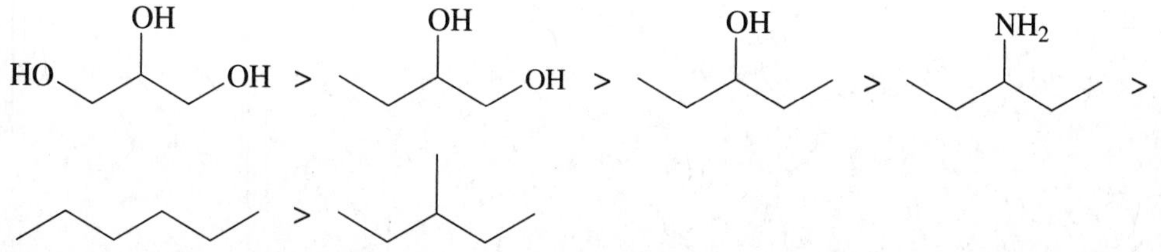

23. **a.** $CH_3CH_2CH_2CH_2CH_2CH_2Br$ > $CH_3CH_2CH_2CH_2CH_2Br$ > $CH_3CH_2CH_2CH_2Br$

b. $CH_3CH_2CH_2CH_2CH_2CH_2CH_2CH_2CH_3$ > $CH_3CH_2CH_2CH_2CH_2CH_2CH_2CH_3$ >

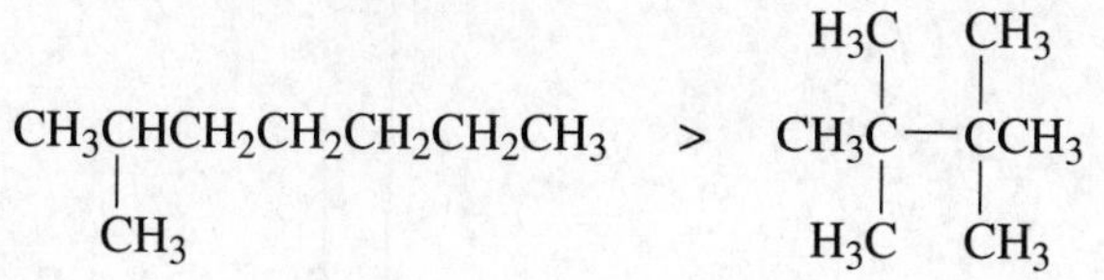

c. $CH_3CH_2CH_2CH_2CH_2OH$ > $CH_3CH_2CH_2CH_2OH$ > $CH_3CH_2CH_2CH_2Cl$ > $CH_3CH_2CH_2CH_2CH_3$

24. Figure 3.2 in the text shows that even number alkanes have higher melting points than odd number alkanes. Therefore, since more energy is required to disrupt the crystal lattice of even numbered alkanes, we can conclude that even number alkanes pack more tightly.

25. **a.** $HOCH_2CH_2CH_2OH$ > $CH_3CH_2CH_2OH$ > $CH_3CH_2CH_2CH_2OH$ > $CH_3CH_2CH_2CH_2Cl$

b.

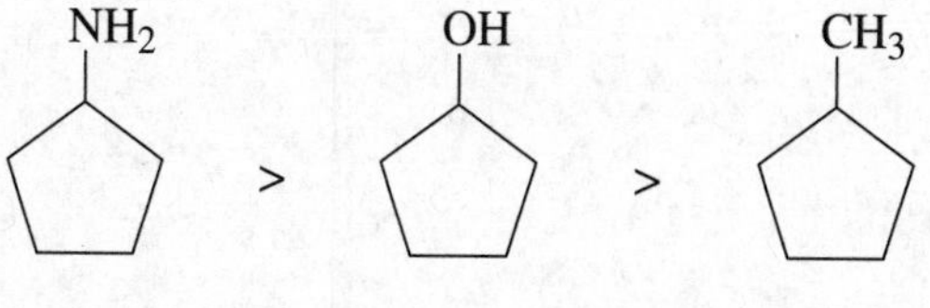

The amine is more soluble than the alcohol in water, because the amine has two hydrogens that can form hydrogen bonds with water and the alcohol has only one.

26. Because cyclohexane is a nonpolar compound, it will have the lowest solubility in the most polar solvent, which, of the solvents given, is ethanol.

$CH_3CH_2CH_2CH_2CH_2OH$	$CH_3CH_2OCH_2CH_3$	CH_3CH_2OH	$CH_3CH_2CH_2CH_2CH_2CH_3$
1-pentanol	diethyl ether	ethanol	hexane

27. Hexethal would be expected to be the more effective sedative because it is more nonpolar than barbital since hexethal has a hexyl group in place of the ethyl group of barbital. Being more nonpolar, hexethal will be better able to penetrate the nonpolar membrane of a cell.

28. **a.** 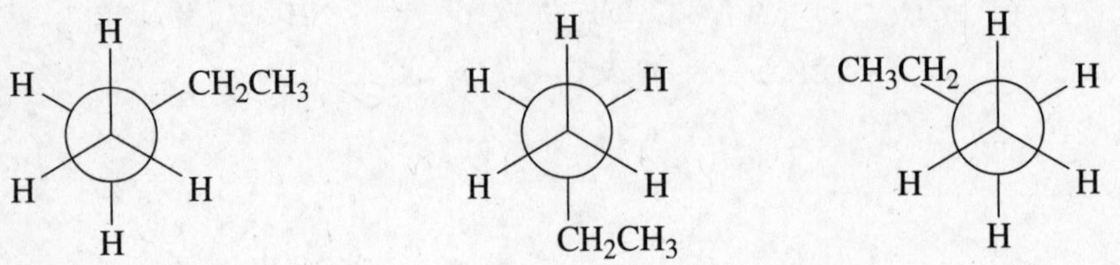
H
H
CH2CH3
H
H
H
H
H
H
H
H
CH2CH3
H
CH3CH2
H
H
H
H

b. yes **c.** yes

29. **a.** 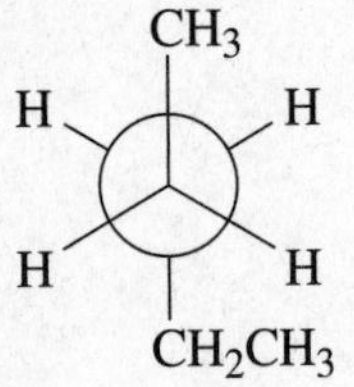
CH3
H
H
H
H
CH2CH3

b. 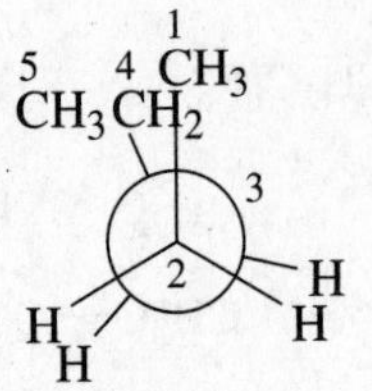
1
5
4 CH3
CH3 CH2
3
2
H
H
H
H

30. 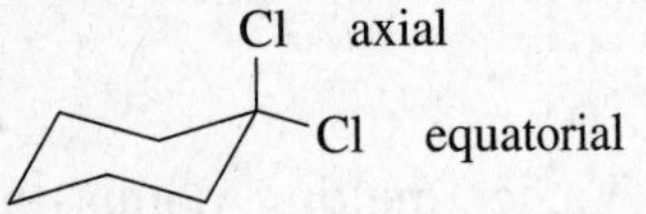
Cl axial
Cl equatorial

31. At any one time, there will be more molecules of isopropylcyclohexane with the substituent in the equatorial position because the isopropyl substituent is larger than the ethyl substituent. Because the isopropyl substituent is larger, the axial confomer of isopropylcyclohexane has more steric strain than the axial conformer of ethylcyclohexane, so the isopropyl group will have a greater preference for the equatorial position.

32. **a.** cis **b.** cis **c.** trans **d.** trans

33. **a.** 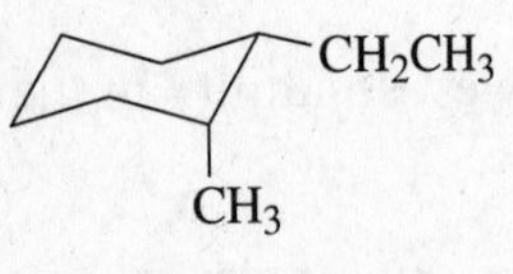
CH2CH3
CH3

b. 
CH2CH3
CH3

c. *trans*-1-Ethyl-2-methylcyclohexane is more stable because both substituents are in equatorial positions, whereas *cis*-1-ethyl-2-methylcyclohexane has one substituent in the equatorial position and the other in the axial position.

34. Both Kekulé and skeletal structures are shown.

a. 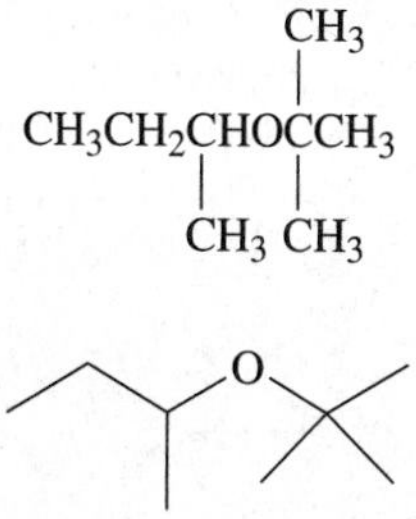
CH3
CH3CH2CHOCCH3
CH3 CH3
O

b. 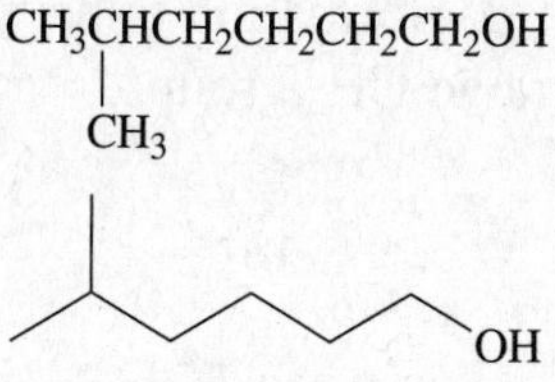
CH3CHCH2CH2CH2CH2OH
CH3
OH

c. 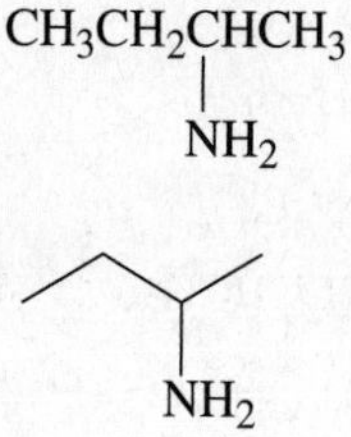
CH3CH2CHCH3
NH2
NH2

d. 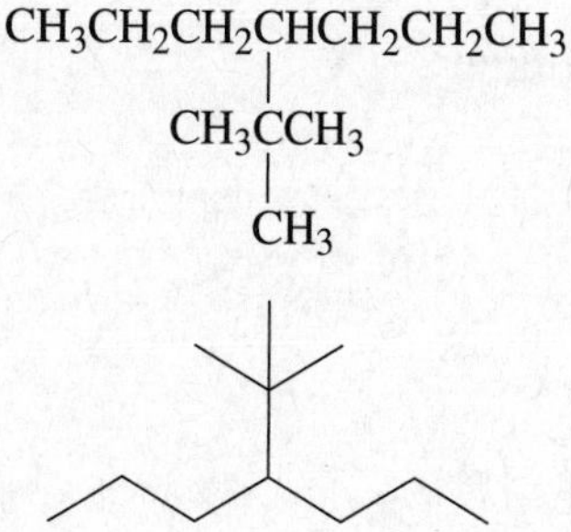
CH3CH2CH2CHCH2CH2CH3
CH3CCH3
CH3

e. 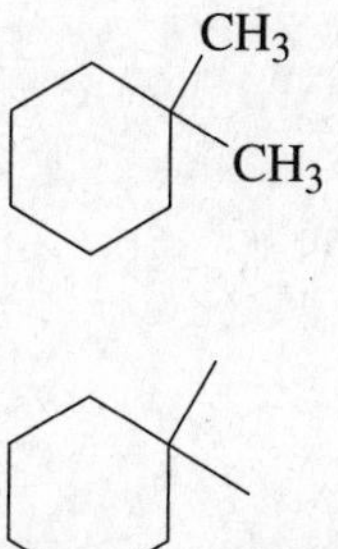
CH3
CH3

f. 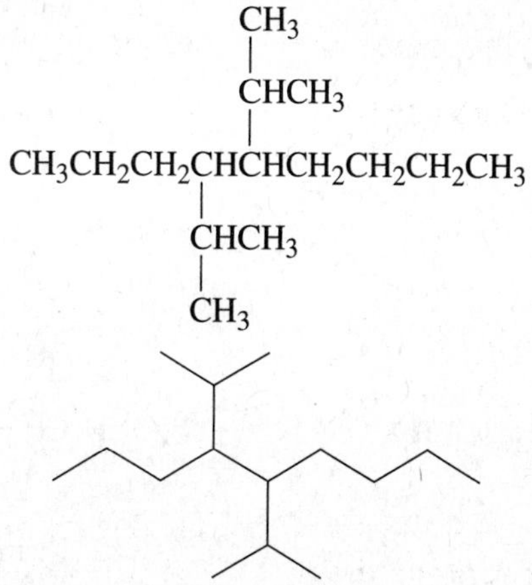
CH3
CHCH3
CH3CH2CH2CHCHCH2CH2CH2CH3
CHCH3
CH3

g. 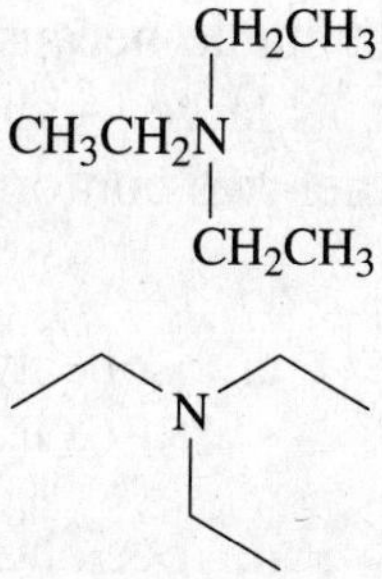
CH2CH3
CH3CH2N
CH2CH3
N

h. 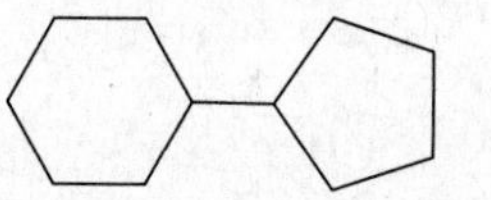

i. 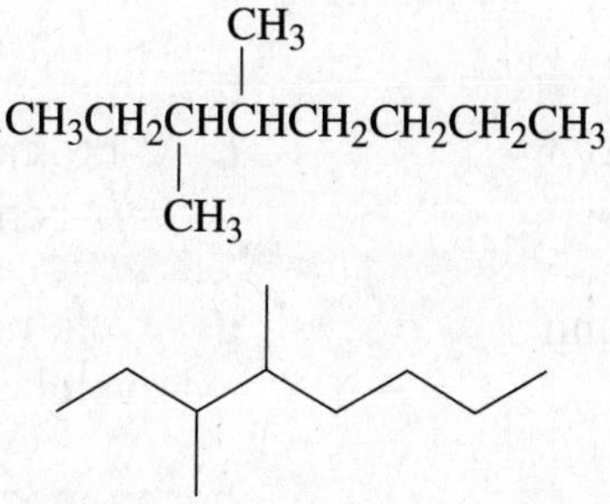
CH3
CH3CH2CHCHCH2CH2CH2CH3
CH3

j. 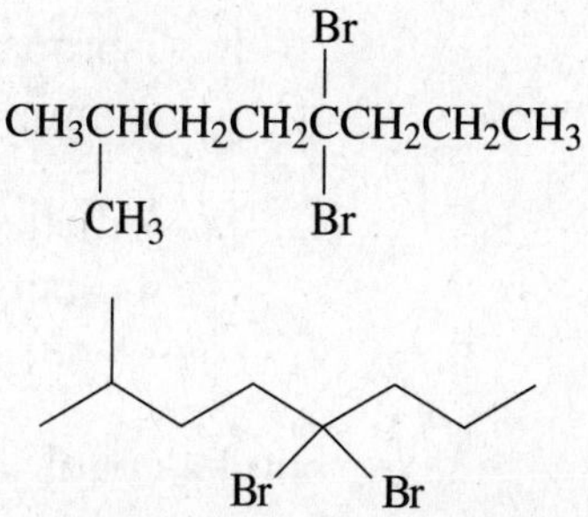
Br
CH3CHCH2CH2CCH2CH2CH3
CH3 Br
Br Br

35. **a.** 5-bromo-2-methyloctane
b. 2,2,6-trimethylheptane
c. 2,3,5-trimethylhexane
d. 3,3-diethylpentane
e. isopropylcyclohexane
f. *N,N*-dimethylcyclohexanamine

36. **a.** 3 **b.** 6 **c.** 3

37. **a.** 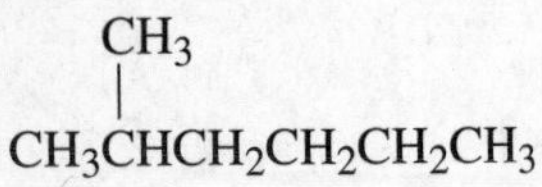

2-methylhexane

b. 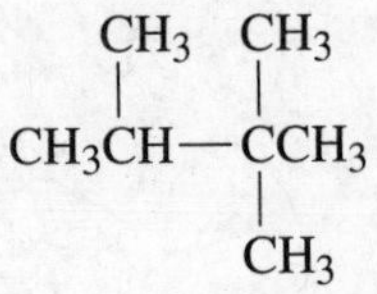

2,2,3-trimethylbutane

38. The first conformation is the most stable because the three substituents are more spread out, so its gauche interactions will not be as large — the Cl in the first conformation is between a CH_3 and an H, whereas the Cl in the other two conformers is between two CH_3 groups.

39. **a.** diethylpropylamine
TERTIARY

b. *sec*-butylisobutylamine
SECONDARY

c. isopentylpropylamine
SECONDARY

d. cyclohexylamine
PRIMARY

40. **a.** 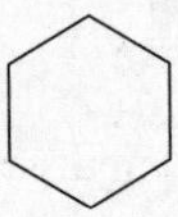

b. 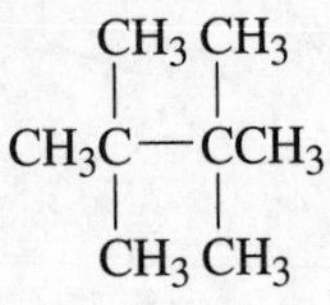

c. 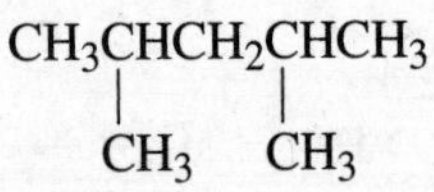

41. **a.** 1-ethoxypropane
ethyl propyl ether

b. 4-methyl-1-pentanol
isohexyl alcohol

c. 2-butanamine
sec-butylamine

d. 2-chlorobutane
sec-butyl chloride

e. 2-methylpentane
isohexane

f. 2-bromo-2-methylbutane
tert-pentyl bromide

g. cyclohexanol
cyclohexyl alcohol

h. bromocyclopentane
cyclopentyl bromide

i. 3-propanamine
isopropylamine

j. 3-methyl-*N*-propyl-1-butanamine
sec-butylethylamine

42. **a.** 1-bromohexane
b. pentyl chloride
c. butyl alcohol
d. hexyl alcohol
e. hexane
f. pentyl alcohol

43. **a.** 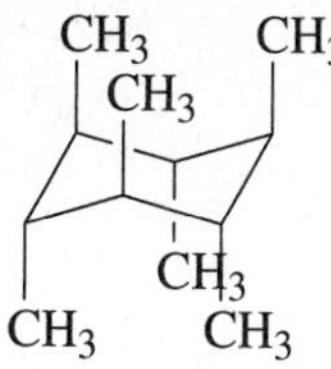 **b.** 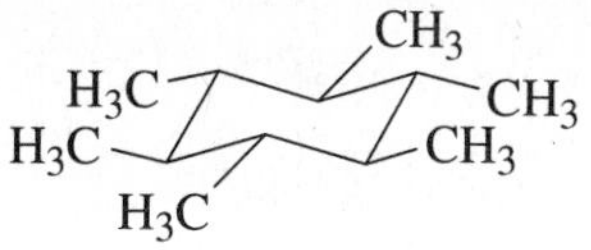

44. Ansaid is more soluble in water. It has a fluoro substituent that can hydrogen bond to water. Motrin has a nonpolar isobutyl substituent in place of Ansaid's fluoro substituent.

45. Only one is named correctly.

a. 3-ethyl-2-methyloctane
b. 4-ethyl-2,2-dimethylheptane
c. 1-bromo-3-methylbutane
d. correct
e. 2,5-dimethylheptane
f. 2,3,3-trimethyloctane

46. B has the highest energy (is the least stable). They are all diaxial-substituted cyclohexanes, so each one has four 1,3-diaxial interactions. Only B has a 1,3-diaxial interaction between CH_3 and Cl, which will be greater than a 1,3-diaxial interaction between CH_3 and H or a 1,3-diaxial interaction between Cl and H.

47. The only one is 2,2,3-trimethylbutane.

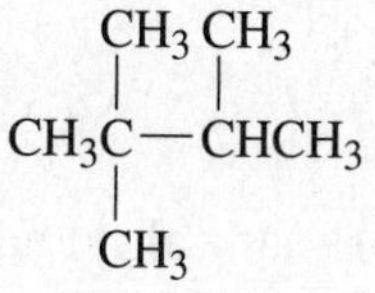 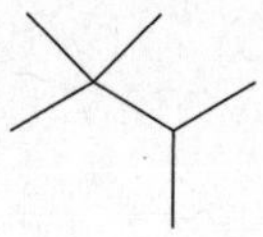

48. **a.** 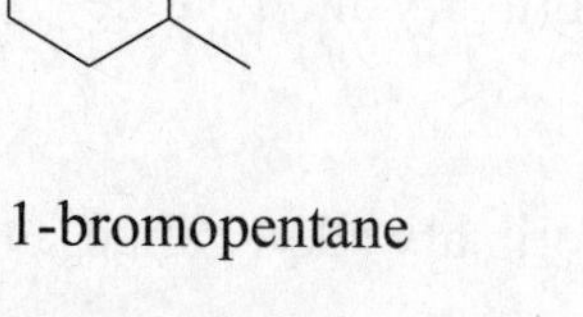

c. 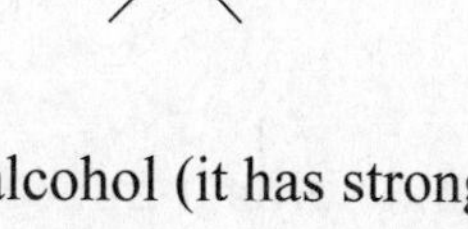

b. 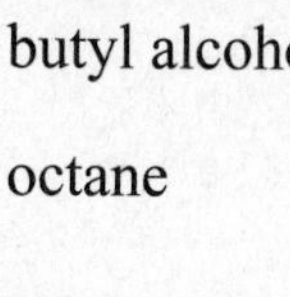

d.

49. **a.** 1-bromopentane
b. butyl alcohol
c. octane
d. isopentyl alcohol (it has stronger hydrogen bonds)
e. hexylamine (it has more hydrogen bonds)

50. Alcohols with low molecular weights are more water soluble than alcohols with high molecular weights because, as a result of having fewer carbons, they have a smaller nonpolar component that has to be dragged into water.

51. **a.**

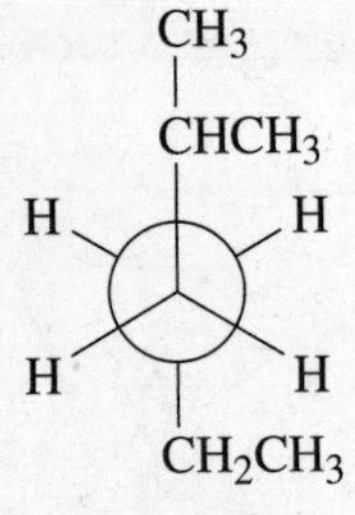

most stable

b.

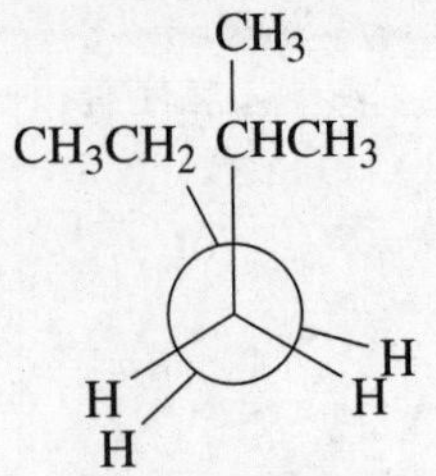

least stable

c. Rotation can occur about all the C—C bonds. There are six carbon-carbon bonds in the compound, so there are five other carbon-carbon bonds, in addition to the C_3-C_4 bond, about which rotation can occur.

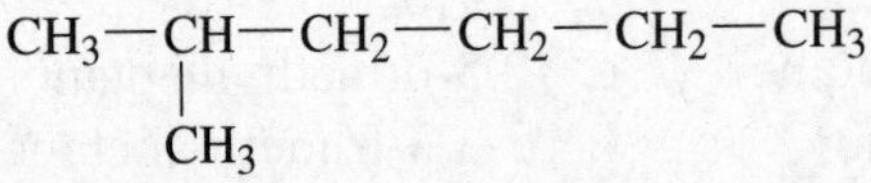

d. Three of the carbon-carbon bonds have staggered conformers that are equally stable because each is bonded to a carbon with three identical substituents.

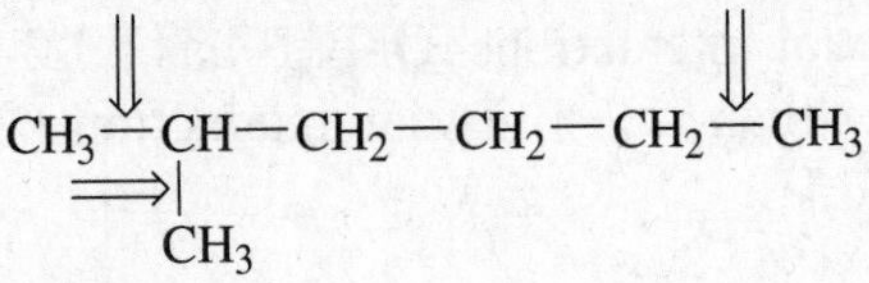

52. **C** and **D** are cis isomers. (In **C,** both substituents are downward pointing; in **D,** both substituents are upward pointing.)

53. **a.** $CH_3CH_2CH_2CH_2CH_2Br$ 1-bromopentane primary alkyl halide

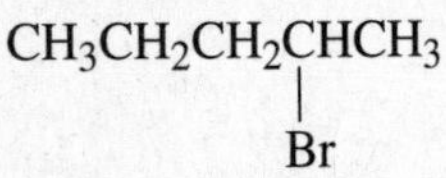 2-bromopentane secondary alkyl halide

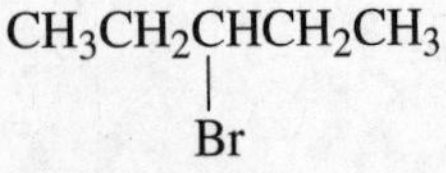 3-bromopentane secondary alkyl halide

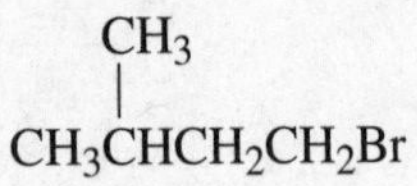 1-bromo-3-methylbutane primary alkyl halide

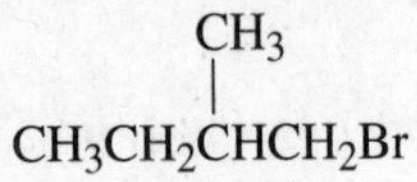 1-bromo-2-methylbutane primary alkyl halide

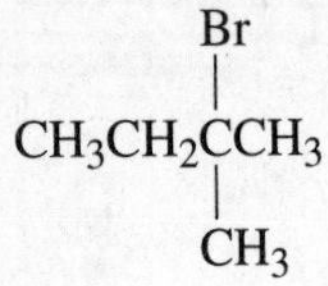 2-bromo-2-methylbutane tertiary alkyl halide

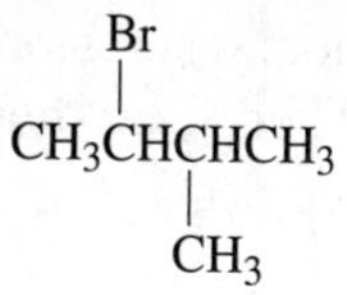

2-bromo-3-methylbutane secondary alkyl halide

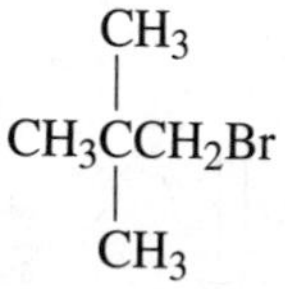

1-bromo-2,2-dimethylpropane primary alkyl halide

b. Four isomers are primary alkyl halides.

c. Three isomers are secondary alkyl halides.

d. One isomer is a tertiary alkyl halide.

54. **a.** butane

b. 1-propanol

c. 5-isopropyl-2-methyloctane

d. 6-chloro-4-ethyl-3-methyloctane

e. 6-isobutyl-2,3-dimethyldecane

55. **a.**

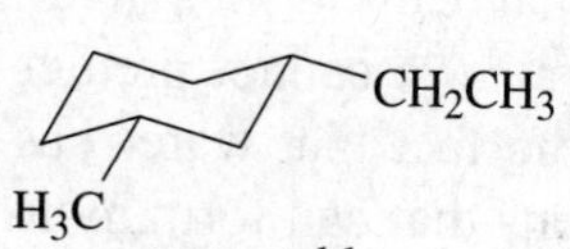

more stable

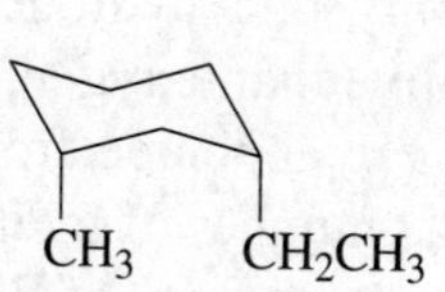

b.

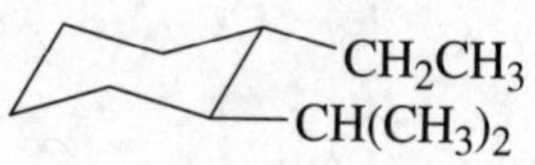

more stable

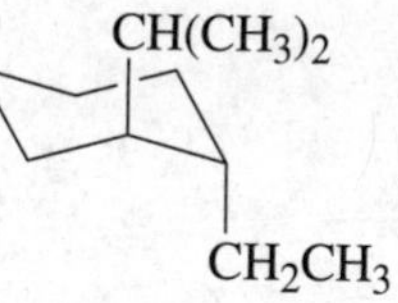

c.

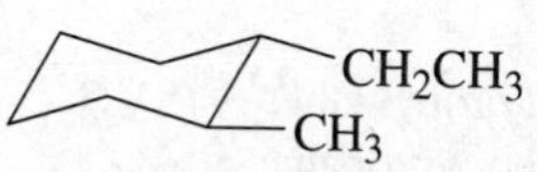

more stable

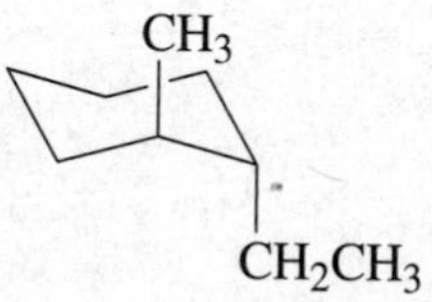

d.

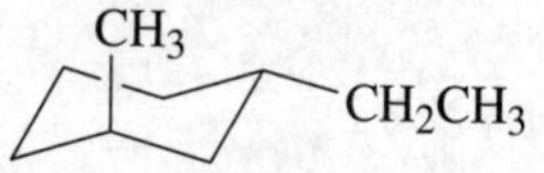

more stable

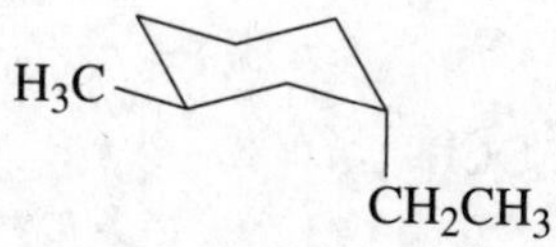

e.

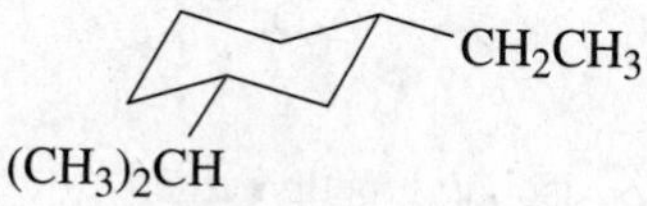

more stable

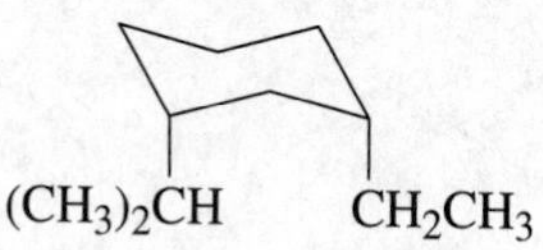

f.

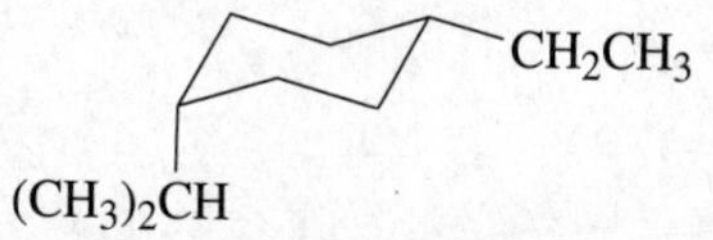

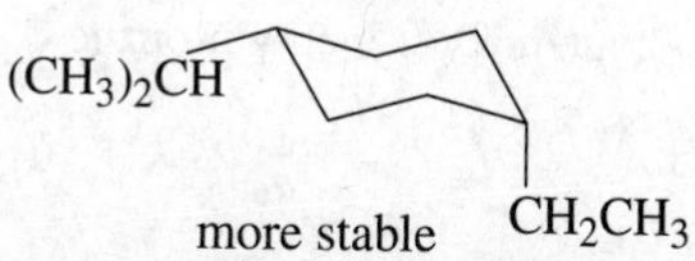

more stable

56. **a.** Each water molecule has two hydrogens that can form hydrogen bonds, whereas each alcohol molecule has only one hydrogen that can form a hydrogen bond. Therefore, there are more hydrogen bonds between water molecules than between alcohol molecules.

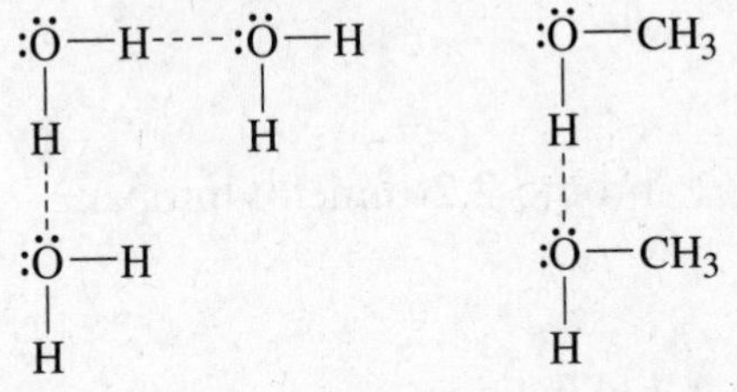

b. Each water molecule has two hydrogens that can form hydrogen bonds, whereas each ammonia has three hydrogens that can form hydrogen bonds. However, oxygen is more electronegative than nitrogen, so the hydrogen bonds between water molecules are stronger than the hydrogen bonds between ammonia molecules. Because the number of hydrogen bonds supports ammonia as having the higher boiling point but the strength of the hydrogen bonds supports water, we could not have predicted which would have the higher boiling point. However, being told that water has the higher boiling point we can conclude that the greater electronegativity of oxygen compared to nitrogen is more important than the number of hydrogens that can form hydrogen bonds.

c. Each water molecule has two hydrogens that can form hydrogen bonds, whereas each molecule of hydrogen fluoride has only one hydrogen that can form a hydrogen bond. However, fluorine is more electronegative than oxygen. Again we cannot predict which will have the higher boiling point, but we can conclude from the fact that water has the higher boiling point that *in this case* the greater number of hydrogens that can form hydrogen bonds is more important than the greater electronegativity of fluorine compared to oxygen.

57. Six ethers have molecular formula $C_5H_{12}O$.

$CH_3OCH_2CH_2CH_2CH_3$

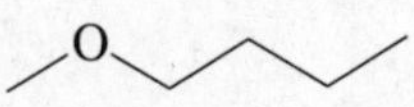

butyl methyl ether
1-methoxybutane

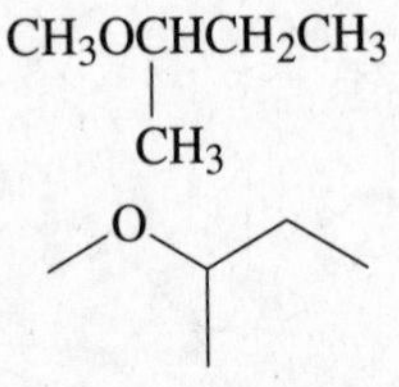

sec-butyl methyl ether
2-methoxybutane

$CH_3CH_2OCH_2CH_2CH_3$

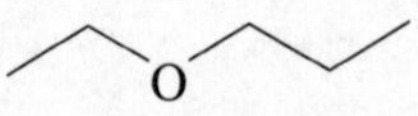

ethyl propyl ether
1-ethoxypropane

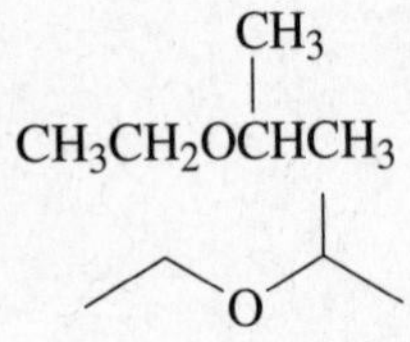

ethyl isopropyl ether
2-ethoxypropane

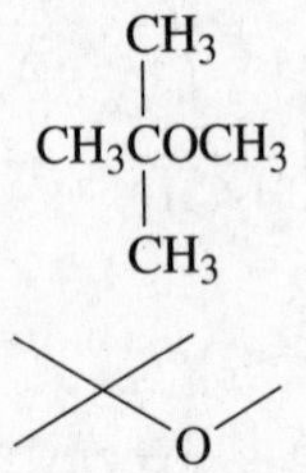

tert-butyl methyl ether
2-methoxy-2-methylpropane

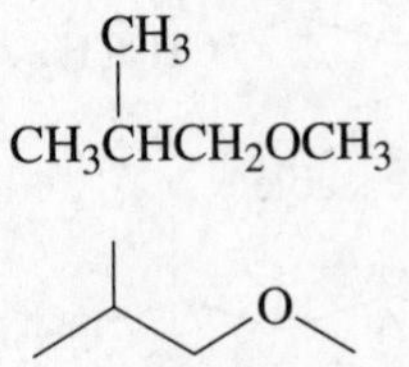

isobutyl methyl ether
1-methoxy-2-methylpropane

58. The most stable conformer has two CH_3 groups in equatorial positions and one in an axial position. (The other conformer would have two CH_3 groups in axial positions and one in an equatorial position.)

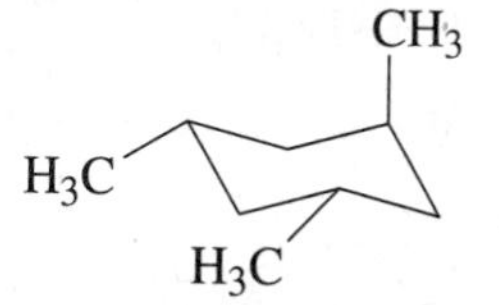
CH3
H3C
H3C

59. **a.** 3-ethyl-2,5-dimethylheptane **b.** 1,4-dichloro-5-methylheptane

60. Dibromomethane does not have constitutional isomers.

If carbon were flat, the two structures shown below would be constitutional isomers, because the Br's would be 90° apart in one compound and 180° apart in the other compound, so they would be different compounds. However, since carbon is tetrahedral the two structures are identical, so dibromomethane does not have constitutional isomers.

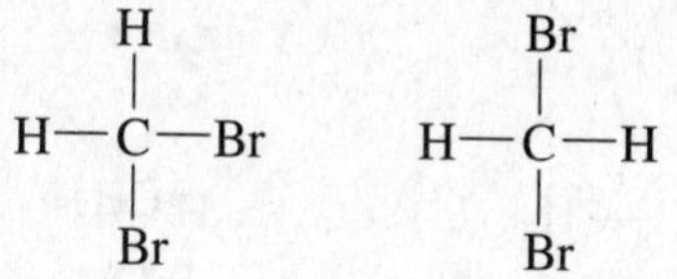
H
H—C—Br
Br
Br
H—C—H
Br

61.

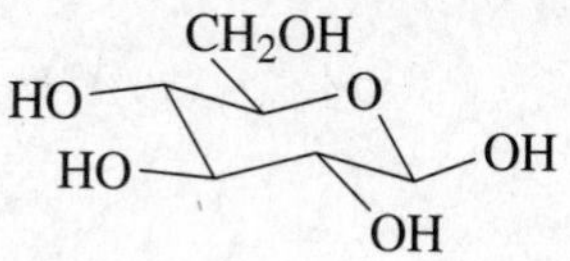
CH2OH
HO
O
HO
OH
OH

62. **#1** $CH_3CH_2CH_2CH_2CH_2CH_2CH_3$
heptane

#4
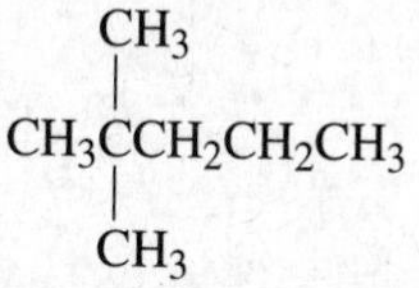
2,2-dimethylpentane

#2
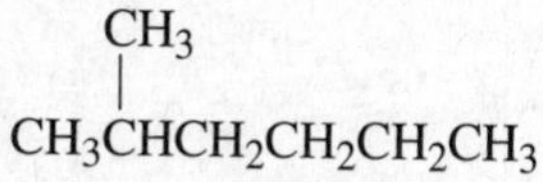
2-methylhexane

#5
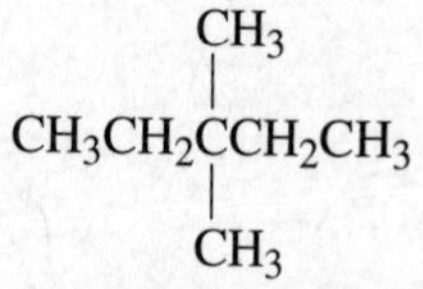
3,3-dimethylpentane

#3
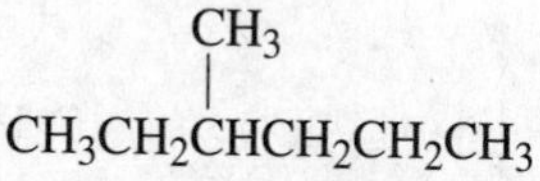
3-methylhexane

#6
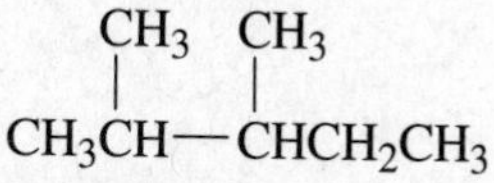
2,3-dimethylpentane

#7 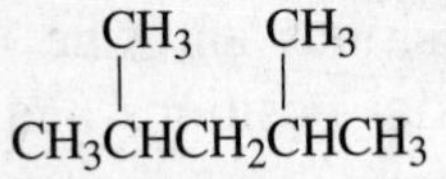

2,4-dimethylpentane

#9 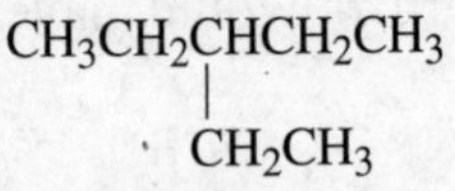

3-ethylpentane

#8 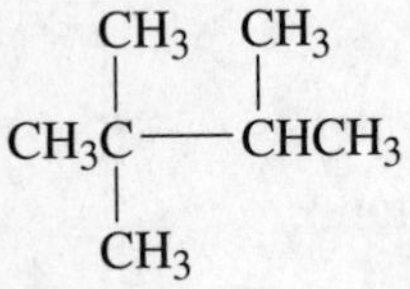

2,2,3-trimethylbutane

63. The most stable conformer has all the substituents in equatorial positions.

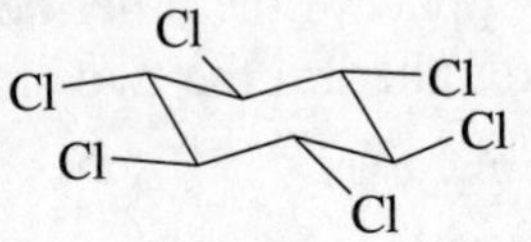

64. **a.**

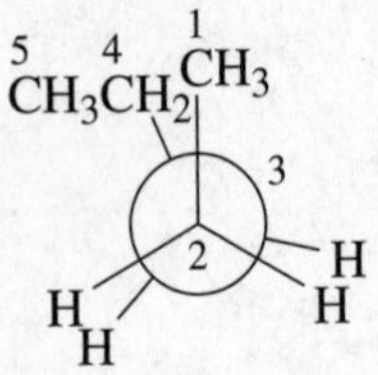

A

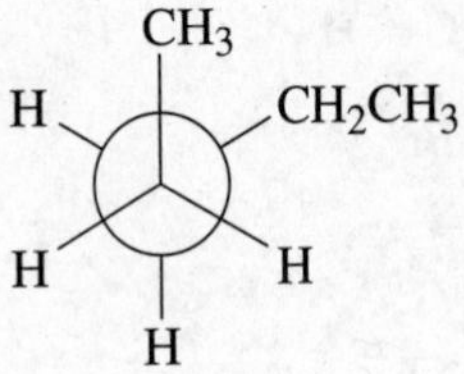

B

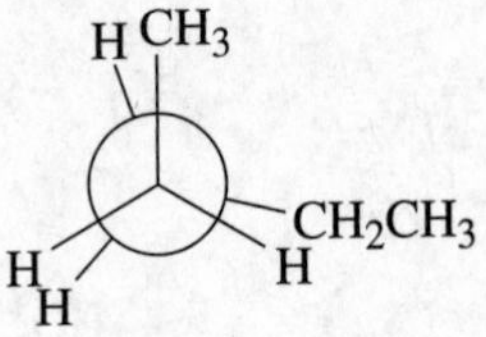

C

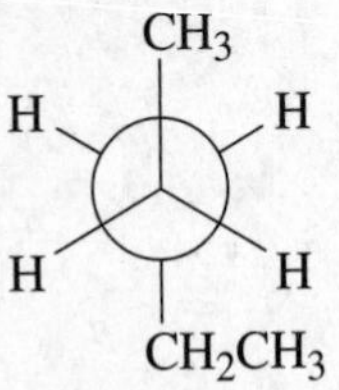

D

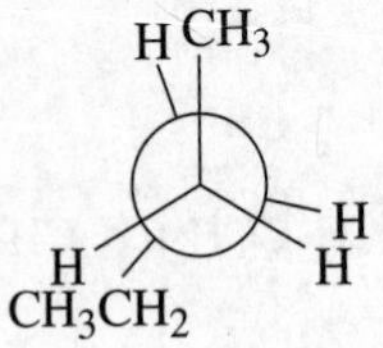

E

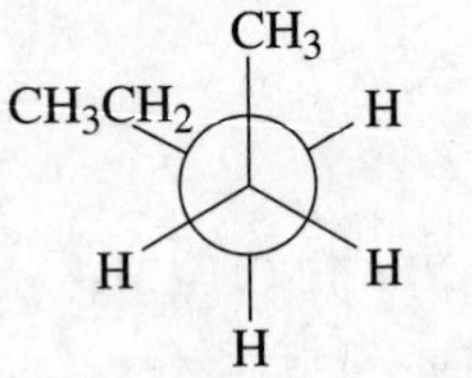

F

b.

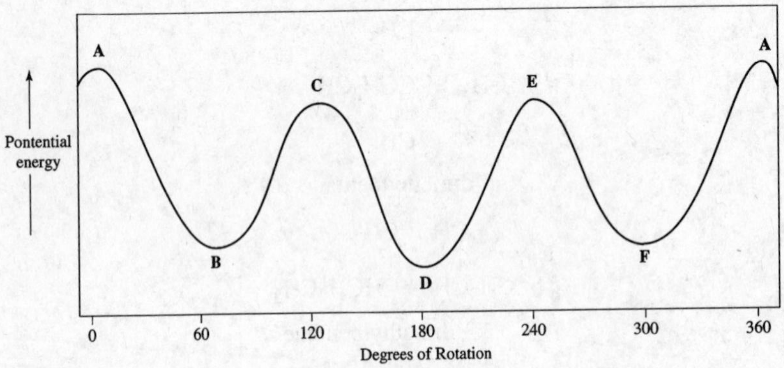
A
C
E
A
B
D
F
Pontential energy
0
60
120
180
240
300
360
Degrees of Rotation

65. **a.**

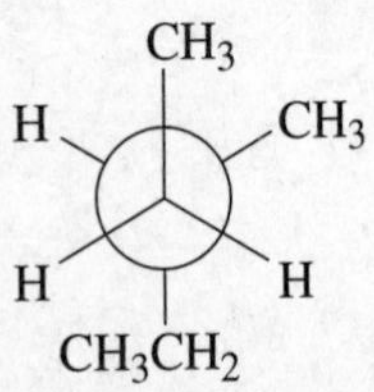

b.

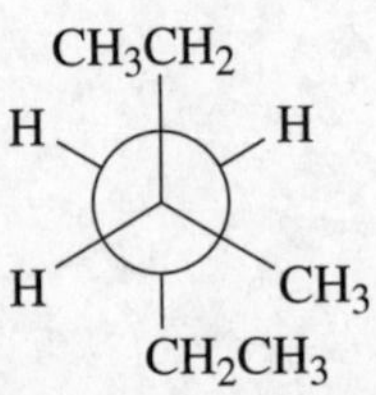

66. **a.** one equatorial and one axial
b. both equatorial and both axial
c. both equatorial and both axial
d. one equatorial and one axial
e. one equatorial and one axial
f. both equatorial and both axial

67. Both *trans*-1,4-dimethylcyclohexane and *cis*-1-*tert*-butyl-3-methylcyclohexane have a conformer with two substituents in the equatorial position and a conformer with two substituents in the axial position. *cis*-1-*tert*-Butyl-3-methylcyclohexane will have a higher percentage of the conformer with two substituents in the equatorial position, because the bulky *tert*-butyl substituent will have a greater preference for the equatorial position.

Chapter 3 Practice Test

1. Name the following compound:

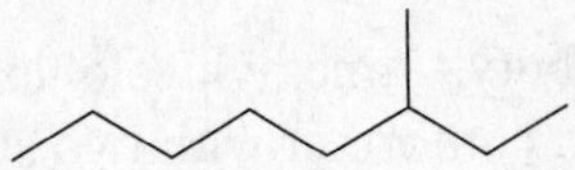

2. Draw a Newman projection for each of the following conformations of hexane considering rotation about the $C_3 — C_4$ bond:

a. the most stable of all the conformations

b. the least stable of all the conformations

3. Give two names for each of the following:

a. $CH_3CH_2CHCH_3$
with Cl on the CH

b. $CH_3CHCH_2CH_2F$
with CH_3 on the CH

c.

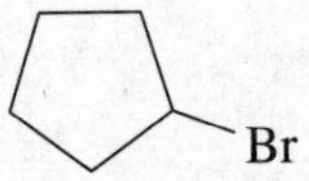

4. Label the three compounds in each set in order of decreasing boiling point. (Label the highest boiling compound #1, the next #2, and the lowest boiling #3.)

a. $CH_3CH_2CH_2CH_2CH_2Br$ $\quad$ $CH_3CH_2CH_2Br$ $\quad$ $CH_3CH_2CH_2CH_2Br$

b. $CH_3CH_2CH_2CH_2CH_3$ $\quad$ $CH_3CH_2CH_2CH_2OH$ $\quad$ $CH_3CH_2CH_2CH_2Cl$

c. $CH_3C(CH_3)_2—C(CH_3)_2CH_3$ $\quad$ $CH_3CH_2CH_2CH_2CH_2CH_2CH_2CH_3$ $\quad$ $CH_3CH(CH_3)CH_2CH_2CH_2CH_2CH_3$

5. Give the systematic name for each of the following:

a. $CH_3CHCH_2CH_2CHCH_2CH_3$
with CH_3 on the first CH and Br on the second CH

b. $CH_3CH_2CHCH_3$
with $CH_2CH_2CH_2CH_3$ on the CH

c.

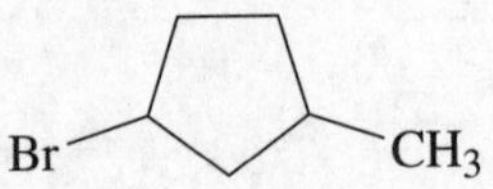

d. $CH_3CHCHCH_2CH_2CH_2Cl$
with Cl on the first CH and CH_2CH_3 on the second CH

6. Draw the other chair conformer.

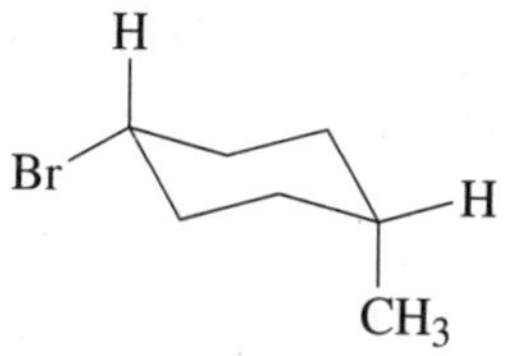
H
Br
H
CH3

7. Draw the most stable conformer of *trans*-1-isopropyl-3-methylcyclohexane.

8. Which of the following has:

a. the higher boiling point: diethyl ether or butyl alcohol?

b. the greater solubility in water: butyl alcohol or pentyl alcohol?

c. the higher boiling point: hexane or isohexane?

d. the higher boiling point: propylamine or ethylmethylamine?

e. the greater solubility in water: ethyl alcohol or ethyl chloride?

9. Give two names for each of the following:

a. $CH_3CH_2\underset{\underset{CH_3}{|}}{C}HBr$ **b.** $CH_3\underset{\underset{CH_3}{|}}{C}HCH_2CH_2OCH_3$ **c.** $CH_3\underset{\underset{CH_3}{|}}{C}HCH_2CH_3$

10. Which is more stable:

a. a staggered conformation or an eclipsed conformation?

b. the chair conformer of methylcyclohexane with the methyl group in the axial position or the chair conformer of methylcyclohexane with the methyl group in the equatorial position?

c. cyclohexane or cyclobutane?

11. Give the structure of the following:

a. a secondary alkyl bromide that has three carbons.

b. a secondary amine that has three carbons.

c. an alkane with no secondary hydrogens.

d. a constitutional isomer of butane.

e. three compounds with molecular formula C_3H_8O.

SPECIAL TOPIC I

An Exercise in Drawing Curved Arrows "Pushing Electrons"

This is an extension of what you learned about drawing curved arrows in Section 4.7 on pages 97–99 of the text. Working through these problems will take only a little of your time. It will, however, be time well spent, because curved arrows will be used throughout the course and it is important that you are comfortable with this notation. (You will not encounter some of the reaction steps shown in this exercise for weeks or even months, so do not worry about why the chemical changes take place.)

Chemists use curved arrows to show how electrons move as covalent bonds break and/or new covalent bonds form. The tail of the arrow is positioned at the point where the electrons are in the reactant, and the head of the arrow points to where these same electrons end up in the product.

In the following reaction step, the bond between bromine and a carbon of the cyclohexane ring breaks and both electrons in the bond end up with bromine in the product. Thus, **the arrow starts at the electrons that carbon and bromine share in the reactant,** and **the head of the arrow points at bromine** because this is where the two electrons end up in the product.

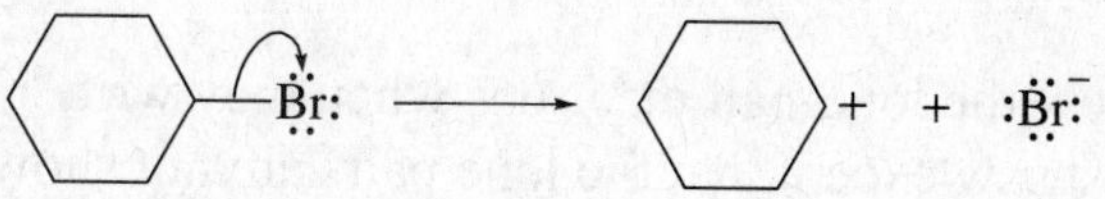

Notice that the carbon of the cyclohexane ring is positively charged in the product because it has lost the two electrons it was sharing with bromine. The bromine is negatively charged in the product because it has gained the electrons that it shared with carbon in the reactant. The fact that two electrons move in this example is indicated by the two barbs on the arrowhead.

Notice that the arrow always starts at a bond or at a lone pair. It does not start at a negative charge.

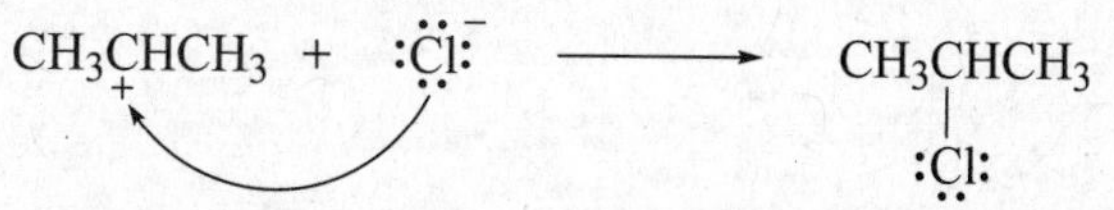

In the following reaction step, a bond is being formed between the oxygen of water and a carbon of the other reactant. The arrow starts at one of the lone pairs of the oxygen and points at the atom (the carbon) that will share the electrons in the product. The oxygen in the product is positively charged because the electrons that oxygen had to itself in the reactant are now being shared with carbon. The carbon that was positively charged in the reactant is not charged in the product, because it has gained a share in a pair of electrons.

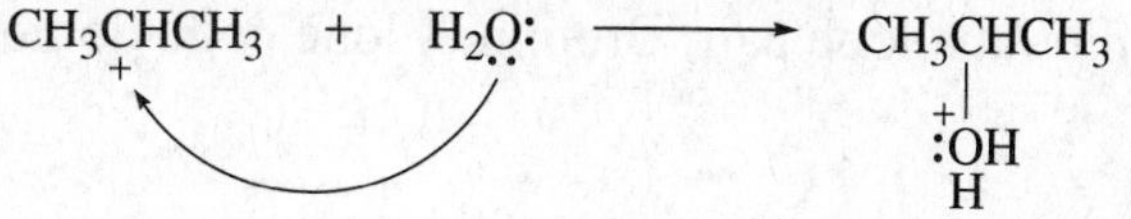

Problem 1.   Draw curved arrows to show the movement of the electrons in the following reaction steps. (You will find the answers immediately after Problem 10.)

a. $CH_3CH_2\underset{\underset{CH_3}{|}}{\overset{\overset{CH_3}{|}}{C}}$—Br: $\longrightarrow$ $CH_3CH_2\underset{\underset{CH_3}{|}}{\overset{\overset{CH_3}{|}}{C}}^+$ + :Br:⁻ 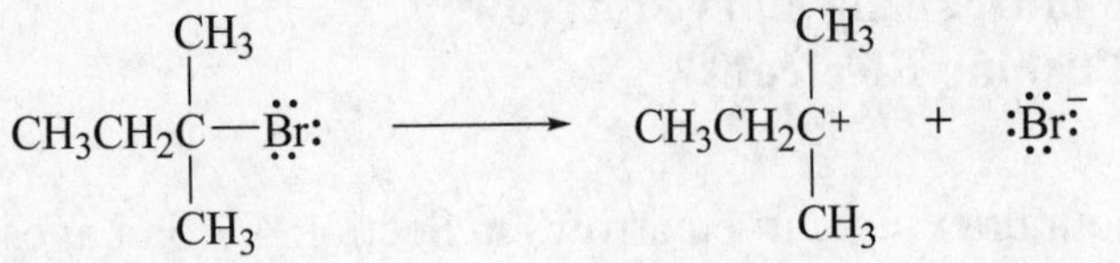

b. $\longrightarrow$ + + :Cl:⁻ 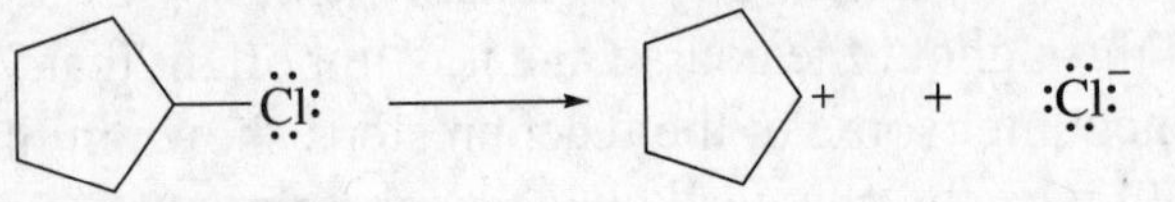

c. $\longrightarrow$ + + H₂Ö: 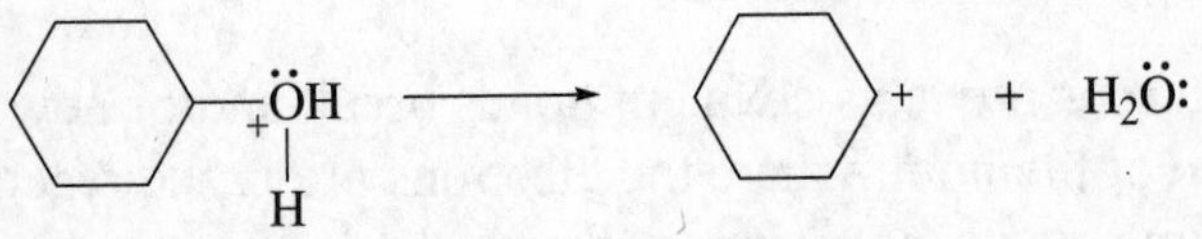

d. CH_3CH_2—MgBr $\longrightarrow$ $CH_3\ddot{C}H_2$ + ⁺MgBr 

e. $CH_3CH_2\underset{+}{C}HCH_3$ + :Br:⁻ $\longrightarrow$ $CH_3CH_2\underset{\underset{:Br:}{|}}{C}HCH_3$ 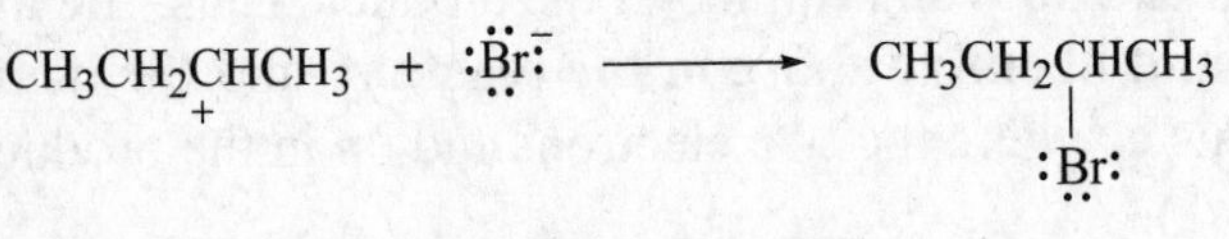

Frequently chemists do not show the lone-pair electrons when they write reactions. The following are the same reaction steps you just saw except that the lone pairs are not shown.

Problem 2.   Draw curved arrows to show the movement of the electrons.

a. $CH_3CH_2\underset{\underset{CH_3}{|}}{\overset{\overset{CH_3}{|}}{C}}$—Br $\longrightarrow$ $CH_3CH_2\underset{\underset{CH_3}{|}}{\overset{\overset{CH_3}{|}}{C}}^+$ + Br⁻ 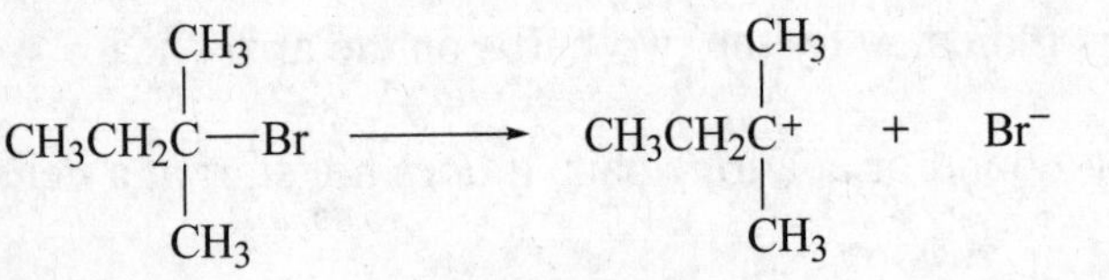

b. $\longrightarrow$ + + Cl⁻ 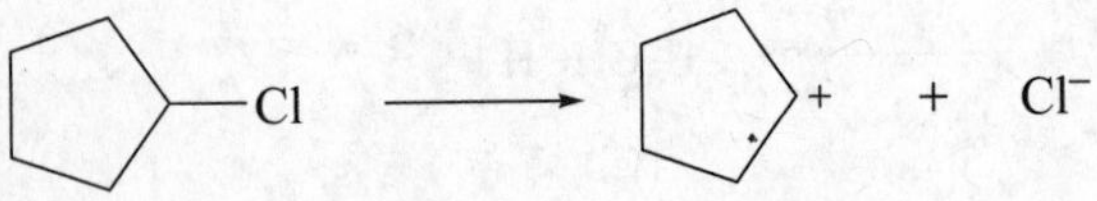

c. $\longrightarrow$ + + H₂O 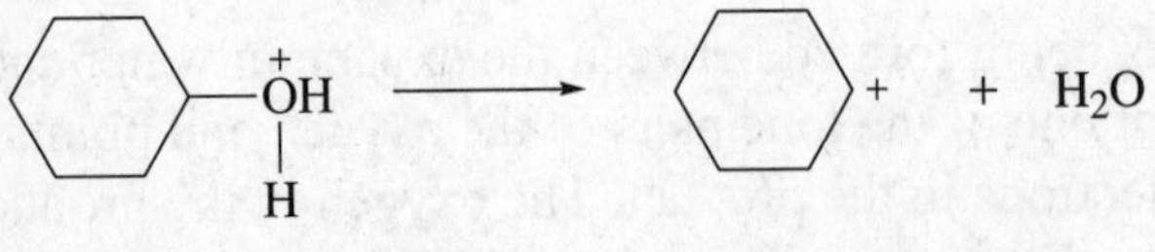

d. CH_3CH_2—MgBr $\longrightarrow$ $CH_3\overset{-}{C}H_2$ + ⁺MgBr 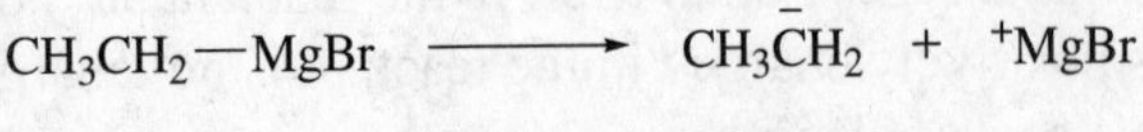

The lone-pair electrons on Br⁻ in example **e** have to be shown in the reactant because an arrow can start only at a bond or at a lone pair. Bromine's lone pairs do not have to be shown in the product.

e. $CH_3CH_2\underset{+}{C}HCH_3$ + :Br:⁻ $\longrightarrow$ $CH_3CH_2\underset{\underset{Br}{|}}{C}HCH_3$ 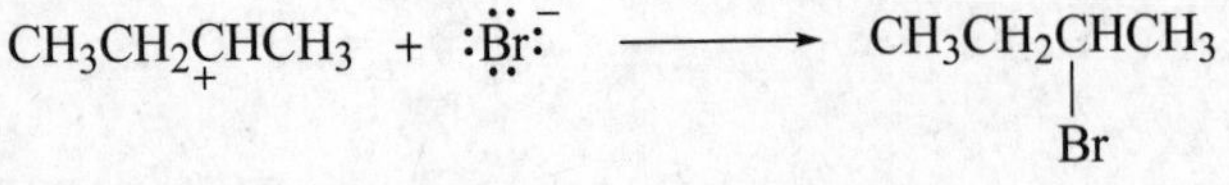

Many reaction steps involve both bond breaking and bond formation. In the following examples, the electrons in the bond that breaks are the same as the electrons in the bond that forms, so only one arrow is needed to show how the electrons move. As in the previous examples, the arrow starts at the point where the electrons are in the reactant, and the head of the arrow points to where these same electrons end up in the product (at the carbon atom in the first example, and between the carbons in the next example). Notice that the atom that loses a share in a pair of electrons (C in the first example, H in the second) ends up with a positive charge.

$$
\underset{\underset{CH_3}{|}}{\overset{\overset{CH_3}{|}}{CH_3\overset{+}{C}-CHCH_3}} \longrightarrow \underset{\underset{CH_3}{|}}{\overset{\overset{CH_3}{|}}{CH_3\overset{+}{C}-CHCH_3}}
$$

$$
\underset{H}{\overset{+}{CH_2-CHCH_3}} \longrightarrow CH_2{=}CHCH_3 \;+\; H^+
$$

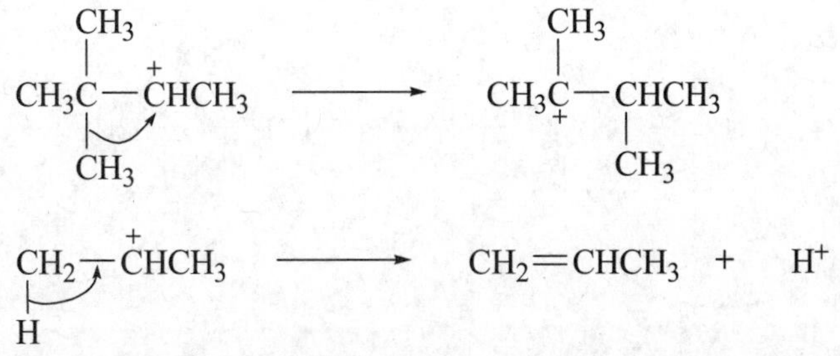

Frequently, the electrons in the bond that breaks are not the same as the electrons in the bond that forms. In such cases, two or more arrows are needed to show the movement of the electrons. In each of the following examples, look at the arrows that illustrate how the electrons move. Notice how the movement of the electrons allows you to determine both the structure of the products and the charges on the products.

$$
CH_3CH{=}CH_2 \;+\; H-\ddot{B}r\colon \longrightarrow CH_3\overset{+}{C}HCH_3 \;+\; \colon\!\ddot{B}\!\ddot{r}\colon^{-}
$$

$$
H\ddot{O}\colon^{-} \;+\; CH_3CH_2-\ddot{B}r\colon \longrightarrow CH_3CH_2-\ddot{O}H \;+\; \colon\!\ddot{B}\!\ddot{r}\colon^{-}
$$

$$
\underset{\underset{CH_3}{|}}{CH_3-\overset{\overset{\colon\ddot{O}H}{|}}{C}-\ddot{C}l\colon} \longrightarrow CH_3-\overset{\overset{+\ddot{O}H}{\|}}{C}-CH_3 \;+\; \colon\!\ddot{C}\!\ddot{l}\colon^{-}
$$

$$
+ \;Br^- \;+\; H_3\overset{+}{O}\colon
$$

$$
CH_3-\overset{\overset{+\ddot{O}H}{\|}}{C}-CH_3 \;+\; H_2O\colon \longrightarrow \underset{\underset{\underset{H}{|}}{\overset{+}{\colon}OH}}{CH_3-\overset{\overset{\colon\ddot{O}H}{|}}{C}-CH_3}
$$

$$
CH_3CH_2-\underset{\underset{H}{|}}{\overset{+}{\ddot{O}}}H \;+\; H_2O\colon \longrightarrow CH_3CH_2-\ddot{O}H \;+\; H_3\overset{+}{O}\colon
$$

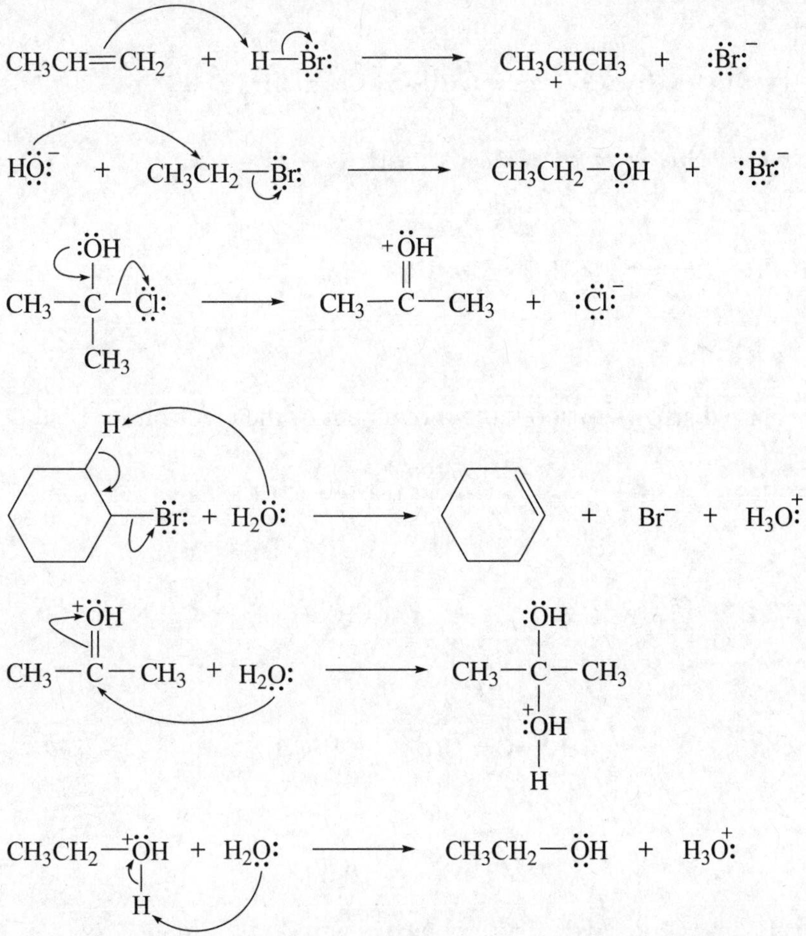

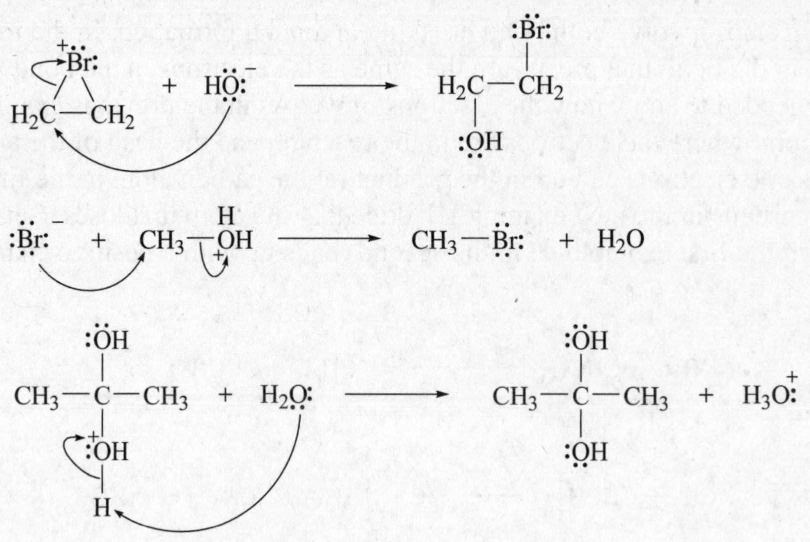

$$\overset{+}{\ddot{B}r:} \quad + \quad H\ddot{O}:^- \quad \longrightarrow \quad H_2C-CH_2$$
$$H_2C-CH_2 \qquad\qquad\qquad\qquad\qquad :OH , \quad :\ddot{B}r:$$

$$:\ddot{B}r:^- \quad + \quad CH_3-\overset{H}{\underset{+}{O}}H \quad \longrightarrow \quad CH_3-\ddot{B}r: \quad + \quad H_2O$$

$$\underset{\overset{+}{\underset{H}{:OH}}}{\overset{:\ddot{O}H}{CH_3-C-CH_3}} + H_2\ddot{O}: \quad \longrightarrow \quad \underset{:OH}{\overset{:\ddot{O}H}{CH_3-C-CH_3}} + H_3\overset{+}{O}:$$

Problem 3. Draw curved arrows to show the movement of the electrons that result in formation of the given product.

a. $\underset{CH_3}{\overset{:\ddot{O}H}{CH_3-\underset{H}{\overset{|}{C}}-\overset{+}{O}H}} \longrightarrow CH_3-\overset{+\ddot{O}H}{\overset{\|}{C}}-CH_3 + H_2O$ 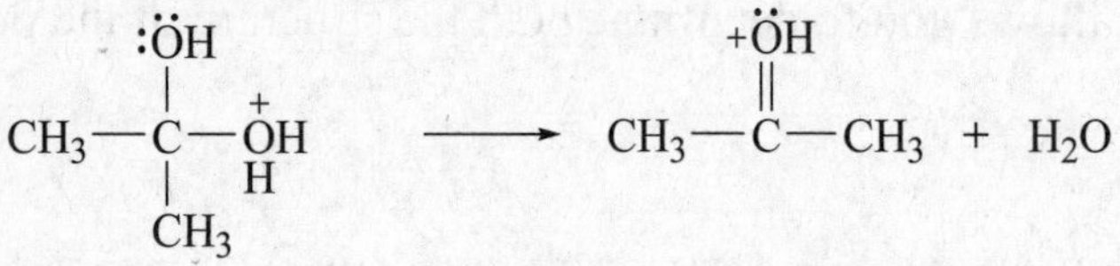

b. $CH_3CH_2CH=CH_2 + H-Cl \longrightarrow CH_3CH_2\overset{+}{C}H-CH_3 + Cl^-$

c. $CH_3CH_2-Br + \ddot{N}H_3 \longrightarrow CH_3CH_2-\overset{+}{N}H_3 + Br^-$

d. $\underset{H}{\overset{CH_3}{CH_3\overset{|}{C}-\overset{+}{C}HCH_3}} \longrightarrow \overset{CH_3}{CH_3\overset{|}{\underset{+}{C}}-CH_2CH_3}$ 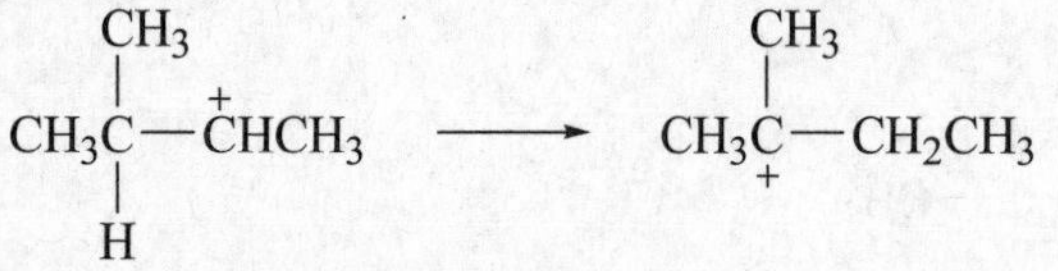

Problem 4. Draw curved arrows to show the movement of the electrons.

a. $CH_3CH=CHCH_3 + H-\underset{H}{\overset{+}{\ddot{O}}}-H \longrightarrow CH_3\overset{+}{C}H-CH_2CH_3 + H_2\ddot{O}:$ 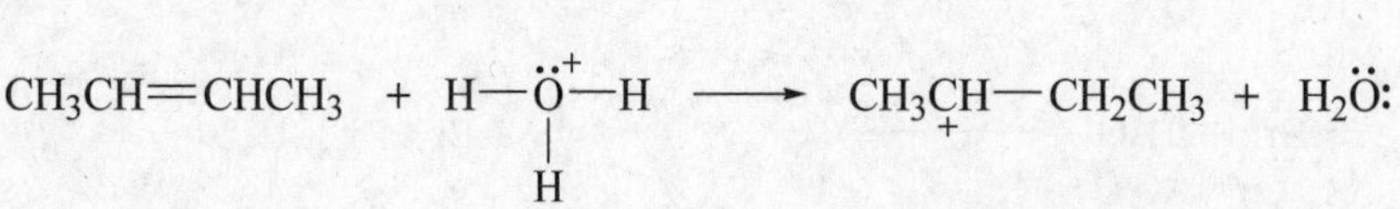

b. $CH_3CH_2CH_2CH_2-\ddot{C}l: + {}^-\ddot{C}\equiv N \longrightarrow CH_3CH_2CH_2CH_2-C\equiv N + :\ddot{C}l:^-$

c. $\underset{OH}{\overset{:\ddot{O}H}{CH_3-\underset{H}{\overset{|}{C}}-\overset{+}{\ddot{O}}CH_3}} \longrightarrow CH_3-\overset{+\ddot{O}H}{\overset{\|}{C}}-OH + CH_3\ddot{O}H$ 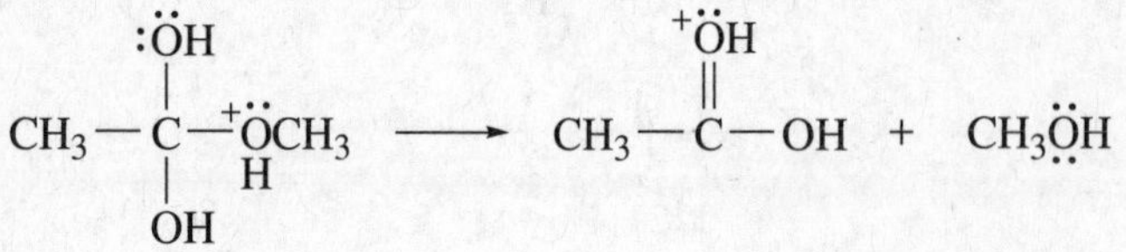

d. $CH_3-\overset{\ddot{O}:}{\overset{\|}{C}}-H + CH_3-MgBr \longrightarrow \underset{CH_3}{CH_3-\overset{:\ddot{O}:^-}{\overset{|}{C}}-H} + {}^+MgBr$ 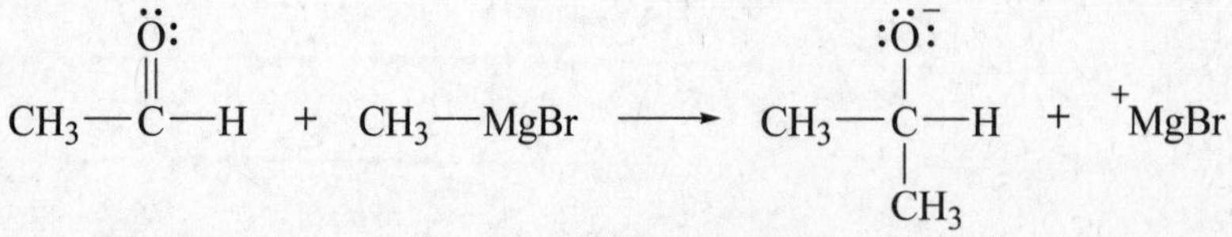

Problem 5. Draw curved arrows to show the movement of the electrons.

a. 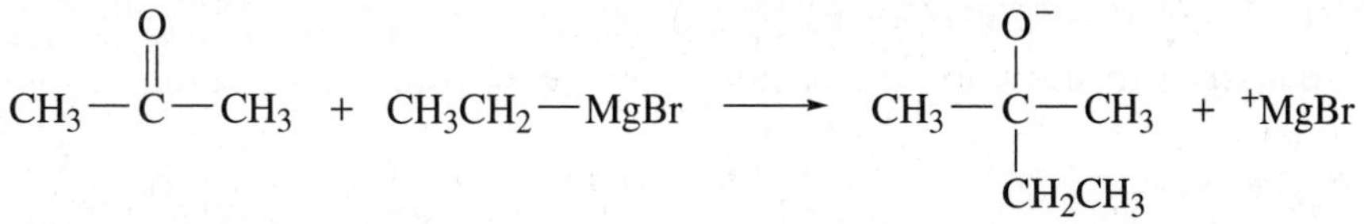

b. $CH_3CH_2CH_2-Br + CH_3\ddot{O}:^- \longrightarrow CH_3CH_2CH_2-OCH_3 + Br^-$

c. 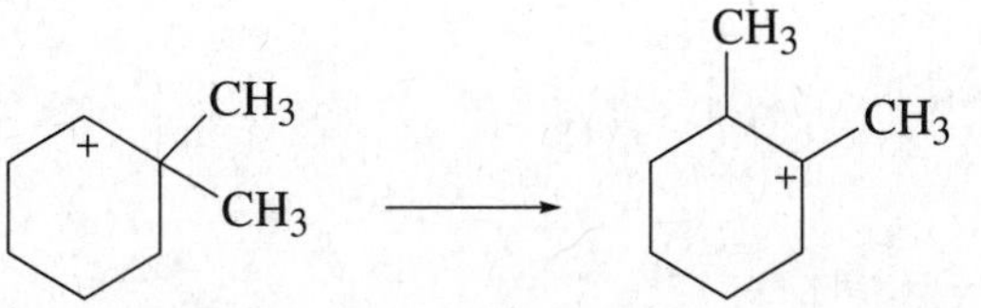

d. 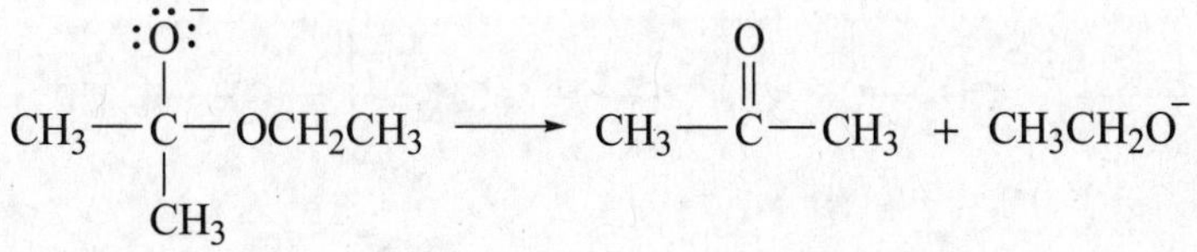

Problem 6. Draw curved arrows to show the movement of the electrons.

a. 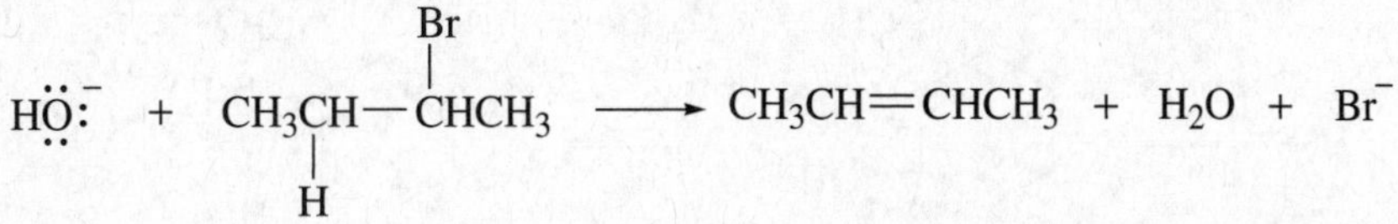

b. $CH_3CH_2C\equiv C-H + {}^-:\ddot{N}H_2 \longrightarrow CH_3CH_2C\equiv \ddot{C}^- + \ddot{N}H_3$

c. 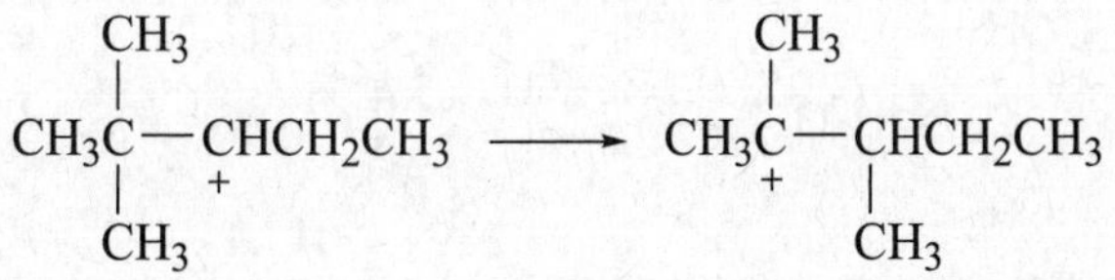

d. 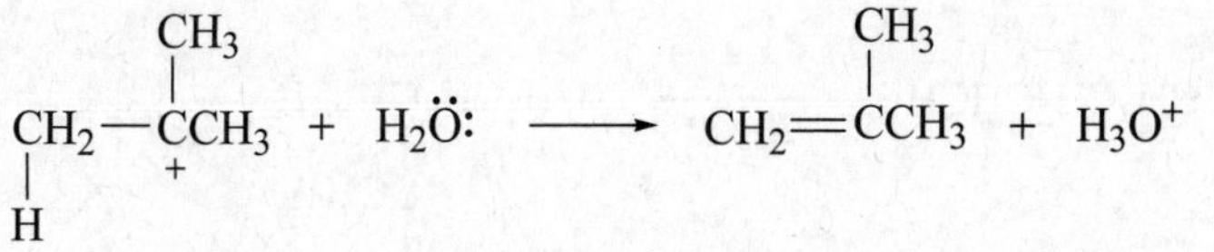

Problem 7. Draw curved arrows to show the movement of the electrons.

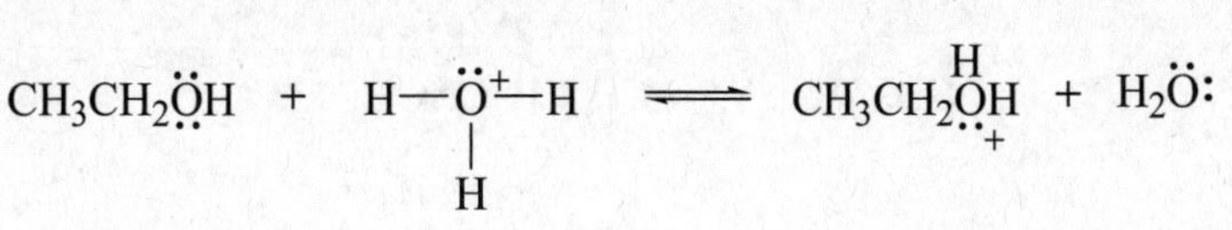

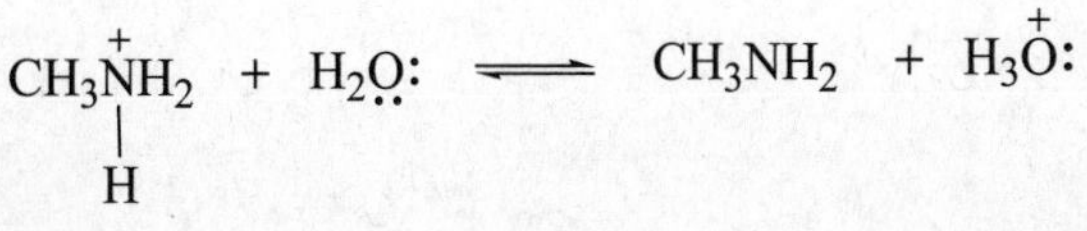

Curved arrows are used to show the movement of electrons in each step of a reaction.

Problem 8.   Draw curved arrows to show the movement of the electrons in each step of the following reaction sequences.

a. $CH_3CH{=}CH_2$ + $H{-}\overset{\cdot\cdot}{Br}{:}$ $\longrightarrow$ $CH_3\underset{+}{CH}{-}CH_3$ + $:\overset{\cdot\cdot}{Br}{:}^-$ $\longrightarrow$ $CH_3CH{-}CH_3$ 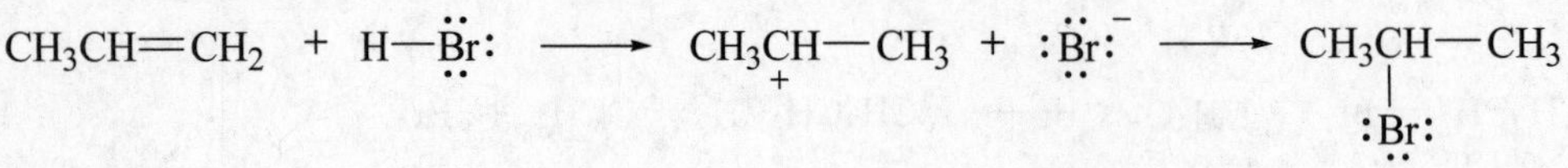
$\qquad\qquad\qquad\qquad\qquad\qquad\qquad\qquad\qquad\qquad\qquad\qquad\qquad\qquad\qquad\quad :\overset{\cdot\cdot}{Br}{:}$

b. $CH_3\overset{\displaystyle CH_3}{\underset{\displaystyle CH_3}{C}}{-}Cl$ $\rightleftharpoons$ $CH_3\overset{\displaystyle CH_3}{\underset{\displaystyle CH_3}{C}}{+}$ + Cl⁻ $\xrightarrow{\ddot{N}H_3}$ $CH_3\overset{\displaystyle CH_3}{\underset{\displaystyle CH_3}{C}}\overset{+}{-}NH_3$ 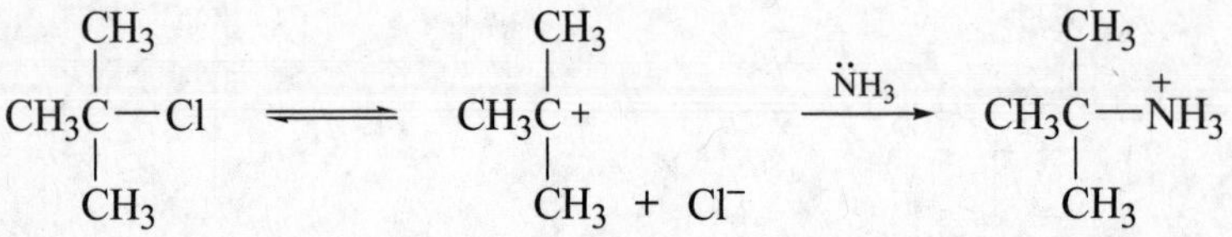

c. $CH_3{-}\overset{\displaystyle \overset{\cdot\cdot}{O}{:}}{C}{-}Cl$ + $H\overset{\cdot\cdot}{\underset{\cdot\cdot}{O}}{:}^-$ $\longrightarrow$ $CH_3{-}\overset{\displaystyle :\overset{\cdot\cdot}{O}{:}^-}{\underset{\displaystyle :OH}{C}}{-}Cl$ $\longrightarrow$ $CH_3{-}\overset{\displaystyle \overset{\cdot\cdot}{O}{:}}{C}{-}OH$ + Cl⁻ 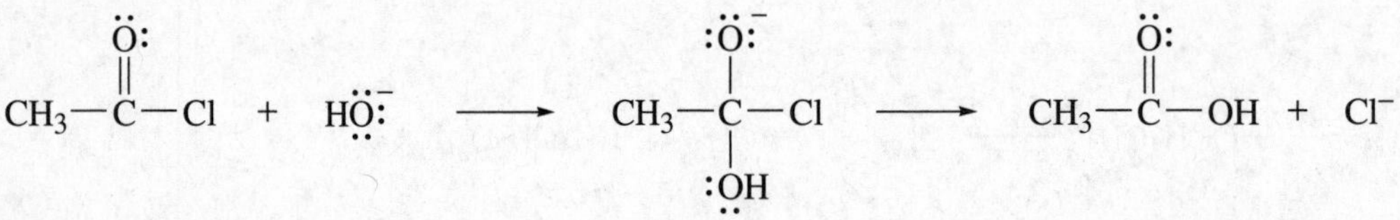

Problem 9.   Draw curved arrows to show the movement of the electrons in each step of the following reactions.

a. 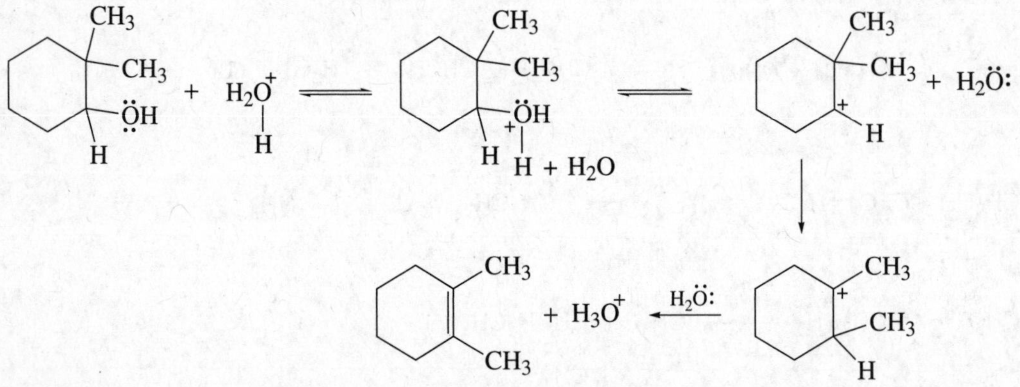

b. $CH_3CH_2CH{=}CH_2$ $\xrightarrow{\overset{\displaystyle H}{\underset{\displaystyle }{CH_3\overset{+}{O}H}}}$ $CH_3CH_2\underset{+}{CH}CH_3$ $\xrightarrow{CH_3\overset{\cdot\cdot}{O}H}$ $CH_3CH_2CHCH_3$ 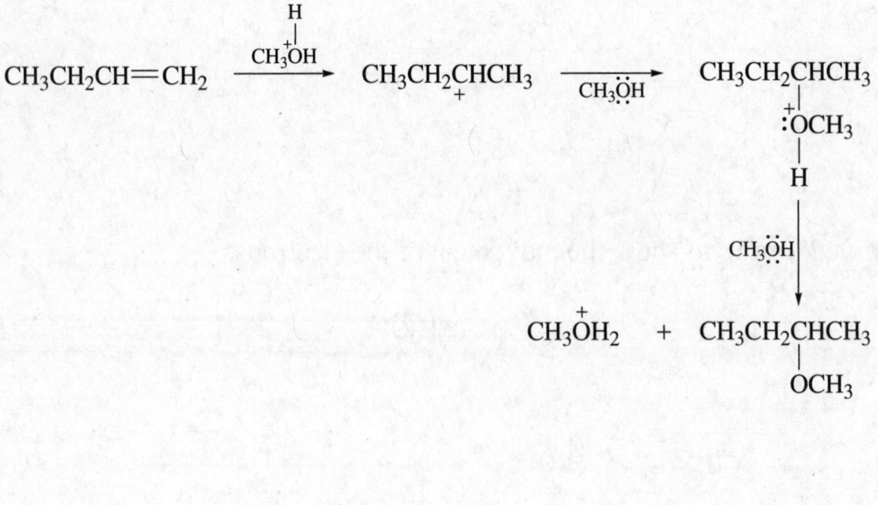
$\qquad\qquad\qquad\qquad\qquad\qquad\qquad\qquad\qquad\qquad\qquad\qquad\qquad\qquad\qquad\quad \overset{+}{:}OCH_3$
$\qquad\qquad\qquad\qquad\qquad\qquad\qquad\qquad\qquad\qquad\qquad\qquad\qquad\qquad\qquad\quad H$
$\qquad\qquad\qquad\qquad\qquad\qquad\qquad\qquad\qquad\qquad\qquad\qquad\qquad\qquad\quad CH_3\overset{\cdot\cdot}{O}H$
$CH_3\overset{+}{O}H_2$ + $CH_3CH_2CHCH_3$
$\qquad\qquad\qquad\qquad\quad OCH_3$

Problem 10. Use what the curved arrows tell you about electron movement to determine the product of each reaction step.

a. 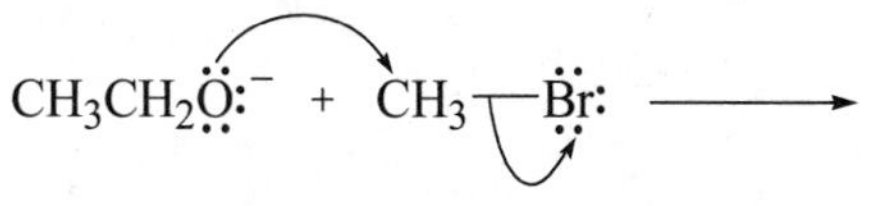

b. 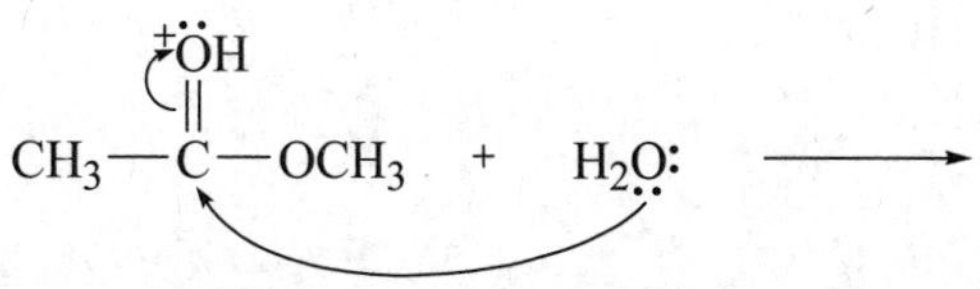

c. 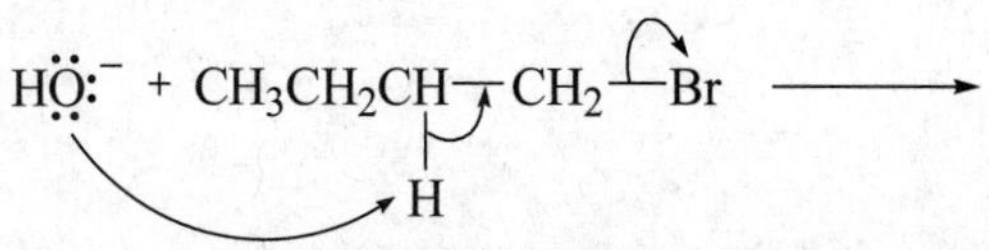

d. 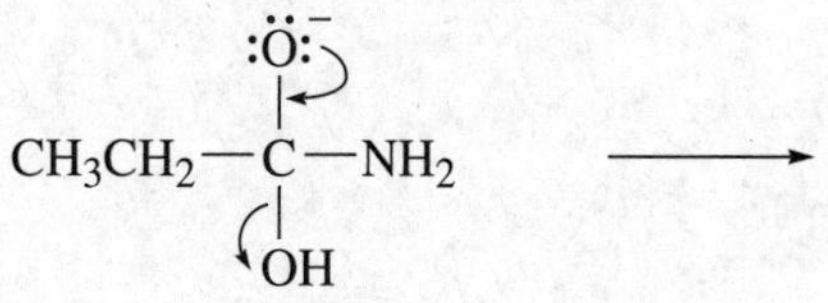

e. 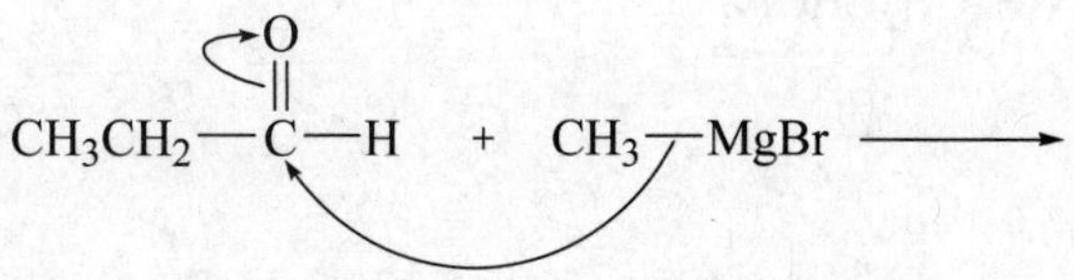

f. 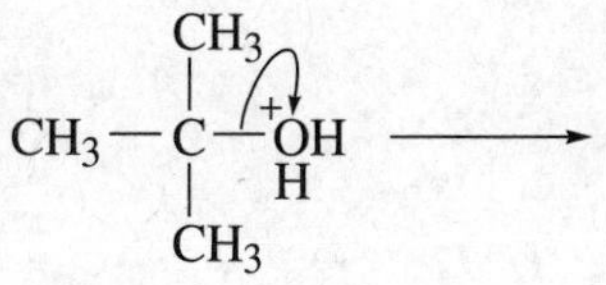

g. 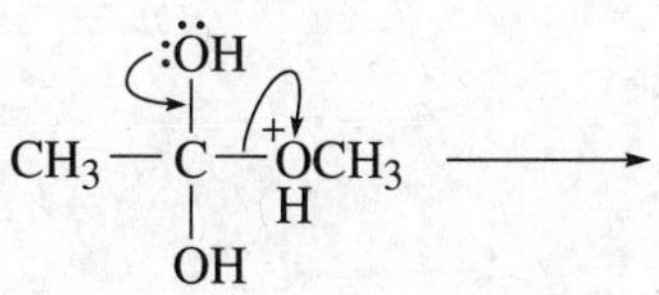

Answers to Electron Pushing Problems

Problem 1

a. 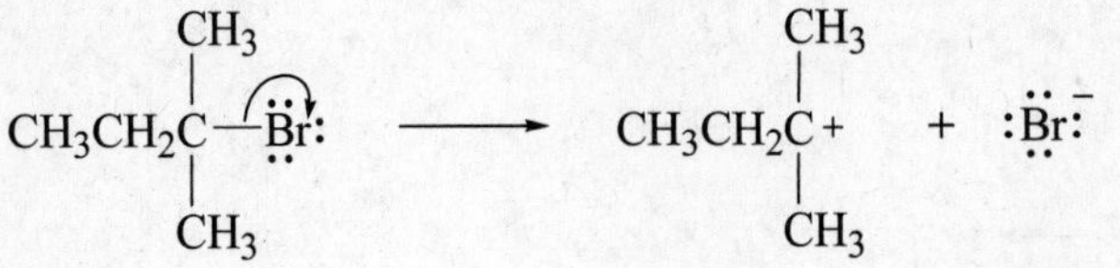

b. 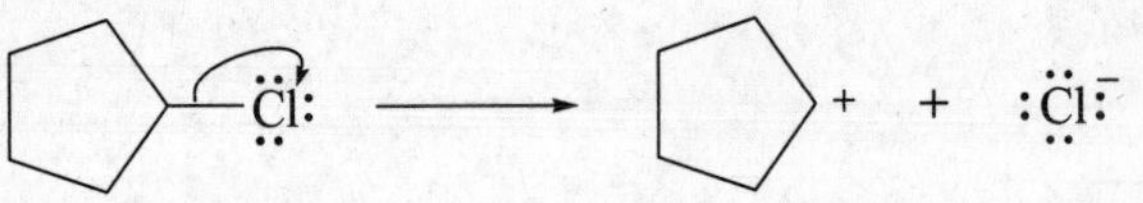

c. 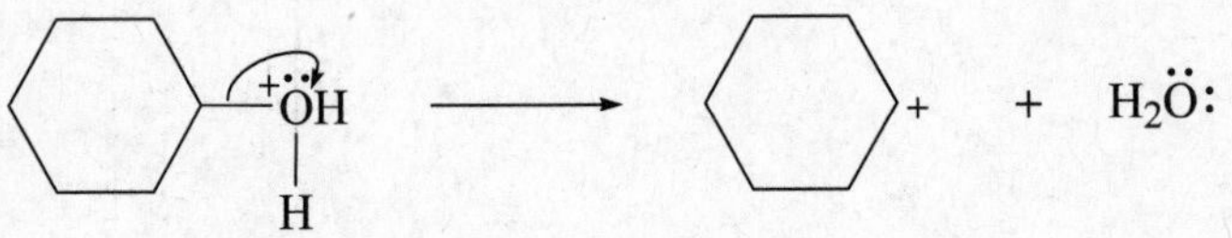

d. 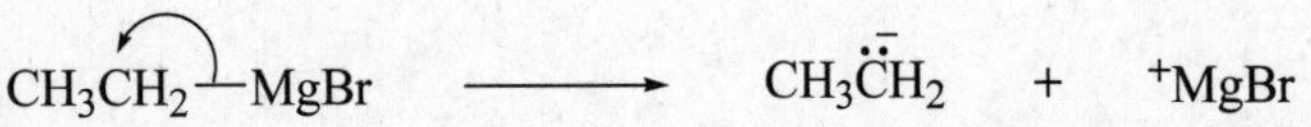

e. 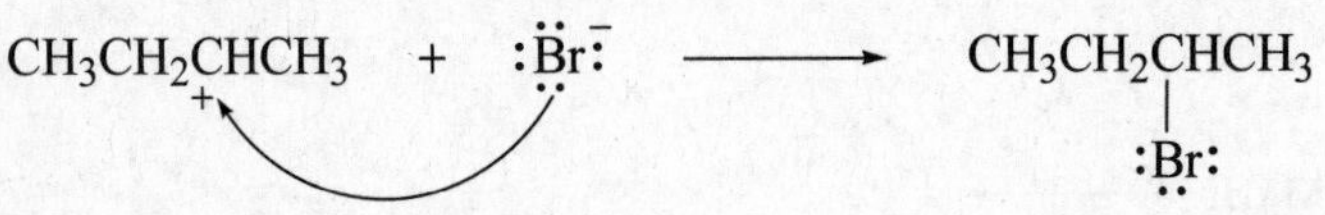

Problem 2

a. 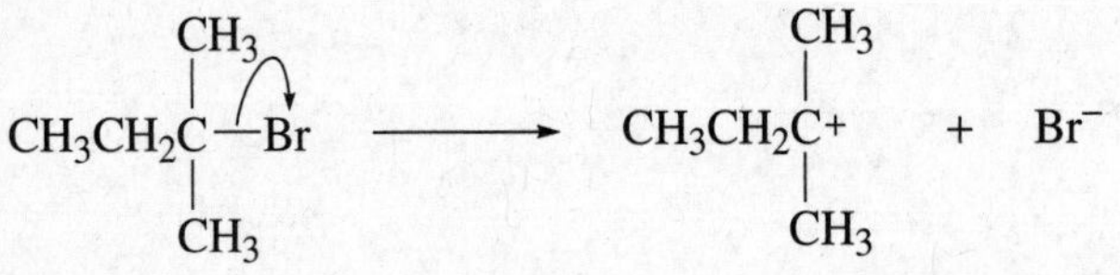

b. 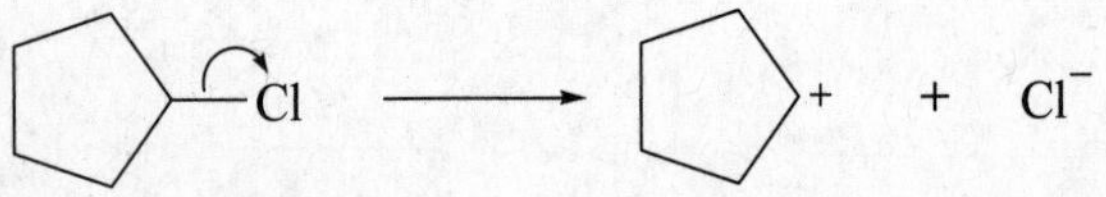

c. 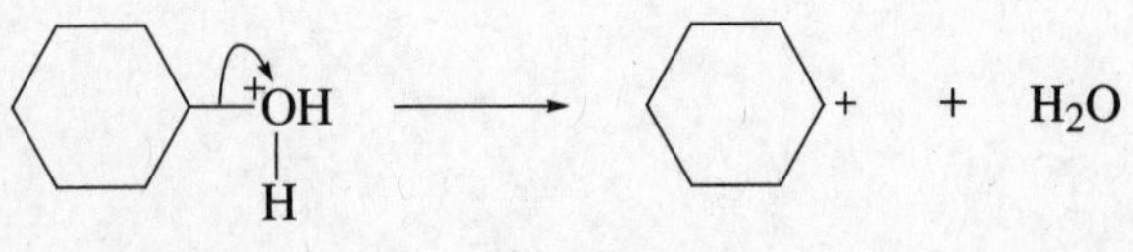

d. 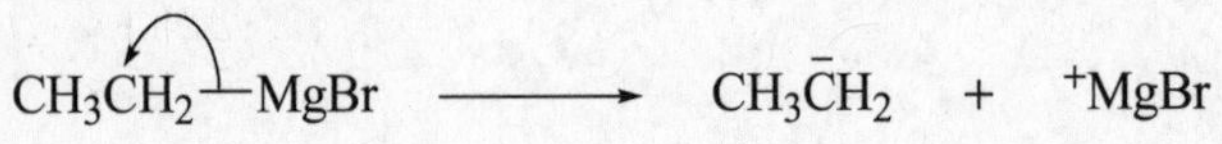

Problem 3

a. 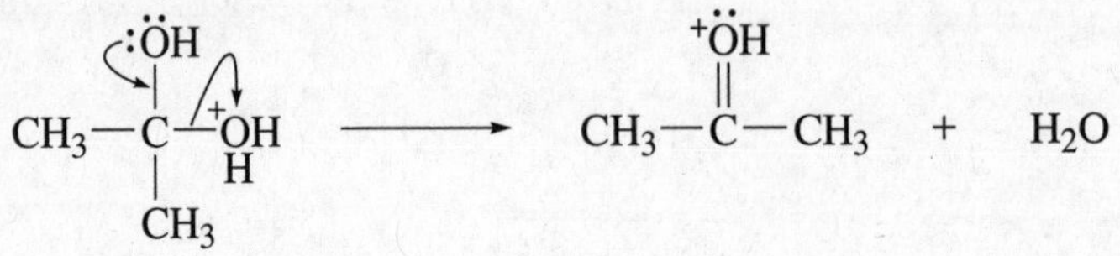

b. 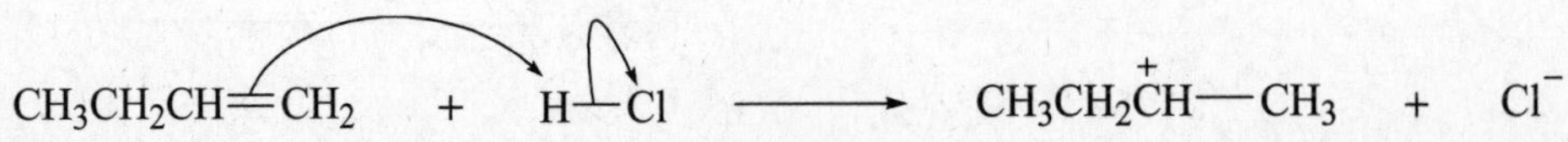

c. 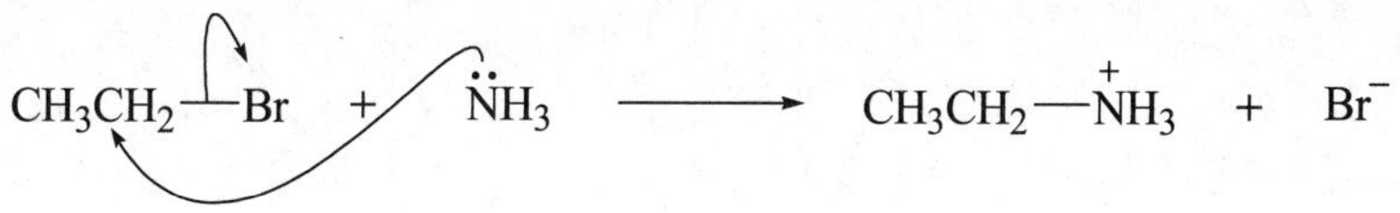

d. 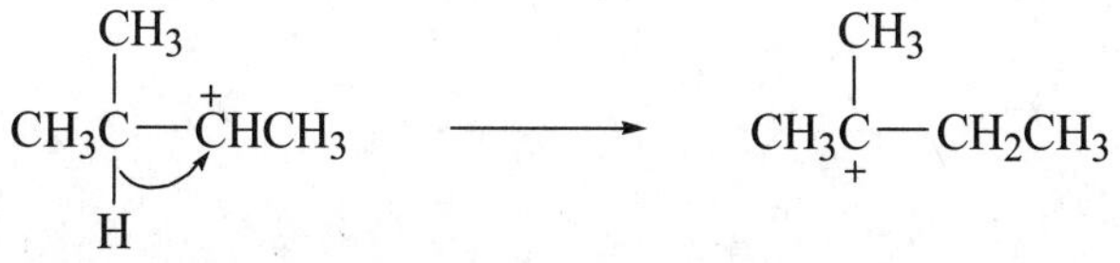

Problem 4

a. 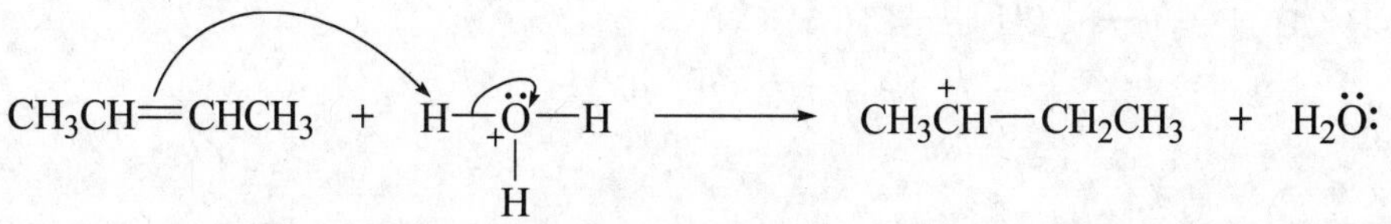

b. 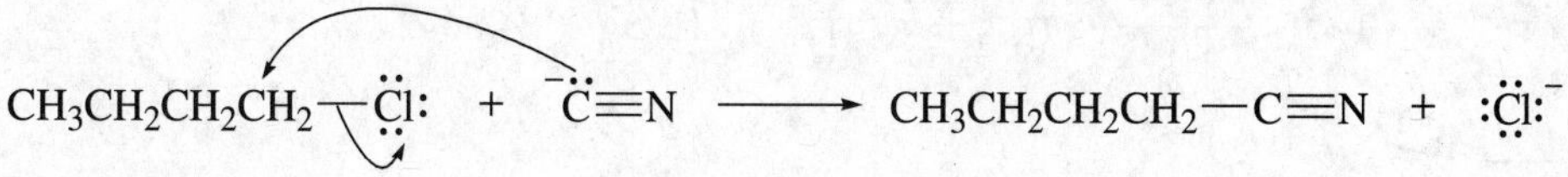

c. 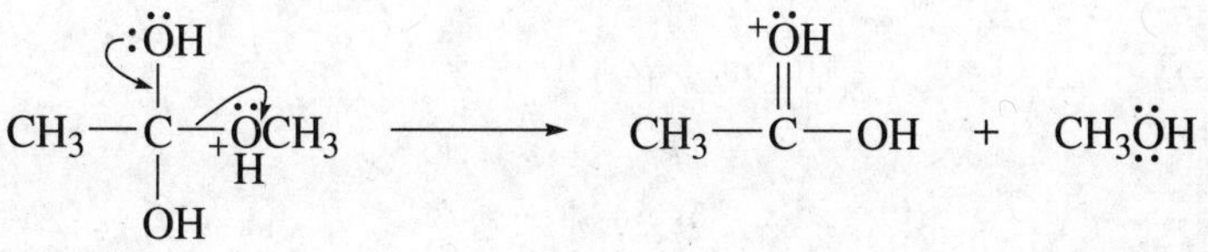

d. 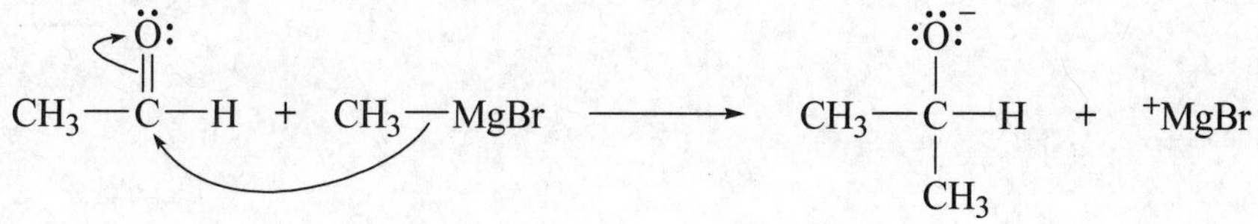

Problem 5

a. 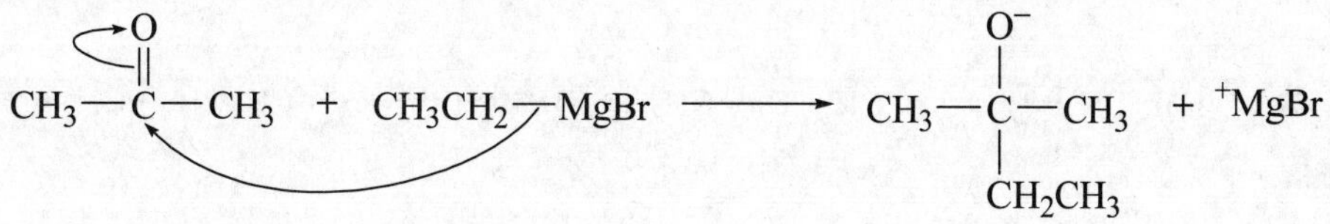

b. 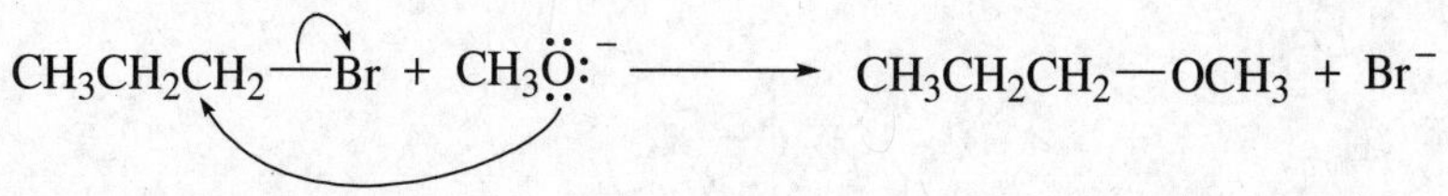

c. 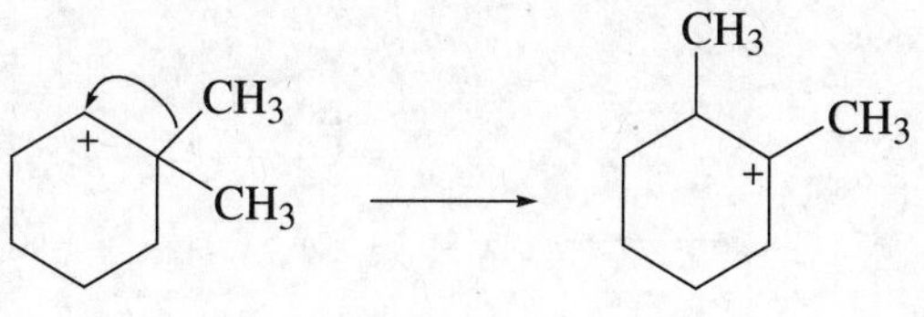

d. 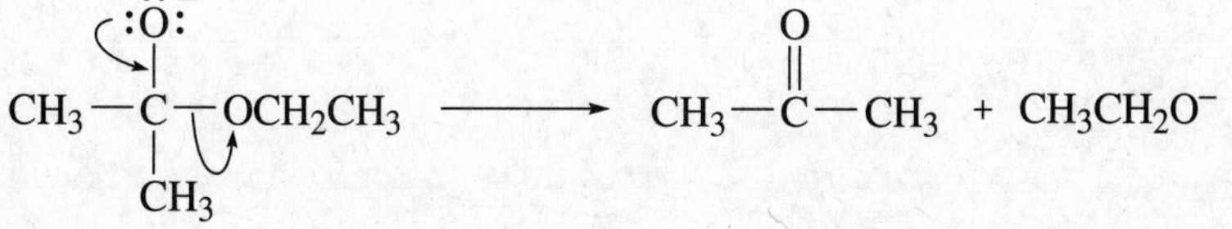

Problem 6

a. 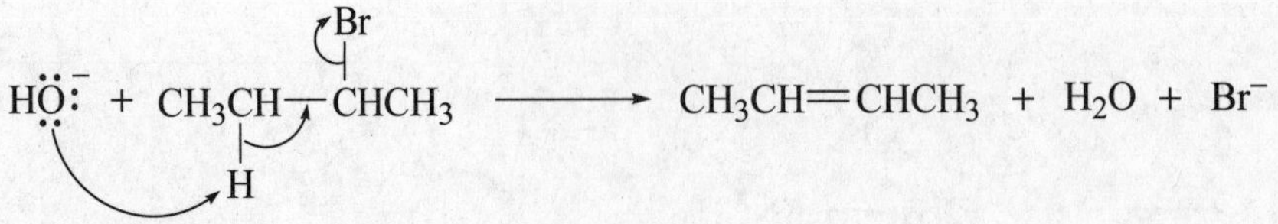
HO: + CH3CH—CHCH3 ⟶ CH3CH=CHCH3 + H2O + Br−
Br
H

b. 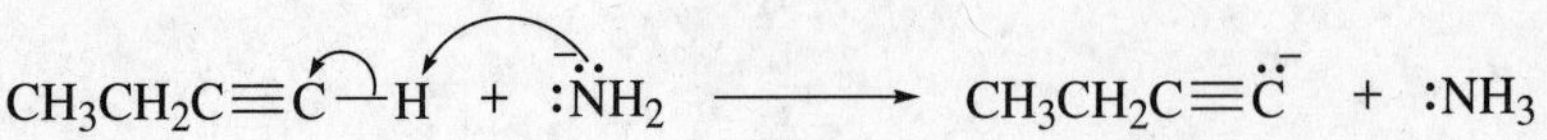
CH3CH2C≡C—H + :NH2 ⟶ CH3CH2C≡C + :NH3

c. 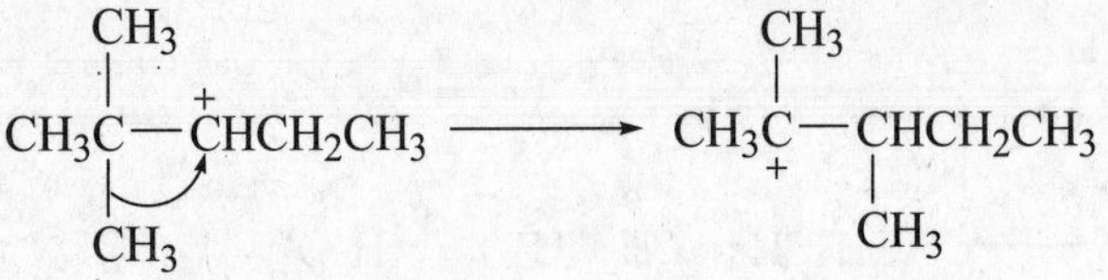
CH3C—CHCH2CH3 ⟶ CH3C—CHCH2CH3
CH3
CH3

d. 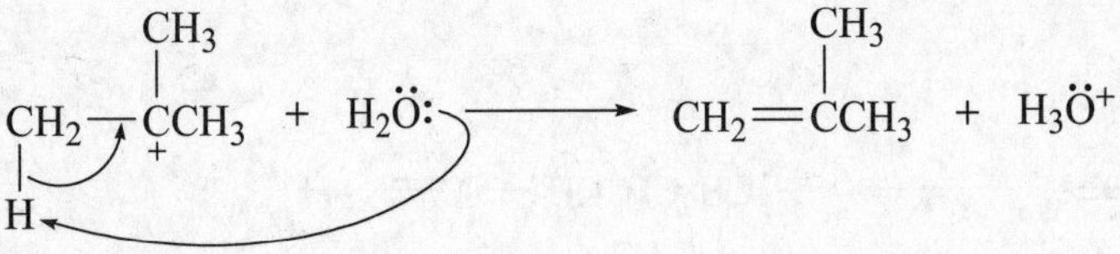
CH2—CCH3 + H2O: ⟶ CH2=CCH3 + H3O+
CH3
H

Problem 7

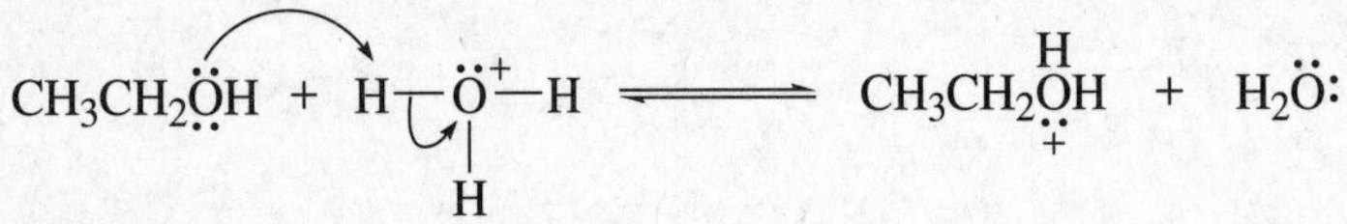
CH3CH2OH + H—O—H ⇌ CH3CH2OH + H2O:
H

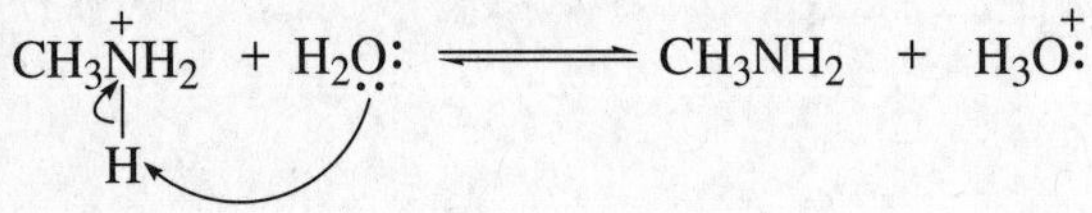
CH3NH2 + H2O: ⇌ CH3NH2 + H3O:
H

Problem 8

a. 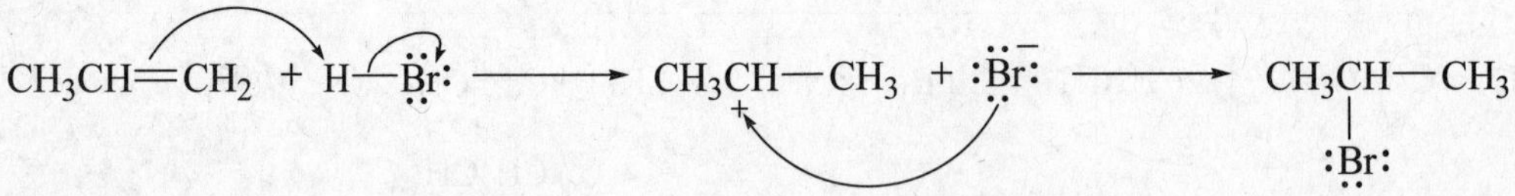
CH3CH=CH2 + H—Br: ⟶ CH3CH—CH3 + :Br: ⟶ CH3CH—CH3
:Br:

b. 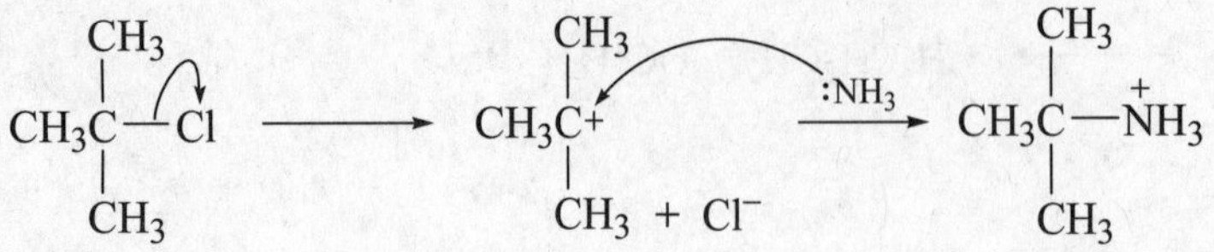
CH3C—Cl ⟶ CH3C+ ⟶ CH3C—NH3
:NH3
CH3 + Cl−

c. 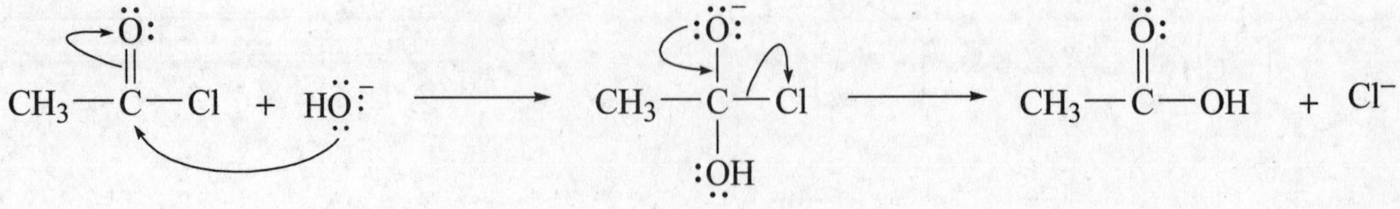
CH3—C—Cl + HO: ⟶ CH3—C—Cl ⟶ CH3—C—OH + Cl−
:OH

Problem 9

a.

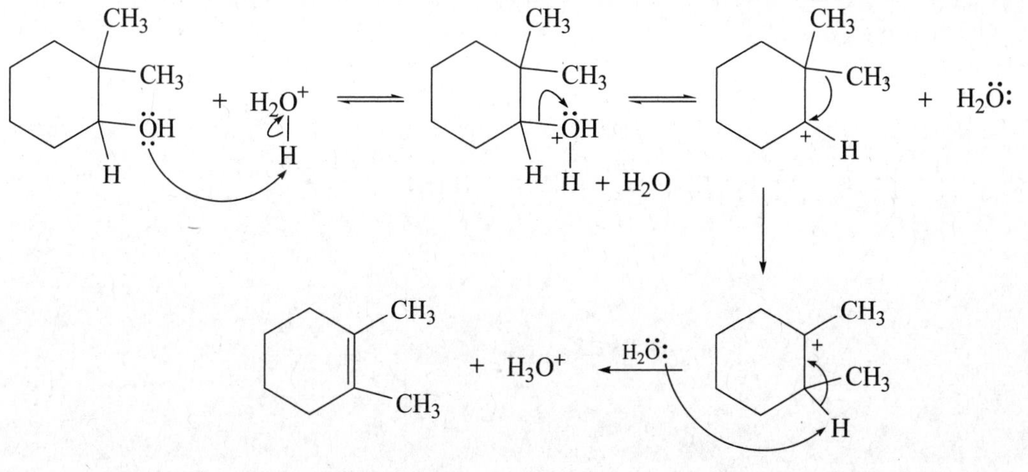
CH3
CH3
ÖH
H
+ H2O+
H
CH3
CH3
+ÖH
H
H + H2O
CH3
CH3
+
H
+ H2Ö:
CH3
+
CH3
H
H2Ö:
CH3
CH3
+ H3O+

b.

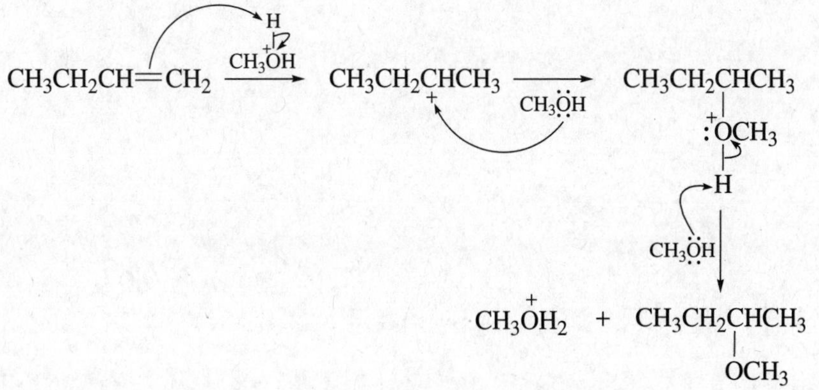
CH3CH2CH=CH2
H
CH3ÖH
+
CH3CH2CHCH3
+
CH3ÖH
CH3CH2CHCH3
+
:OCH3
H
CH3ÖH
+
CH3OH2 + CH3CH2CHCH3
OCH3

Problem 10

a.

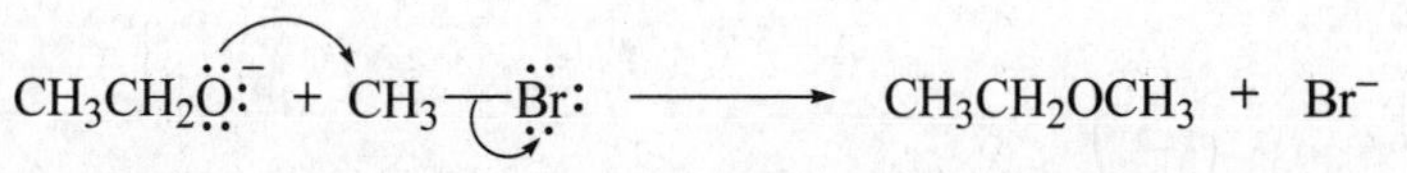
CH3CH2Ö:− + CH3—Br:
CH3CH2OCH3 + Br−

b.

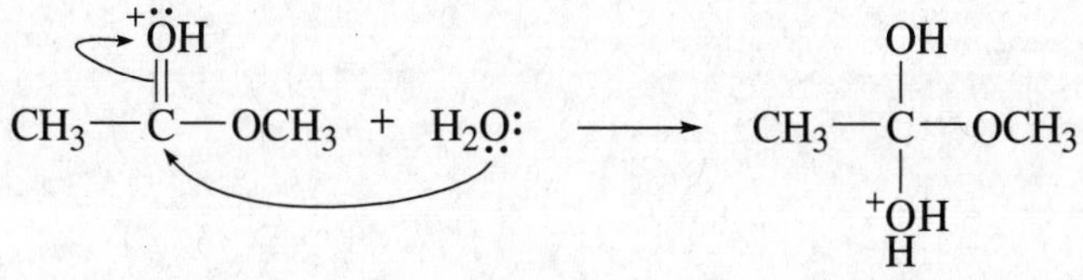
+ÖH
CH3—C—OCH3 + H2O:
OH
CH3—C—OCH3
+OH
H

c.

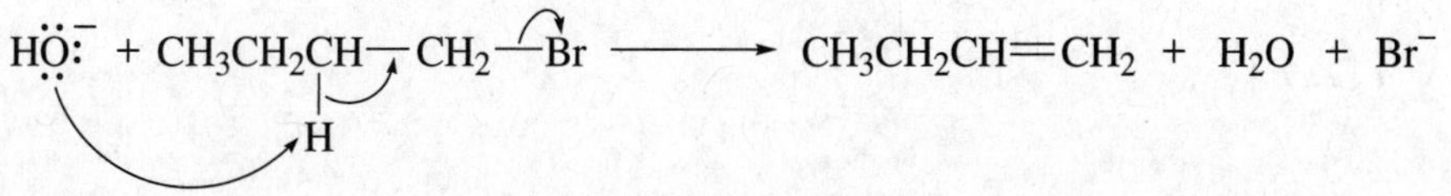
HÖ:− + CH3CH2CH—CH2—Br
H
CH3CH2CH=CH2 + H2O + Br−

d.

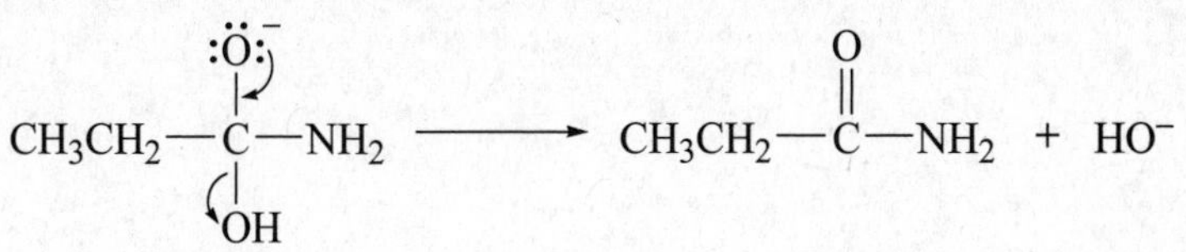
:Ö:−
CH3CH2—C—NH2
OH
O
CH3CH2—C—NH2 + HO−

e.

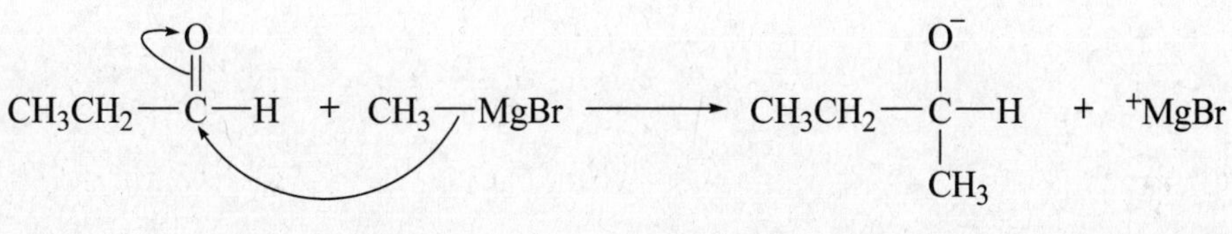
O
CH3CH2—C—H + CH3—MgBr
O−
CH3CH2—C—H + +MgBr
CH3

f. 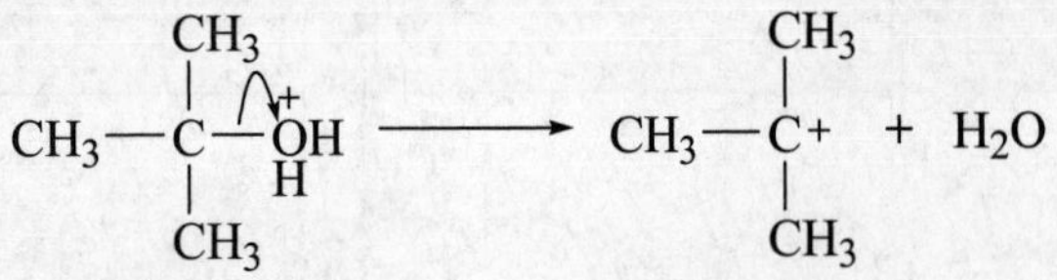

g. 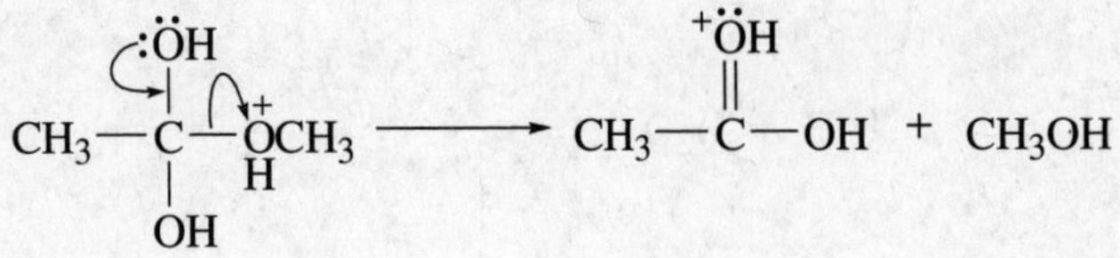

CHAPTER 4

Alkenes: Structure, Nomenclature, Stability, and an Introduction to Reactivity

1. **a.** C_5H_8 **b.** C_4H_6 **c.** $C_{10}H_{16}$

2. **a.** Solved in the text. **b.** 4 **c.** 1

3. **a.**

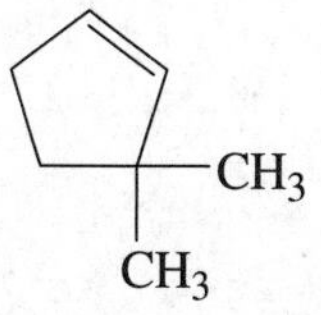

c. $CH_3CH_2OCH{=}CH_2$

b.

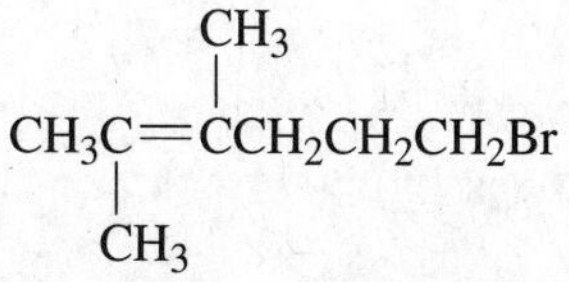

d. $CH_2{=}CHCH_2OH$

4. **a.** 4-methyl-2-pentene **c.** 1-bromo-4-methyl-3-hexene

b. 2-chloro-3,4-dimethyl-3-hexene **d.** 1,5-dimethylcyclohexene

5. **a.** Solved in the text. **b.** 4 **c.** 4 **d.** 6

6. **a.** 1 and 3

b. **1.**

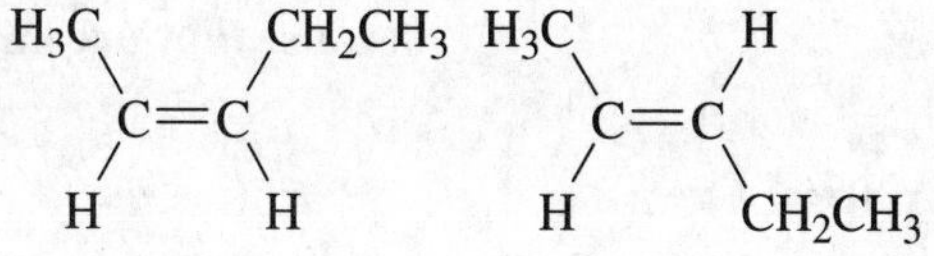

cis trans

3.

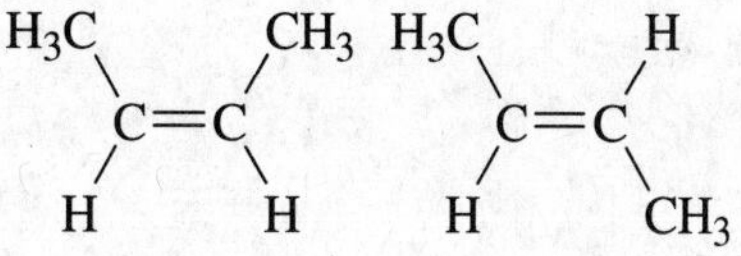

cis trans

7. **a.** –I > –Br > –OH > $–CH_3$

b. –OH > $–CH_2Cl$ > $–CH{=}CH_2$ > $–CH_2CH_2OH$

8. **a.**

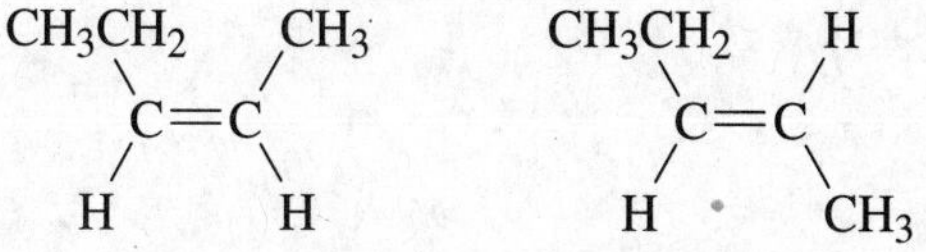

Z *E*

b.

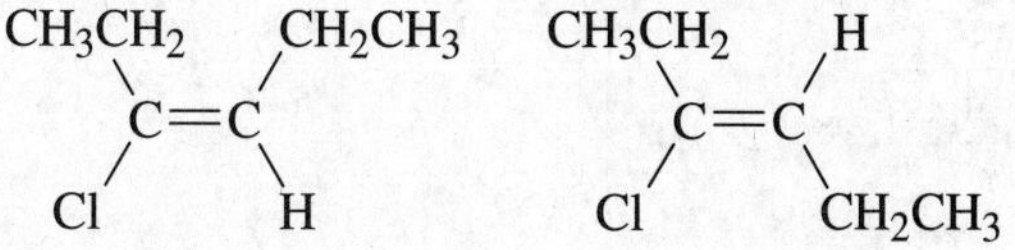

E *Z*

c.

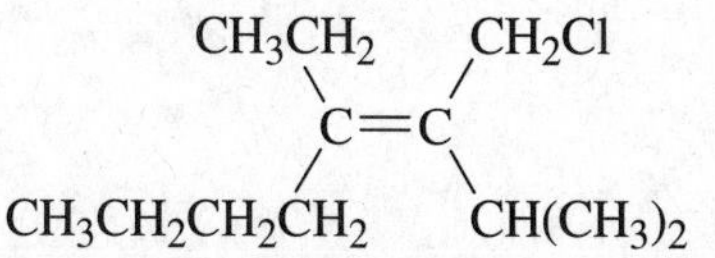

E

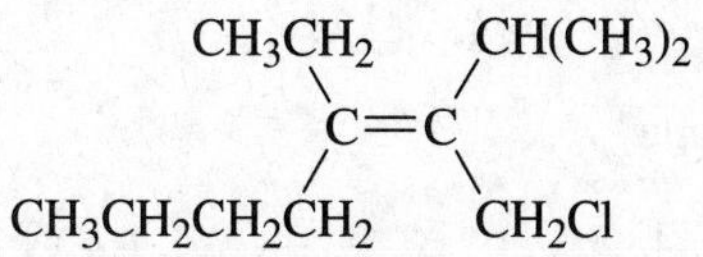

Z

9.   **a.** *Z*   **b.** *E*

10. 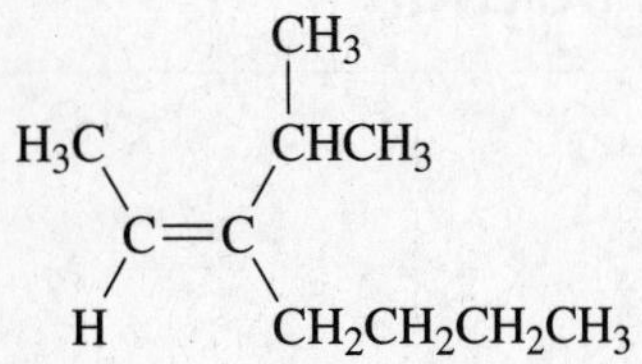

$$
\begin{array}{c}
\underset{\text{CH}_3}{\overset{\displaystyle \text{CH}_3}{\underset{|}{}}} \\
\overset{\displaystyle \text{H}_3\text{C}}{}\qquad \overset{\displaystyle \text{CHCH}_3}{} \\
\text{C}{=}\text{C} \\
\overset{\displaystyle \text{H}}{}\qquad \overset{\displaystyle \text{CH}_2\text{CH}_2\text{CH}_2\text{CH}_3}{}
\end{array}
$$

11.   **a.** 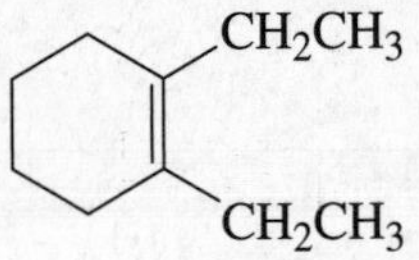

This alkene is the most stable because it has the greatest number of alkyl substituents bonded to the sp^2 carbons.

b. 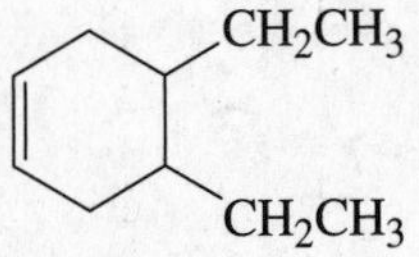

This alkene is the least stable because it has the fewest number of alkyl substituents bonded to the sp^2 carbons.

12. 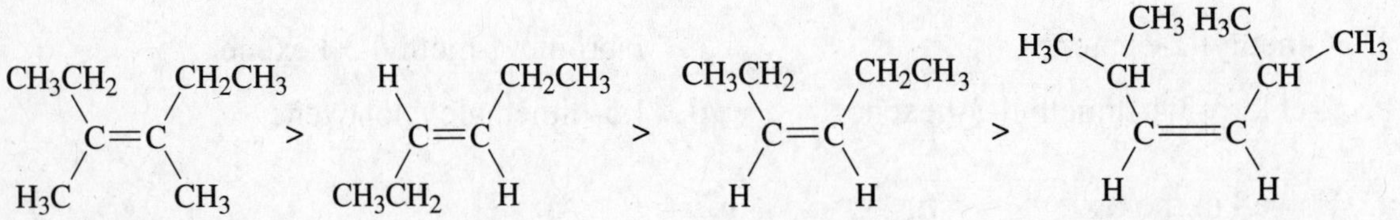

cis-3,4-dimethyl-3-hexene	*trans*-3-hexene	*cis*-3-hexene	*cis*-2,5-dimethyl-3-hexene
4 substituents	**2 trans substituents**	**2 cis substituents**	**2 cis substituents that**
most stable			**cause greater steric strain**
			because they are larger

13.   **a.** AlCl_3 + $:\text{NH}_3$ $\rightleftharpoons$ $\text{Cl}_3\overset{-}{\text{Al}}{-}\overset{+}{\text{N}}\text{H}_3$

        electrophile   nucleophile

b. $\text{H}{-}\ddot{\text{B}}\text{r}:$ + $\text{H}\ddot{\text{O}}:^{-}$ $\rightleftharpoons$ $:\ddot{\text{B}}\text{r}:^{-}$ + $\text{H}_2\ddot{\text{O}}:$

        electrophile   nucleophile

14.   nucleophiles: H^{-} CH_3O^{-} $\text{CH}_3\text{C}{\equiv}\text{CH}$ NH_3

        electrophiles: $\text{CH}_3\overset{+}{\text{C}}\text{HCH}_3$

15.   **a.** $\text{CH}_3\overset{\displaystyle \text{O}}{\overset{\|}{\text{C}}}{-}\text{O}{-}\text{H}$ + $\text{H}\ddot{\text{O}}:^{-}$ $\longrightarrow$ $\text{CH}_3\overset{\displaystyle \text{O}}{\overset{\|}{\text{C}}}{-}\text{O}^{-}$ + H_2O 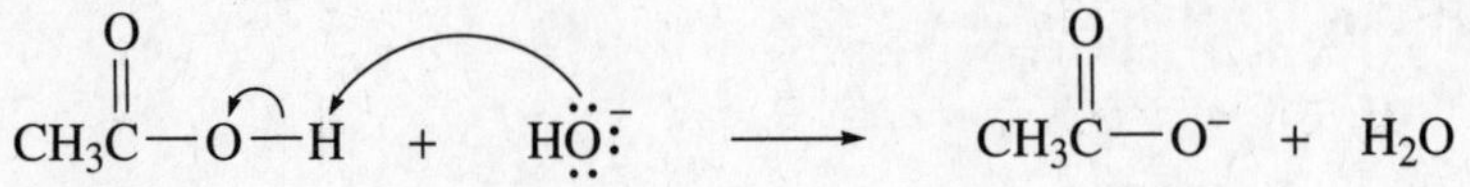

        electrophile   nucleophile

b. 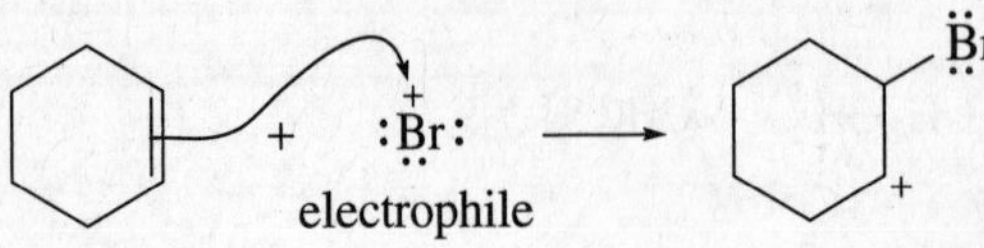

        nucleophile   electrophile

c. 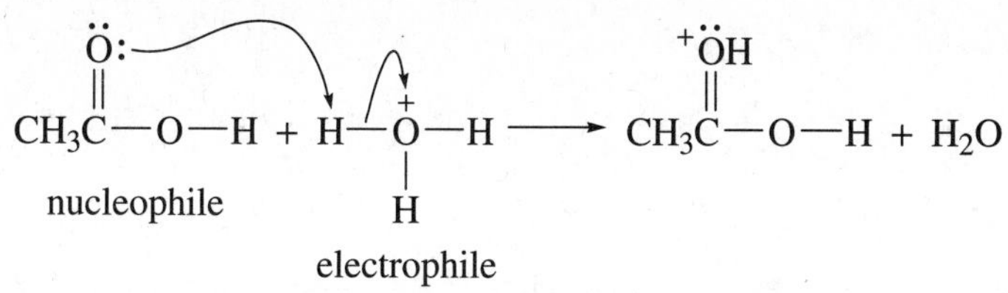
nucleophile
electrophile

d. 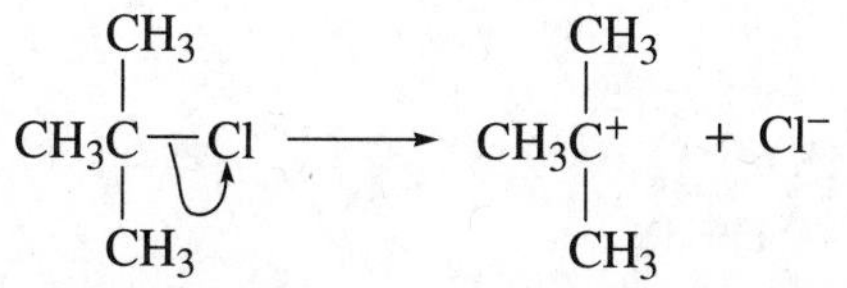

16. The reactants are labeled (above). Note that part d. has only one reactant, so the reaction does not involve the reaction of an electrophile with a nucleophile; it involves dissociation of a bond.

17. **a.**

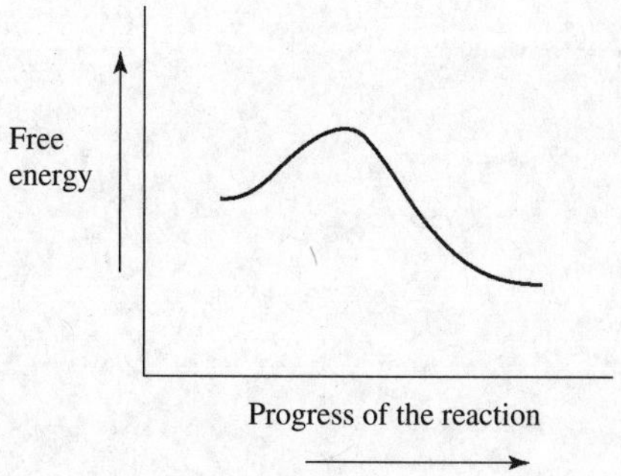
Free energy
Progress of the reaction

b.

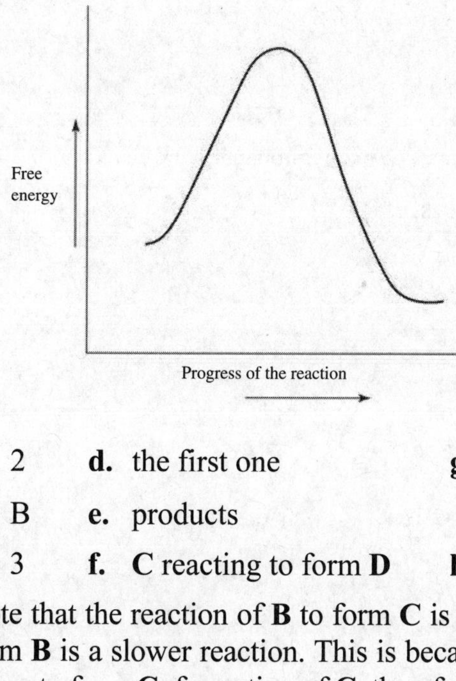
Free energy
Progress of the reaction

c.

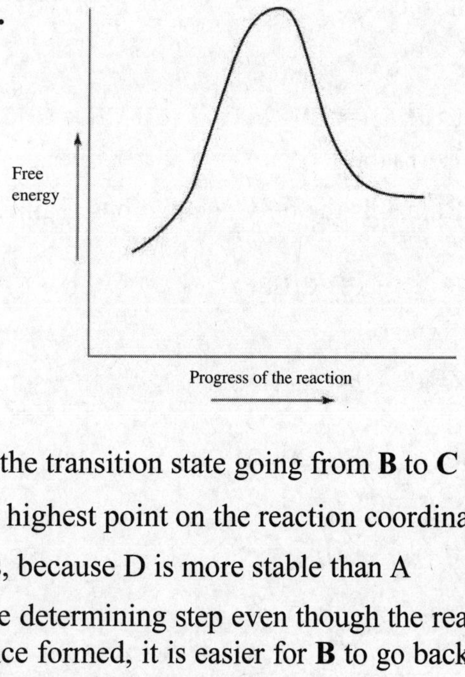
Free energy
Progress of the reaction

18. **a.** 2 **d.** the first one **g.** **B** (the transition state going from **B** to **C** is at the highest point on the reaction coordinate)

b. B **e.** products

c. 3 **f.** **C** reacting to form **D** **h.** yes, because D is more stable than A

Note that the reaction of **B** to form **C** is the rate determining step even though the reaction of **A** to form **B** is a slower reaction. This is because once formed, it is easier for **B** to go back to **A** than to go on to form **C**; formation of **C**, therefore, is the rate-determining step.

19. **a.** 4-methylcyclohexene **c.** 1-methyl-1,3-cyclopentadiene

b. 1-ethyl-1,4-cyclohexadiene

20. If the number of carbons is 30, $C_nH_{2n+2} = C_{30}H_{62}$.

A compound with molecular formula $C_{30}H_{50}$ is missing 12 hydrogens. Because it has no rings, squalene has 6 π bonds (12/2 = 6).

21. **a.** 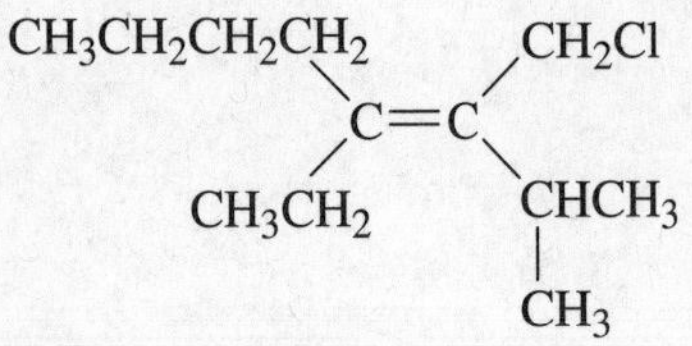 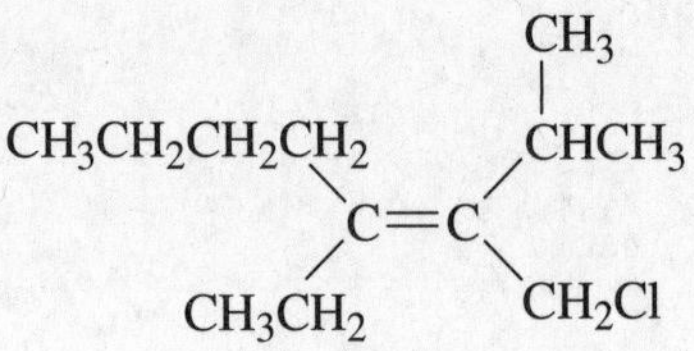

Z *E*

b. 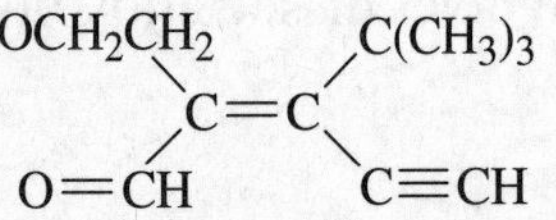 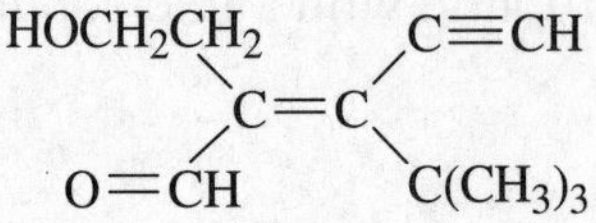

Z *E*

22. **a.** 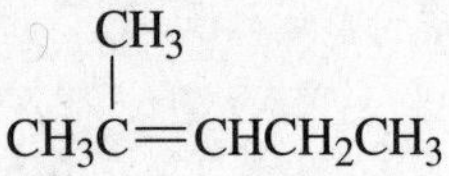

One hydrogen is attached to the sp^2 carbons

b. 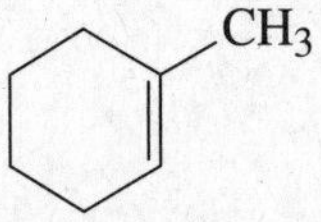

One hydrogen is attached to the sp^2 carbons

23. **a.** $CH_3CH_2CH{=}CH_2$ 1-butene; $CH_3CH{=}CHCH_3$ 2-butene; 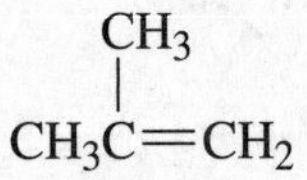 2-methylpropene

b. 2-Butene is the only one that has *E* and *Z* isomers.

24. **a.** 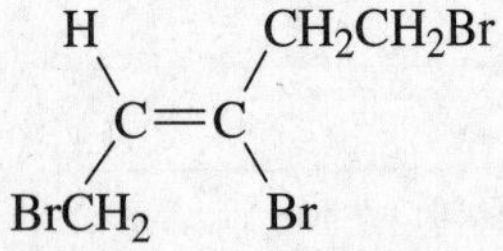

d. $CH_2{=}CHBr$

b. 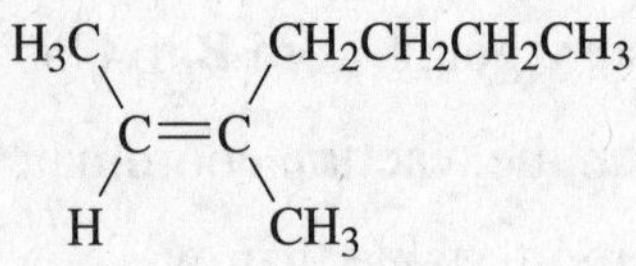

e. 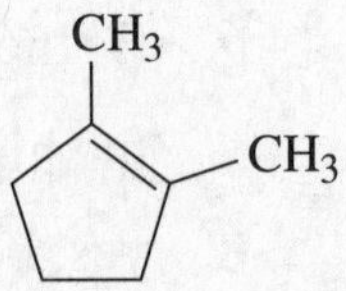

c. 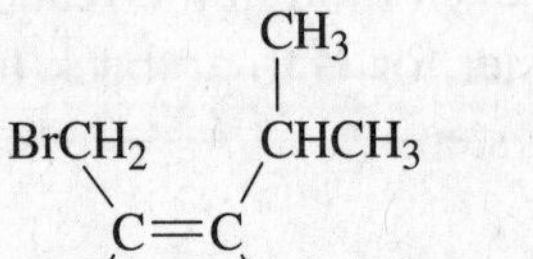

f. $CH_2{=}CHCH_2NHCH_2CH{=}CH_2$

25. **a.** 3 **b.** 13

26. **a.** (*E*)-3-methyl-3-hexene

b. *trans*-8-methyl-4-nonene or (*E*)-8-methyl-4-nonene

c. *trans*-9-bromo-2-nonene or (*E*)-9-bromo-2-nonene

d. 2,4-dimethyl-1-pentene

e. 2-ethyl-1-pentene

f. *cis*-2-pentene or (*Z*)-2-pentene

27.

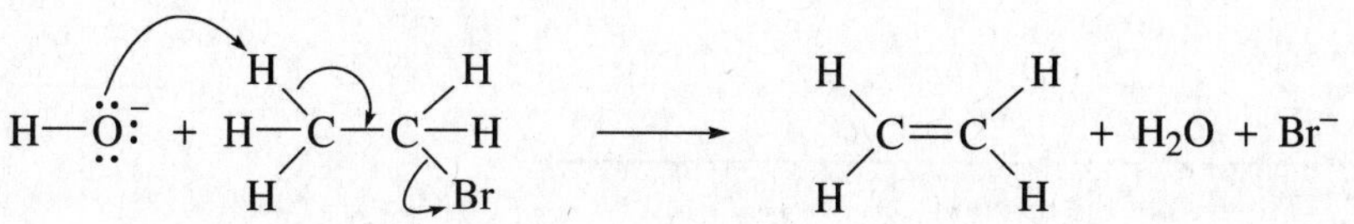

28.

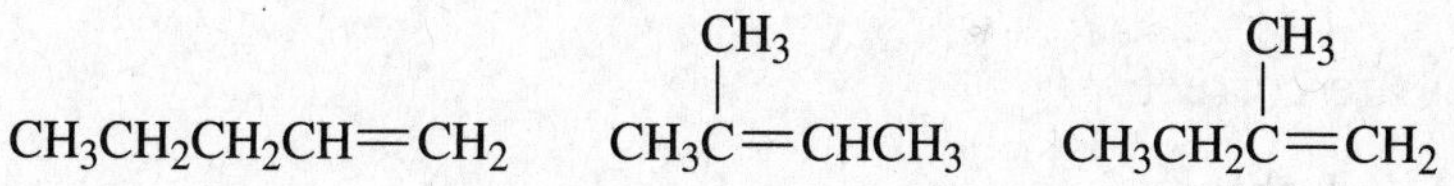

29. **a.** *Z* **b.** *E* **c.** *E* **d.** *Z*

30.

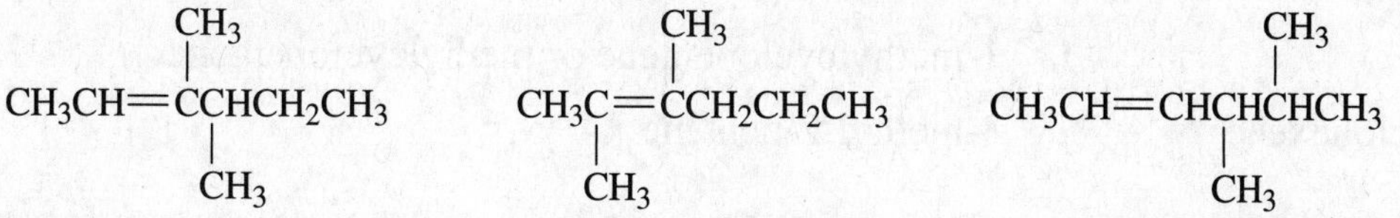

3,4-dimethyl-2-hexene
3 alkyl substituents

2,3-dimethyl-2-hexene
4 alkyl substituents
most stable

4,5-dimethyl-2-hexene
2 alkyl substituents
least stable

31. **a.** $-CH{=}CH_2 > -CH(CH_3)_2 > -CH_2CH_2CH_3 > -CH_3$

b. $-OH > -NH_2 > -CH_2OH > -CH_2NH_2$

c. $-Cl > -C({=}O)CH_3 > -CN > -CH{=}CH_2$

32. **a.** C_5H_8 **b.** C_8H_{10}

33. **a.** 3,8-dibromo-4-nonene

b. (*Z*)-4-ethyl-3,7-dimethyl-3-octene

c. 1,5-dimethylcyclopentene

d. 3-ethyl-2-methyl-2-heptene

34.

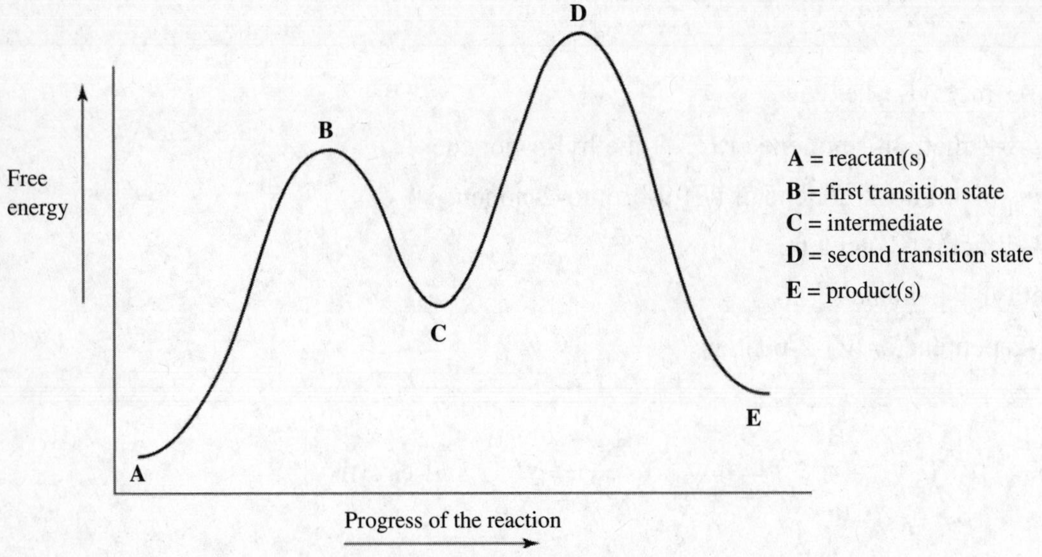
D
B
Free
energy
A = reactant(s)
B = first transition state
C = intermediate
D = second transition state
E = product(s)
C
E
A
Progress of the reaction

35. Only two of Molly Kule's names are correct.

a. 2-pentene
b. correct
c. 3-methyl-1-hexene
d. 3-heptene
e. 4-ethylcyclohexene
f. 2-chloro-3-hexene
g. correct
h. 2-methyl-1-hexene
i. 1-methylcyclopentene or methylcyclopentene
j. 3-methyl-2-pentene

36. **a.** rings + π bonds = 1 **b.** rings + π bonds = 2 **c.** rings + π bonds = 2

a.
$CH_3CH{=}CH_2$

b.
$CH_3C{\equiv}CH$

$CH_2{=}C{=}CH_2$

c.
$HC{\equiv}CCH_2CH_3$

$CH_3C{\equiv}CCH_3$

$CH_2{=}CHCH{=}CH_2$

$CH_2{=}C{=}CHCH_3$

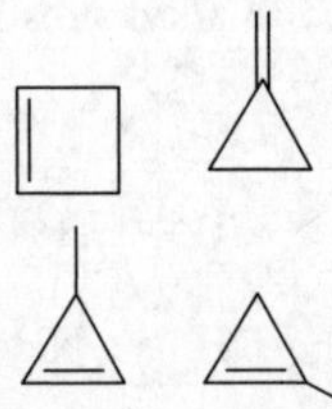

37. k_1 is the rate constant for A being converted to B, k_2 for B being converted to C, k_{-1} for B being converted to A and k_{-2} for C being converted to B

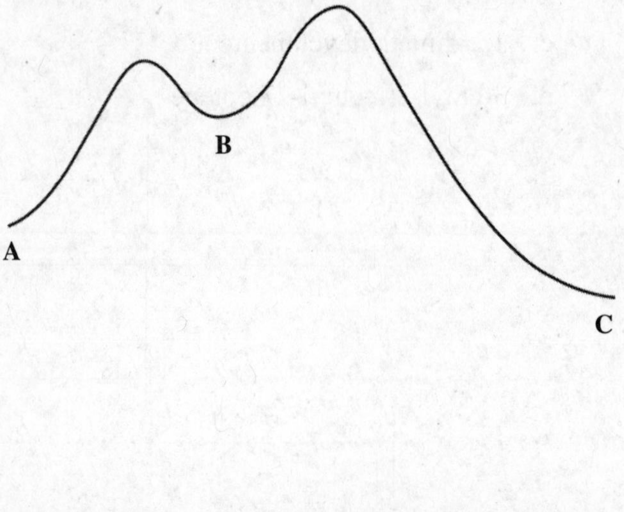
B
A
C

a. one

b. two

c. the second step (k_2) (In this particular diagram, $k_2 > k_1$: if you had made the transition state for the second step a lot higher, you could have had a diagram in which $k_1 > k_2$.)

d. the second step (k_{-1}), (that is, B forming A)

e. the second step in the reverse direction (k_{-1}), (that is, B forming A)

f. B to C, since its transition state has the highest energy

g. C forming B

38. The rate constant for a reaction can be increased by **decreasing** the stability of the reactant or by **increasing** the stability of the transition state.

39. 3,7,11-trimethyl-(1,3*E*,6*E*,10)-dodecatetraene

The configuration of the double bond at the 1-position and at the 10-position is not specified because stereoisomers are not possible at those positions, because there are **two hydrogens** on C-1 and **two methyl groups** on C-11.

40. $CH_2{=}CHCH_2CH_2CH_2CH_3$ — 1-hexene

$CH_3CH{=}CHCH_2CH_2CH_3$ — 2-hexene

$CH_3CH_2CH{=}CHCH_2CH_3$ — 3-hexene

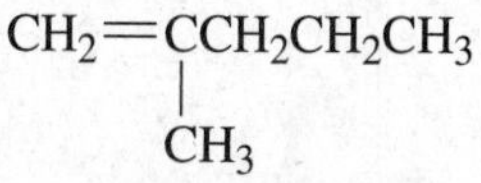
2-methyl-1-pentene

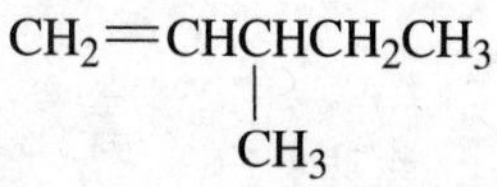
3-methyl-1-pentene

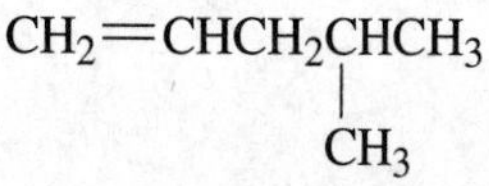
4-methyl-1-pentene

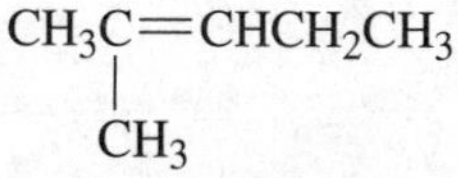
2-methyl-2-pentene

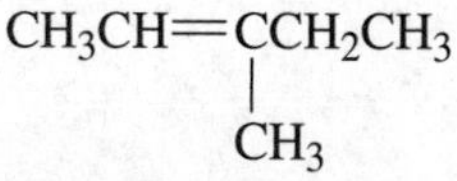
3-methyl-2-pentene

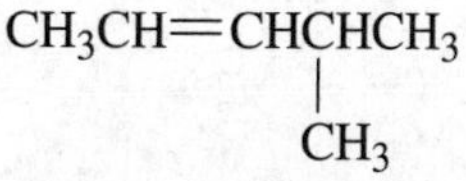
4-methyl-2-pentene

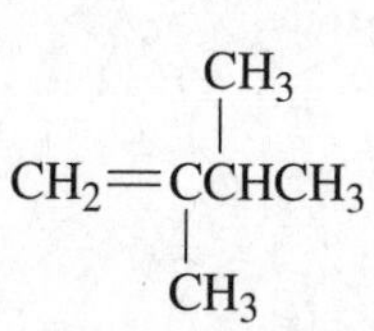
2,3-dimethyl-1-butene

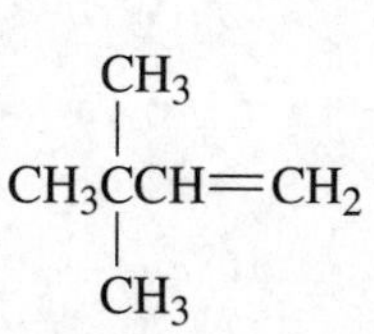
3,3-dimethyl-1-butene

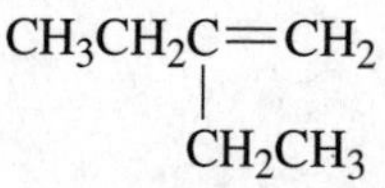
2-ethyl-1-butene

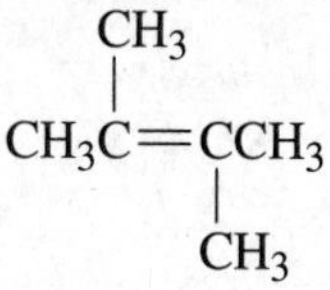
2,3-dimethyl-2-butene

a. Of the compounds shown above, the following have *E* and *Z* isomers:

2-hexene, 3-hexene, 3-methyl-2-pentene, 4-methyl-2-pentene

b. 2,3-Dimethyl-2-butene is the most stable; it is the only one that has no hydrogens bonded to the sp^2 carbons.

41. Tamoxifen is an *E* isomer.

Chapter 4 Practice Test

1. Name the following compounds:

a. $CH_3CH_2\underset{\displaystyle CH_3}{\underset{|}{C}}HCH_2CH{=}CH_2$

c. $CH_3CH_2CH{=}CHCH_2CH_2\underset{\displaystyle CH_2CH_3}{\underset{|}{C}}HCH_3$

b. 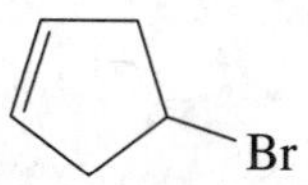

d. 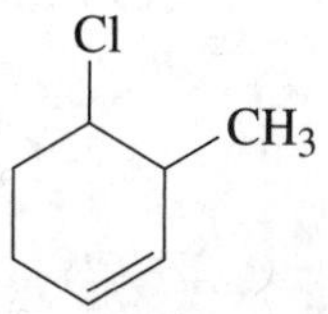

2. Label the following substituents in order of decreasing priority in the *E,Z* system of nomenclature. Label the highest priority **#1.**

$-\overset{\displaystyle O}{\overset{\|}{C}}CH_3$ $\quad -CH{=}CH_2 \quad -Cl \quad -C{\equiv}N$

3. Correct the incorrect names.

a. 3-pentene **c.** 2-ethyl-2-butene

b. 2-vinylpentane **d.** 2-methylcyclohexene

4. Which member of each pair is more stable?

a. 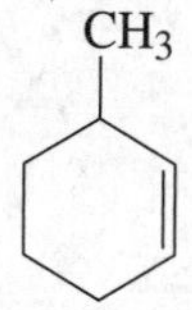 or 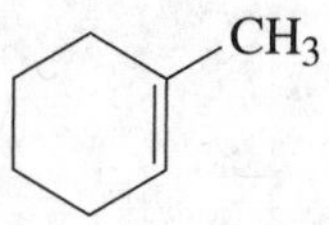

b. 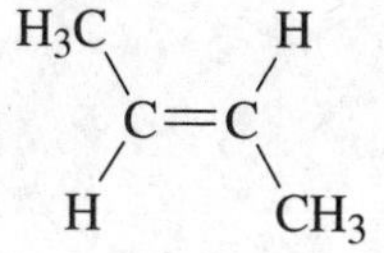 or 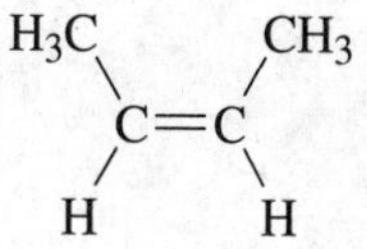

5. Which of the following have cis-trans isomers?

a. 1-pentene **c.** 2-bromo-3-hexene

b. 4-methyl-2-hexene **d.** 2-methyl-2-hexene

6. How many π bonds and/or rings does a hydrocarbon have if it has a molecular formula of C_8H_8?

7. Using curved arrows show the movement of electrons in the following reaction:

$$CH_3CH{=}CH_2 + H{-}\ddot{\underset{..}{Cl}}: \rightleftharpoons CH_3\overset{}{\underset{+}{C}}H{-}CH_3 + :\ddot{\underset{..}{Cl}}:^- \longrightarrow CH_3\underset{\displaystyle :\underset{..}{Cl}:}{\underset{|}{C}}H{-}CH_3$$

8. Indicate whether each of the following statements is true or false:

a. Increasing the energy of activation, increases the rate of the reaction. T F

b. An exergonic reaction is one with a $-\Delta G°$. T F

c. An alkene is an electrophile. T F

d. The higher the energy of activation, the more slowly the reaction will take place. T F

e. Another name for *trans*-2-butene is *Z*-2-butene. T F

f. The more stable the compound, the greater its concentration at equilibrium. T F

g. 2,3-Dimethyl-2-pentene is more stable than 3,4-dimethyl-2-pentene. T F

9. Do the following compounds have the *E* or the *Z* configuration?

a. 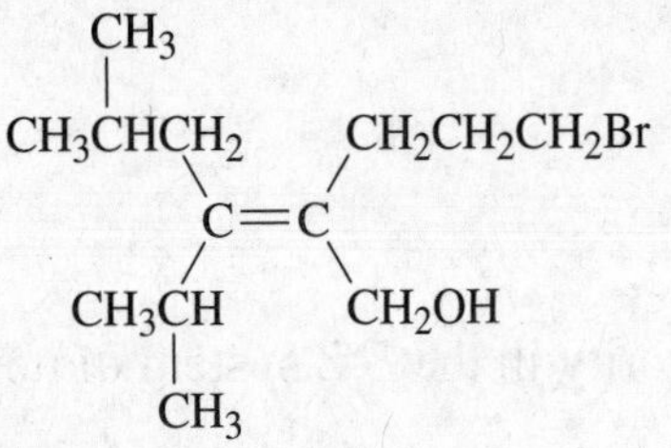
CH3
CH3CHCH2
CH2CH2CH2Br
C=C
CH3CH
CH2OH
CH3

b. 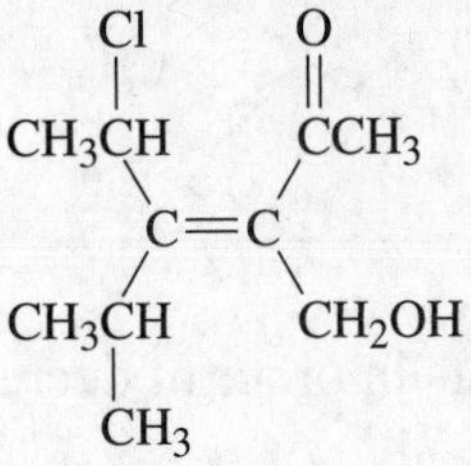
Cl
O
CH3CH
CCH3
C=C
CH3CH
CH2OH
CH3

10. Draw structures for the following:

a. allyl alcohol

b. 3-methylcyclohexene

c. *cis*-3-heptene

d. vinyl bromide

11. Which of the following alkenes is the most stable?

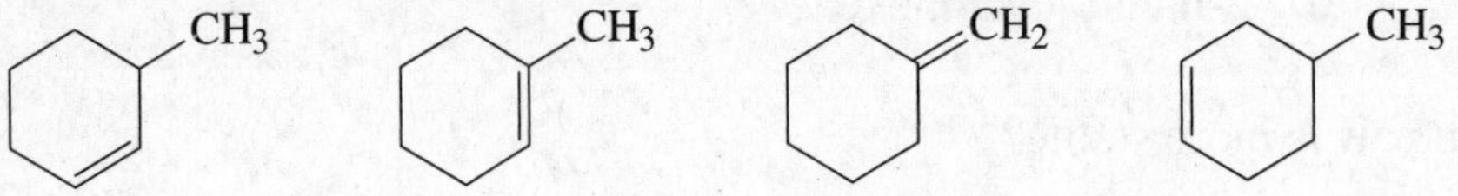
CH3
CH3
CH2
CH3

<u>**CHAPTER 5**</u>

Reactions of Alkenes and Alkynes: An Introduction to Multistep Synthesis

1. 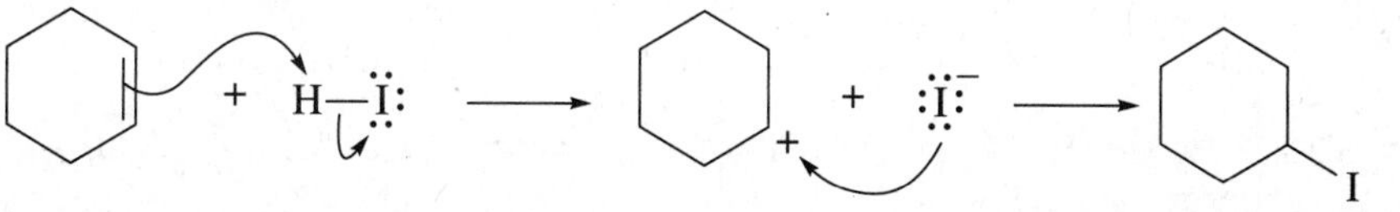

2. The ethyl cation is more stable. The ethyl cation has a methyl group in place of one of the hydrogens of the methyl cation. This methyl substituent can donate electrons toward the positive charge which decreases the concentration of positive charge, thereby making the ethyl carbocation more stable.

3. 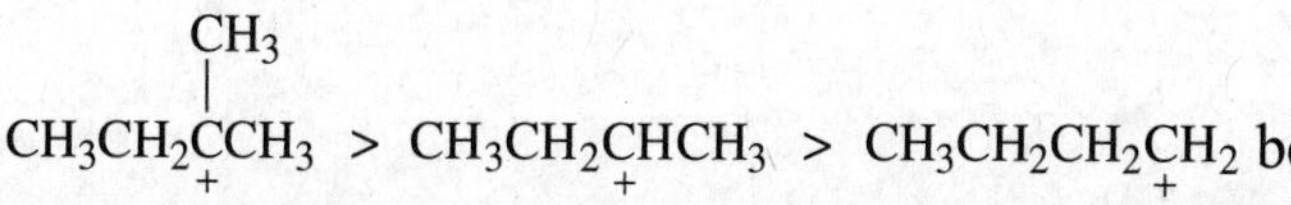 because a tertiary carbocation is more stable than a secondary carbocation, which is more stable than a primary carbocation.

4. **a.** 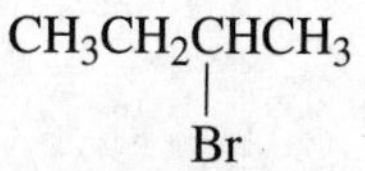 **c.** 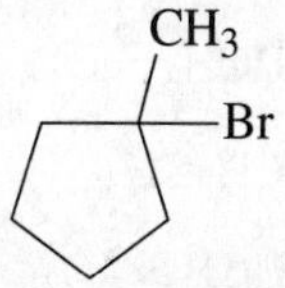 **e.** 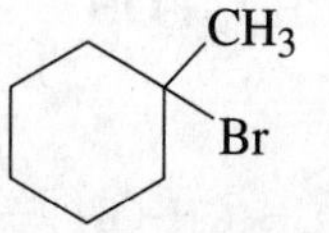

b. 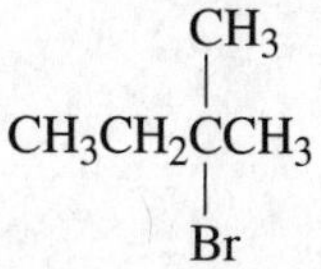 **d.** 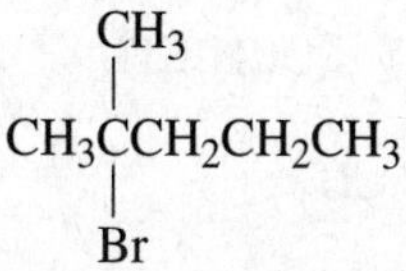 **f.** 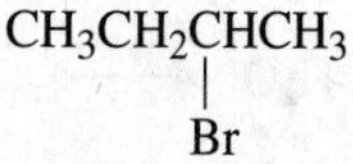

5. **a.** 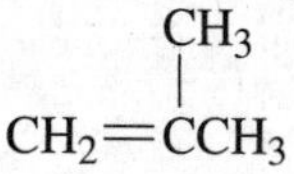 **c.** 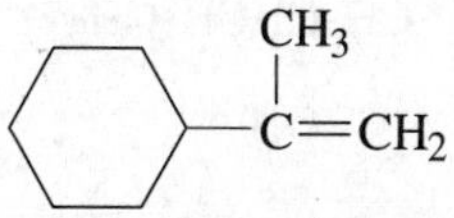

b. 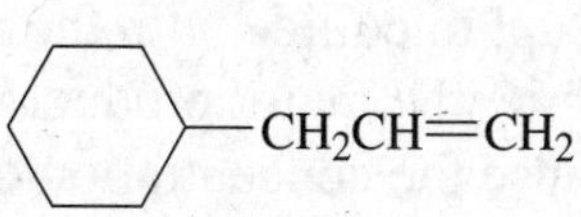 **d.** 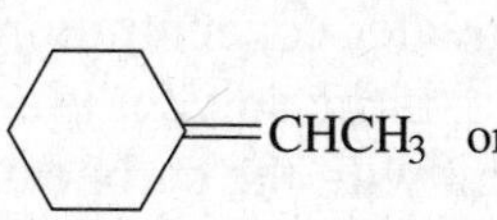 or 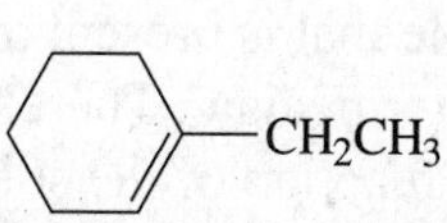

6. **a.** 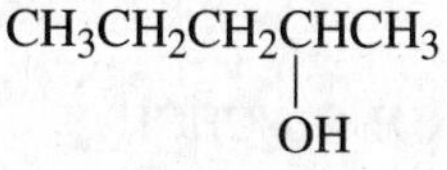 **c.** 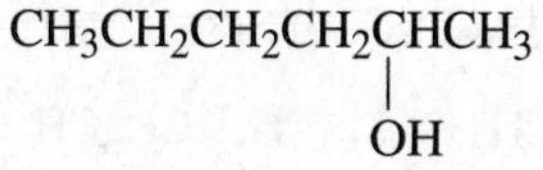 and 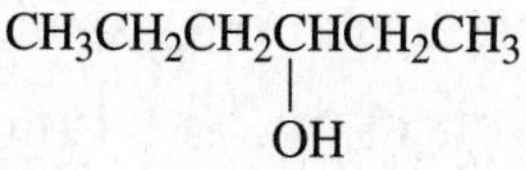

b. 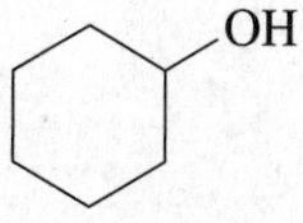 **d.** 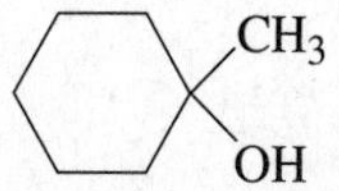

7. **a.** 1. 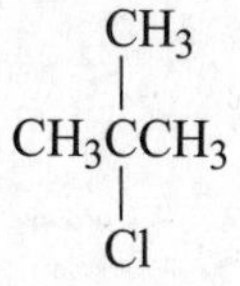 2. 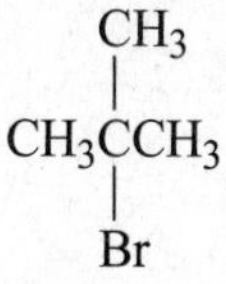 3. 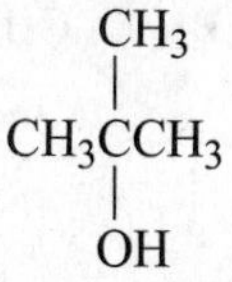 4. 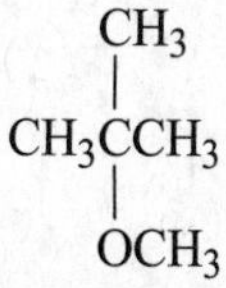

b. The first step in all the reactions is addition of an electrophilic proton (H^+) to the carbon of the CH_2 group.

The *tert*-butyl carbocation is formed as an intermediate in each of the reactions.

c. The nucleophile that adds to the *tert*-butyl carbocation is different in each reaction.

In reactions #3 and #4, there is a third step—a proton is lost from the group that was the nucleophile in the second step of the reaction.

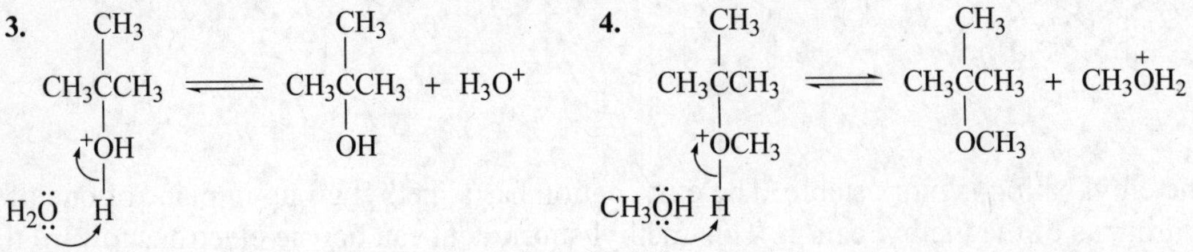

8.

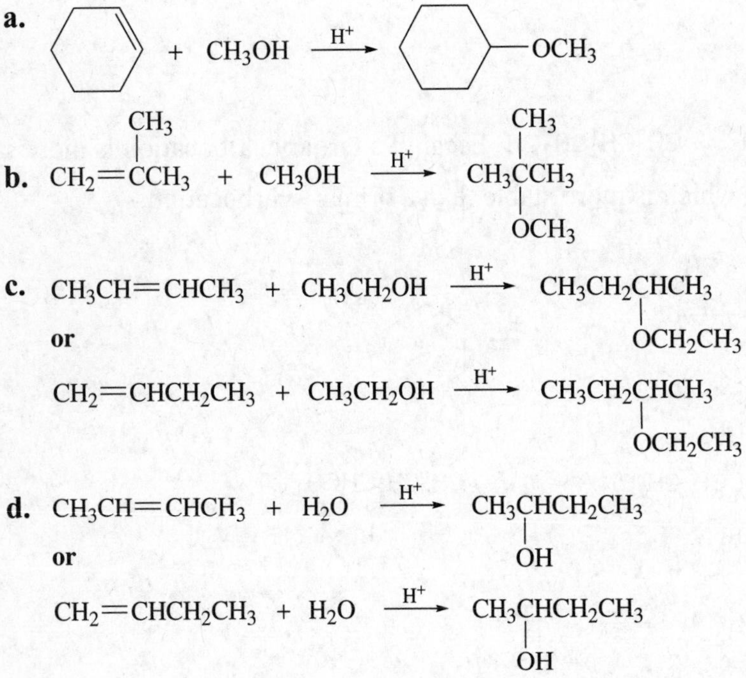

9. The nucleophile that is present in greater concentration is more apt to collide with the carbocation intermediate. Therefore, if the solvent is a nucleophile, the major product will be formed by the reaction of the solvent with the carbocation, because the concentration of the solvent is much greater than the concentration of the other nucleophile. (For example, in "a" the concentration of H_2O is much greater than the concentration of Cl^-.)

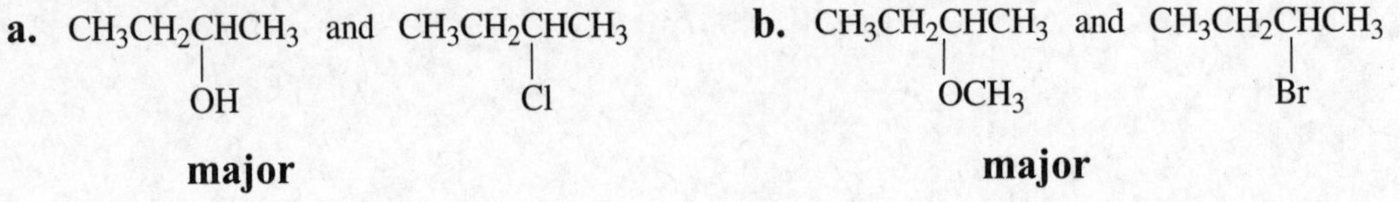

10.

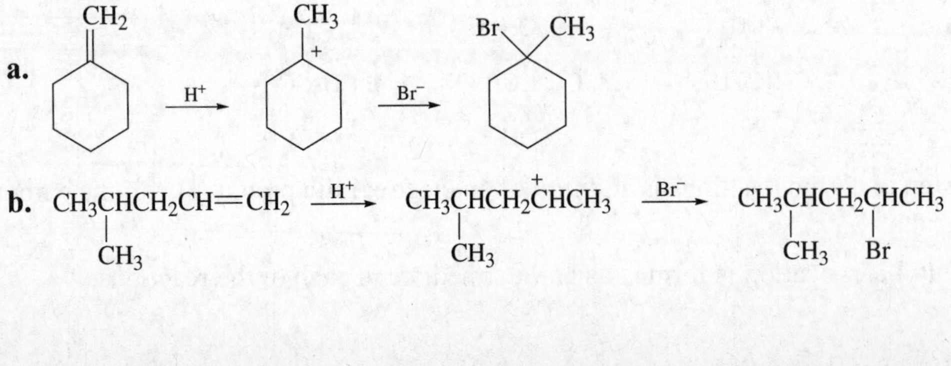

c.

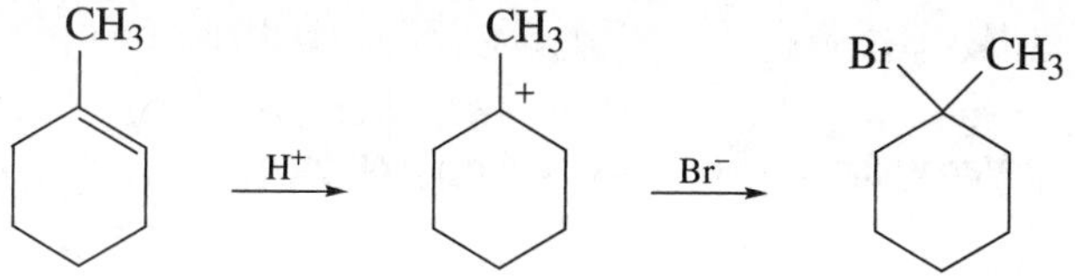
CH3
H+
CH3
+
Br−
Br
CH3

d.

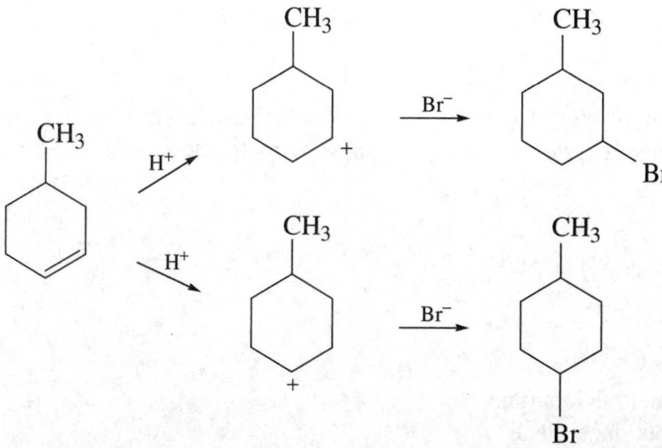
CH3
H+
H+
CH3
+
CH3
+
Br−
Br−
CH3
Br
CH3
Br

1-Bromo-3-methylcyclohexane and 1-bromo-4-methylcyclohexane will be obtained in approximately equal amounts because in each case the intermediate is a secondary carbocation.

11. The general molecular formula of a cyclic alkyne is C_nH_{2n-4}.

Because a compound has two fewer hydrogens for every ring and π bond, a compound with one ring and 2π bonds (one triple bond) would have 6 fewer hydrogens than the C_nH_{2n+2} formula.

12. The molecular formula of a noncyclic hydrocarbon with 14 carbons and no π bonds is $C_{14}H_{30}$ (C_nH_{2n+2}).

Because a hydrocarbon has two fewer hydrogens for every ring and π bond, a hydrocarbon with one ring and 4π bonds (2 triple bonds) would have 10 fewer hydrogens than the C_nH_{2n+2} formula. Thus, the molecular formula is $C_{14}H_{20}$.

13. **a.** $ClCH_2CH_2C{\equiv}CCH_2CH_3$ **b.**

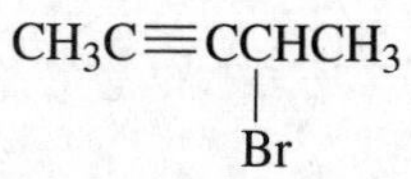
CH3C≡CCHCH3
Br

c.

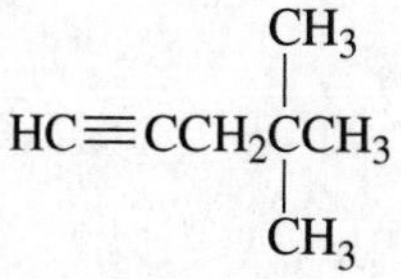
CH3
HC≡CCH2CCH3
CH3

14. **a.** 4-methyl-1-pentyne **b.** 2-hexyne

15. $HC{\equiv}CCH_2CH_2CH_2CH_3$ — 1-hexyne, butylacetylene

$CH_3C{\equiv}CCH_2CH_2CH_3$ — 2-hexyne, methylpropylacetylene

$CH_3CH_2C{\equiv}CCH_2CH_3$ — 3-hexyne, diethylacetylene

$CH_3CH_2\underset{\underset{CH_3}{|}}{C}HC{\equiv}CH$ — 3-methyl-1-pentyne, *sec*-butylacetylene

$CH_3\underset{\underset{CH_3}{|}}{C}HCH_2C{\equiv}CH$ — 4-methyl-1-pentyne, isobutylacetylene

$CH_3\underset{\underset{CH_3}{|}}{C}HC{\equiv}CCH_3$ — 4-methyl-2-pentyne, isopropylmethylacetylene

$CH_3\overset{\overset{CH_3}{|}}{\underset{\underset{CH_3}{|}}{C}}C{\equiv}CH$ — 3,3-dimethyl-1-butyne, *tert*-butylacetylene

16. **a.** 5-bromo-2-pentyne **c.** 5-methyl-3-heptyne
b. 6-bromo-2-chloro-4-octyne **d.** 3-ethyl-1-hexyne

17.
a. sp^2–sp^2 **d.** sp–sp^3 **g.** sp^2–sp^3
b. sp^2–sp^3 **e.** sp–sp **h.** sp–sp^3
c. sp–sp^2 **f.** sp^2–sp^2 **i.** sp^2–sp

18. **a.** $CH_2{=}\underset{\underset{Br}{|}}{C}CH_3$ **c.** $CH_3CH_2\overset{\overset{Br}{|}}{\underset{\underset{Br}{|}}{C}}CH_3$

b. $CH_3\overset{\overset{Br}{|}}{\underset{\underset{Br}{|}}{C}}CH_3$ **d.** $CH_3\overset{\overset{Br}{|}}{\underset{\underset{Br}{|}}{C}}CH_2CH_2CH_3 + CH_3CH_2\overset{\overset{Br}{|}}{\underset{\underset{Br}{|}}{C}}CH_2CH_3$

19. Because the alkyne is not symmetrical, two ketones will be obtained.

$$CH_3CH_2C{\equiv}CCH_2CH_2CH_3 + H_2O \xrightarrow{H_2SO_4} CH_3CH_2\overset{\overset{O}{\|}}{C}CH_2CH_2CH_2CH_3 + CH_3CH_2CH_2\overset{\overset{O}{\|}}{C}CH_2CH_2CH_3$$

20. **a.** $CH_3C{\equiv}CH$ **b.** $CH_3CH_2C{\equiv}CCH_2CH_3$ **c.** 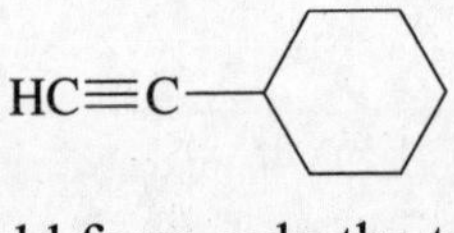

The best answer for "b" is 3-hexyne, because it would form only the target ketone. 2-Hexyne would form two different ketones, so only half of the product would be the target ketone.

21. Note that both of the enols can exist as *E* and *Z* isomers.

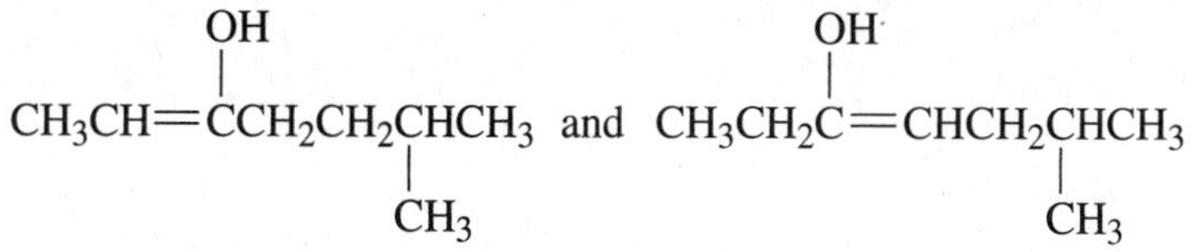
OH
CH3CH=CCH2CH2CHCH3 and CH3CH2C=CHCH2CHCH3
CH3
OH
CH3

22. **a.** $CH_3C{\equiv}CCH_3 \xrightarrow[\text{Lindlar catalyst}]{H_2}$

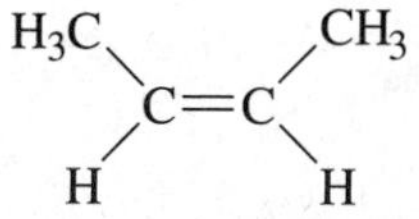
H3C
CH3
C=C
H
H

2-butyne

b. $CH_3CH_2CH_2CH_2C{\equiv}CH \xrightarrow[\text{Lindlar catalyst}]{H_2} CH_3CH_2CH_2CH_2CH{=}CH_2$

1-hexyne

23. **a.** $CH_3CH_2CH{=}CH_2$ or $CH_3CH{=}CHCH_3$

1-butene 2-butene

b. $CH_3CH_2CH_2CH{=}CH_2$ or $CH_3CH_2CH{=}CHCH_3$

1-pentene 2-pentene

c.

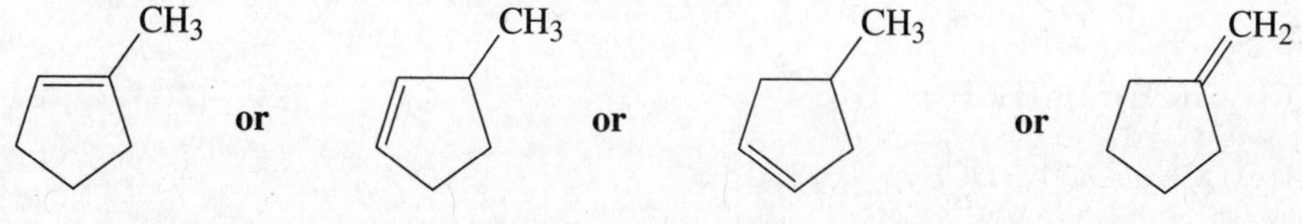
CH3
or
CH3
or
CH3
or
CH2

1-methylcyclopentene 3-methylcyclopentene 4-methylcyclopentene methylenecyclopentene

24. The reaction of sodium amide with an alkane will not favor products because the carbanion that would be formed is a stronger base than the amide ion. (Recall that the equilibrium favors reaction of the strong and formation of the weak; Section 2.5)

$$CH_3CH_3 + \bar{N}H_2 \rightleftharpoons CH_3\bar{C}H_2 + NH_3$$

CH_3CH_3	$\bar{N}H_2$	$CH_3\bar{C}H_2$	NH_3
$pK_a > 60$			$pK_a = 36$
weaker acid	weaker base	stronger base	stronger acid

25. Remember: the weaker the acid, the stronger its conjugate base.

a. $CH_3CH_2CH_2\bar{C}H_2 > CH_3CH_2CH{=}\bar{C}H > CH_3CH_2C{\equiv}C^-$

b. $\bar{N}H_2 > CH_3C{\equiv}C^- > CH_3CH_2O^- > F^-$

26. The electronegativities of carbon atoms decrease in the order: $sp > sp^2 > sp^3$.
The more electronegative the carbon atom, the less stable it will be with a positive charge.

a. $CH_3\overset{+}{C}H_2$ **b.** $H_2C{=}\overset{+}{C}H$

27. Solved in the text.

28. **a.** $HC\equiv CH \xrightarrow[2.\ CH_3CH_2CH_2Br]{1.\ \bar{N}H_2} CH_3CH_2CH_2C\equiv CH$

b. product of **a** $\xrightarrow[HgSO_4]{H_2O,\ H_2SO_4} CH_3CH_2CH_2\overset{O}{\overset{\|}{C}}CH_3$

c. $HC\equiv CH \xrightarrow[2.\ CH_3Br]{1.\ \bar{N}H_2} CH_3C\equiv CH \xrightarrow[Na/NH_3\ liq]{H_2/\text{Lindlar catalyst or}} CH_3CH{=}CH_2$

d. product of **c** $\xrightarrow{HBr} CH_3\underset{Br}{\underset{|}{C}}HCH_3$

e. 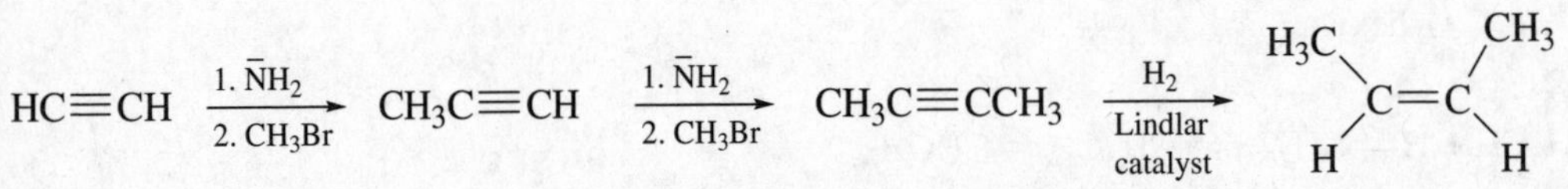

f. $HC\equiv CH \xrightarrow[2.\ CH_3Br]{1.\ \bar{N}H_2} CH_3C\equiv CH \xrightarrow{\text{excess HCl}} CH_3\overset{Cl}{\overset{|}{\underset{Cl}{\underset{|}{C}}}}CH_3$

29. $-CH_2\underset{OCH_3}{\underset{|}{C}}HCH_2\underset{OCH_3}{\underset{|}{C}}HCH_2\underset{OCH_3}{\underset{|}{C}}HCH_2\underset{OCH_3}{\underset{|}{C}}H-$

30. **a.** $CH_2{=}CHCl$ **b.** $CH_2{=}\underset{\underset{OCH_3}{|}}{\underset{C=O}{\underset{|}{C}}}CH_3$ **c.** $CF_2{=}CF_2$

31. 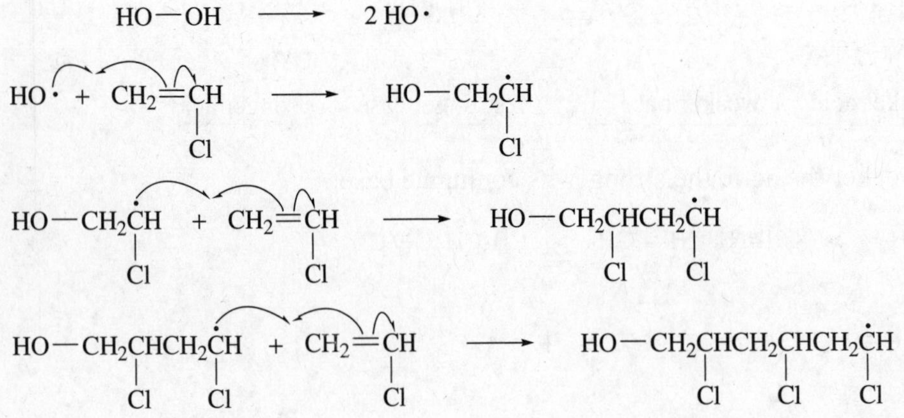

32. Since branching increases the flexibility of the polymer, beach balls are made from more highly branched polyethylene.

33. **a.** 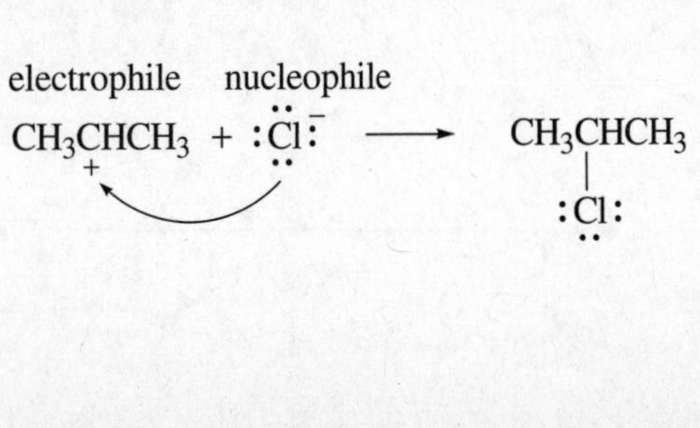
electrophile
nucleophile

b. 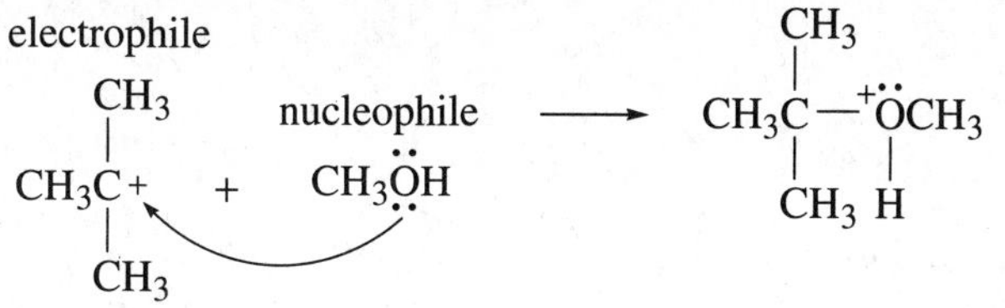
electrophile
nucleophile

34. **a.** 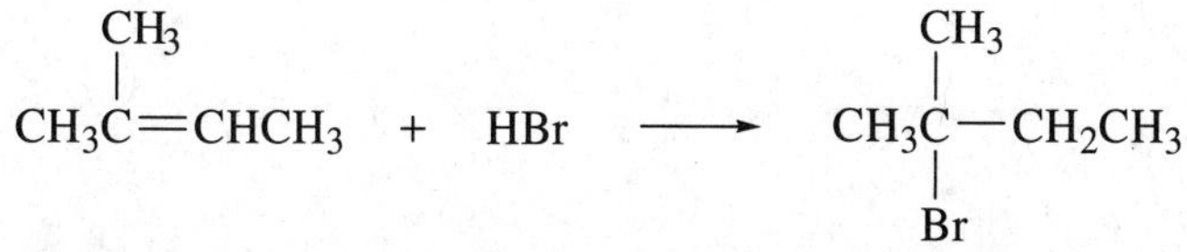

b. 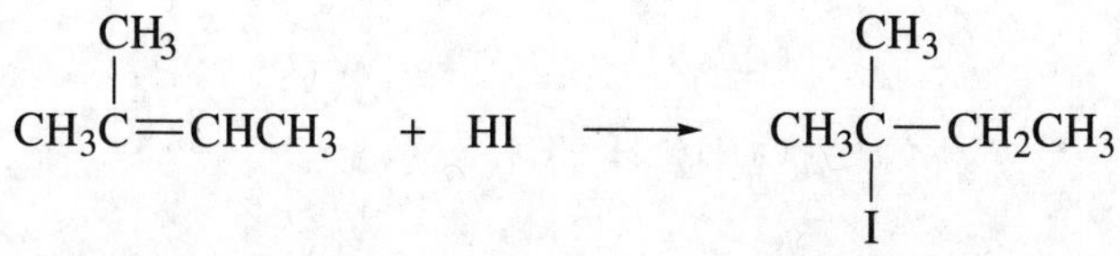

c. 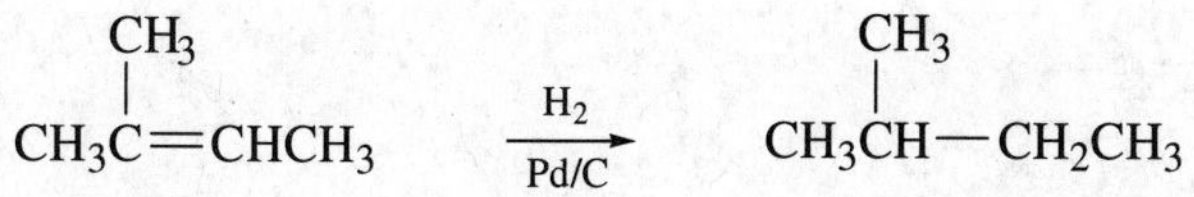
H_2
Pd/C

d. 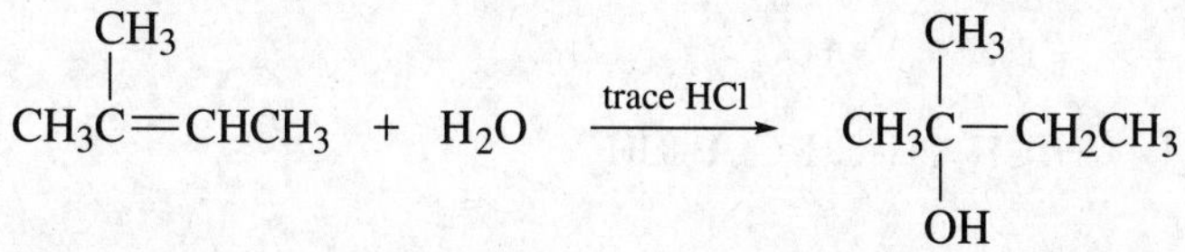
trace HCl

e. 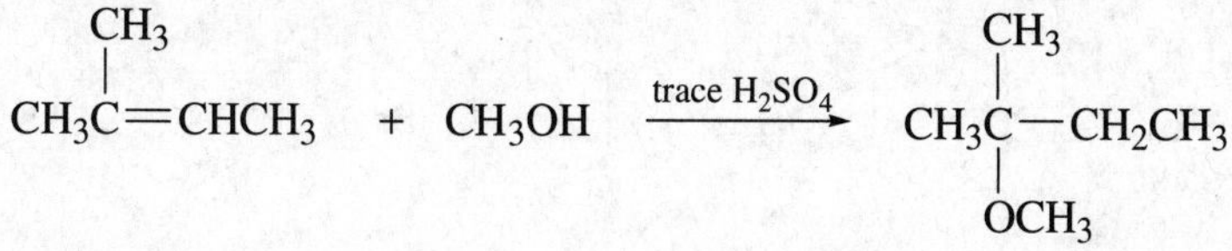
trace H_2SO_4

f. 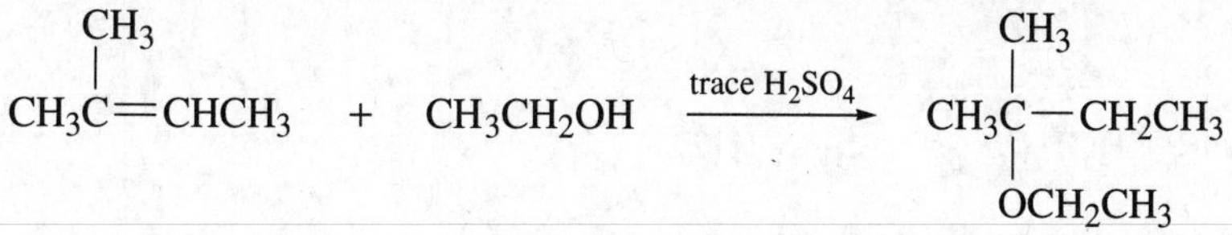
trace H_2SO_4

35. **a.** 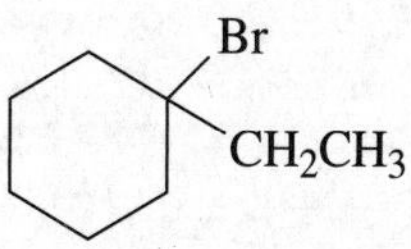 **b.** 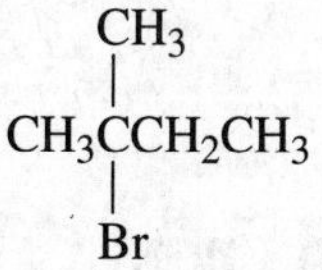 **c.** 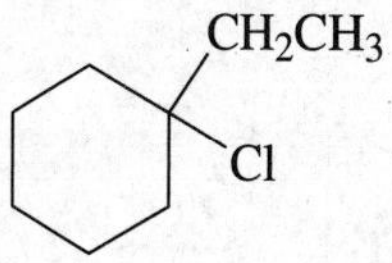 **d.** 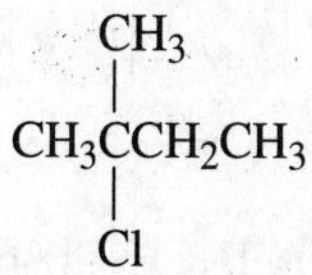

36. **a.** 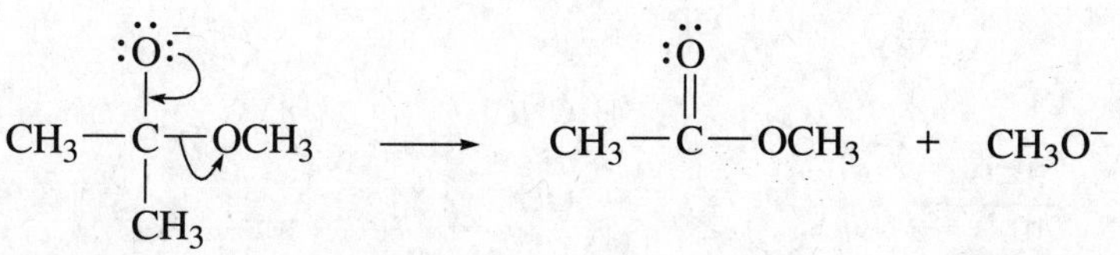

b. 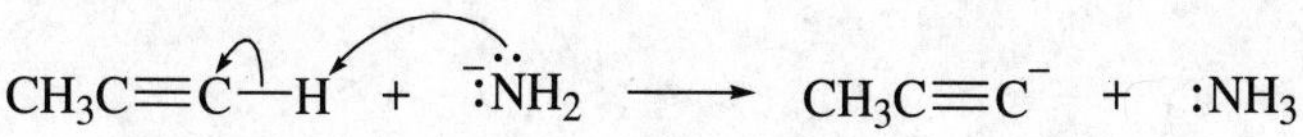

c. 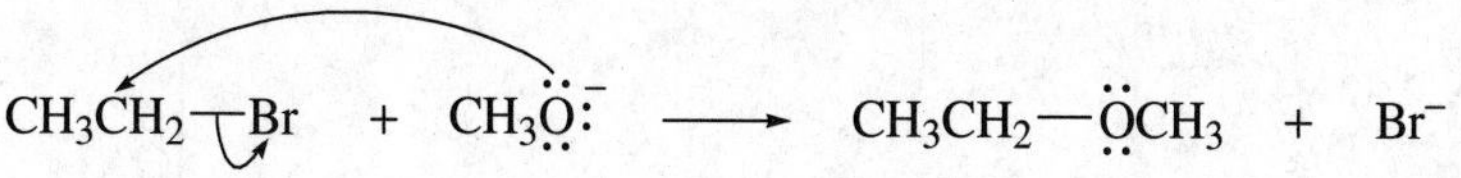

37.

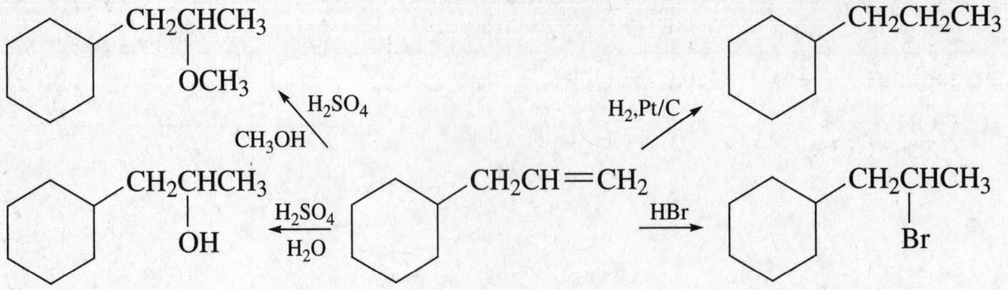

38. **a.** $CH_3C(OH){=}CH_2$

b. $CH_3CH_2C(OH){=}CHCH_2CH_3$ and $CH_3CH{=}C(OH)CH_2CH_2CH_3$

c. 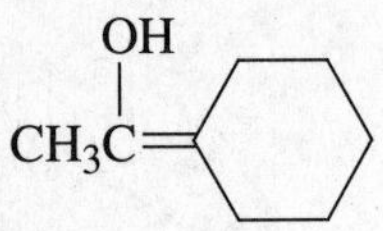 and 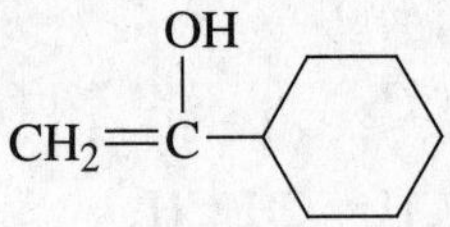

39. $CH_3CH_2CH_2C(=O)CH_2CH_2CH_3$ and $CH_3CH_2CH_2CH_2C(=O)CH_2CH_3$

40. **a.** 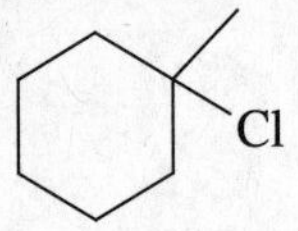

b. no reaction without an acid catalyst

c. 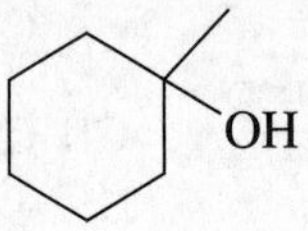

d. 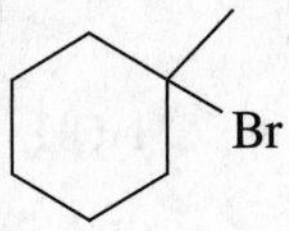

e. 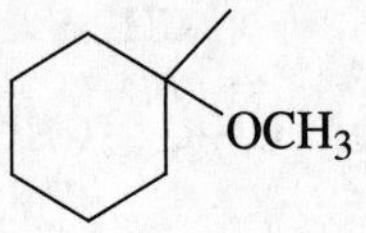

41. **a.** $CH_3\overset{+}{C}(CH_3)CH_3$ **b.** $CH_3\overset{+}{C}HCH_3$ **c.** $CH_3\overset{+}{C}H_2$

42. **a.** 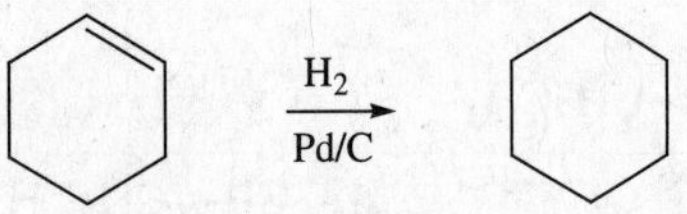

b. $CH_3CH_2CH_2CH{=}CH_2 \xrightarrow{HCl} CH_3CH_2CH_2CH(Cl)CH_3$

c. 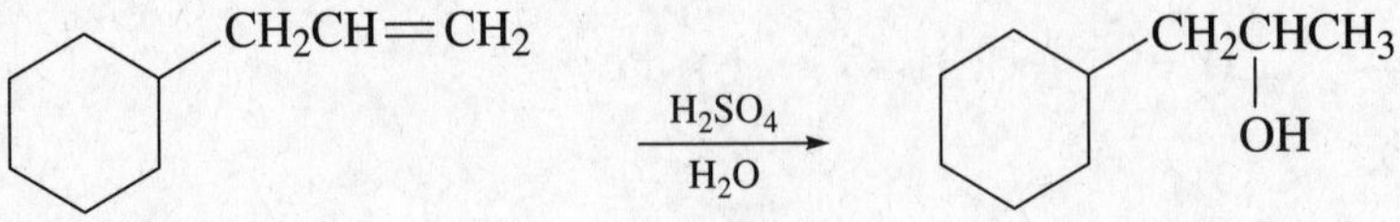

43. 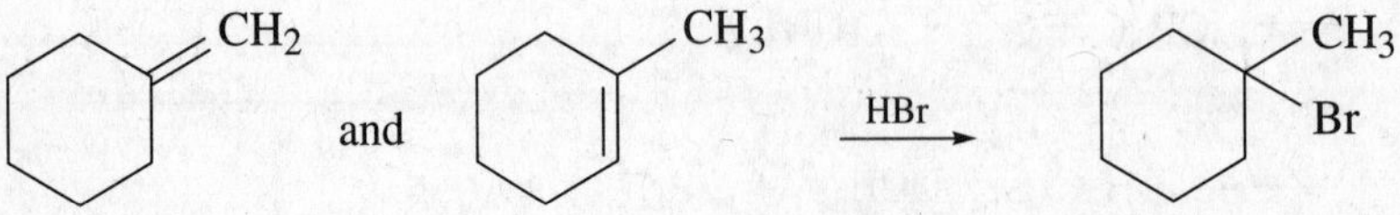

44. To determine their relative rates of hydration, the rate constant of each alkene is divided by the smallest rate constant of the series (3.51×10^{-8}).

propene	$= 4.95 \times 10^{-8}/3.51 \times 10^{-8} = 1.41$
(*Z*)-2-butene	$= 8.32 \times 10^{-8}/3.51 \times 10^{-8} = 2.37$
(*E*)-2-butene	$= 3.51 \times 10^{-8}/3.51 \times 10^{-8} = 1$
2-methyl-2-butene	$= 2.15 \times 10^{-4}/3.51 \times 10^{-8} = 6.12 \times 10^{3}$
2,3-dimethyl-2-butene	$= 3.42 \times 10^{-4}/3.51 \times 10^{-8} = 9.74 \times 10^{3}$

a. Both compounds form the same carbocation, but since (*Z*)-2-butene is less stable than (*E*)-2-butene, (*Z*)-2-butene has a smaller free energy of activation.

b. 2-Methyl-2-butene reacts faster because it forms a tertiary carbocation in the rate-limiting step, while (*Z*)-2-butene forms a less stable secondary carbocation.

c. Both compounds form tertiary carbocations. However, 2,3-dimethyl-2-butene has two sp^2 carbons that a proton can be attached to, but 2-methyl-2-butene has only one. Therefore, 2,3-dimethyl-2-butene will have a greater number of productive collisions with a proton.

45. **a.**

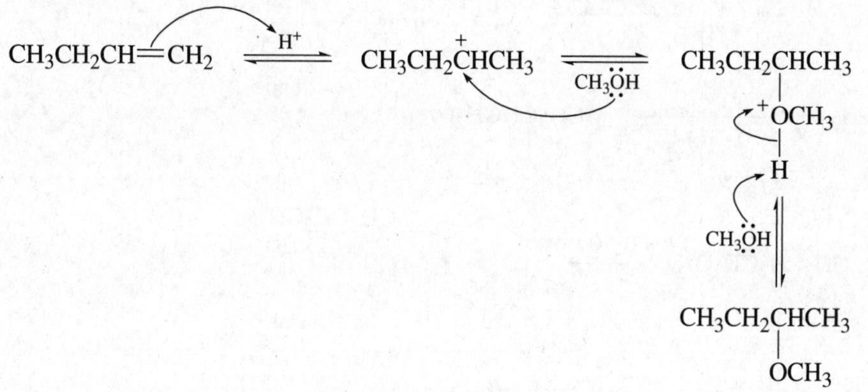
CH3CH2CH=CH2
H+
CH3CH2CHCH3
CH3OH
CH3CH2CHCH3
OCH3
H
CH3OH
CH3CH2CHCH3
OCH3

b. the first step

c. H^+

d. 1-butene

e. the *sec*-butyl cation

f. methyl alcohol

46. As long as the pH is greater than –2.4 and less than 15.9, more than 50% of 2-propanol will be in its neutral, nonprotonated form.

$$\overset{+}{ROH_2} \underset{}{\overset{pK_a = \sim -2.4}{\rightleftharpoons}} ROH \overset{pK_a = \sim 15.9}{\rightleftharpoons} RO^-$$

Because when the pH = pK_a, half the compound is in its acid form and half is in its basic form. Therefore at a pH less than –2.4, more than half of the compound will be in its positively charged protonated form.

At a pH greater than 15.9, more than half of the compound will exist as the negatively charged anion.

Thus at a pH between –2.4 and 15.9, more than half of the compound will exist in the neutral nonprotonated form.

47. **a.**

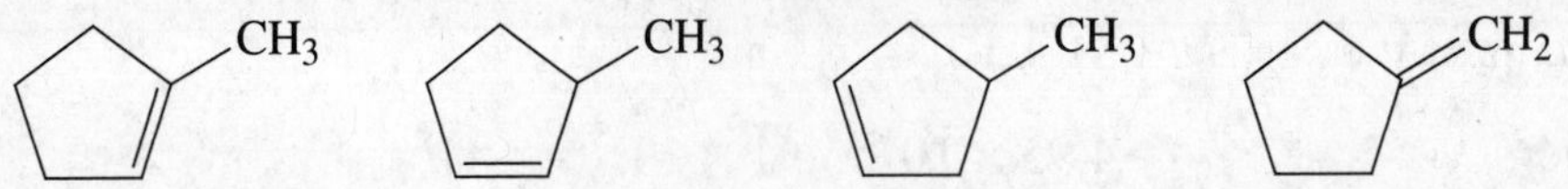

b.

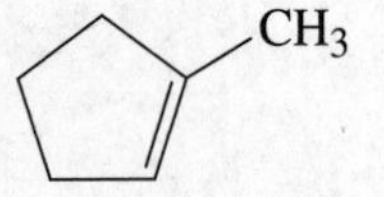

48. **a.** $CH_3CH{=}CH_2 \xrightarrow[CH_3OH]{H_2SO_4} CH_3CHCH_3$ (with OCH_3 on the central carbon)

b.

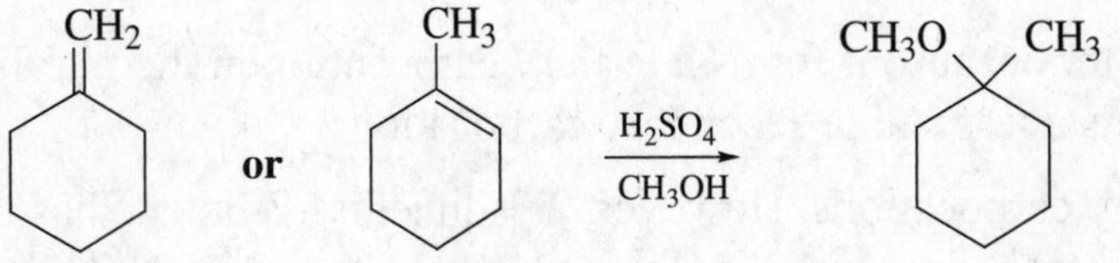

c.

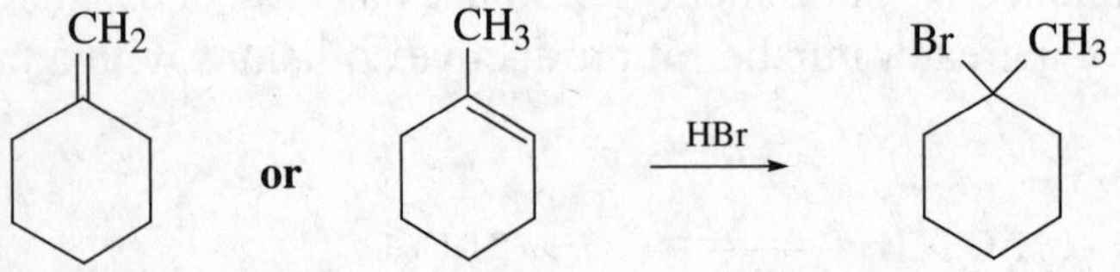

d. $CH_3CH{=}CHCH_3 \xrightarrow[CH_3CH_2OH]{H_2SO_4} CH_3CHCH_2CH_3$ (with OCH_2CH_3 on C-2)

e.

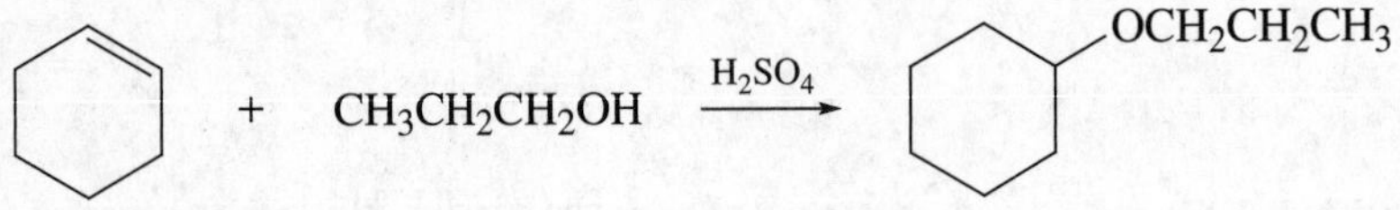

f. $CH_3C{=}CHCH_3$ (with CH_3 on C-2) $\xrightarrow[H_2O]{H_2SO_4} CH_3CCH_2CH_3$ (with CH_3 and OH on C-2)

49. **a.** $CH_3C{\equiv}CCH_2CH_2CH_3$

b. $CH_3CH_2C{\equiv}CCHCH_2CH_2CH_3$ (with CH_2CH_3 on C-5)

c. $BrC{\equiv}CCH_2CH_2CH_3$

d. $CH_3C{\equiv}CCH_2CHCHCH_3$ (with CH_3 on C-5 and CH_3 on C-6)

50. **a.** $CH_3CH_2CCH_3$ (with Cl and Cl on C-2)

b. $CH_3CH_2CH_2CH_2CCH_2CH_3$ (with Cl and Cl on C-3)

c. $CH_3CH_2CH_2CCH_2CH_2CH_3$ (with Cl and Cl on C-4) + $CH_3CH_2CCH_2CH_2CH_2CH_3$ (with Cl and Cl on C-3)

51. **a.** 5-bromo-2-hexyne **c.** 5,5-dimethyl-2-hexyne

b. 5-methyl-2-octyne **d.** 6-chloro-2-methyl-3-heptyne

52. **a.**

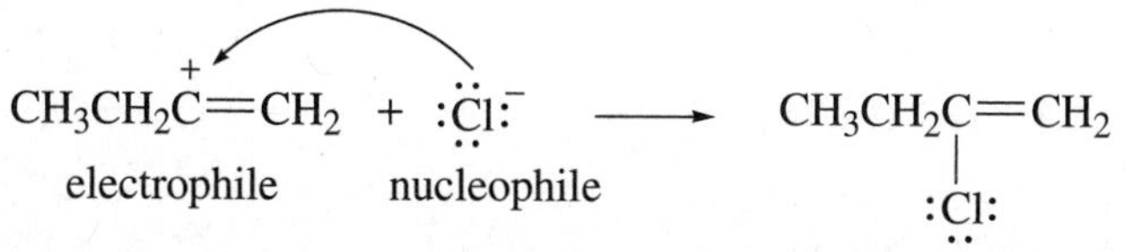
electrophile
nucleophile

b.

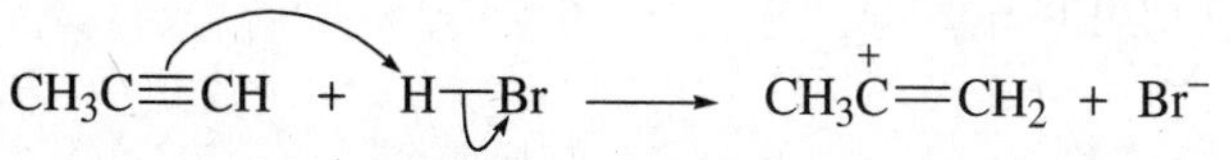

nucleophile electrophile

c.

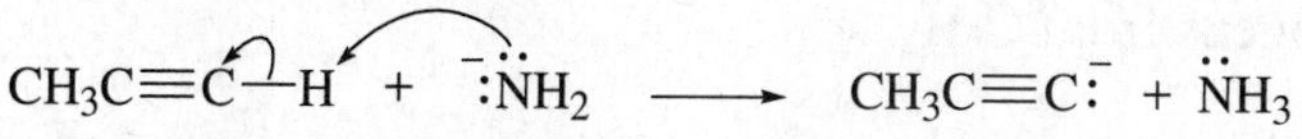
electrophile
nucleophile

53. He named only one compound correctly.

a. 4-methyl-2-hexyne **c.** correct

b. 7-bromo-3-heptyne **d.** 2-pentyne

54. $HC{\equiv}CCH_2CH_2CH_2CH_2CH_3$ 1-heptyne, pentylacetylene

$CH_3C{\equiv}CCH_2CH_2CH_2CH_3$ 2-heptyne, butylmethylacetylene

$CH_3CH_2C{\equiv}CCH_2CH_2CH_3$ 3-heptyne, ethylpropylacetylene

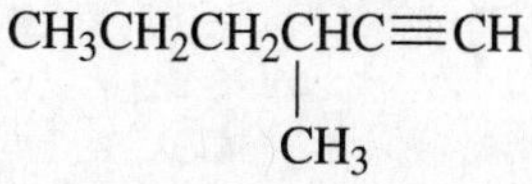

3-methyl-1-hexyne

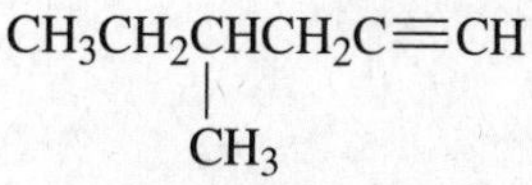

4-methyl-1-hexyne

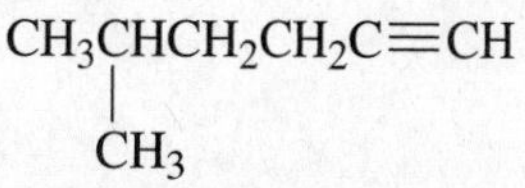

5-methyl-1-hexyne
isopentylacetylene

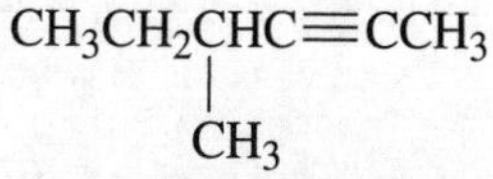

4-methyl-2-hexyne
sec-butylmethylacetylene

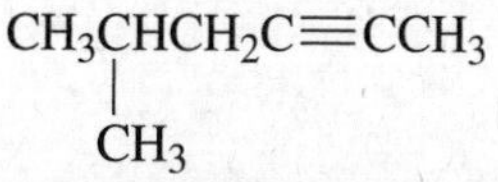

5-methyl-2-hexyne
isobutylmethylacetylene

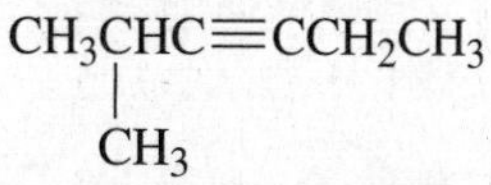

2-methyl-3-hexyne
ethylisopropylacetylene

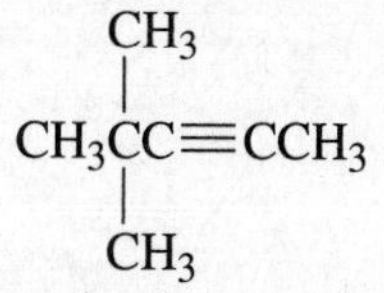

4,4-dimethyl-2-pentyne
tert-butylmethylacetylene

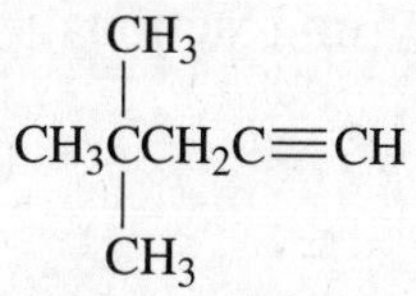

4,4-dimethyl-1-pentyne

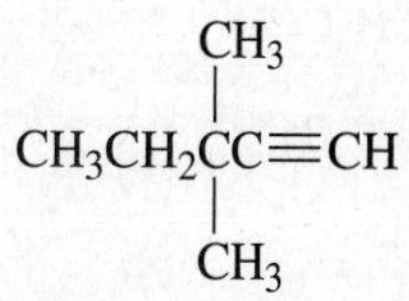

3,3-dimethyl-1-pentyne

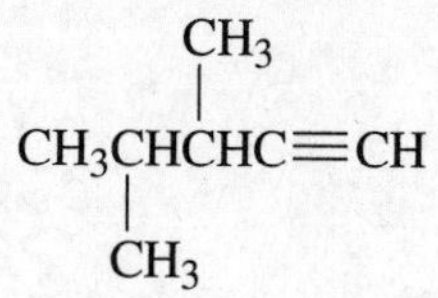

3,4-dimethyl-1-pentyne

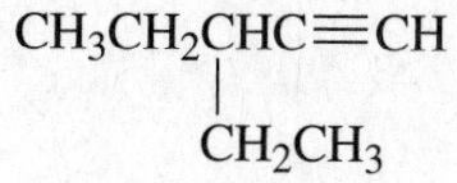

3-ethyl-1-pentyne

55.   **a.** $CH_3CH_2C{\equiv}CCH_2CH_3$ $\xrightarrow[\text{Lindlar catalyst}]{H_2}$ 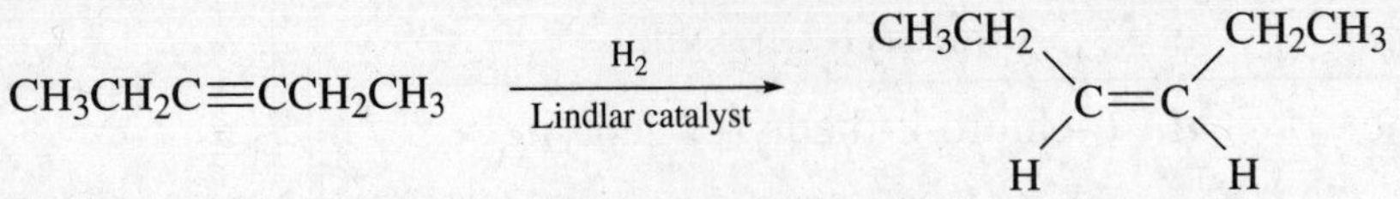

$$\underset{H}{\overset{CH_3CH_2}{\diagdown}}C=C\underset{H}{\overset{CH_2CH_3}{\diagup}}$$

  b. $CH_3CH_2C{\equiv}CCH_2CH_3$ $\xrightarrow[\text{Pt/C}]{H_2}$ $CH_3CH_2CH_2CH_2CH_2CH_3$

56.   The molecular formula of the hydrocarbon is $C_{32}H_{56}$.

C_nH_{2n+2} = $C_{32}H_{66}$

With one triple bond, two double bonds, and one ring, the degree of unsaturation is 5. Therefore, the compound is missing 10 hydrogens from C_nH_{2n+2}.

57.   **a.** $CH_2{=}CCH_3$ 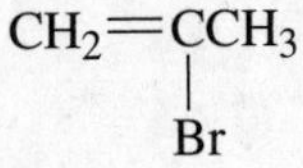
             |
             Br

  b. $CH_3\overset{\overset{\displaystyle Br}{|}}{\underset{\underset{\displaystyle Br}{|}}{C}}CH_3$ 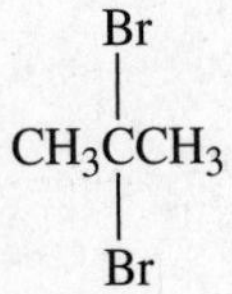

  c. $CH_3\overset{\overset{\displaystyle O}{\|}}{C}CH_3$ 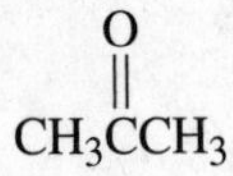

  d. $CH_3CH_2CH_3$

  e. $CH_3CH{=}CH_2$

  f. $CH_3C{\equiv}C^-$

  g. $CH_3C{\equiv}CCH_2CH_2CH_2CH_2CH_3$

58.   **a.** $CH_3CH{=}CCH_3$ 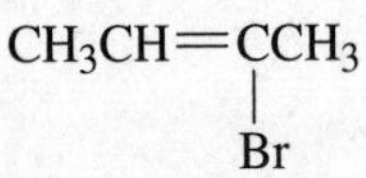
             |
             Br

  b. $CH_3CH_2\overset{\overset{\displaystyle Br}{|}}{\underset{\underset{\displaystyle Br}{|}}{C}}CH_3$ 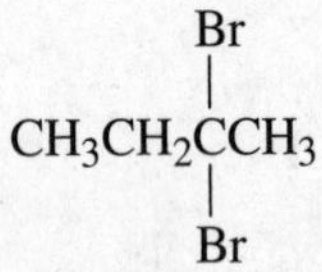

  c. $CH_3\overset{\overset{\displaystyle O}{\|}}{C}CH_2CH_3$ 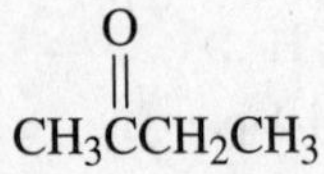

  d. $CH_3CH_2CH_2CH_3$

  e. 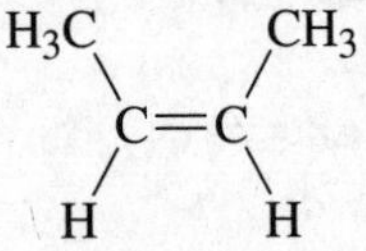
$$\underset{H}{\overset{H_3C}{\diagdown}}C=C\underset{H}{\overset{CH_3}{\diagup}}$$

  f.   no reaction

  g.   no reaction

59. 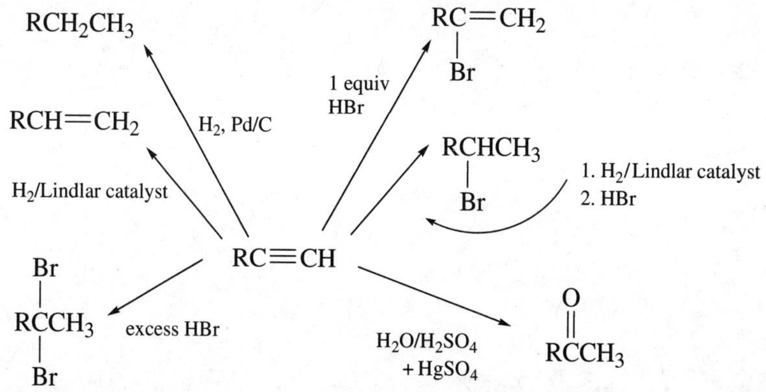

60.

a. 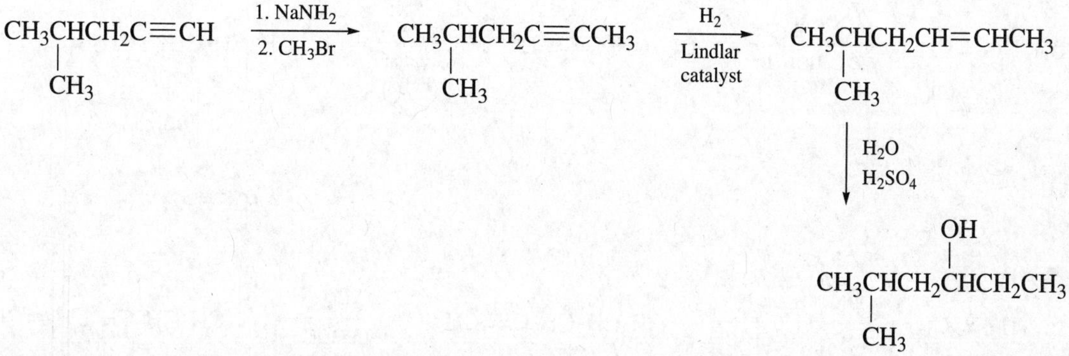

b. 5-Methyl-2-hexanol also will be obtained because in the third step of the synthesis, the proton can add to either of the sp^2 carbons.

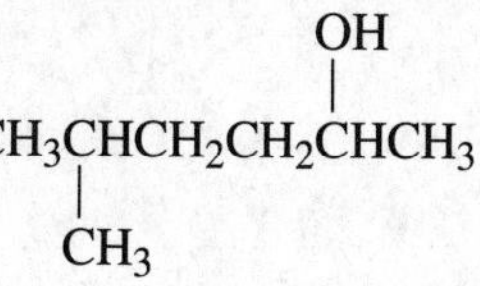

61. Three are named correctly.

a. 3-heptyne

b. 5-methyl-3-heptyne

c. correct

d. 6,7-dimethyl-3-octyne

e. correct

f. correct

62. only **b** is a pair of keto-enol tautomers.

63.

a. $HC\equiv CH$ $\xrightarrow[\text{2. CH}_3\text{Br}]{\text{1. NaNH}_2}$ $HC\equiv CCH_3$ $\xrightarrow[\text{HgSO}_4]{\text{H}_2\text{O, H}_2\text{SO}_4}$ $CH_3\overset{\overset{\displaystyle O}{\|}}{C}CH_3$ 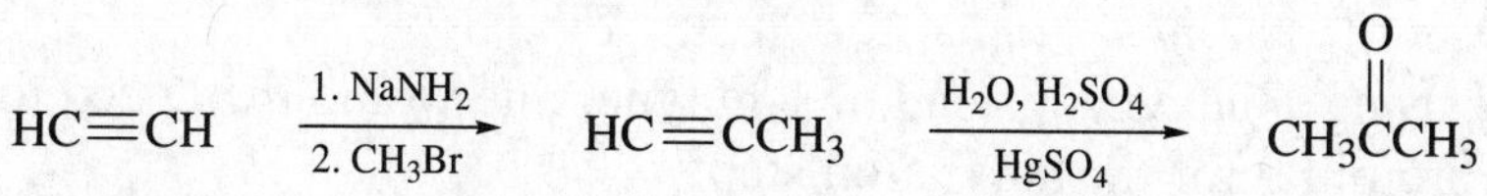

b. $HC\equiv CH$ $\xrightarrow[\text{2. CH}_3\text{Br}]{\text{1. NaNH}_2}$ $HC\equiv CCH_3$ $\xrightarrow[\text{2. CH}_3\text{CH}_2\text{Br}]{\text{1. NaNH}_2}$ $CH_3C\equiv CCH_2CH_3$ 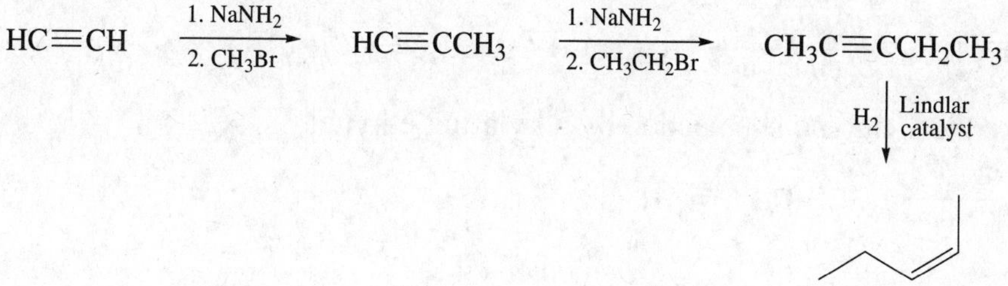

c. $HC\equiv CH$ $\xrightarrow[\text{2. CH}_3\text{Br}]{\text{1. NaNH}_2}$ $HC\equiv CCH_3$ $\xrightarrow[\text{2. CH}_3\text{CH}_2\text{Br}]{\text{1. NaNH}_2}$ $CH_3C\equiv CCH_2CH_3$ 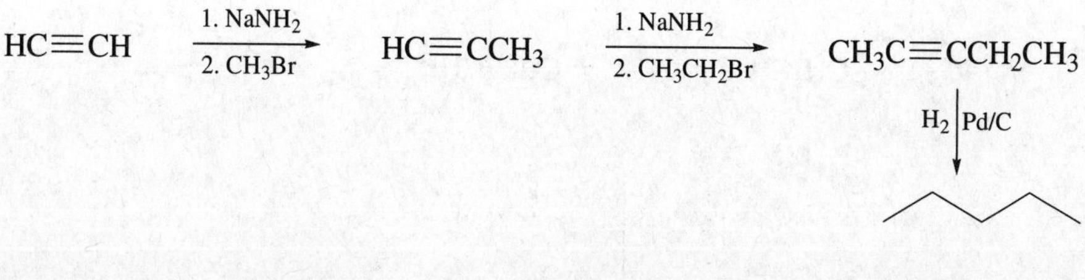

64. **a.** $CH_3CH_2C(=O)CH_3$ **b.** $CH_3CH_2CH_2C(=O)CH_3$ **c.** 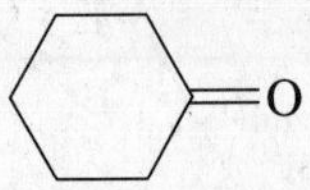 **d.** 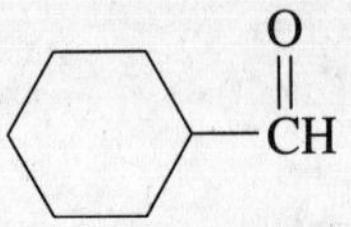

65. **a.** $HC\equiv CH \xrightarrow[\text{2. } CH_3CH_2CH_2CH_2Br]{\text{1. } NaNH_2} CH_3CH_2CH_2CH_2C\equiv CH$

$\xrightarrow{H_2O,\ H_2SO_4,\ HgSO_4} CH_3CH_2CH_2CH_2C(=O)CH_3$

b. $HC\equiv CH \xrightarrow[\text{2. } CH_3CH_2Br]{\text{1. } NaNH_2} CH_3CH_2C\equiv CH \xrightarrow[\text{Lindlar catalyst}]{H_2} CH_3CH_2CH{=}CH_2$

$\xrightarrow{HBr} CH_3CH_2CH(Br)CH_3$

c. $HC\equiv CH \xrightarrow[\text{2. } CH_3CH_2CH_2Br]{\text{1. } NaNH_2} CH_3CH_2CH_2C\equiv CH \xrightarrow[\text{Lindlar catalyst}]{H_2} CH_3CH_2CH_2CH{=}CH_2$

$\xrightarrow{H_2O,\ H_2SO_4} CH_3CH_2CH_2CH(OH)CH_3$

d. 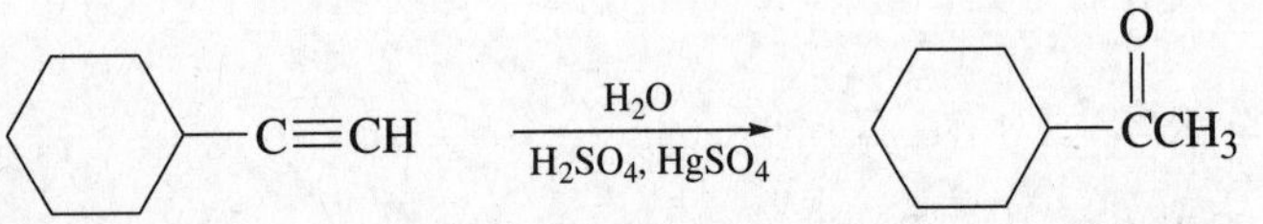
C≡CH
H2O
H2SO4, HgSO4
O
CCH3

66. The base used to remove a proton must be stronger than the base that is formed as a result of proton removal. A terminal alkyne has a pK_a ~ 25. Therefore, the base used to remove a proton from a terminal alkyne must be a stronger base than the terminal alkyne. In other words, any base whose conjugate acid has a pK_a greater than 25 can be used.

67. She can make 3-octyne by using 1-hexyne instead of 1-butyne. She would then need to use ethyl bromide (instead of butyl bromide) for the alkylation step:

$CH_3CH_2CH_2CH_2C\equiv CH \xrightarrow[\text{2. } CH_3CH_2Br]{\text{1. } NaNH_2} CH_3CH_2CH_2CH_2C\equiv CCH_2CH_3$

Or she could make the 1-butyne she needed by alkylating ethyne:

$HC\equiv CH \xrightarrow[\text{2. } CH_3CH_2Br]{\text{1. } NaNH_2} CH_3CH_2C\equiv CH$

$\xrightarrow[\text{2. } CH_3CH_2CH_2CH_2Br]{\text{1. } NaNH_2} CH_3CH_2CH_2CH_2C\equiv CCH_2CH_3$

68. **a.** 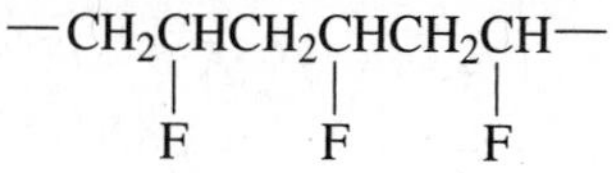

b. 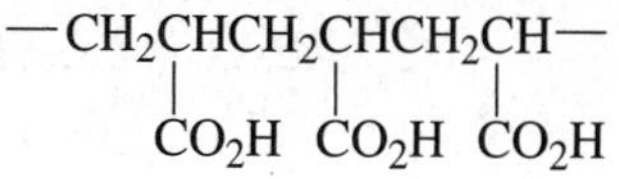

69. **a.** $CH_2{=}CHCH_2CH_3$

b. 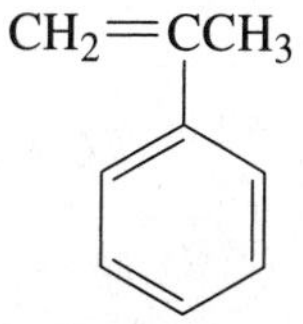

70. 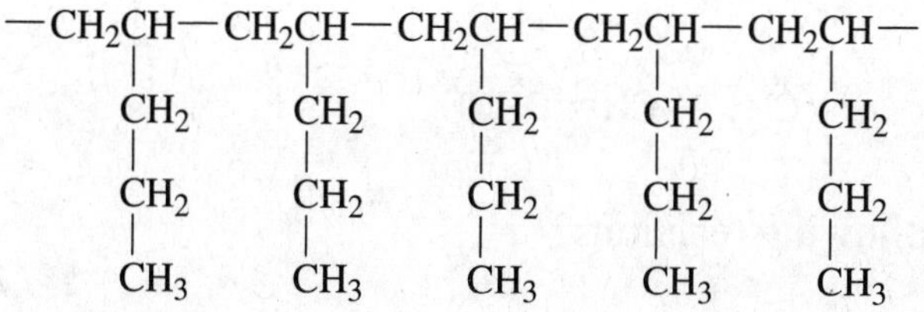

71. When BF_3 and H_2O are used to generated the H^+ electrophile, a negatively charged counter ion is not formed.

$$BF_3 + H_2\ddot{O}: \rightleftharpoons F_3\overset{-}{B}{:}\overset{+}{\ddot{O}}H_2 \rightleftharpoons FB_3{:}\ddot{O}H + H^+$$

When HCl is used to generated the H^+ electrophile, a negatively charged counter ion is formed.

$$HCl \rightleftharpoons H^+ + :\ddot{Cl}:^-$$

Because a negatively charged counter ion can act as a chain terminator, BF_3 and H_2O are used as the source of H^+.

Chapter 5 Practice Test

1. Which member of each pair is more stable?

a. $CH_3\underset{+}{C}HCH_2CH_3$ or $CH_3\underset{+}{\overset{CH_3}{\overset{|}{C}}}CH_3$ **b.** $CH_3CH_2\overset{+}{C}H_2$ or $CH_3\overset{+}{C}HCH_3$

2. Which would be a better compound to use as a starting material for the synthesis of 2-bromopentane?

$CH_3CH_2CH_2CH{=}CH_2$ or $CH_3CH_2CH{=}CHCH_3$

3. What is the major product of each of the following reactions?

a. $CH_2{=}\overset{CH_3}{\overset{|}{C}}CH_2CH_3 + HBr \longrightarrow$

b. $CH_3CH_2CH{=}CH_2 + HCl \longrightarrow$

4. Indicate how each of the following compounds could be synthesized using an alkene as one of the starting materials:

a. $CH_3\underset{\underset{CH_3}{|}}{\overset{CH_3}{\overset{|}{C}}}CH_2CH_3$ **b.** $CH_3\underset{\underset{OH}{|}}{\overset{CH_3}{\overset{|}{C}}}CH_2CH_3$

5. Draw the enol tautomer of the following compound:

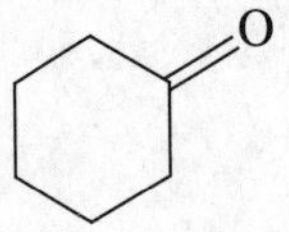

6. Indicate whether each of the following statements is true or false:

a. 1-Butyne is more acidic than 1-butene. T F

b. An sp^2 carbon is more electronegative than an sp^3 carbon. T F

c. Water is a stronger acid than ammonia. T F

d. The reaction of 1-butene with HCl will form 1-chlorobutane as the major product. T F

7. What reagents could be used to convert the given starting material into the desired product?

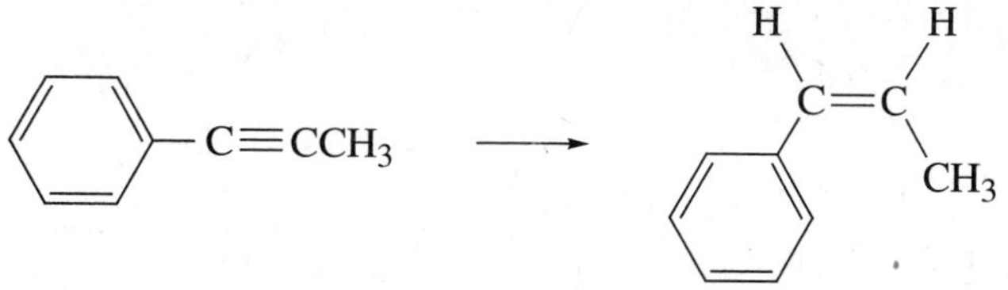
C≡CCH3
H
H
C=C
CH3

8. Give the systematic name for the following compound:

$CH_3CH(CH_3)C{\equiv}CCH_2CH_2Br$

9. What would be the best alkyne to use for the synthesis of the following ketone?

$CH_3CH_2CH_2C(=O)CH_3$

10. Rank the following compounds in order of decreasing acidity:

(Label the most acidic compound #1.)

NH_3 $CH_3C{\equiv}CH$ CH_3CH_3 H_2O $CH_3CH{=}CH_2$

11. Give an example of a ketone that has two enol tautomers.

12. Show how the target compounds could be prepared from the given starting materials.

a. $CH_3CH_2C{\equiv}CH \longrightarrow CH_3CH_2CH_2CH_2CH_2CH_3$

b. $CH_3CH_2C{\equiv}CH \longrightarrow CH_3CH_2C(=O)CH_2CH_2CH_3$

CHAPTER 6
Isomers and Stereochemistry

Do the following exercises using molecular models.

1. Build the enantiomers of 2-bromobutane.
   a. Try to superimpose them.
   b. Show that they are mirror images.
   c. Which one is (*R*)-2-bromobutane?

2. Build the four stereoisomers of 2-bromo-3-chlorobutane.

3. Build the three stereoisomers of 2,3-dibromobutane.

4. Build the four stereoisomers of 2,3-dibromopentane.
   Why does 2,3-dibromopentane have four stereoisomers, whereas 2,3-dibromobutane has only three stereoisomers?

5. Build (*S*)-2-pentanol.

6. Build (*S*)-1-bromo-2-methylbutane. Substitute the Br^- with a Cl^- to form 1-chloro-2-methylbutane. What is the configuration of your model of 1-chloro-2-methylbutane?

Answers to Problems

1. **a.** $CH_3CH_2CH_2OH$ CH_3CHOH (with CH_3 on the CH) $CH_3CH_2OCH_3$

   b. There are seven constitutional isomers with molecular formula $C_4H_{10}O$.

   $CH_3CH_2CH_2CH_2OH$ CH_3CHCH_2OH (with CH_3 on the CH) CH_3COH (with CH_3 above and CH_3 below on the C) $CH_3CHCH_2CH_3$ (with OH on the CH)

   $CH_3CH_2OCH_2CH_3$ $CH_3OCH_2CH_2CH_3$ CH_3OCHCH_3 (with CH_3 on the CH)

2. **a.**
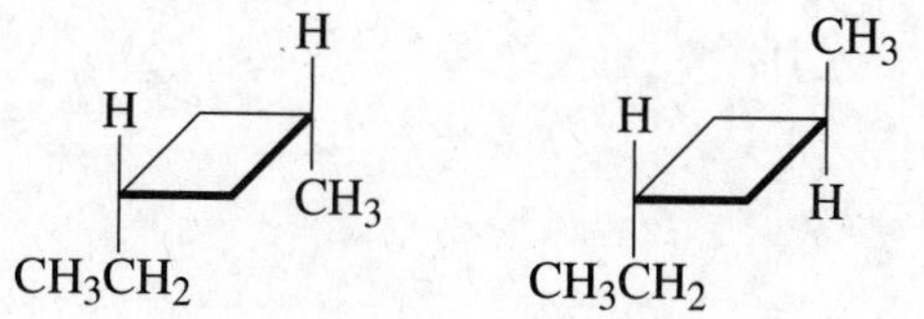

   c.
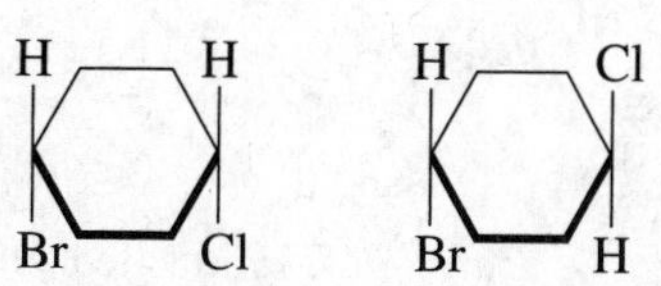

   b.
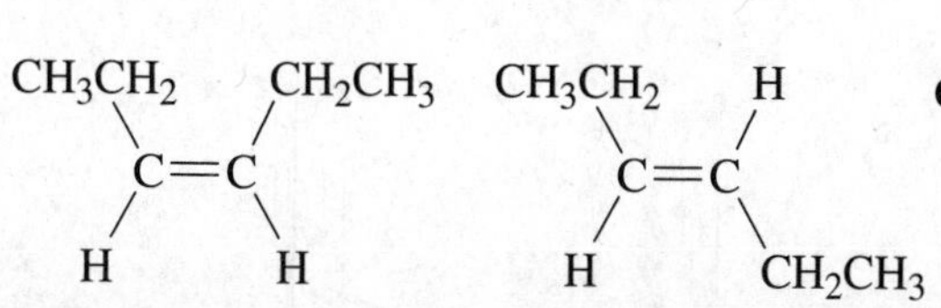

   d.
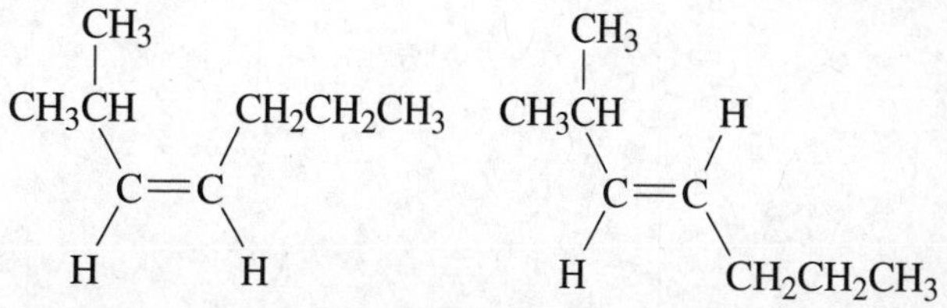

3. **a.** F, G, J, L, N, P, Q, R, S, Z

b. A, C, D, H, I, M, O, T, U, V, W, X, Y

Whether B, E, and K are chiral or achiral depends on how they are drawn. For example, if B is drawn with two equal loops, it is achiral; if the loops differ in size, it is chiral.

4. **a**, **c**, and **f** have asymmetric centers.

5. Solved in the text.

6. **a**, **c**, and **f**, because in order to be able to exist as a pair of enantiomers, the compound must have an asymmetric center. (See the answer to Problem 4.)

7. Draw the first enantiomer with the groups in any order you want. Then draw the second enantiomer by drawing the mirror image of the first enantiomer.

a.

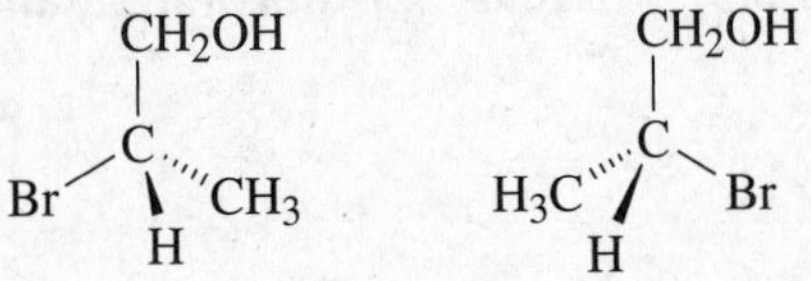

c.

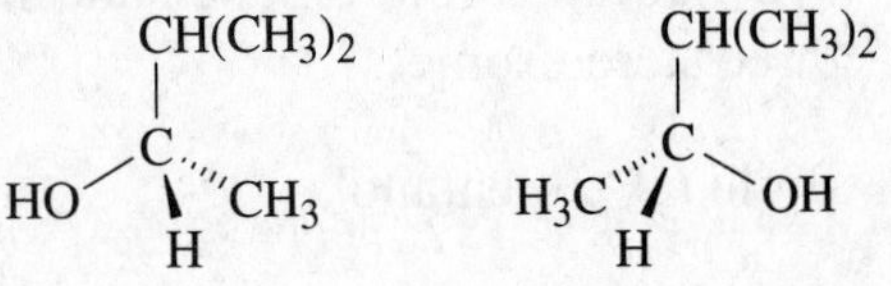

b.

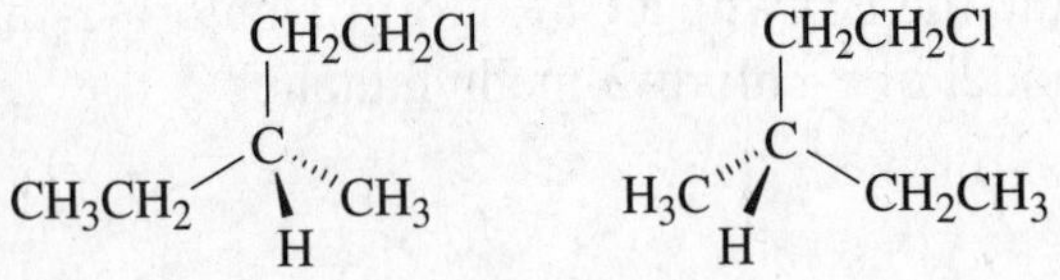

8. Solved in the text.

9. **A**, **B**, and **C** are identical.

10. **a.** $-CH_2OH$ (1) $-CH_3$ (3) $-CH_2CH_2OH$ (2) $-H$ (4)

b. $-CH{=}O$ (2) $-OH$ (1) $-CH_3$ (4) $-CH_2OH$ (3)

c. $-CH(CH_3)_2$ (2) $-CH_2CH_2Br$ (3) $-Cl$ (1) $-CH_2CH_2CH_2Br$ (4)

d. $-CH{=}CH_2$ (2) $-CH_2CH_3$ (3) $-C{\equiv}CH$ (1) $-CH_3$ (4)

C is attached to 2 C's (for $-CH{=}CH_2$)

C is attached to 3 C's (for $-C{\equiv}CH$)

11. **a.** Switch a pair of substituents so that the H is on a dotted bond. The configuration of the compound with the switched pair is *S*. Therefore, the compound given in the question has the *R* configuration.

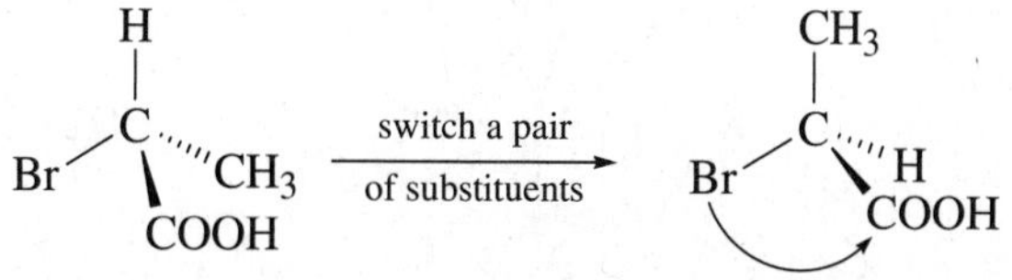
H
C
Br
CH3
COOH
switch a pair
of substituents
CH3
C
Br
H
COOH

R *S*

b. Remember that the arrow can bypass #4 but not #3. The compound has the *R* configuration.

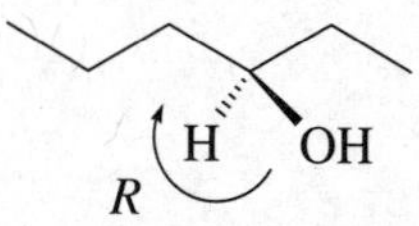
H
OH
R

12. **a.** (*R*)-2-bromobutane **b.** (*R*)-1,3-dichlorobutane

13. **a.** enantiomers **b.** enantiomers

14. Your answer might be correct yet not look like the answers shown here. If you can get the answer shown here by interchanging **two** pairs of groups bonded to an asymmetric center, then your answer is correct. If you get the answer shown here by interchanging **one** pair of groups bonded to an asymmetric center, then your answer is not correct.

a.

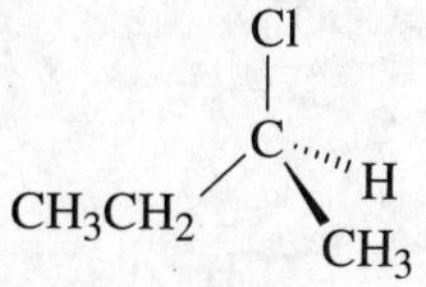
Cl
C
CH3CH2
H
CH3

b.

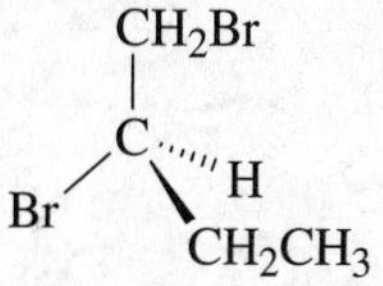
CH2Br
C
Br
H
CH2CH3

15. **a.** levorotatory **b.** dextrorotatory

16. **a.** *S* **b.** *R* **c.** *R* **d.** *S*

17.

$$\text{observed specific rotation} = \frac{\text{observed rotation}}{\text{concentration} \times \text{length}}$$

$$[\alpha] = \frac{\alpha}{\text{concentration} \times \text{length}} = \frac{+13.4^\circ}{\frac{2}{50} \times 2\ \text{dm}} = \frac{+13.4^\circ}{0.08} = +168$$

18. **a.** -24 **b.** 0

19. From the data given, you cannot determine what the configuration of naproxen is.

20. **a.** enantiomers

b. identical compounds (Therefore, they are not isomers.)

c. diastereomers

21. **a.** Find the sp^3 carbons that are bonded to four different substituents; these are the asymmetric centers. Cholesterol has eight asymmetric centers. They are indicated by arrows.

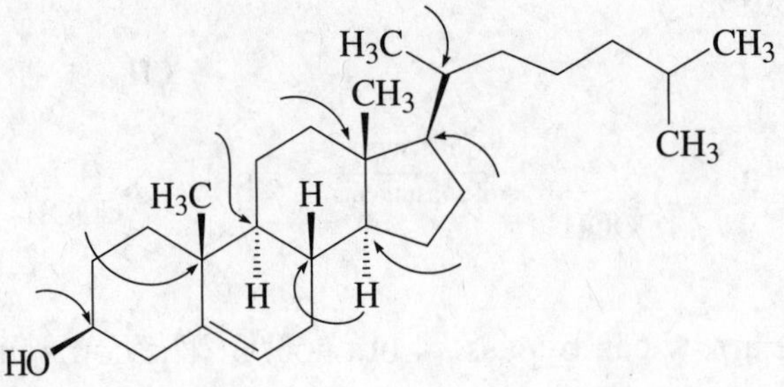

b. $2^8 = 256$

22. **a.** Leucine has one asymmetric center.

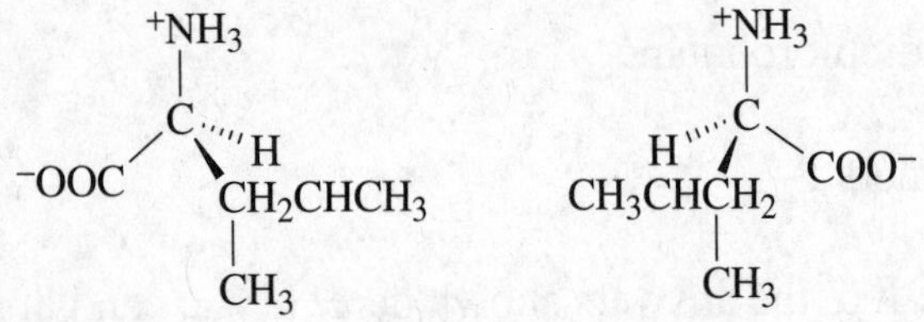

enantiomers

b. Isoleucine has two asymmetric centers.

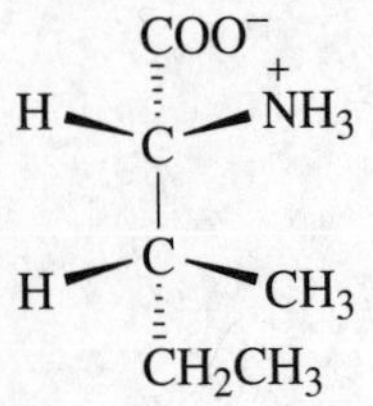
1

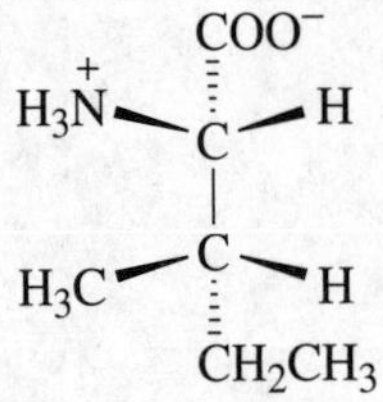
2

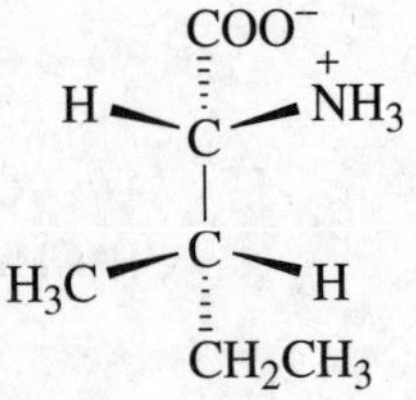
3

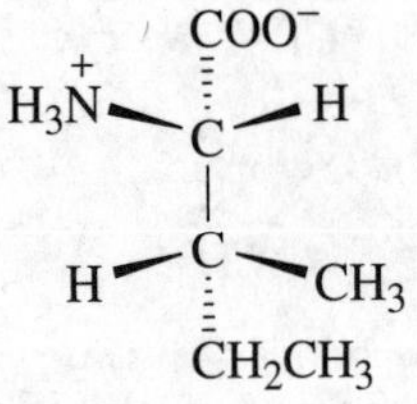
4

enantiomers: 1 and 2, 3 and 4

distereomers: 1 and 3, 1 and 4, 2 and 3, 2 and 4

23. **a.** disastereomers **b.** enantiomers **c.** identical **d.** constitutional isomers

24. **B**

A does not have a stereoisomer that is a meso compound, because its two asymmetric centers are not bonded to the same four groups.

C does not have a stereoisomer that is a meso compound, because it does not have asymmetric centers.

25. **a.**

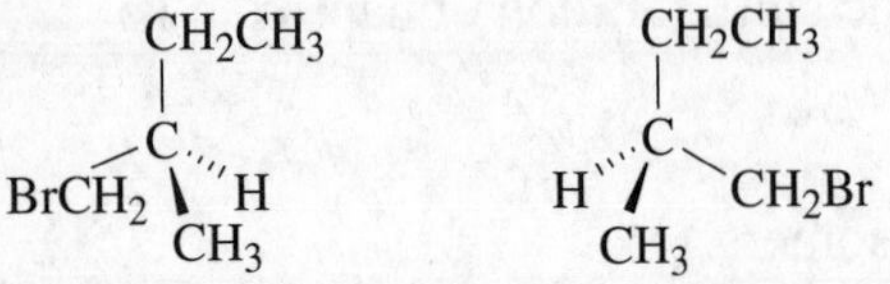

b. This compound does not have any stereoisomers because it does not have an asymmetric center.

c.

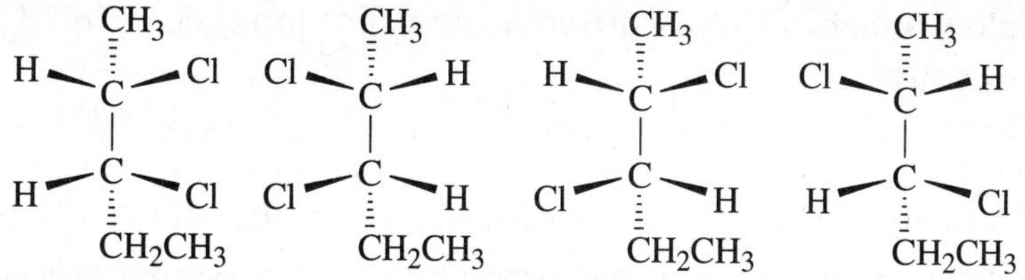

d.

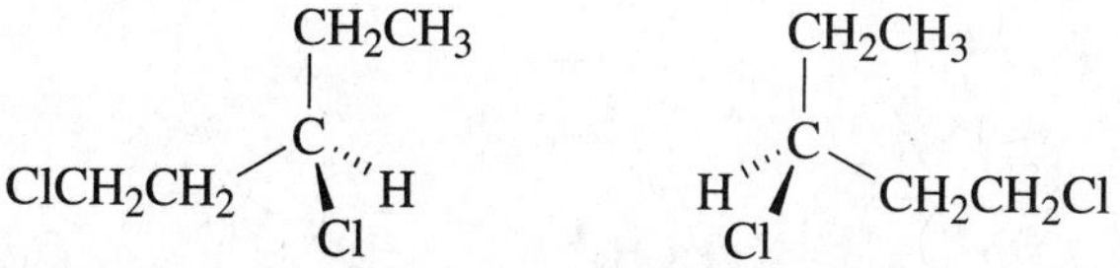

26. Using (+)-limonene, add a hatched wedge, making sure that the two lines in the plane of the paper are adjacent to one another. Put the H on the hatched wedge. The arrow drawn from the group with the highest priority to the next highest priority is clockwise. Therefore, (+)-limonene is the *R* enantiomer, the one found in oranges.

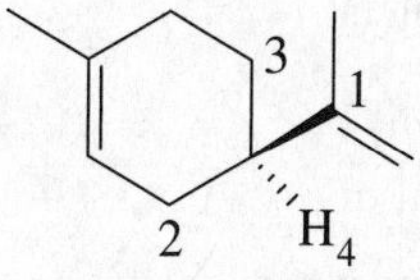

You could have started with (−)-limonene. If you did, you would have found that (−)-limonene was the *S* enantiomer, the one found in lemons.

27. Only the stereoisomers of the major product of each reaction are shown.

a.

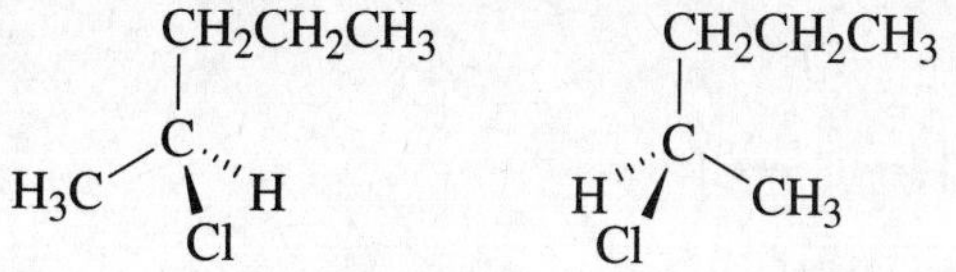

c.

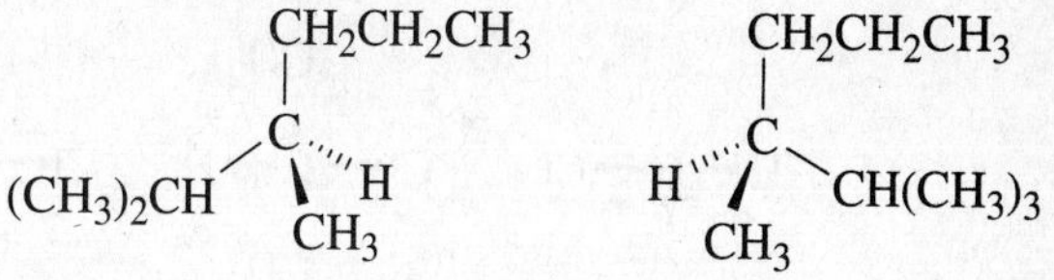

b.

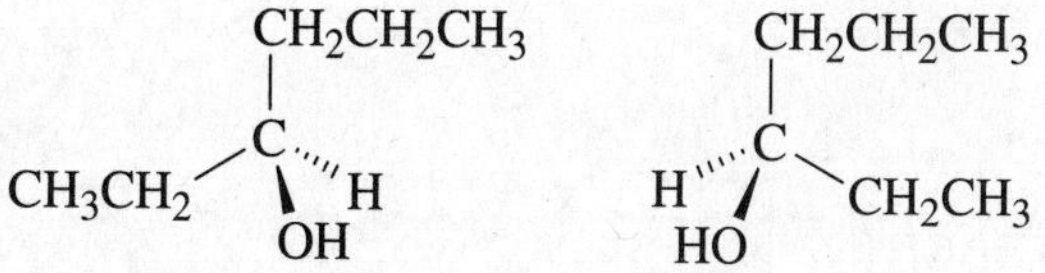

d.

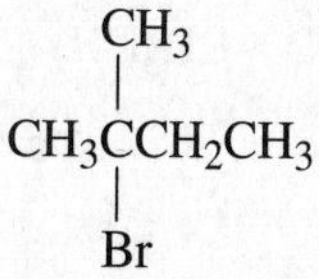

This compound does not have any stereoisomers because it does not have an asymmetric center.

28. **a.**

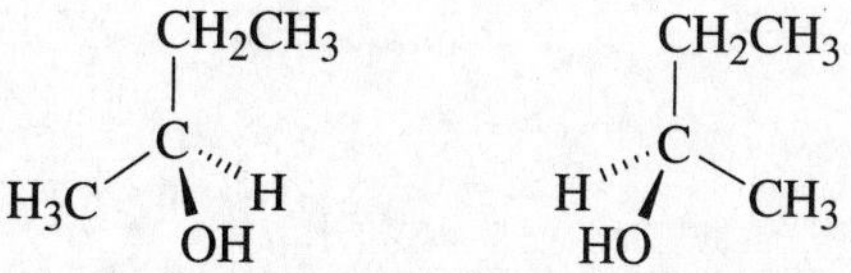

c.

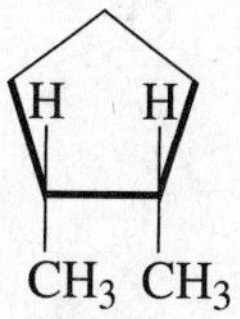

b.

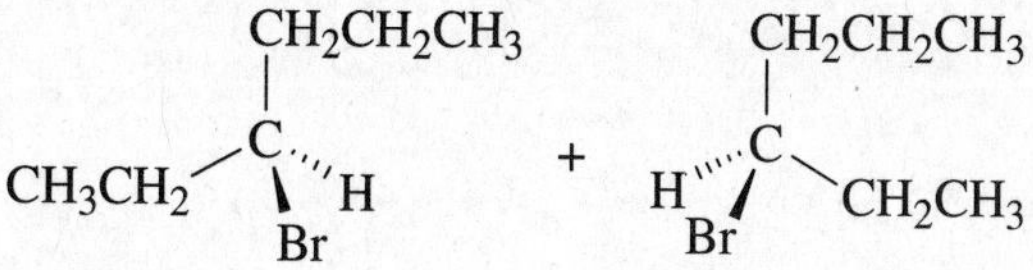

29. **a.** (*R*)-malate and (*S*)-malate (A product with one asymmetric center would be formed from a reactant with no asymmetric centers. Thus, the product would be a racemic mixture.)

b. (*R*)-malate and (*S*)-malate (A product with one asymmetric center would be formed from a reactant with no asymmetric centers. Thus, the product would be a racemic mixture.)

30. An asymmetric center is an atom attached to four different atoms (or groups). CHFBrCl is the only one with an asymmetric center.

31.

$CH_3CH{=}CHCH_2CH_3$
2 stereoisomers
[cis and trans]

$CH_2{=}CHCH_2CH_2CH_3$
no stereoisomers

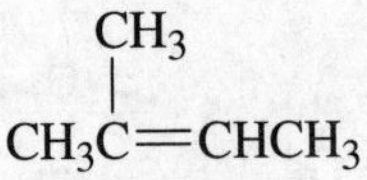
no stereoisomers

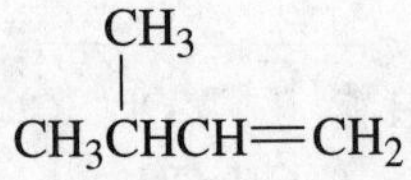
no stereoisomers

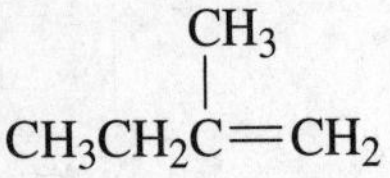
no stereoisomers

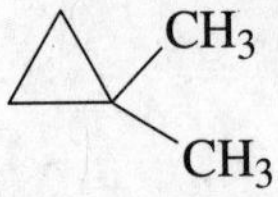
no stereoisomers

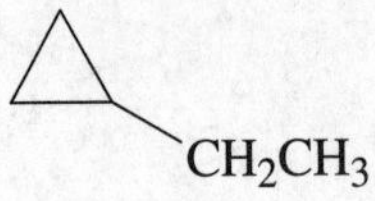
no stereoisomers

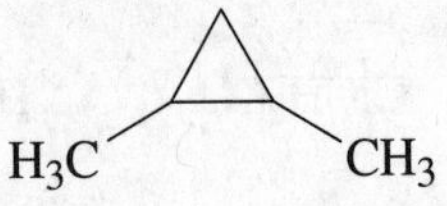
3 stereoisomers
[Cis is a meso compound.]
[Trans is a pair of enantiomers.]

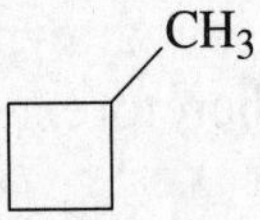
no stereoisomers

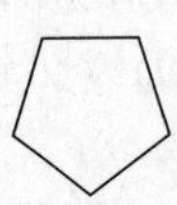
no stereoisomers

32. **a.**

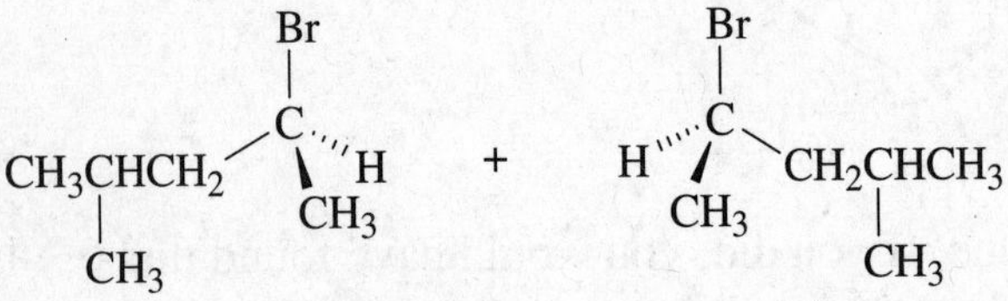

b.

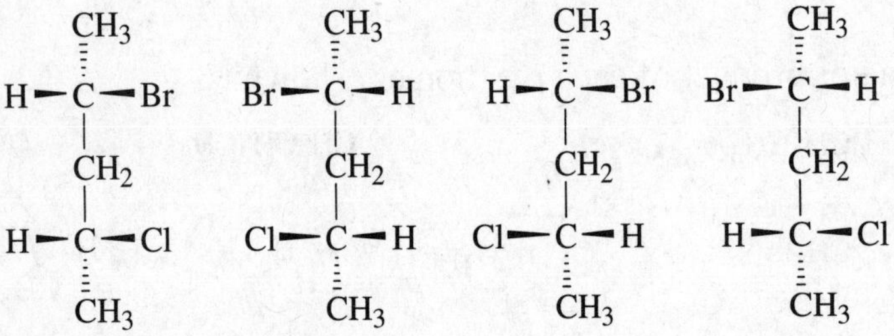

c.

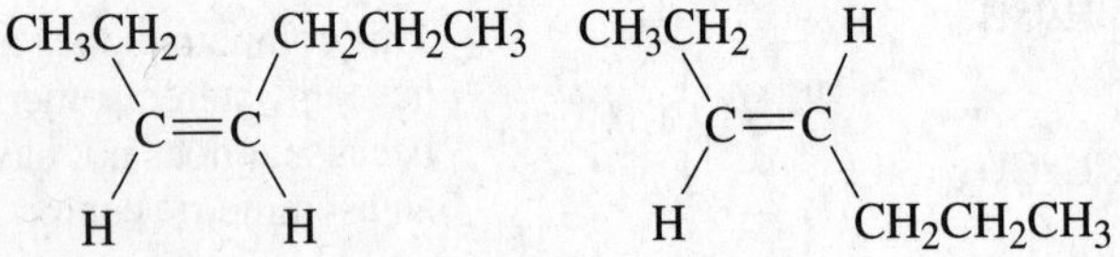

d.

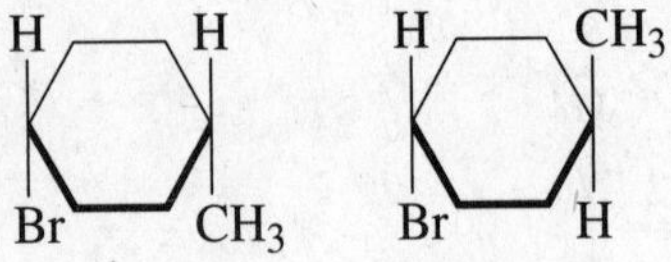

e.

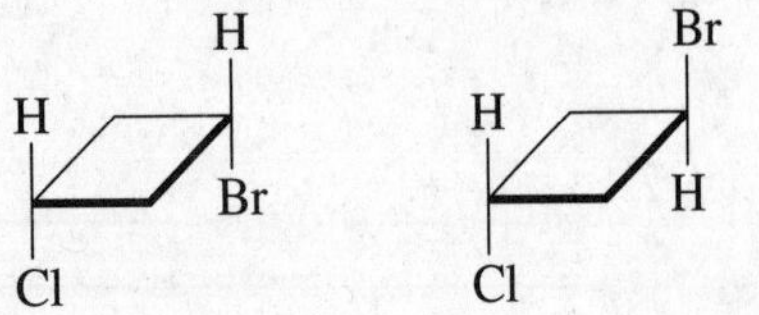

f.

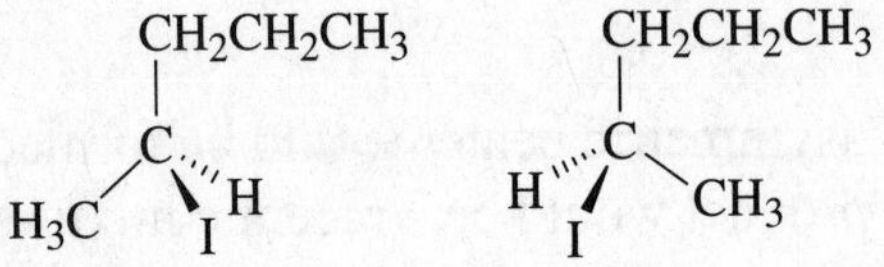

g. 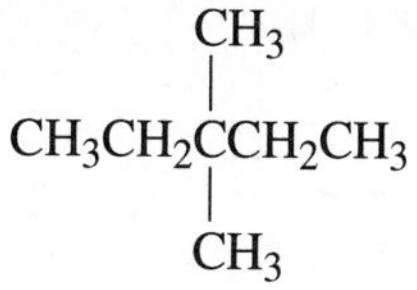

No isomers are possible for this compound because it does not have an asymmetric center.

h. 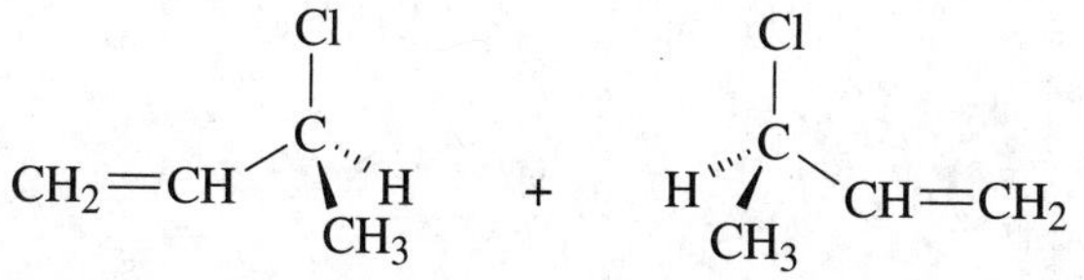

33. a. (*E*)-1-bromo-2-chloro-2-fluoro-1-iodoethene
 b. (*S*)-2,3-dimethylpentane
 c. (*Z*)-2-bromo-1-chloro-1-fluoroethene
 d. (*S*)-1-chloro-3-methylpentane

34. Mevacor has eight asymmetric centers.

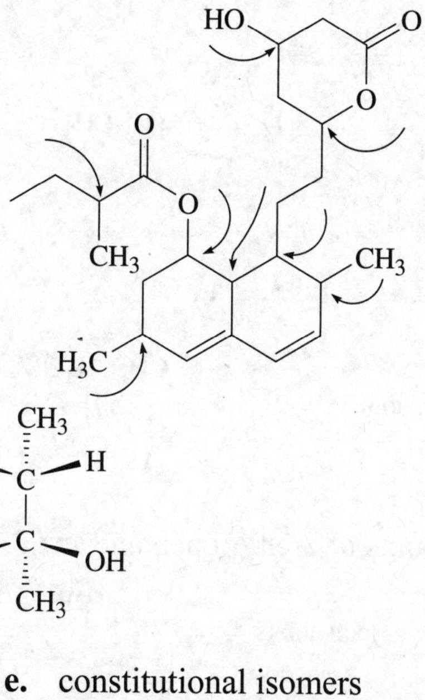

35. 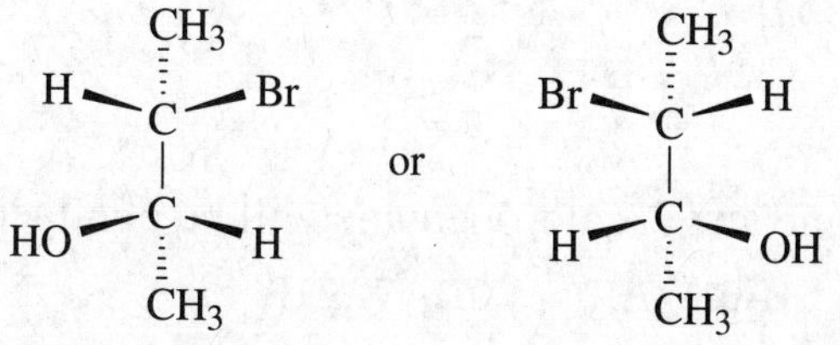

36. a. diastereomers
 b. identical
 c. constitutional isomers
 d. diastereomers
 e. constitutional isomers
 f. enantiomers
 g. identical
 h. diastereomers

Diastereomers are stereoisomers that are not enantiomers. Therefore, cis-trans isomers are diastereomers.

37. a. 1. 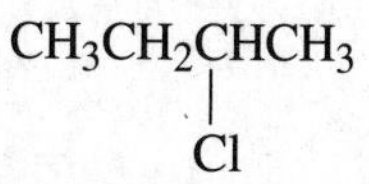 stereoisomers = 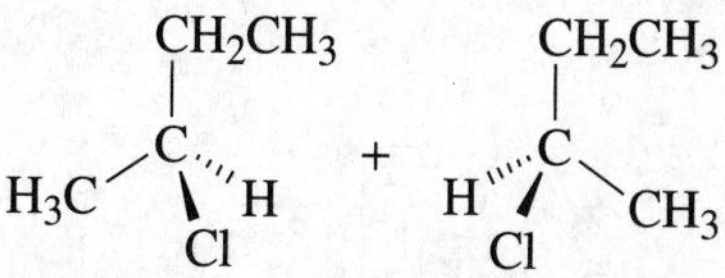

2. $CH_3CH_2CHCH_3$
$\quad\quad\quad|$
$\quad\quad\;\,OH$

stereoisomers =

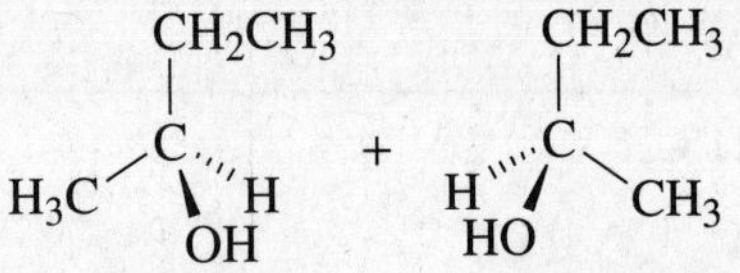

3. $CH_3CH_2CH_2CH_3$

4. $CH_3CH_2CHCH_3$
$\quad\quad\quad|$
$\quad\quad\;\,OCH_3$

stereoisomers =

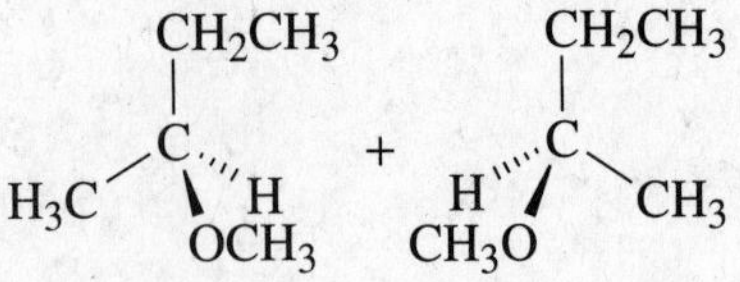

b. The products would be the same if the reactant had been the trans isomer.

38. **a** and **c**

39.

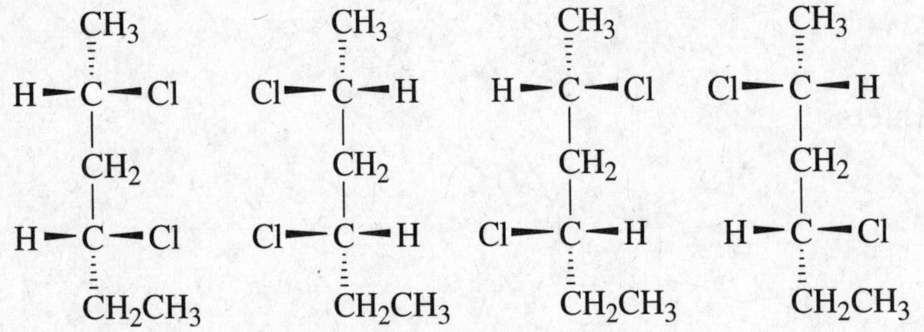

1 **2** **3** **4**

enantiomers = 1 and 2; 3 and 4

diastereomers = 1 and 3; 1 and 4; 2 and 3; 2 and 4

40. **a.** $CH_3CH_2CHCH_3$
$\quad\quad\quad|$
$\quad\quad\;\,Cl$

stereoisomers =

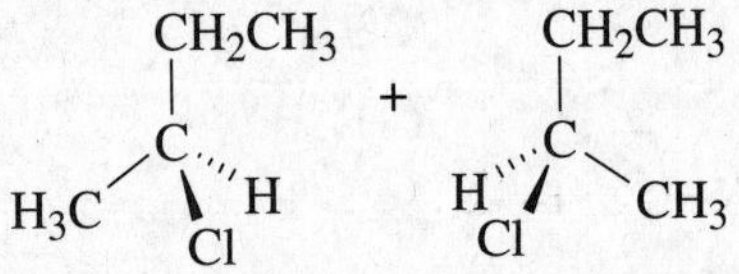

b. Approximately equal amounts of 2-chloropentane and 3-chloropentane will be obtained.

$CH_3CH_2CH_2CHCH_3$
$\quad\quad\quad\quad|$
$\quad\quad\quad\;\,Cl$

stereoisomers =

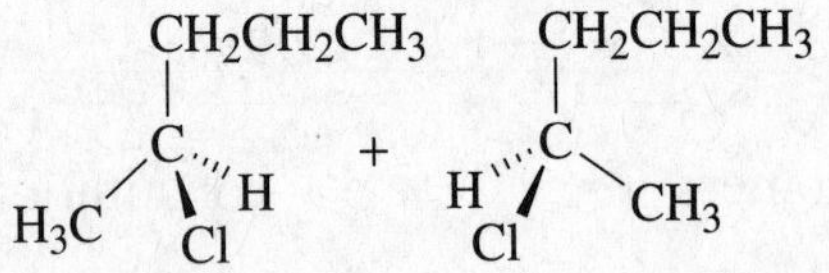

$CH_3CH_2CHCH_2CH_3$
$\quad\quad\quad|$
$\quad\quad\;\,Cl$

This product does not have an asymmetric center, so it does not have stereoisomers.

c.

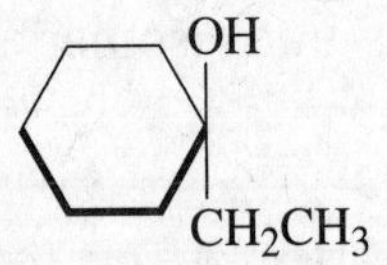

This product does not have an asymmetric center, so it does not have stereoisomers.

41. $$[\alpha] = \frac{\alpha}{l \times c} = \frac{-1.8^\circ}{[2.0\,\text{dm}][0.15\,\text{g/mL}]} = -6.0$$

42. Butaclamol has three asymmetric centers that are carbon atoms. It also has an asymmetric center that is a nitrogen atom (it too has 4 different substituents attached to it, one of which is a lone pair). Thus, butaclamol has a total of four asymmetric atoms.

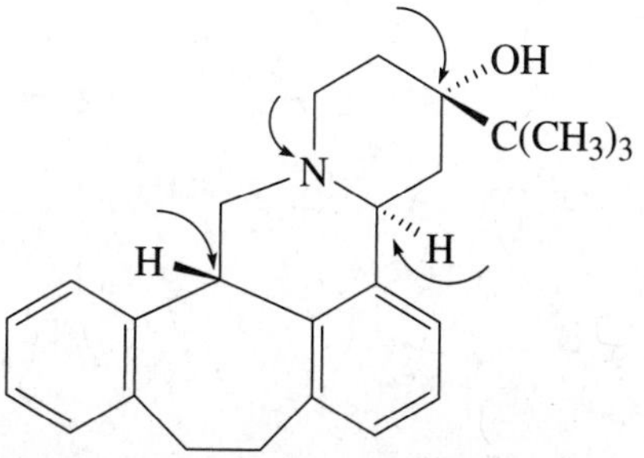
OH
C(CH3)3
N
H
H

43. **a, b, c, e** (think of a British car and an American car), and **h** are chiral.

d, f, and **g** are each superimposable on its mirror image. These, therefore, are achiral.

44. *R* and *S* are related to (+) and (–) in that if one configuration (say, *R*) is (+), the other one is (–).

Because some compounds with the *R* configuration are (+) and some are (–), there is no way to determine whether a particular *R* enantiomer is (+) or (–) without putting the compound in a polarimeter or going to the library to see if someone else has previously determined how it rotates the plane of polarization of polarized light.

45. **a.**

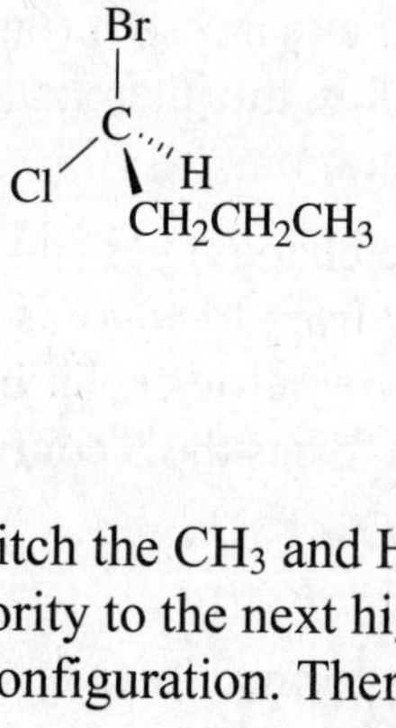
Br
C
Cl
H
CH2CH2CH3

b.

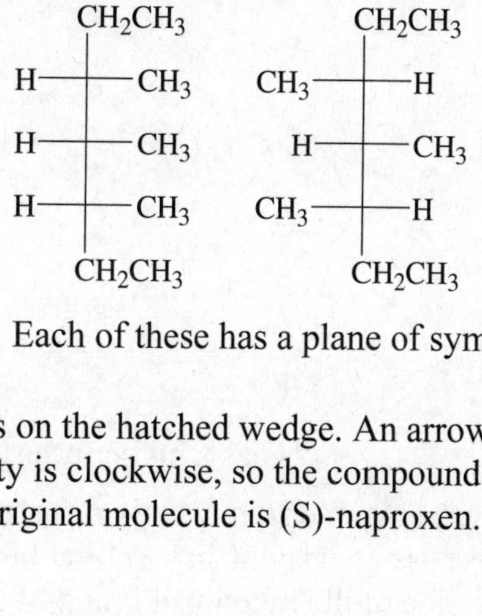
CH2CH3
H
CH3
H
CH3
H
CH3
CH2CH3
CH2CH3
CH3
H
H
CH3
CH3
H
CH2CH3

Each of these has a plane of symmetry, so they are achiral.

46. Switch the CH_3 and H so the H is on the hatched wedge. An arrow drawn from the highest priority to the next highest priority is clockwise, so the compound with the switched pair has the *R* configuration. Therefore, the original molecule is (S)-naproxen.

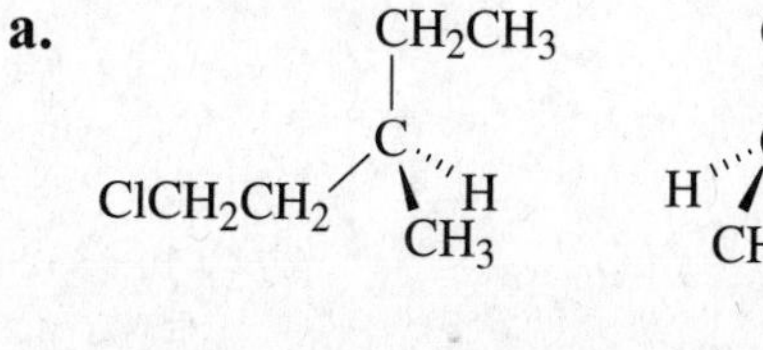
H
CH3
C
COH
O
CH3O
(*S*)-naproxen

3
CH3
H
4
2
C
1
COH
O
CH3O
(*R*)-naproxen

47. **a.**

CH2CH3
C
ClCH2CH2
H
CH3
CH2CH3
C
H
CH2CH2Cl
CH3

b.

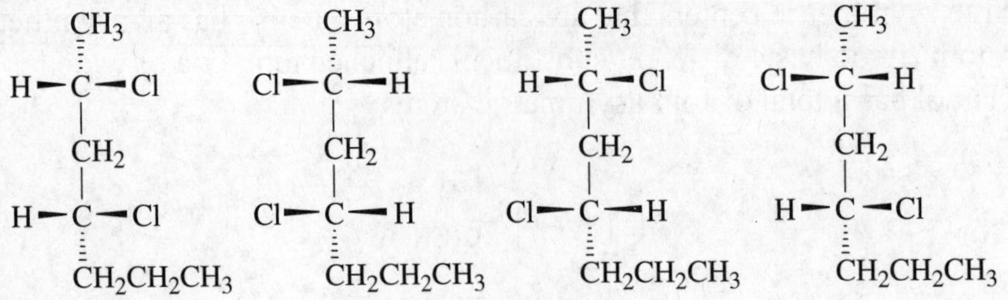

c.

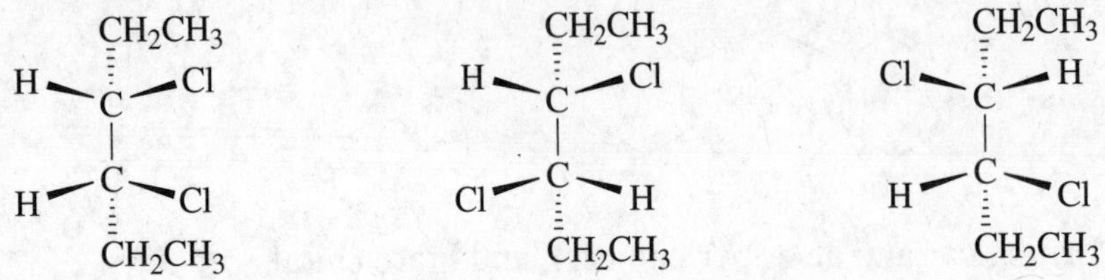

a meso compound

48. **a.**

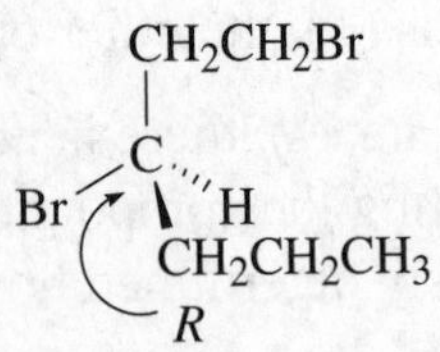

b.

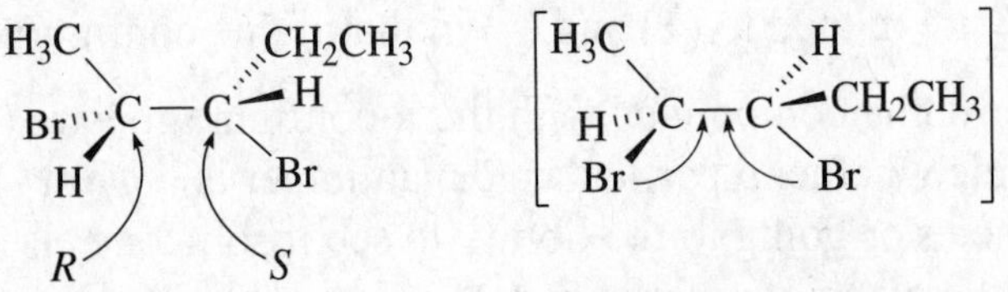

After interchanging H and Br in order to have H attached by a dashed bond, the arrow from #1 to #2 is counterclockwise, so the asymmetric center before the interchange has the *R* configuration.

After interchanging H and CH_2CH_3, the arrow from #1 to #2 is clockwise, so the asymmetric center before the interchange has the *S* configuration.

49. **a.** (*S*)-citric acid (^{14}C has a higher priority than ^{12}C)

b. The reaction is catalyzed by an enzyme. Only one stereoisomer is formed in an enzyme-catalyzed reaction because an enzyme has a chiral binding site which allows reagents to be delivered to only one side of the functional group of the compound.

c. The product of the reaction will be achiral because if it doesn't have a ^{14}C, the two CH_2COOH groups will be identical so it will not have an asymmetric center.

50.

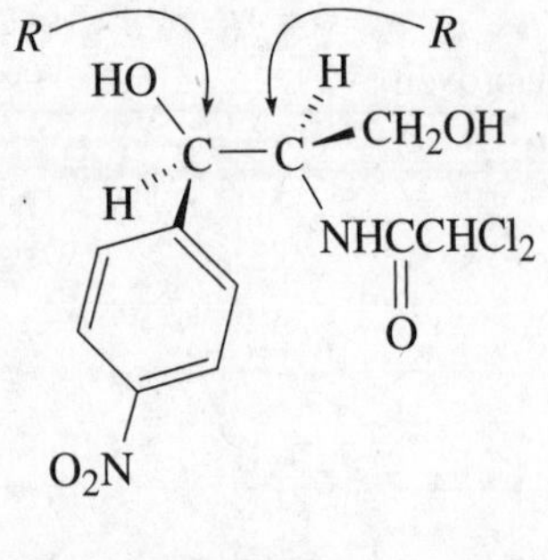

51. **a.** Because there are two asymmetric centers, there are four possible stereoisomers.

b.

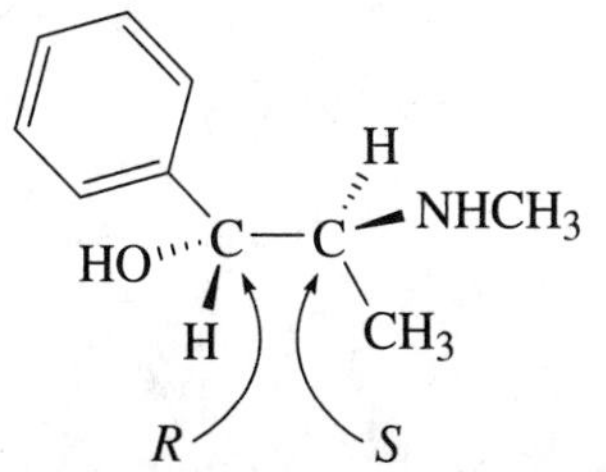

52. **a.** If you rotate one of the structures 180° (keeping it in the same plane), you can see that they are identical.

b. Because the two molecules have the opposite configuration at both asymmetric centers, they are enantiomers.

53. Because of the double bond, the compound has cis and trans isomers. Because of the asymmetric center, the cis and trans isomers each has a pair of enantiomers.

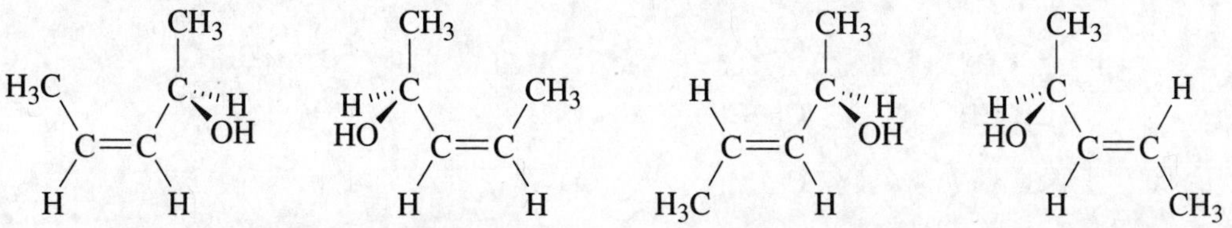

enantiomers for the cis stereoisomer

enantiomers for the trans stereoisomer

54.

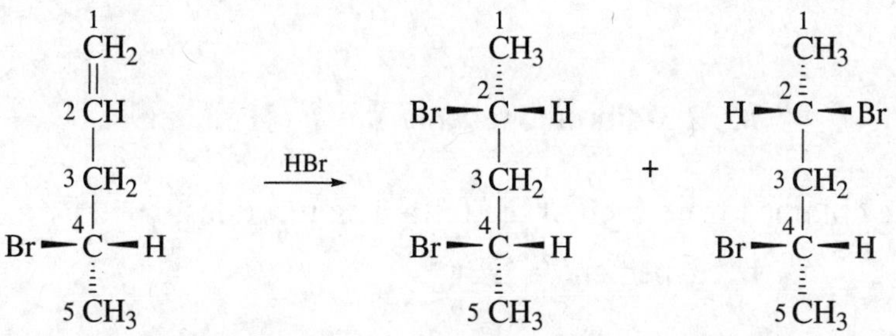

(*S*)-4-bromo-1-pentene

optically inactive meso compound

optically active

The configuration at C-4 does not change, because no bonds to this carbon are broken.

The new asymmetric center at C-2 can have either the *R* or the *S* configuration.

The product with C-2 in the *R* configuration has a plane of symmetry; therefore, it is achiral (optically inactive).

The product with C-2 in the *S* configuration does not have a plane of symmetry; therefore, it is chiral (optically active).

55. **a.** diastereomers **b.** identical **c.** constitutional isomers **d.** diastereomers

Chapter 6 Practice Test

1. Are the following compounds identical or a pair of enantiomers?

a.

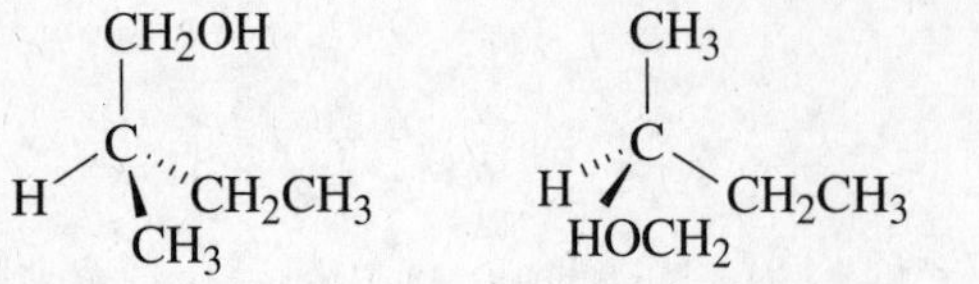

b.

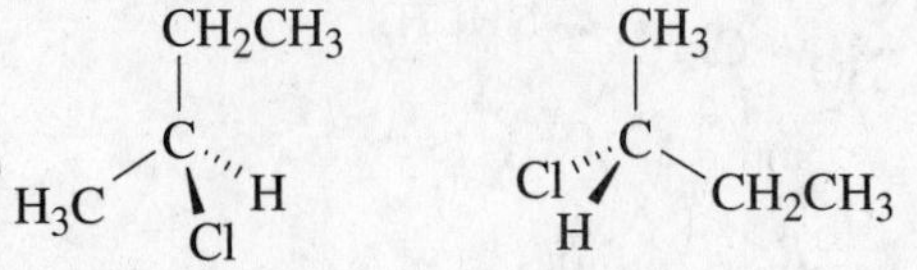

2. 30 mL of a solution containing 2.4 g of a compound rotates the plane of polarization of polarized light –0.48° in a polarimeter with a 2.0 decimeter sample tube. What is the specific rotation of the compound?

3. Which are meso compounds?

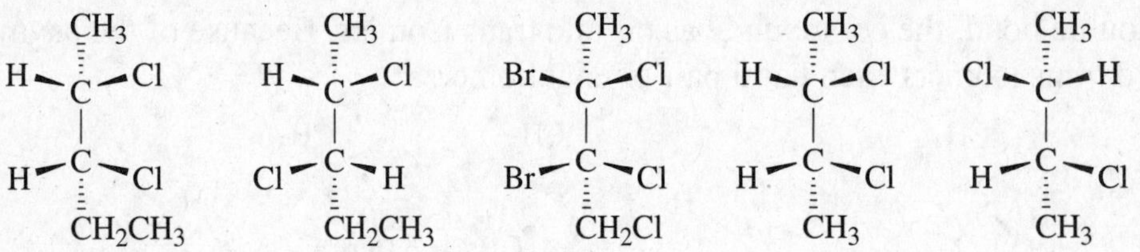

A B C D E

4. Draw the constitutional isomers that have molecular formula C_4H_9Cl.

5. Draw the stereoisomers of:

a. 2,3-dibromobutane **b.** 2,3-dibromopentane

6. What stereoisomers would be obtained from each of the following reactions:

a. 1-butene + HCl **b.** 2-pentene + HBr

7. What stereoisomers would be obtained from each of the following reactions?

a.

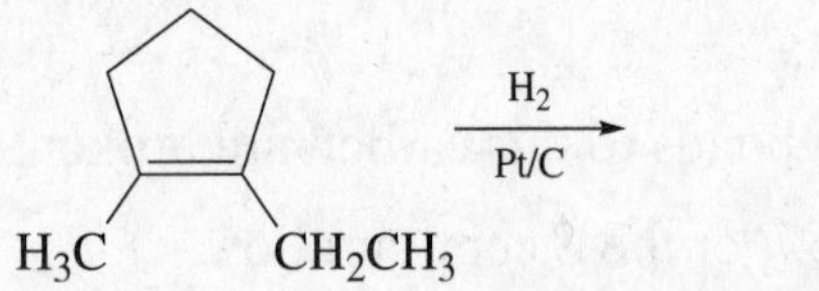

b.

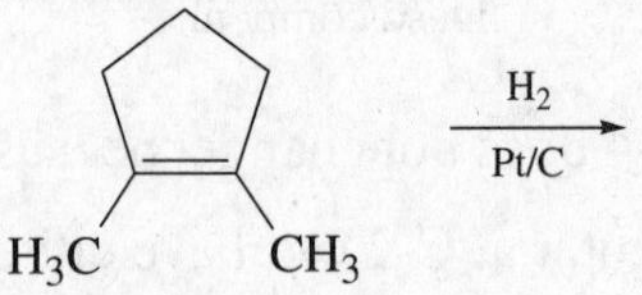

8. Which of the following have the *R* configuration?

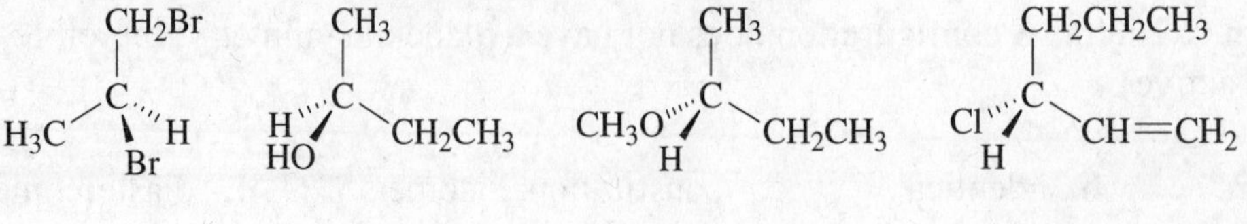

9. Draw a diastereomer of the following compound:

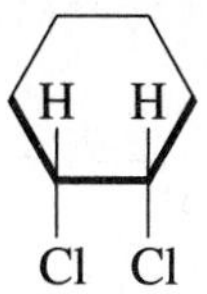
H H
Cl Cl

10. Indicate whether each of the following statements is true or false:

a. Diastereomers have the same melting points. T F

b. Meso compounds do not rotate polarized light. T F

c. 2,3-Dichloropentane has a stereoisomer that is a meso compound. T F

d. All chiral compounds with the *R* configuration are dextrorotatory. T F

e. A compound with three asymmetric centers can have a maximum of nine stereoisomers. T F

<u>**CHAPTER 7**</u>

Delocalized Electrons and Their Effect on Stability, Reactivity, and p*K*a: Ultraviolet and Visible Spectroscopy

1. **a**, **b**, **e**, **f**, and **g** have delocalized electrons.

2. **a.**

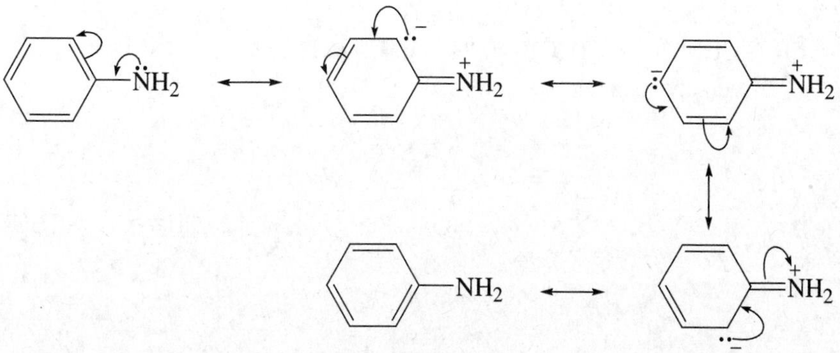

b. Because the lone-pair electrons on nitrogen are localized electrons, the only resonance structures are the two resonance structures of the benzene ring.

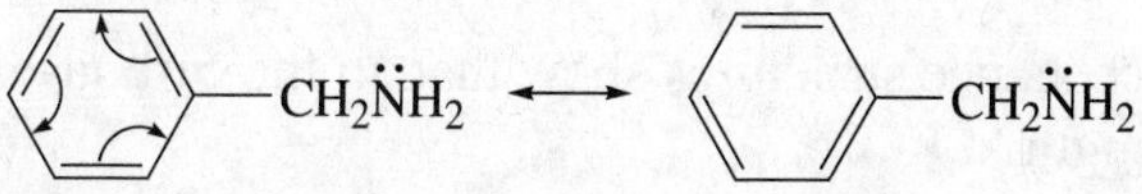

e.

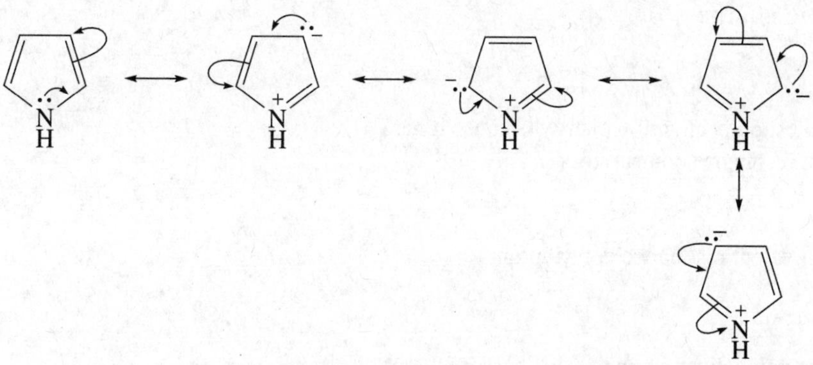

f.

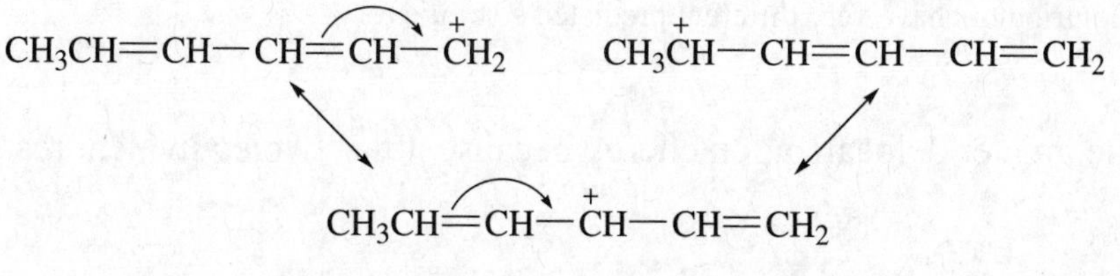

g.

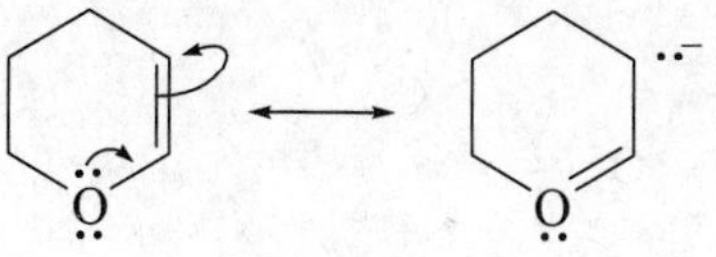

3. The resonance contributor that makes the greatest contribution to the hybrid is labeled "**A**". "**B**" contributes less to the hybrid than "**A**".

a. Solved in the text.

b. 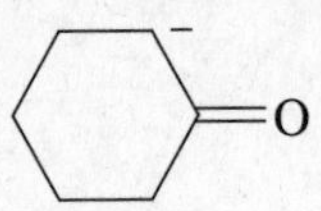 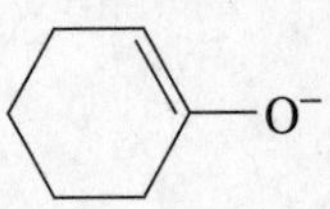

B **A**

c. $CH_3—\overset{+}{C}H—CH{=}CH—CH_3$ $CH_3CH{=}CH—\overset{+}{C}H—CH_3$

equally stable

4. **a.** $CH_3\overset{\delta+}{C}{\cdots}CH{\cdots}\overset{\delta+}{C}HCH_3$ with CH_3 on the first C **b.** 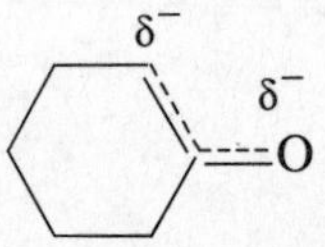 **c.** $CH_3\overset{\delta+}{C}H{\cdots}CH{\cdots}\overset{\delta+}{C}HCH_3$

5. 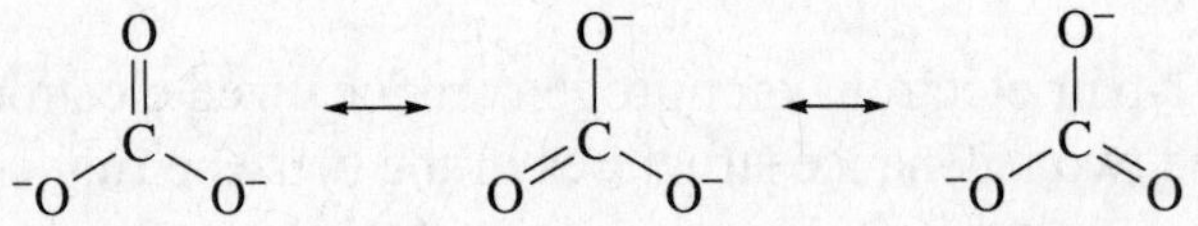

a. The contributing resonance structures show that all the carbon-oxygen bonds in the carbonate ion are the same length.

b. Because the two negative charges are shared equally by three oxygens, each oxygen has two thirds of a negative charge.

6. 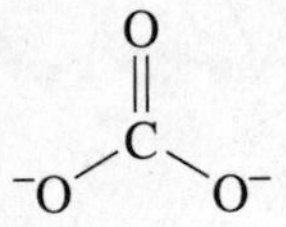 has the greatest delocalization energy because it has 3 equivalent resonance contributors

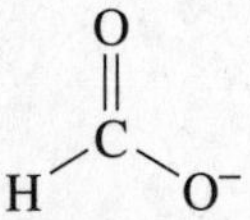 has two equivalent resonance contributors

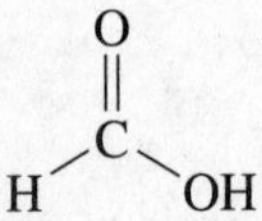 has the least delocalization energy because its two resonance contributors have very different predicted stabilities

7. The acetate ion has the greater delocalization energy because it has two equivalent resonance contributors.

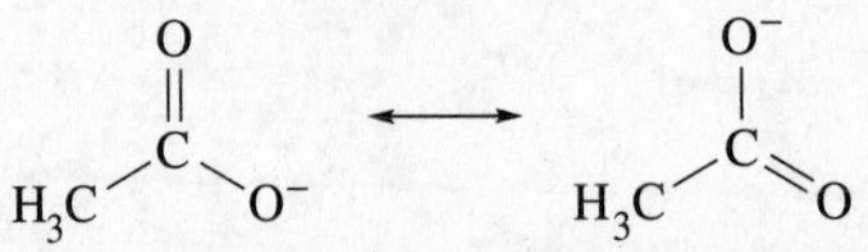

8. 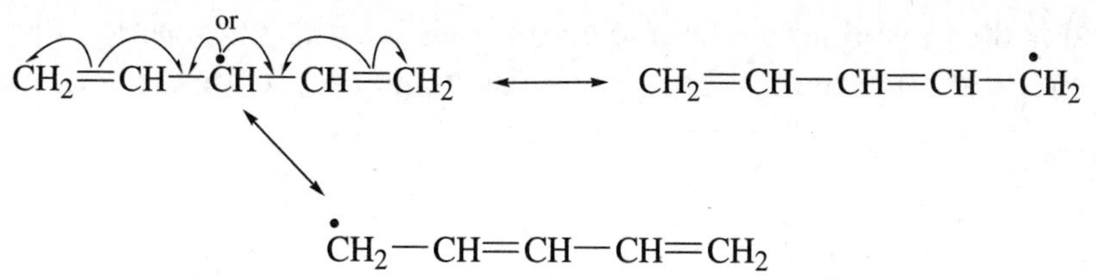

9. The six starred carbons in rhodopsin share the unpaired electrons.

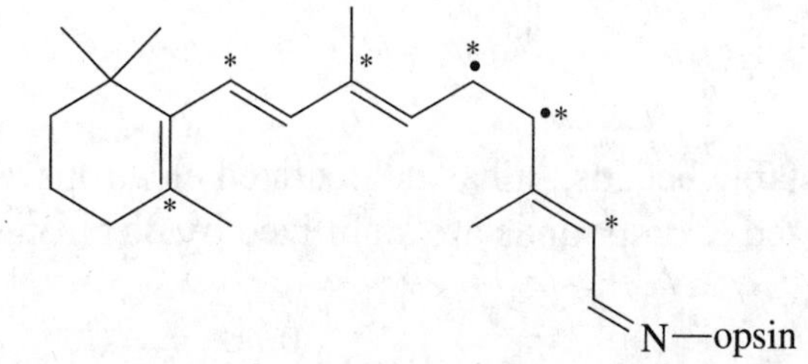
N—opsin

10. **a.** $CH_3CH{=}CH\overset{+}{C}HCH_3$

secondary allylic
(the other one is primary allylic)

b. 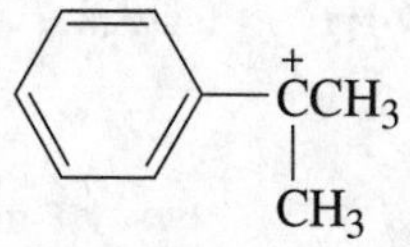

tertiary benzylic
(the other one is secondary benzylic)

11. **a.** 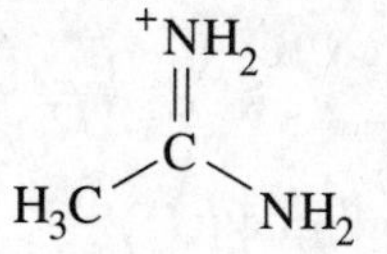

This one is more stable because the positive charge is on an N rather than on a more electronegative O.

b. 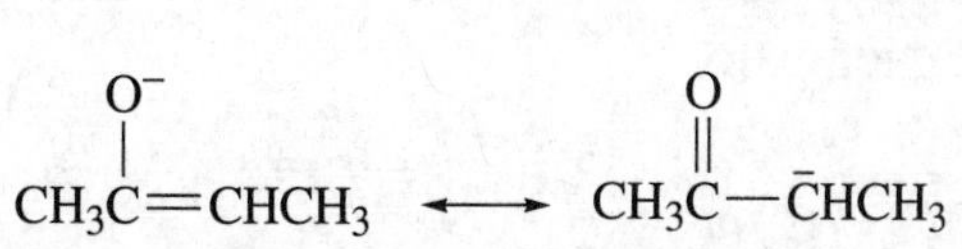

This one is more stable because only in this compound is the negative charge delocalized.

12. **a.** Solved in the text.

b. The contributing resonance structures show that there are two sites that could be protonated.

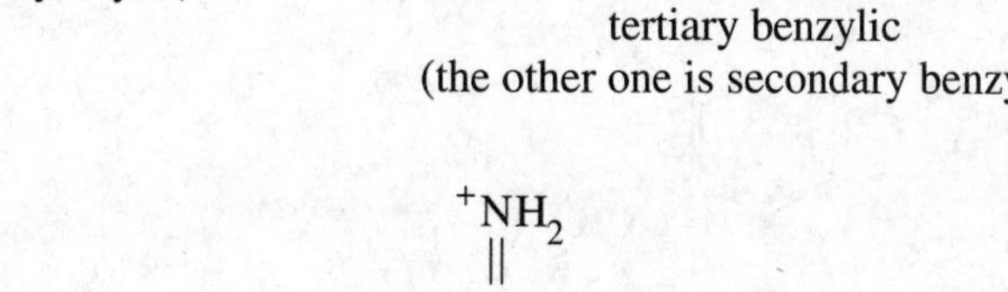

13. **a.** 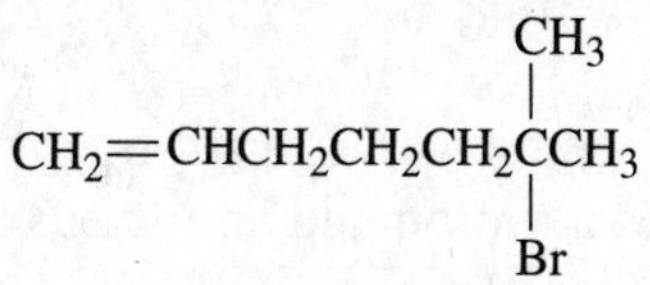

This is the major product because the intermediate that leads to this product is a tertiary carbocation.

b. 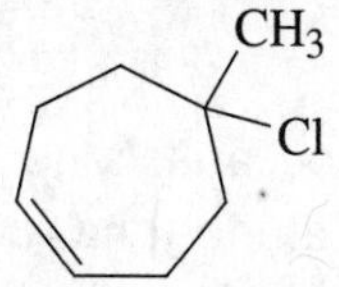

This is the major product because the intermediate that leads to this product is a tertiary carbocation.

14. The indicated double bond is the most reactive in an electrophilic substitution reaction because addition of an electrophile to this double bond forms the most stable carbocation (a tertiary allylic carbocation).

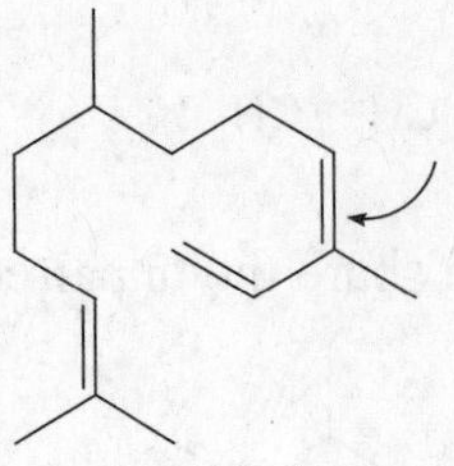

15. 2,4-Heptadiene is more stable because it has delocalized electrons. All the electrons in 2,5-heptadiene are localized. Compounds are stabilized by electron delocalization.

16. **a.**

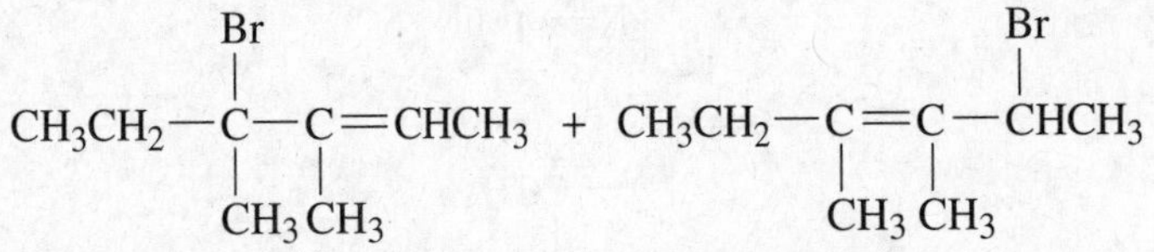
Br
CH3CH2—C—C=CHCH3 + CH3CH2—C=C—CHCH3
CH3 CH3
Br
CH3 CH3

1,2-addition product **1,4-addition product**

b.

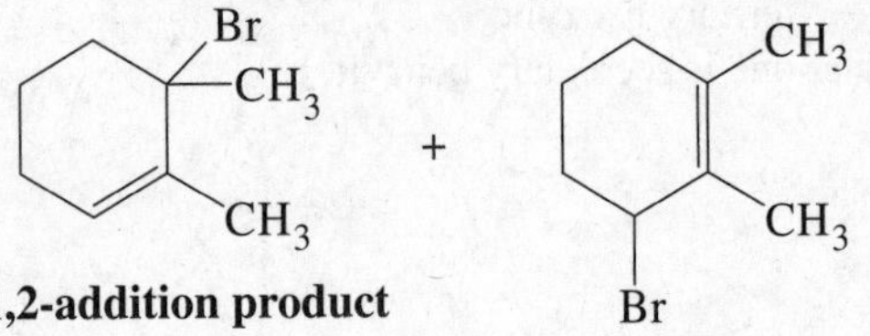
Br
CH3
CH3
+
CH3
CH3
Br

1,2-addition product **1,4-addition product**

17. In each case, the compound shown is the stronger acid because the negative charge that results when the compound loses a proton can be delocalized. Electron delocalization is not possible for the other compound in each pair.

a. $CH_3CH{=}CHOH \longrightarrow H^+ + CH_3CH{=}CHO^- \longleftrightarrow CH_3\bar{C}HCH{=}O$

b.

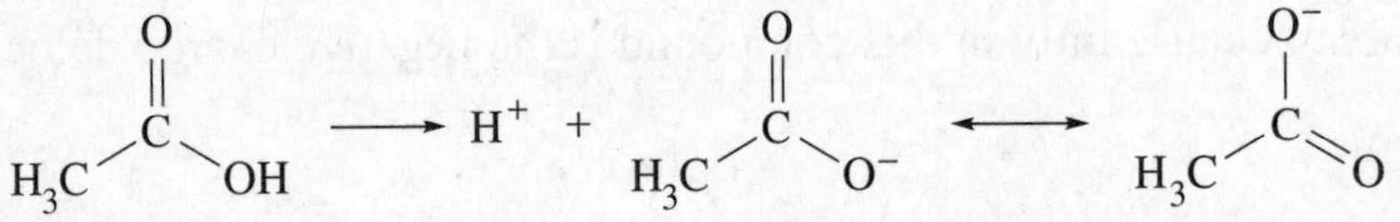
O
H3C C OH
H+ +
O
H3C C O−
O−
H3C C O

c. $CH_3CH{=}CHOH \longrightarrow H^+ + CH_3CH{=}CHO^- \longleftrightarrow CH_3\bar{C}HCH{=}O$

d. $CH_3CH{=}CH\overset{+}{N}H_3 \longrightarrow H^+ + CH_3CH{=}CHNH_2 \longleftrightarrow CH_3\bar{C}HCH{=}\overset{+}{N}H_2$

18. **a.** Ethylamine is a stronger base because when the lone pair on the nitrogen in aniline is protonated, it can no longer be delocalized into the benzene ring, so it is less apt to accept a proton.

b. Ethoxide ion is a stronger base because a negatively charged oxygen is a stronger base than a neutral nitrogen (Recall that an alcohol has a $pK_a \sim 15$ and a protonated amine has a $pK_a \sim 10$, and the weaker the acid, the stronger is its conjugate base.).

c. Ethoxide ion is a stronger base because when the phenolate ion is protonated, the pair of electrons that is protonated can no longer be delocalized into the benzene ring, so it is less apt to accept a proton.

19. The carboxylic acid is the most acidic because its conjugate base has greater delocalization energy than does the conjugate base of phenol, and the more stable the base, the stronger is its conjugate acid. The alcohol is the least stable because, unlike the negative charge on the conjugate base of phenol, the negative charge on the conjugate base of the alcohol cannot be delocalized.

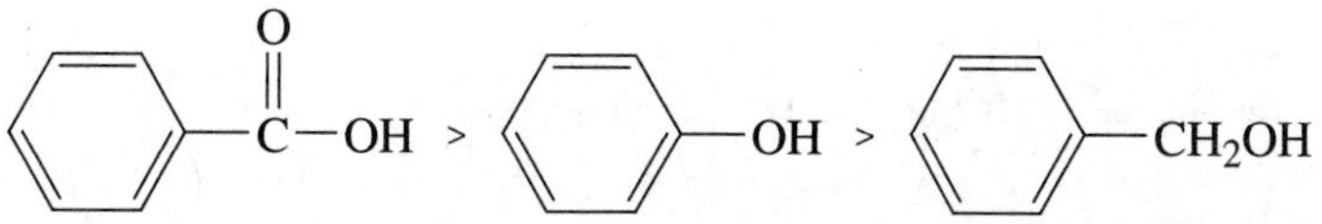
O
C—OH
>
OH
>
CH₂OH

20.

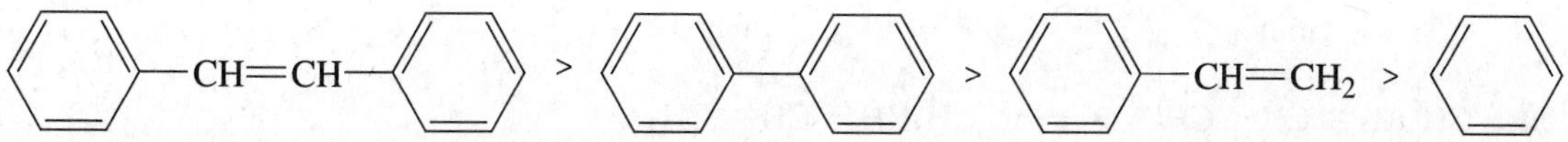
CH=CH
>
>
CH=CH₂
>

21. **a.** Blue is the result of absorption of light of a longer wavelength than light that produces purple.

The compound on the right has an additional group [$N(CH_3)_2$] that will cause it to absorb at a longer wavelength. Therefore, it is the blue compound.

b. They will be the same color at pH = 3 because the $N(CH_3)_2$ group will be protonated and, therefore, will not possess the lone pair that causes the compound to absorb light of a longer wavelength.

22. **a, d, f, h, i, j**

23. **a.**

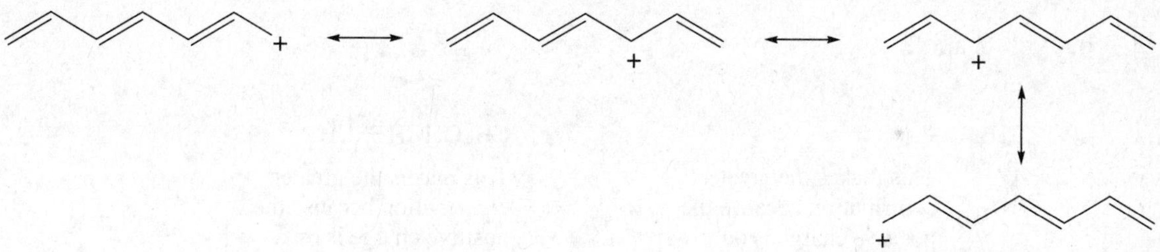

b.

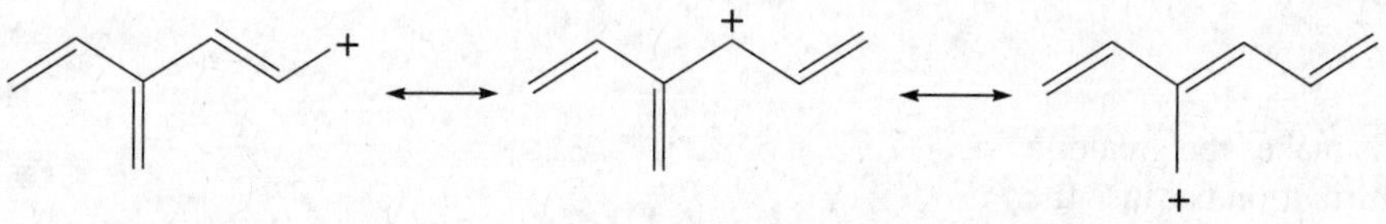

24. **a.** different compounds

b. resonance contributors

c. different compounds

d. different compounds

Notice that in the structures that are different compounds, both atoms and electrons have changed their locations.

In contrast, in structures that are resonance contributors, only the electrons have moved.

25. **a.** **1.** 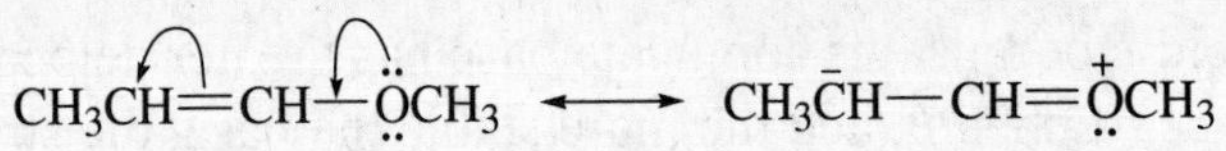

major minor

2. 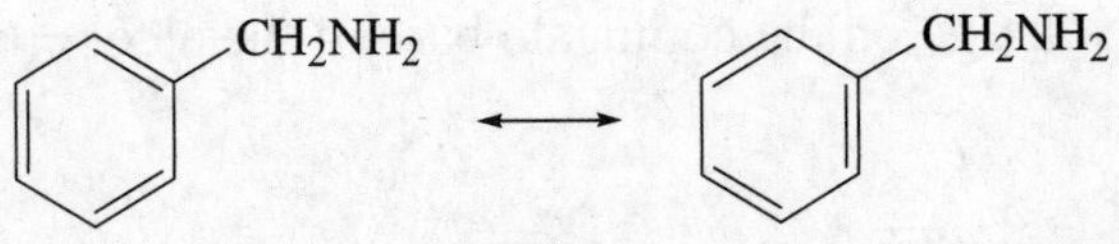

The two resonance contributors have the same stability.

3. 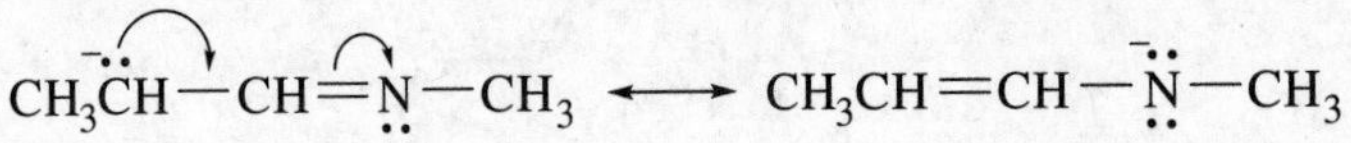

minor major

4. 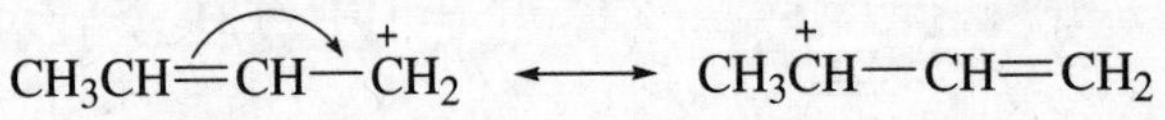

minor major

5. 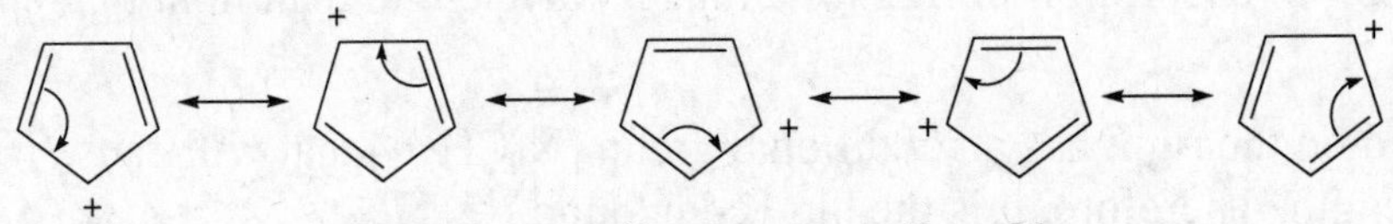

The five contributors are equally stable.

6. 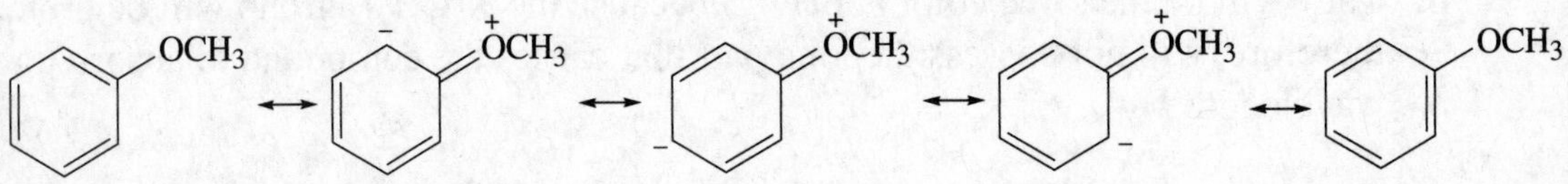

major minor minor minor major

b. yes, **2** and **5**

26. **a.** 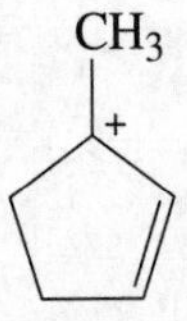

This makes the greater contribution because the positive charge is on a tertiary carbon.

c. $CH_3\overset{+}{C}HCH{=}CH_2$

This makes the greater contribution because the positive charge is on a secondary allylic carbon.

b. 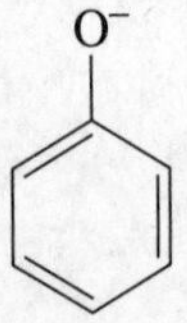

This makes the greater contribution because the negative charge is on an electronegative oxygen.

27. **a.** The resonance contributors show that the carbonyl oxygen has the greater electron density: the carbonyl oxygen has a partial negative charge; the other oxygen (the carboxyl oxygen) has a partial positive charge.

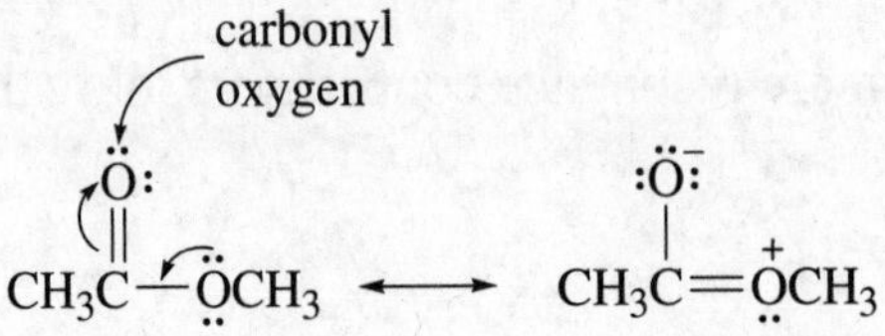
carbonyl
oxygen

b. The compound on the right has the greater electron density on its nitrogen because the compound on the left has a resonance contributor with a positive charge on the nitrogen.

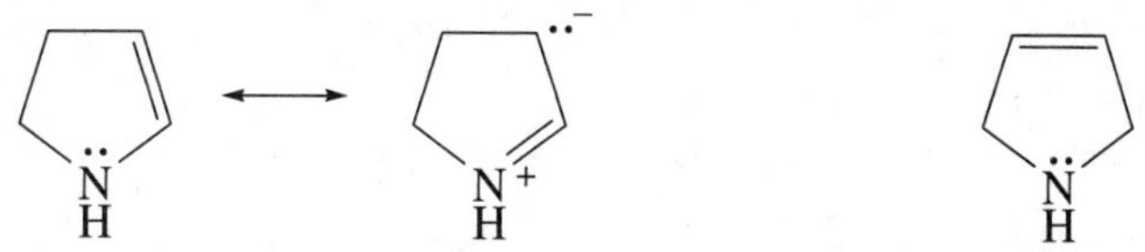

c. The compound with the cyclohexane ring has the greater electron density on its oxygen because the lone pair on the nitrogen can be delocalized onto the oxygen.
There is less delocalization onto oxygen by the lone pair in the compound with the benzene ring because the lone pair can also be delocalized away from the oxygen into the benzene ring.

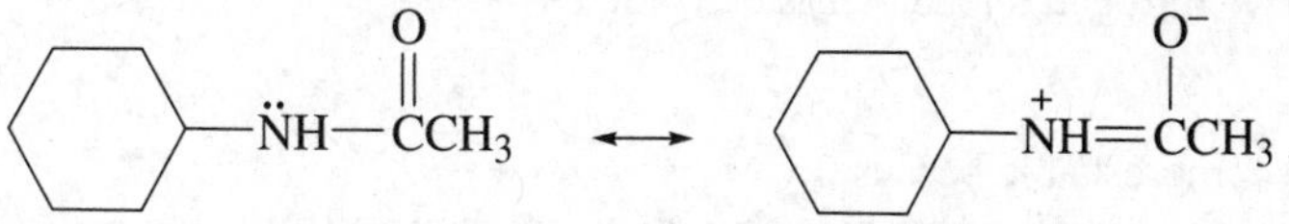

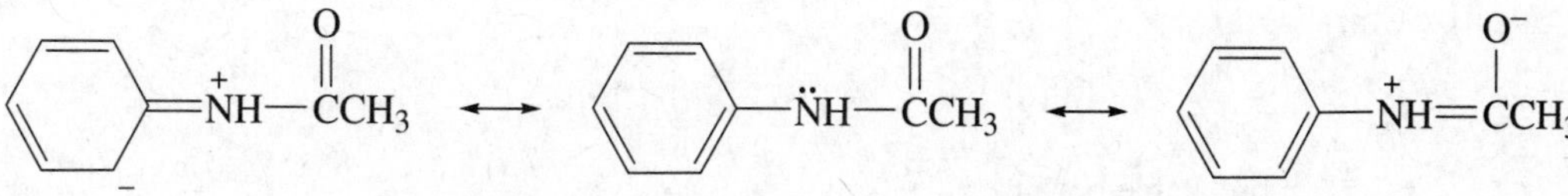

electron delocalization into the benzene ring

electron delocalization onto the carbonyl oxygen

28. The methyl group on benzene can lose a proton easier than the methyl group on cyclohexane because the electrons left behind on the carbon in the former can be delocalized into the benzene ring.

In contrast, the electrons left behind in the other compound are localized on the carbon.

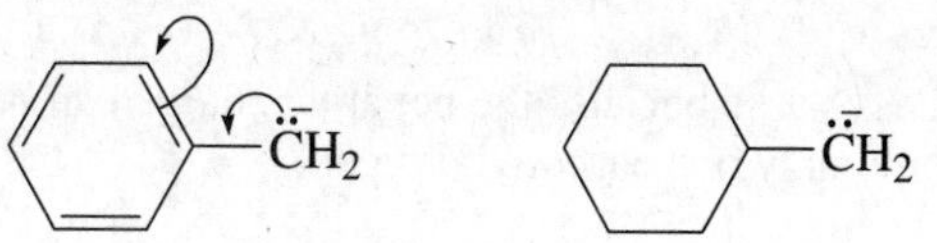

29.

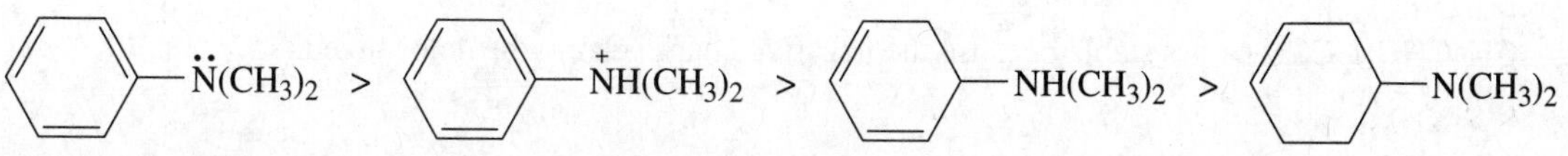

conjugated system has three double bonds and the lone pair on N

conjugated system has three double bonds

conjugated system has two double bonds

this compound has only one double bond

30. The more the electrons that are left behind when the proton is removed can be delocalized, the greater the stability of the base. The more stable the base, the more acidic its conjugate acid.

The bases obtained when a proton is removed are ranked below in order of decreasing stability: the negative charge in the first compound can be delocalized onto two other carbons; the negative charge in the second compound can be delocalized onto one other carbon; the negative charge in the last compound cannot be delocalized.

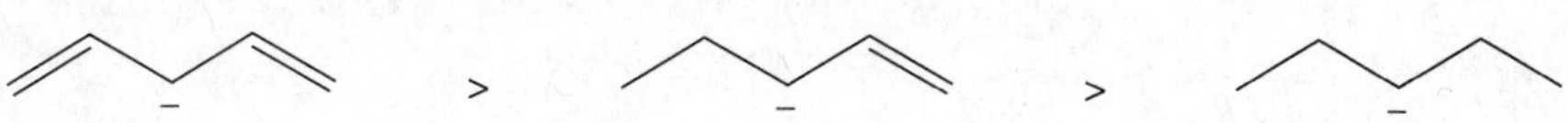

Their conjugate acids are ranked below in order of decreasing acidity:

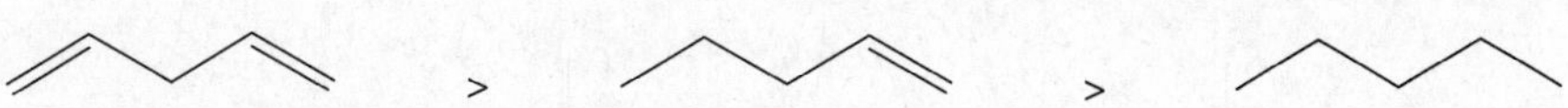

31. **a.** 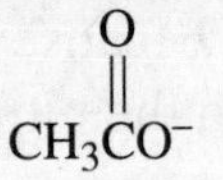 more stable because: the negative charge is shared by 2 oxygens

b. 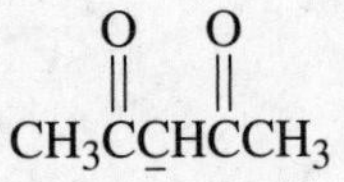 more stable because: the negative charge is shared by a carbon and 2 oxygens

c. 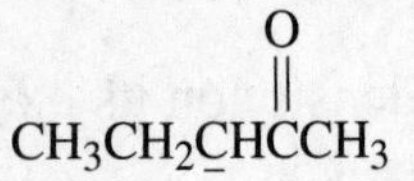 more stable because: the negative charge is shared by a carbon and an oxygen

d. 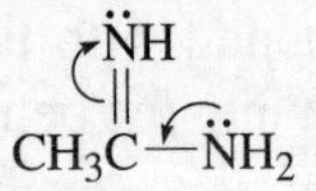 more stable because: nitrogen's lone pair is delocalized

e. 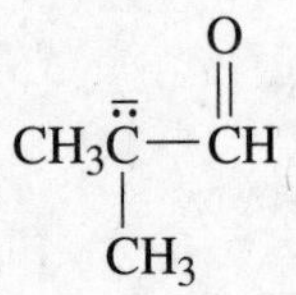 more stable because: the negative charge is shared by a carbon and an oxygen

f. 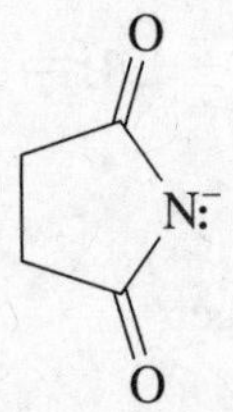 more stable because: the negative charge is shared by a nitrogen and 2 oxygens

32. The stronger base is the less stable base of each pair in Problem 31.

a. $CH_3CH_2O^-$ Less stable because the negative charge cannot be delocalized.

b. 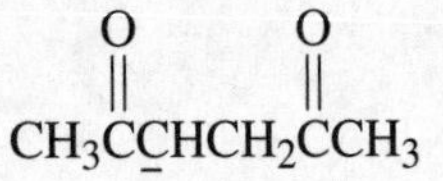 Less stable because the negative charge can be delocalized onto only one carbonyl oxygen.

c. 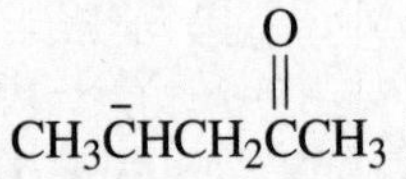 Less stable because the negative charge cannot be delocalized.

d. 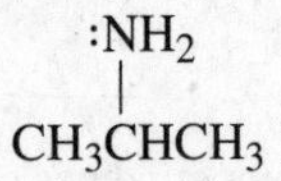 Less stable because nitrogen's lone pair cannot be delocalized.

e. 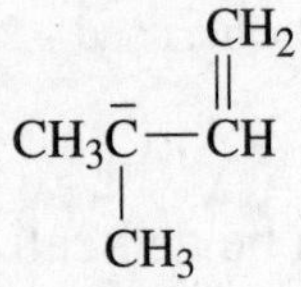 Less stable because the negative charge is shared by two carbons.

f. 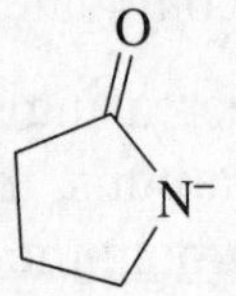 Less stable because the negative charge is shared by a nitrogen and only one oxygen.

33. **a.** 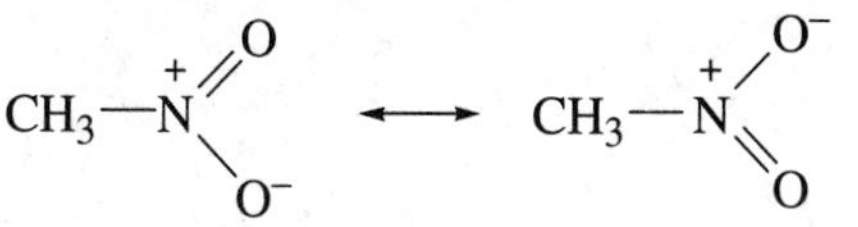
CH3
N
O
O
CH3
N
O
O

the two resonance contributors have the same stability

b. 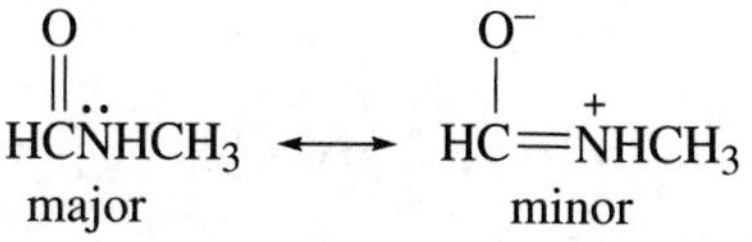
HCNHCH3
major
HC=NHCH3
minor

c. 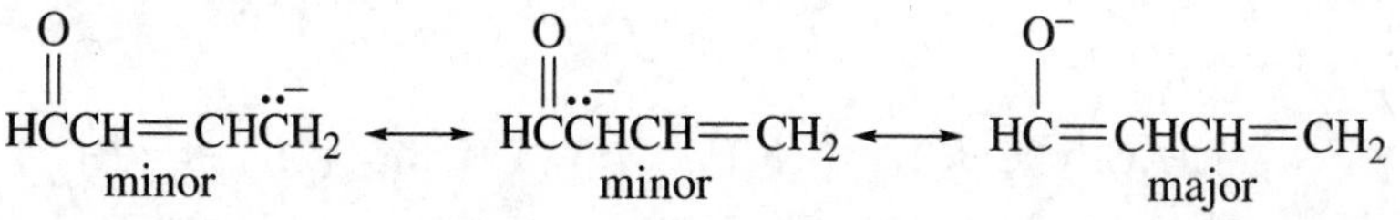
HCCH=CHCH2
minor
HCCHCH=CH2
minor
HC=CHCH=CH2
major

d. 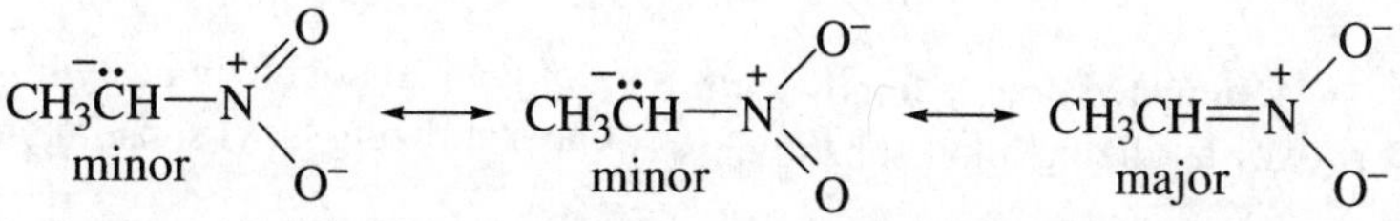
CH3CH—N
minor
CH3CH—N
minor
CH3CH=N
major

e. 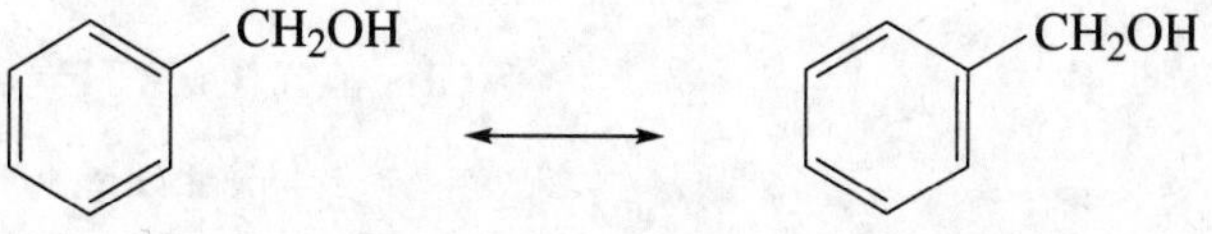
CH2OH
CH2OH

They are equally stable so they contribute equally to the resonance hybrid.

f. 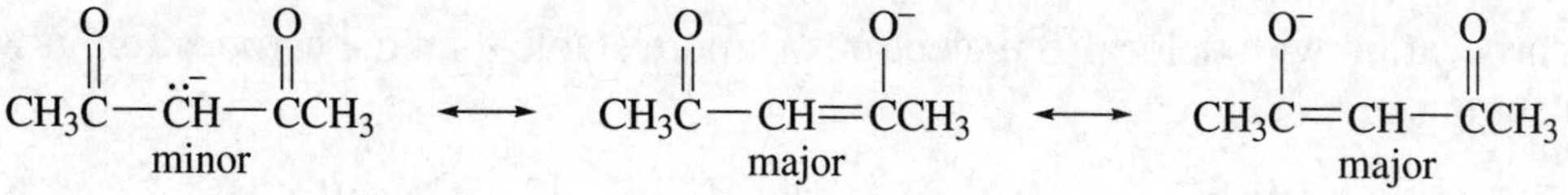
CH3C—CH—CCH3
minor
CH3C—CH=CCH3
major
CH3C=CH—CCH3
major

34. **A** is the most acidic because the electrons left behind when the proton is removed can be delocalized onto two oxygen atoms. **B** is the next most acidic because the electrons left behind when the proton is removed can be delocalized onto one oxygen atom. **C** is the least acidic because the electrons left behind when the proton is removed cannot be delocalized.

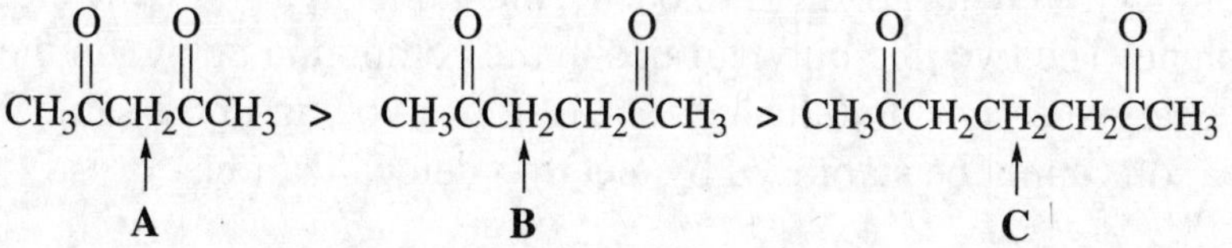
CH3CCH2CCH3
CH3CCH2CH2CCH3
CH3CCH2CH2CH2CCH3
A
B
C

35. The resonance contributor that makes the greatest contribution to the hybrid is labeled "**A.**" "**B**" contributes less to the hybrid than "**A.**"

a. 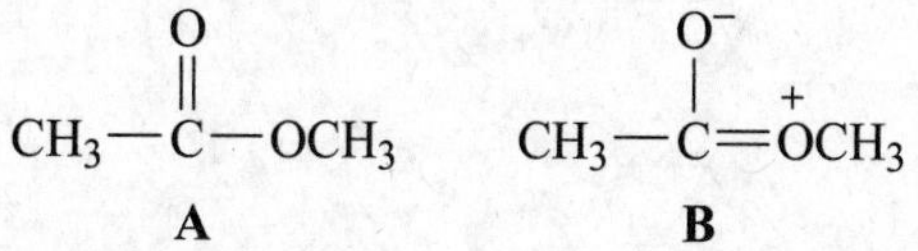
CH3—C—OCH3
A
CH3—C=OCH3
B

b. 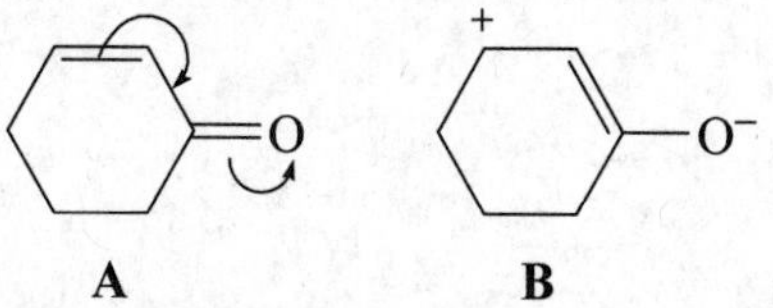
A
B

c. 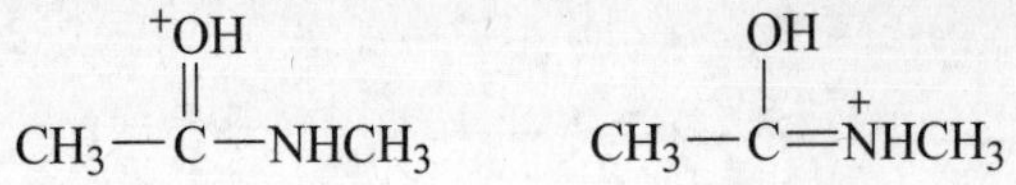

B A

36 **a.** 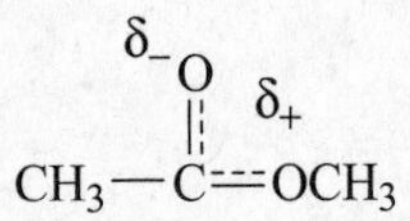 **b.** 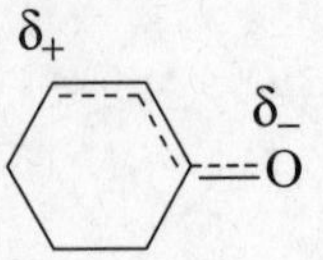 **c.** 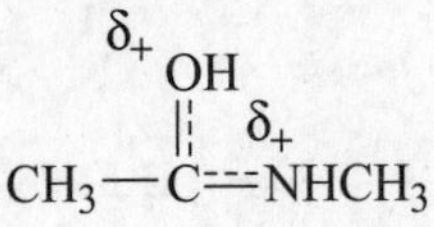

37. 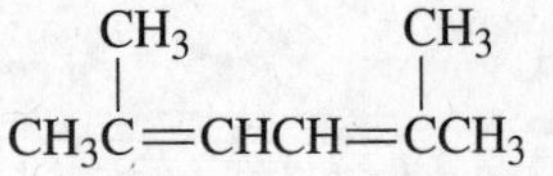 > $CH_3CH{=}CHCH{=}CHCH_3$ > $CH_3CH{=}CHCH{=}CH_2$ >

2,5-dimethyl-2,4-hexadiene
conjugated double bonds with four stabilizing alkyl substituents
most stable

2,4-hexadiene
conjugated double bonds with two stabilizing alkyl substituents

1,3-pentadiene
conjugated double bonds with one stabilizing alkyl substituent

$CH_2{=}CHCH_2CH{=}CH_2$

1,4-pentadiene
isolated double bonds
least stable

38. **a.** The carbocation with delocalized electrons is more stable than the carbocation in which all the electrons are localized.

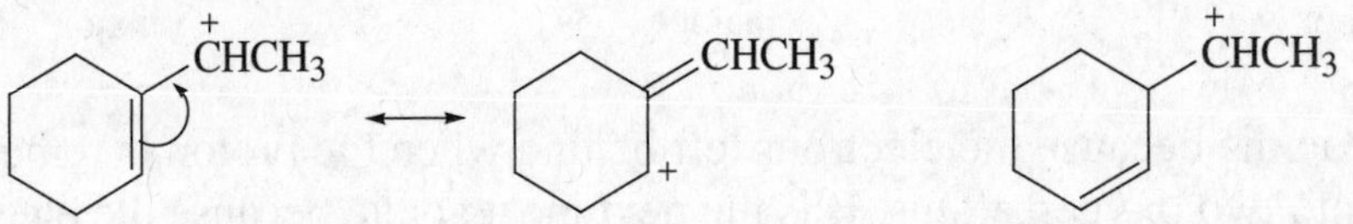

electrons are delocalized — all the electrons are localized

b. In order for electron delocalization to occur, the atoms that share the electrons must be in the same plane. The two *tert*-butyl groups in the compound below on the right, prevent the positively charged carbon and the benzene ring from being in the same plane. Therefore, the carbocation cannot be stabilized by electron delocalization.

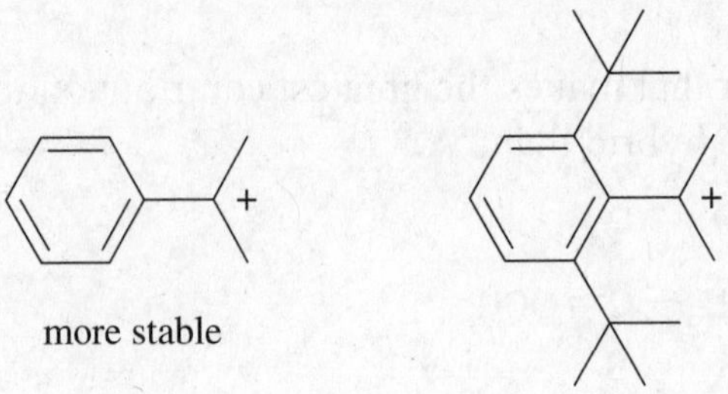

more stable

39. **a.** There are six linear dienes with molecular formula C_6H_{10}.

b. Two are conjugated dienes.

$CH_2{=}CHCH{=}CHCH_2CH_3$

$CH_3CH{=}CHCH{=}CHCH_3$

c. Two are isolated dienes.

$CH_2{=}CHCH_2CH{=}CHCH_3$

$CH_2{=}CHCH_2CH_2CH{=}CH_2$

There are also two dienes with adjacent double bonds.

$CH_2{=}C{=}CHCH_2CH_2CH_3$

$CH_3CH{=}C{=}CHCH_2CH_3$

40. Three addition products would be obtained. First add a proton to the end of the conjugated system. Then draw the resonance contributors. A Br^- will add to each positively charged carbon.

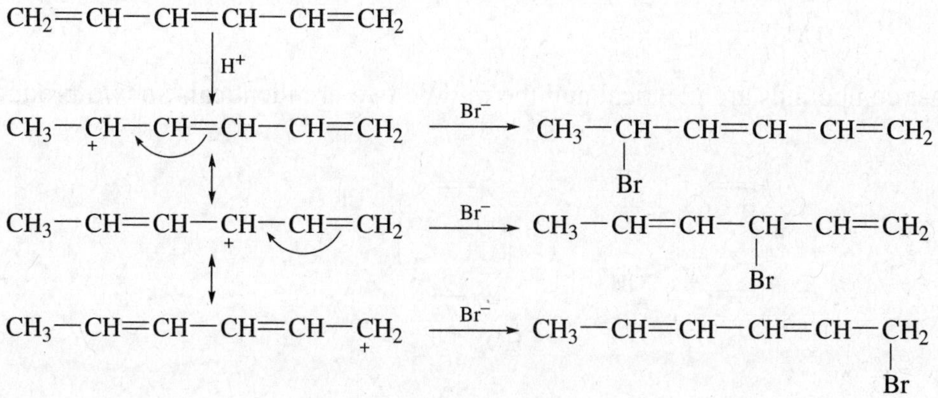

41. **a.**

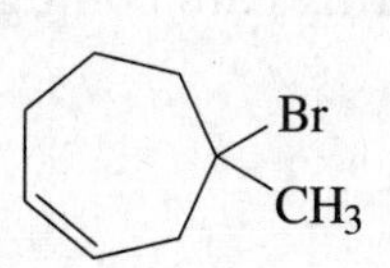

The intermediate that leads to this product is a tertiary carbocation

b.

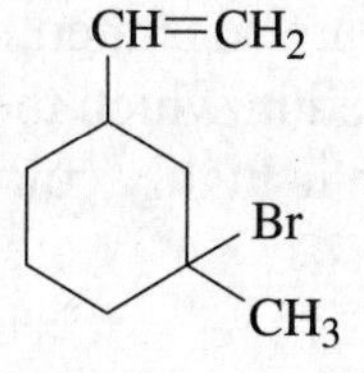

The intermediate that leads to this product is a tertiary carbocation

42. **a.**

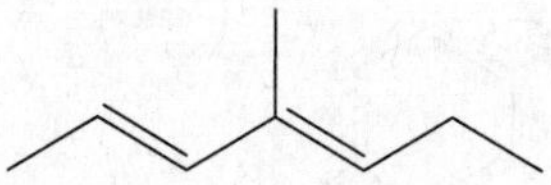

and

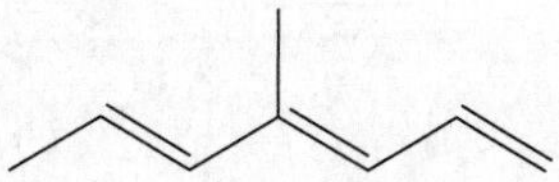

The λ_{max} will be at a longer wavelength because it has more conjugated double bonds.

b.

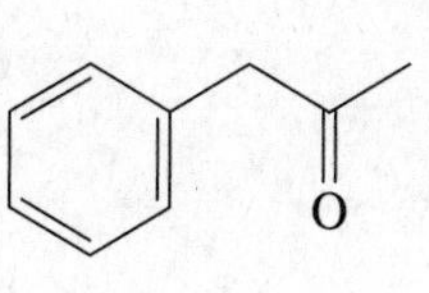

and

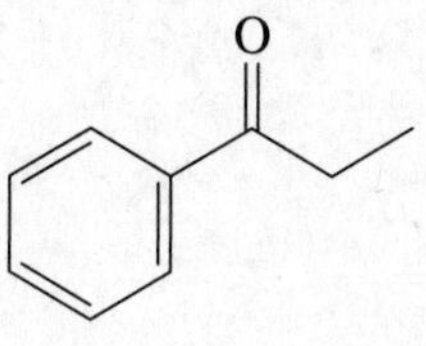

The λ_{max} will be at a longer wavelength because the carbonyl group is conjugated with the benene ring.

43. First draw all the resonance contributors to see which carbons will have a partial positive charge, because hydroxide ion will bond to each of those carbons.

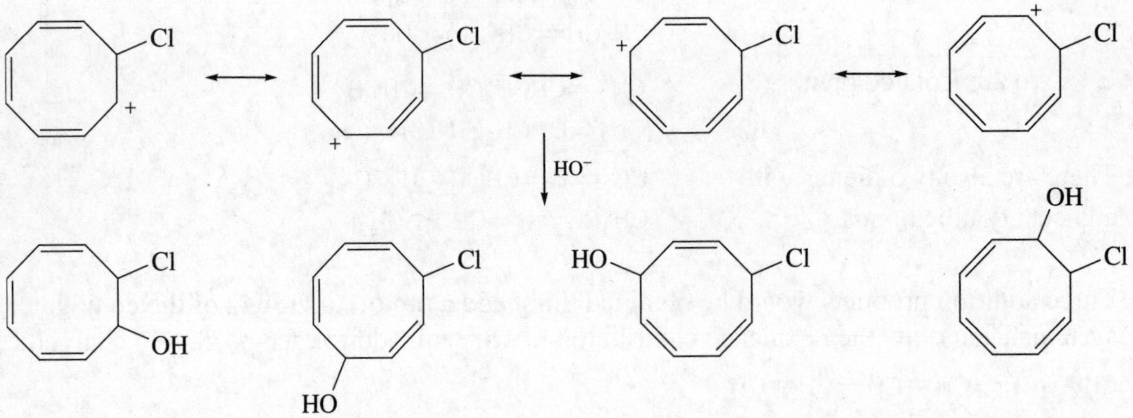

The first and last compounds are identical and the middle two are identical. So two products will be formed.

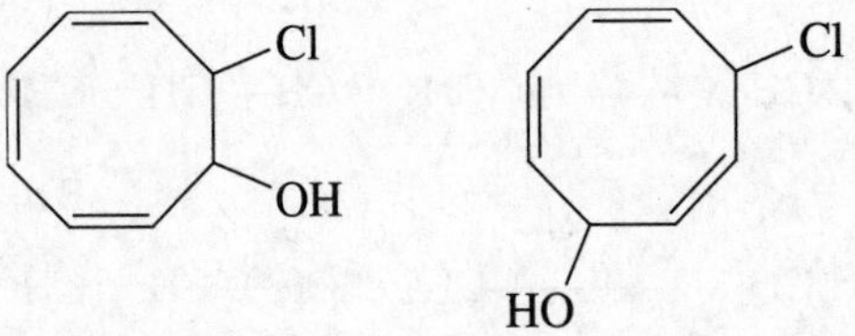

44. Before the addition of acid, the compound is colorless because the benzene rings are not conjugated with each other, since they are separated from each other by two single bonds. In the presence of acid, a carbocation is formed in which the three benzene rings are conjugated with each other. The conjugated carbocation is highly colored.

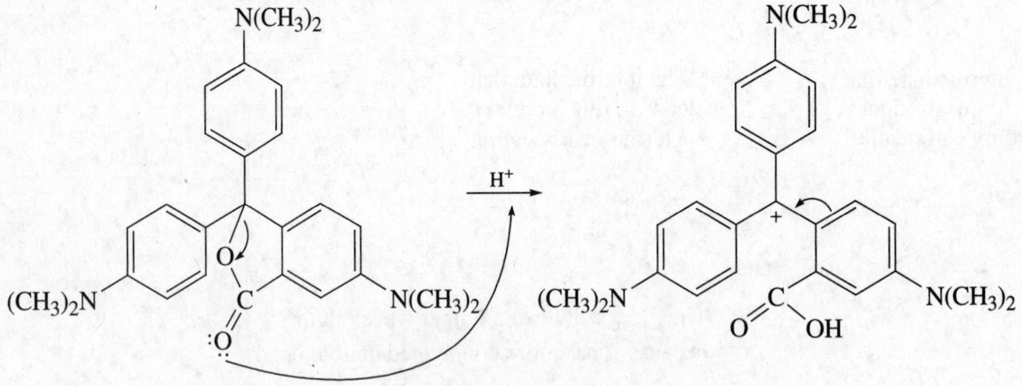

45. **a.**

1.

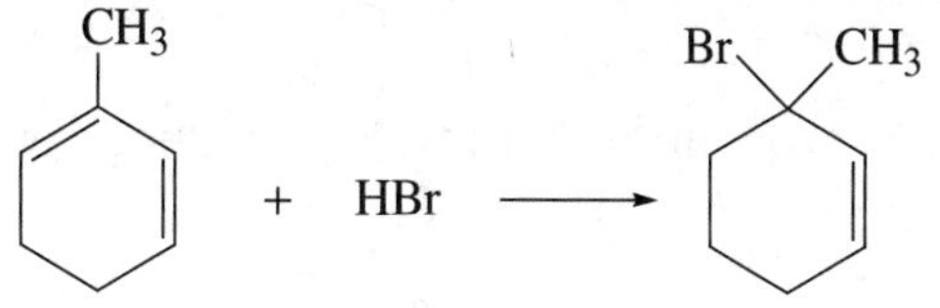
CH3
+ HBr
Br
CH3

or

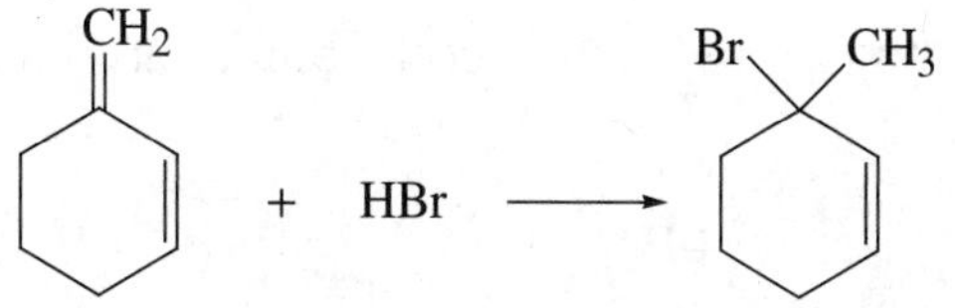
CH2
+ HBr
Br
CH3

2.

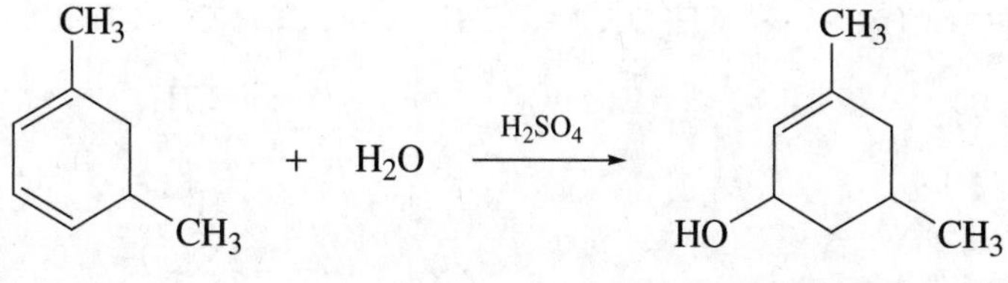
CH3
CH3
+ H2O
H2SO4
HO
CH3

or

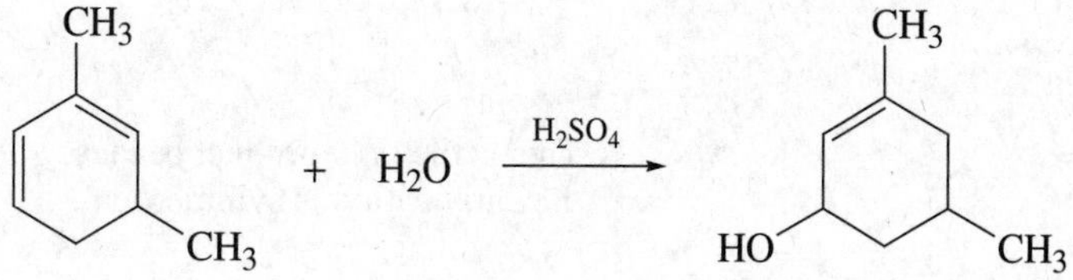
CH3
CH3
+ H2O
H2SO4
HO
CH3

b. **1.**

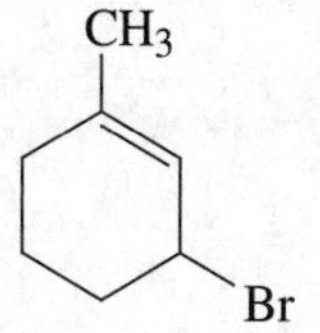
CH3
Br

2.

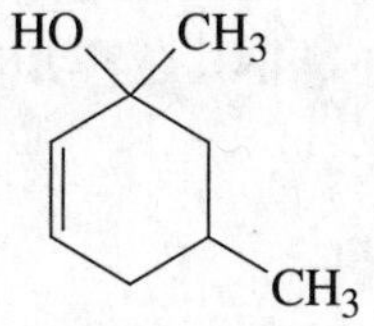
HO
CH3
CH3

46. **a.** Addition of an electrophile to C-1 forms a carbocation with two resonance contributors, a tertiary allylic carbocation and a secondary allylic carbocation.

Addition of an electrophile to C-4 forms a carbocation with two resonance contributors, a tertiary allylic carbocation and a primary allylic carbocation.

Therefore, addition to C-1 results in formation of the more stable carbocation intermediate, and the more stable intermediate leads to the major products.

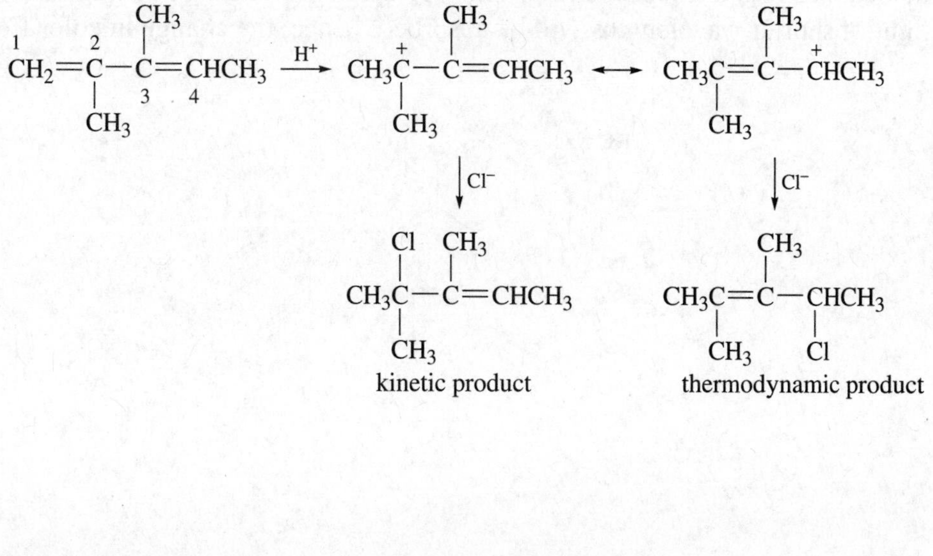
H+
Cl−
Cl−

kinetic product

thermodynamic product

b. Addition of an electrophile to C-1 forms a carbocation with two resonance contributors; both are *tertiary allylic* carbocations.

Addition of an electrophile to C-4 forms a carbocation with two resonance contributors, a *secondary allylic* carbocation and a *primary allylic* carbocation.

Therefore, addition to C-1 results in formation of the more stable carbocation.

"Only" one product is formed, because the two resonance contributors are identical since the carbocation is symmetrical.)

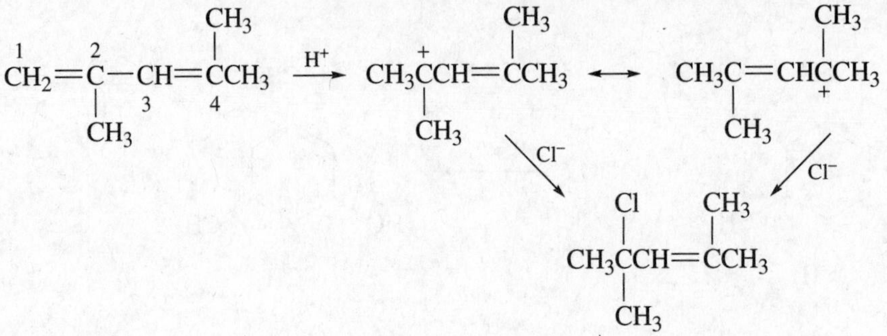

This is the only product because the carbocation is symmetrical.

47. **a.** At pH < 4, the lone pair of the nitrogen of the $N(CH_3)_2$ group of methyl orange is protonated, so it cannot interfere with the fully conjugated system.

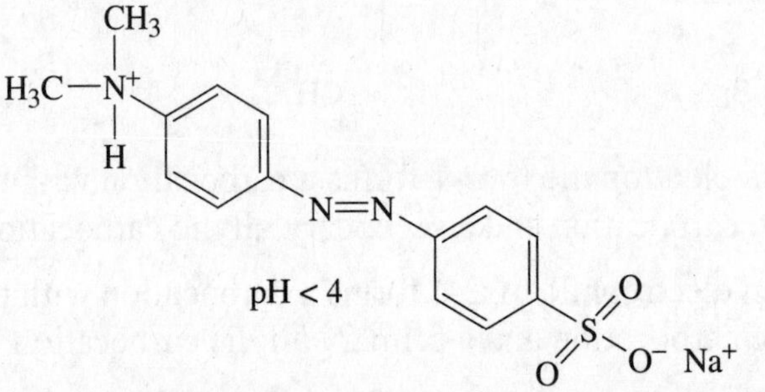
pH < 4

At pH > 4, the nitrogen of the $N(CH_3)_2$ group is not protonated and the lone pair can be delocalized into the benzene ring. This decreases the extent of conjugation and, therefore, light of shorter wavelengths will be absorbed; hence, the change in color from red to yellow.

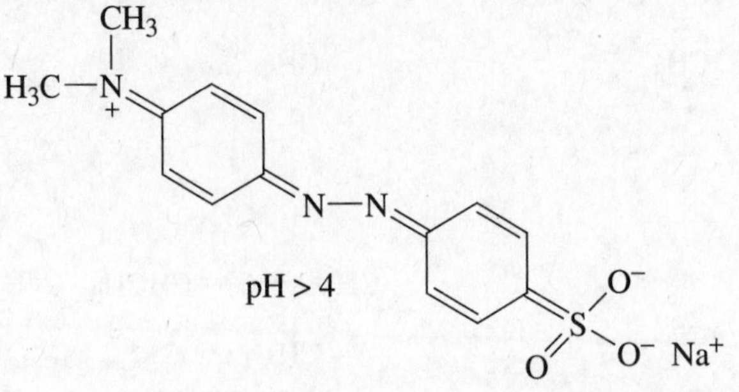
pH > 4

b. In acidic solutions, the three benzene rings of phenolphthalein are isolated from one another.

In basic solutions, as a result of loss of the proton from one of the OH groups, there is a greater degree of conjugation.

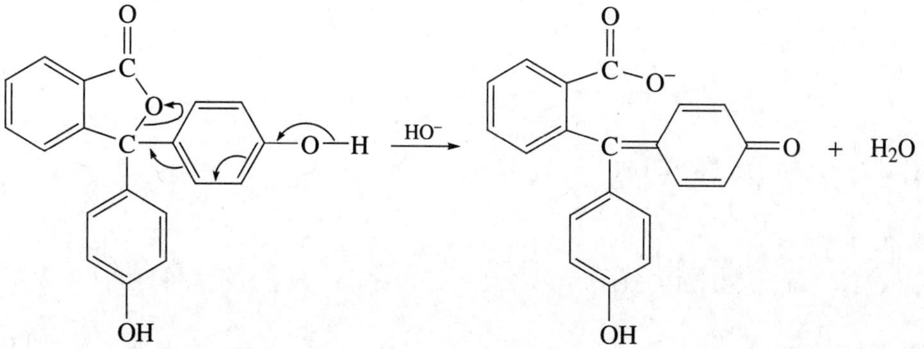
O
C
O
C
O—H
HO⁻
OH
O
C
O⁻
C
O
+ H₂O
OH

Chapter 7 Practice Test

1. For each of the following pairs indicate the one that is the more stable:

a. 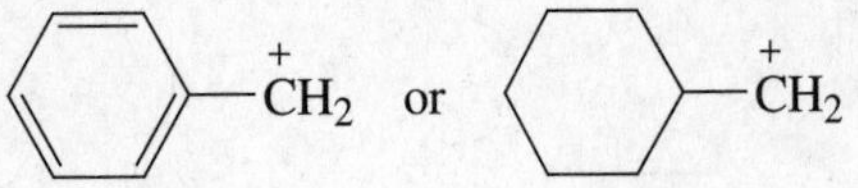

b. $CH_3\bar{C}HCH_3$ or $CH_3\bar{C}HC{\equiv}CH$

c. $CH_3\bar{C}HCH_2\overset{O}{\overset{\|}{C}}CH_3$ or $CH_3\bar{C}H\overset{O}{\overset{\|}{C}}CH_3$

d. $CH_2{=}CH\overset{+}{C}H_2$ or $CH_2{=}CHCH_2\overset{+}{C}H_2$

e. 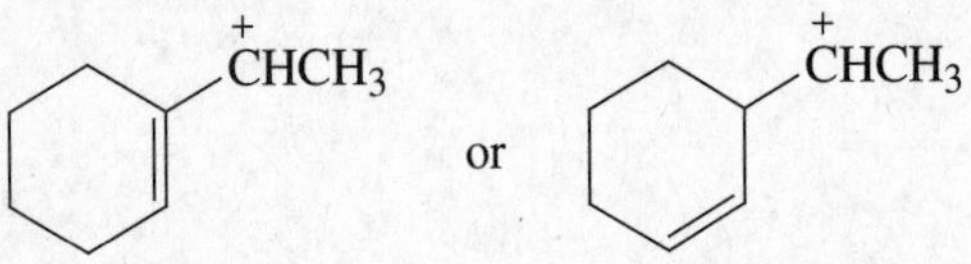

2. Draw resonance contributors for the following:

a. $CH_3CH{=}CH{-}\ddot{\underset{..}{O}}CH_3$

b. $CH_3CH{=}CH{-}CH{=}CH{-}\overset{+}{C}H_2$

c. $\bar{C}H_2{-}CH{=}CH{-}\overset{O}{\overset{\|}{C}}H$

3. Which compounds do not have delocalized electrons?

$CH_3CH_2NHCH{=}CHCH_3$ $\quad$ $CH_3\underset{+}{\overset{CH_3}{\overset{|}{C}}}CH_2CH{=}CH_2$ $\quad$ $CH_2{=}CHCH_2CH{=}CH_2$

$CH_2{=}CH\overset{O}{\overset{\|}{C}}CH_3$ $\quad$ $CH_3CH_2NHCH_2CH{=}CHCH_3$ $\quad$ 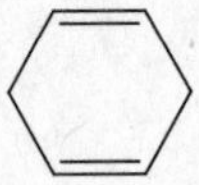

4. Which of the following pairs are resonance contributors?

a. CH_3CH_2OH and CH_3OCH_3

b. 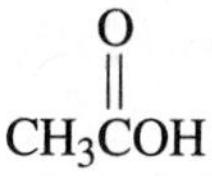 and 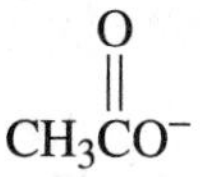

c. 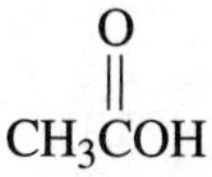 and 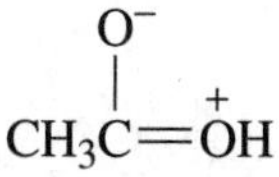

d. 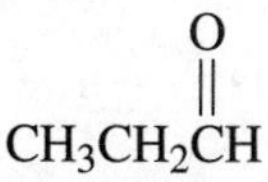 and 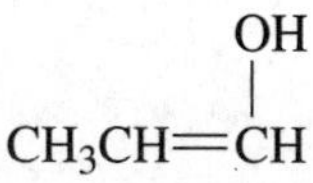

5. Which is a stronger base?

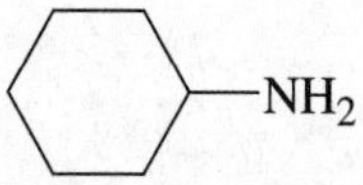 or 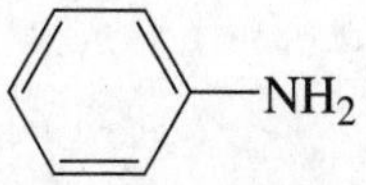

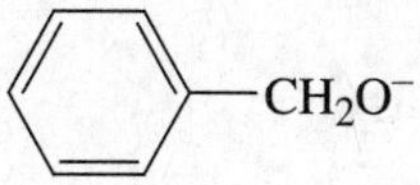 or 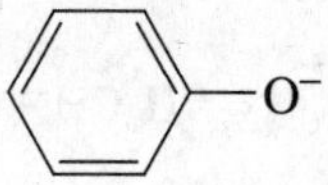

6. Draw resonance contributors for the following:

a. 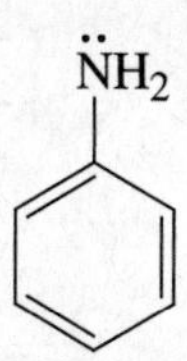

b. 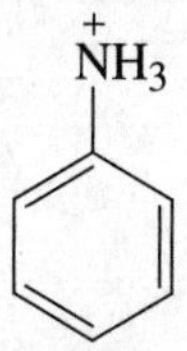

c. 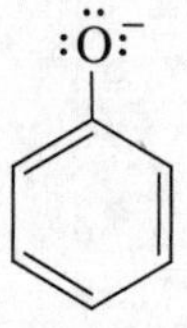

7. For each of the following pairs of compounds, indicate the one that is the more acidic:

a. 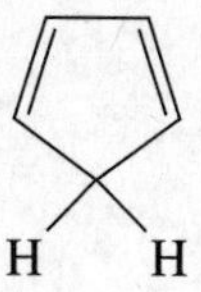 or 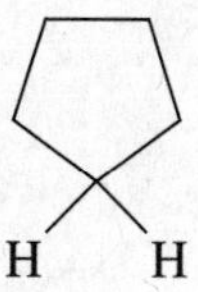

b. $CH_3CH_2CH_3$ or $CH_3CH{=}CH_2$

c. 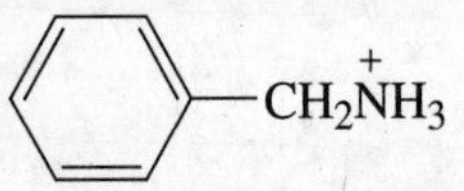 or 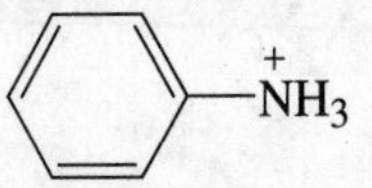

8. Which resonance contributor makes a greater contribution to the resonance hybrid?

a. 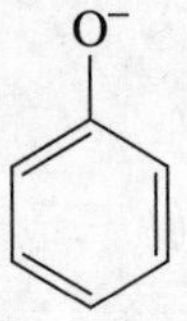 or 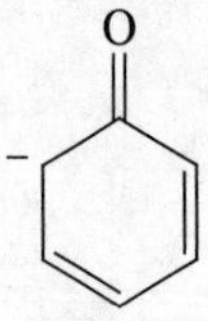

b. 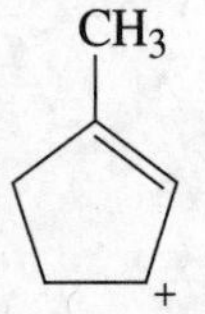 or 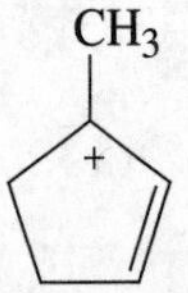

9. Rank the following carbocations in order of decreasing stability:

$CH_3CH{=}CH\overset{+}{C}H_2$ $CH_3CH{=}CH\overset{+}{C}HCH_3$ $CH_3CH{=}CHCH_2\overset{+}{C}H_2$ $CH_3CH{=}CH\overset{+}{C}(CH_3)CH_3$

10. Which compound has the greater λ_{max}? (300 nm is a greater λ_{max} than 250 nm.)

a. 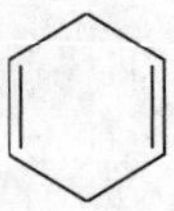 or 

b. 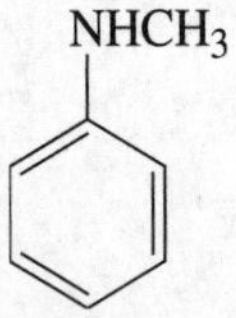 or 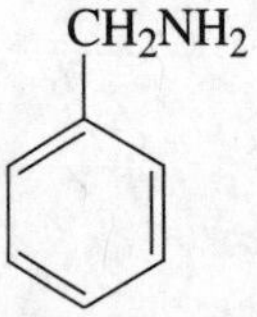

c. 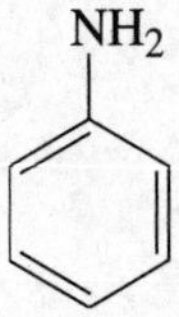 or 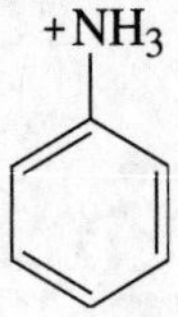

d. 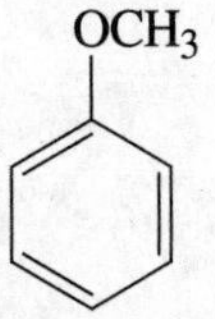 or 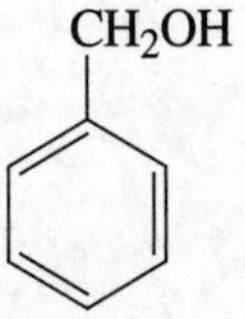

11. Give the four products that could be obtained from the following reaction:

$$CH_2{=}C(CH_3){-}C(CH_3){=}CHCH_3 + HBr \longrightarrow$$

SPECIAL TOPIC II

Drawing Resonance Contributors

On p. 97–99 of the text and in Special Topic I in this *Study Guide and Solutions Manual*, we saw that chemists use curved arrows to show how electrons move when reactants are converted into products. Chemists also use curved arrows when they draw resonance contributors.

We have seen that delocalized electrons are electrons that are shared by more than two atoms. When electrons are shared by more than two atoms, we cannot use solid lines to represent them. For example in the carboxylate ion, a pair of electrons are shared by a carbon and two oxygens. Thus the two oxygen atoms share the negative charge. We show the pair of delocalized electrons by a dotted line spread over the three atoms. We have seen that this structure is called a resonance hybrid.

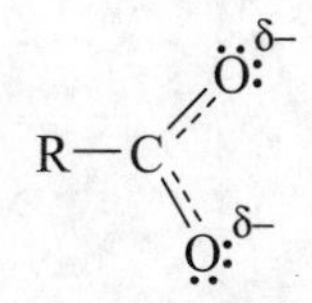

carboxylate ion

Chemists do not like to use dotted lines when drawing structures because, unlike a solid line that represents two electrons, the dotted lines do not specify the number of electrons they represent. Therefore, chemists use structures with localized electrons (solid lines) to approximate the resonance hybrid that has delocalized electrons (dotted lines). These approximate structures are called resonance contributors; a double-headed arrow is drawn between resonance contributors. Curved arrows are used to show the movement of electrons in going from one resonance contributor to the next.

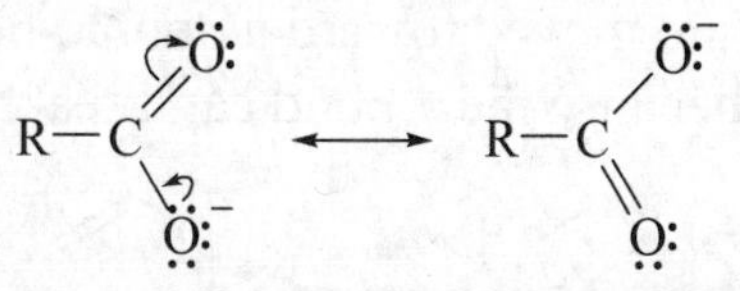

Rules for Drawing Resonance Contributors

Resonance contributors are easy to draw if you remember these three simple rules.

1. Only electrons move; atoms NEVER move.
2. The only electrons that can move are π electrons (electrons in π bonds) and lone-pair electrons.
3. Electrons always are moved toward an sp^2 carbon. Recall that an sp^2 carbon is either a positively charged carbon or a double-bonded carbon.

Moving π electrons toward a positively-charged carbon.

In the following example, π electrons toward a positively-charged carbon. Because the atom has a positive charge, it can accept the electrons. The carbon that is positively charged in the first resonance contributor is neutral in the second resonance contributor because it has received electrons. The carbon in the first resonance contributor that loses its share of the π electrons is positively charged in the second resonance contributor.

We see that the following carbocation has three resonance contributors.

$$CH_2{=}CH{-}CH{=}CH{-}\overset{+}{C}HCH_3 \longleftrightarrow CH_2{=}CH{-}\overset{+}{C}H{-}CH{=}CHCH_3$$

$$\updownarrow$$

$$\overset{+}{C}H_2{-}CH{=}CH{-}CH{=}CHCH_3$$

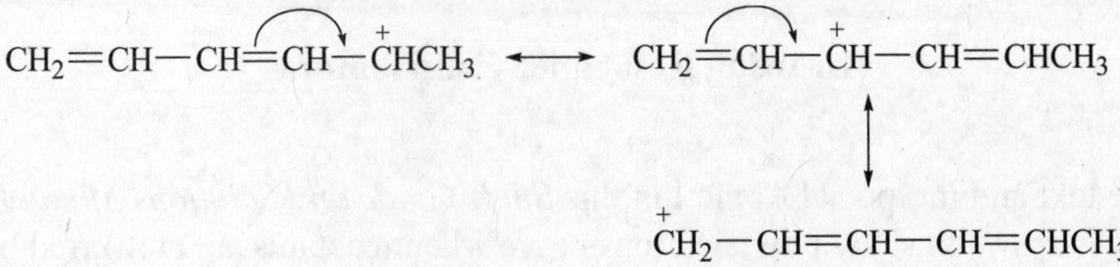

Notice that in going from one resonance contributor to the next, the total number of electrons in the structure does not change. Therefore, each of the resonance contributors must have the same charge.

Problem 1. Draw the resonance contributors for the following carbocation:
(The answers can be found immediately after Problem 11.)

$$CH_3CH{=}CH{-}CH{=}CH{-}CH{=}CH{-}\overset{+}{C}H_2 \longleftrightarrow$$

$$\updownarrow$$

$$\longleftrightarrow$$

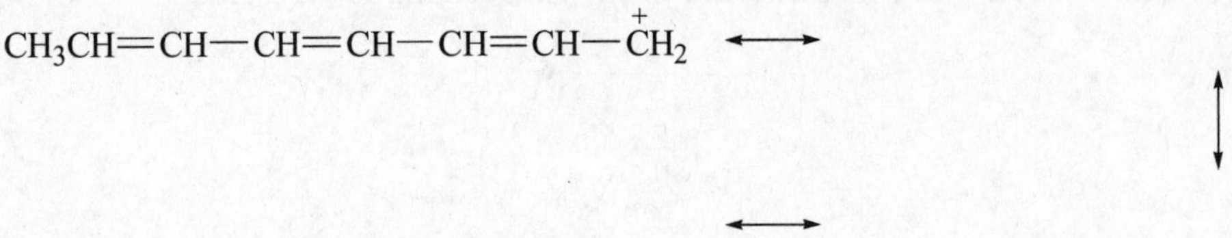

Problem 2. Draw the resonance contributors for the following carbocation:

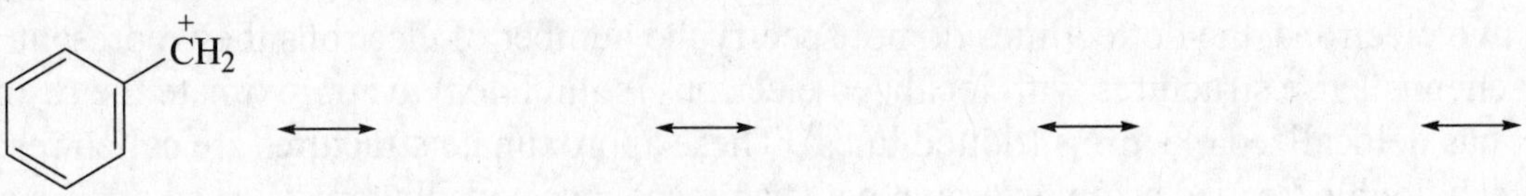

Moving π electrons toward a double-bonded carbon.

In the following example, π electrons are moved toward a double-bonded carbon. The atom to which the electrons are moved can accept them because the π bond can break.

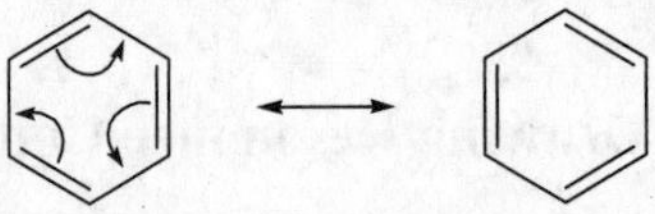

Problem 3. Draw the resonance contributor for the following compound:

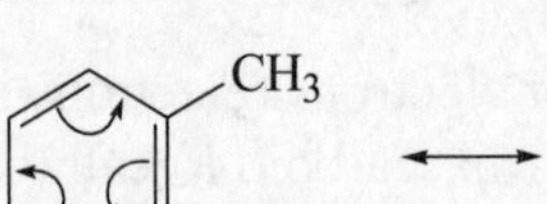

In the next example, the electrons can be moved equally easily to the left (upper pair of arrows) or to the right (bottom pair of arrows). The charges on the end carbons cancel, so there is no charge on any of the carbons in the resonance hybrid.

$$\overset{-}{\ddot{C}}H_2{-}CH{=}CH{-}\overset{+}{C}H_2 \longleftrightarrow CH_2{=}CH{-}CH{=}CH_2 \longleftrightarrow \overset{+}{C}H_2{-}CH{=}CH{-}\overset{-}{\ddot{C}}H_2$$

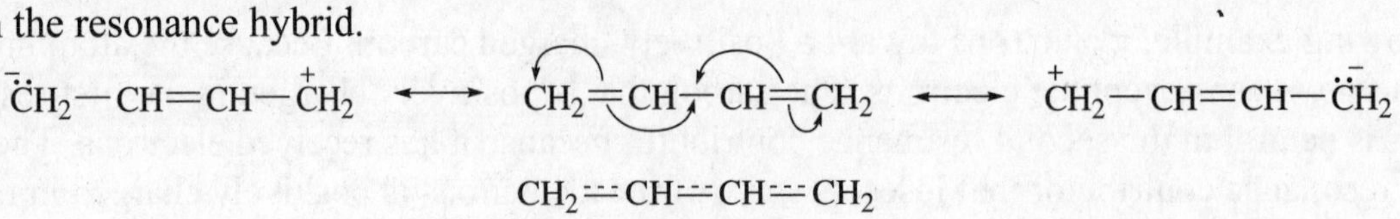

$$CH_2{\dot{=}}CH{\dot{=}}CH{\dot{=}}CH_2$$
resonance hybrid

When electrons can be moved in either of two directions and there is a difference in the electronegativity of the atoms to which they can be moved, always move the electrons toward the more electronegative atom. For example, in the following example the electrons are moved toward oxygen, not toward carbon.

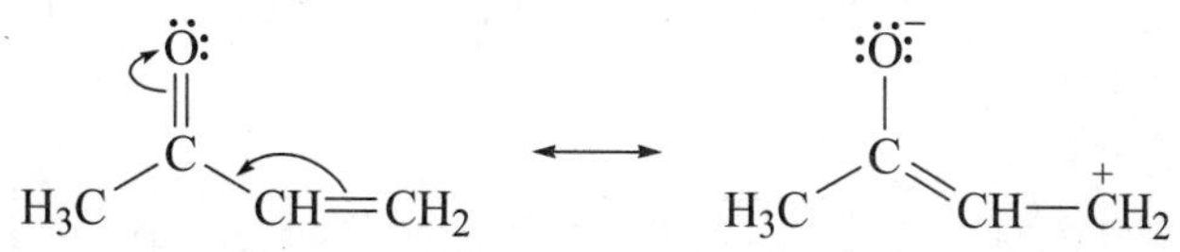

Notice that the first resonance contributor has a net charge of 0. Since the number of electrons in the molecule does not change, the other resonance contributor must have the same net charge. (A net charge of 0 does not mean that there is no charge on any of the atoms; a resonance contributor with a positive charge on one atom and a negative charge on another has a net charge of 0.)

Moving a lone pair toward a double-bonded carbon.

In the following examples, lone-pair electrons are moved toward a double-bonded carbon. Notice that the arrow starts at a pair of electrons, not at a negative charge. In the first example, each of the resonance contributors has a net charge of -1; in the second example, each of the resonance contributors has a net charge of 0.

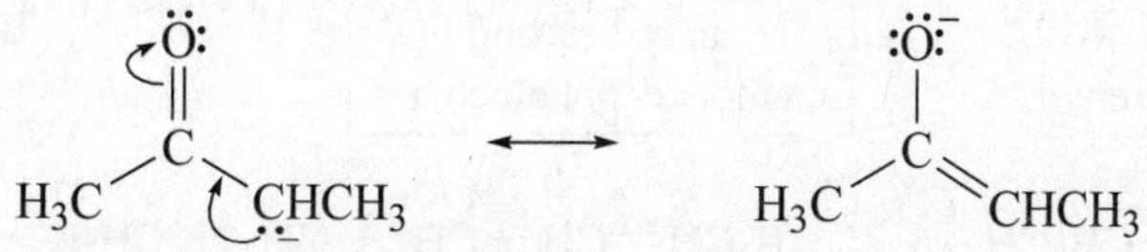

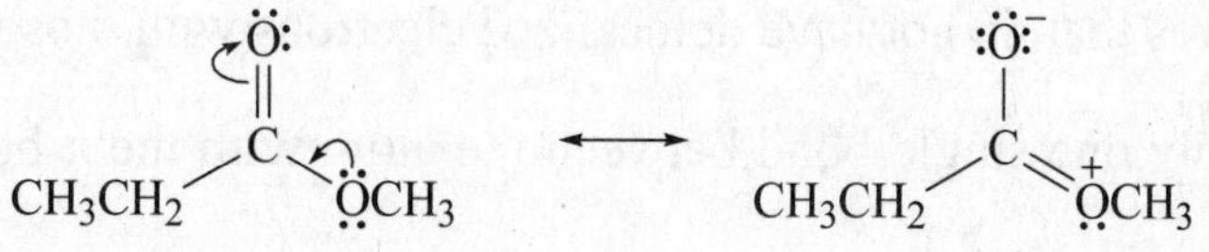

The following species has three resonance contributors. Notice again that the arrow starts at a lone pair, not at a negative charge. The three oxygen atoms share the two negative charges. Therefore, each oxygen atom in the hybrid has 2/3 of a negative charge.

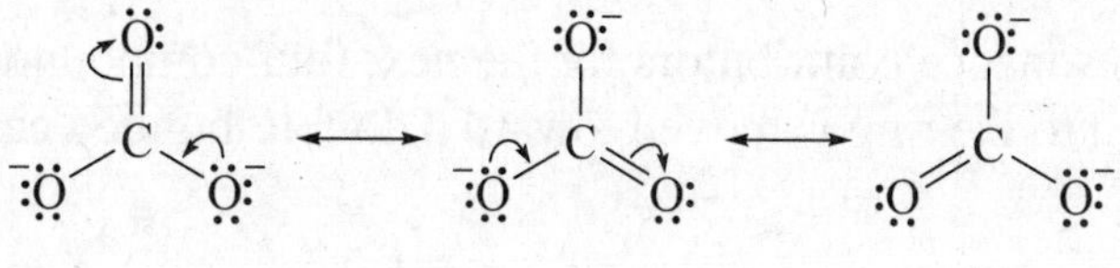

Problem 4. Draw the resonance contributor for the following:

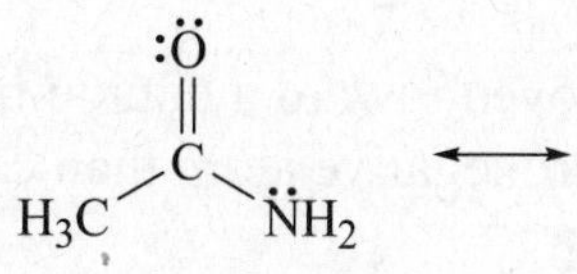

Notice that in the next example the lone-pair electrons are moved away from the most electronegative atom in the molecule. They do so because that is the only way electron delocalization can occur. The π electrons cannot be moved toward the oxygen because the oxygen atom has a complete octet. Recall that electrons can be moved only toward a positive charge or toward a double-bonded carbon.

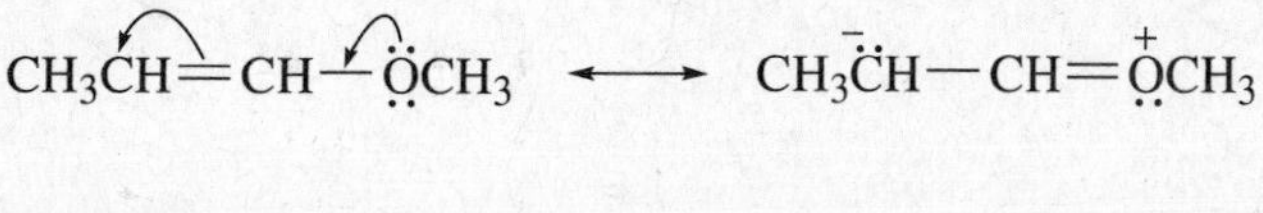

The next compound has five resonance contributors. To get to the second resonance contributor, a lone pair on nitrogen is moved toward a double-bonded carbon. Notice that the first and fifth resonance contributors are not the same; they are similar to the two resonance contributors of benzene.

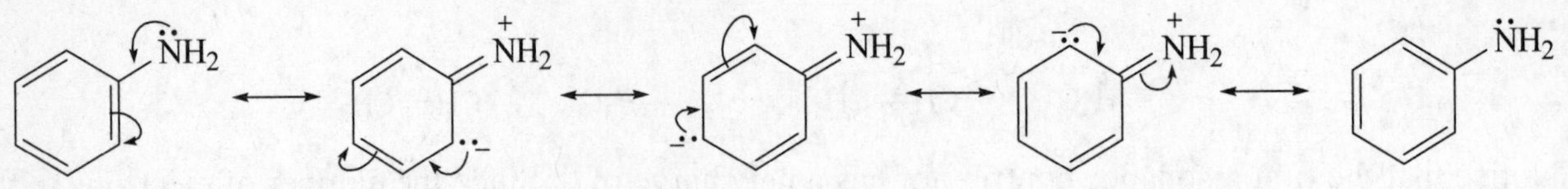

Problem 5. Draw the resonance contributors for the following:

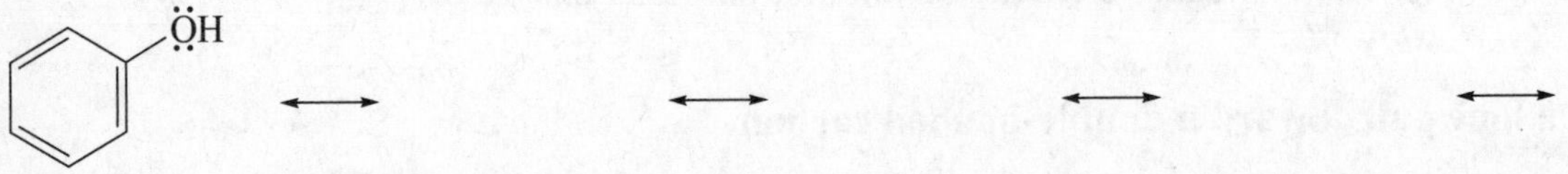

The following species do not have delocalized electrons. Electrons cannot be moved toward an sp^3 hybridized atom because an sp^3 hybridized atom has a complete octet and, therefore, cannot accept any more electrons.

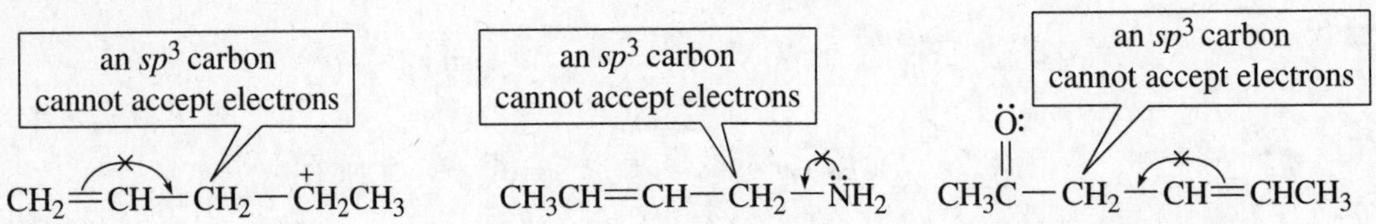
an sp3 carbon cannot accept electrons
an sp3 carbon cannot accept electrons
an sp3 carbon cannot accept electrons

Compare the above structures that do not have delocalized electrons with those that do. Notice that:

1. There can be only one single bond between the atom with the π bond and the atom with the positive charge.
2. There can be only one single bond between the atom with the π bond and the atom with the lone pair.
3. There can be only one single bond between the atoms with the π bonds (that is, the double bonds must be conjugated).

Notice the difference in the resonance contributors for the next four compounds. In the first compound, a lone pair on the atom attached to the ring is moved toward a double-bonded carbon.

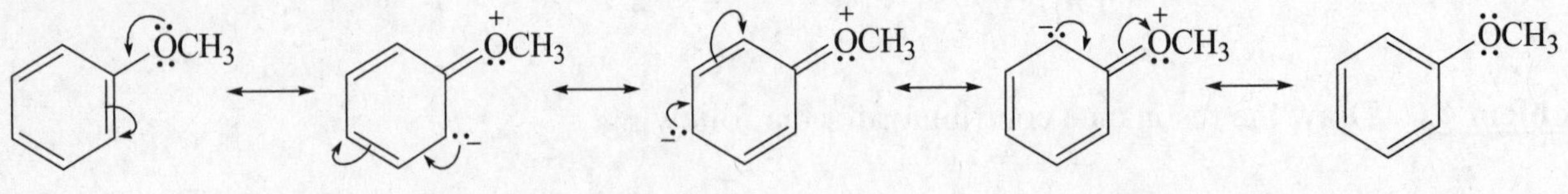

In the next compound, first a π bond is moved toward a double-bonded carbon; the electron movement is toward the oxygen since it is a more electronegative atom than carbon. Then a π bond is moved toward a positive charge.

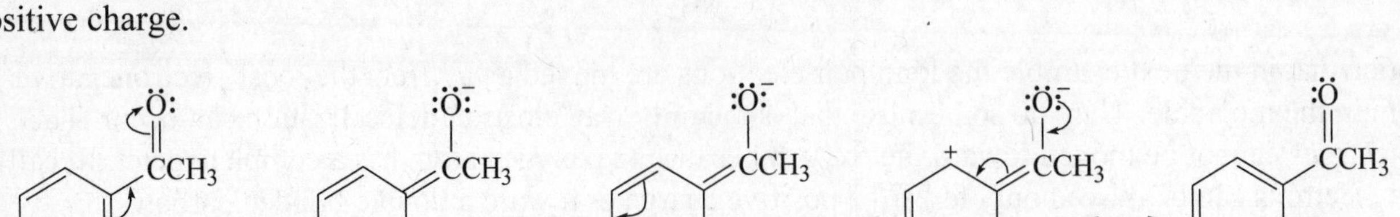

In the next two compounds, the atom attached to the ring has neither a lone pair or a π bond. Therefore, the substituent can neither donate electrons into the ring or accept electrons from the ring. Thus, these compounds have only two resonance contributors—the ones that are similar to the two resonance contributors of benzene.

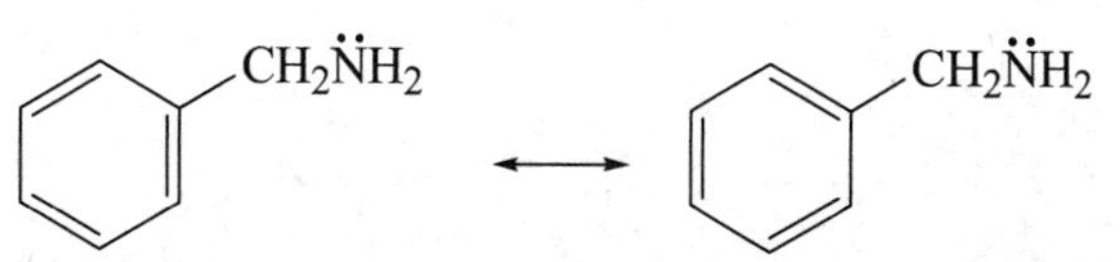
CH₂N̈H₂
CH₂N̈H₂

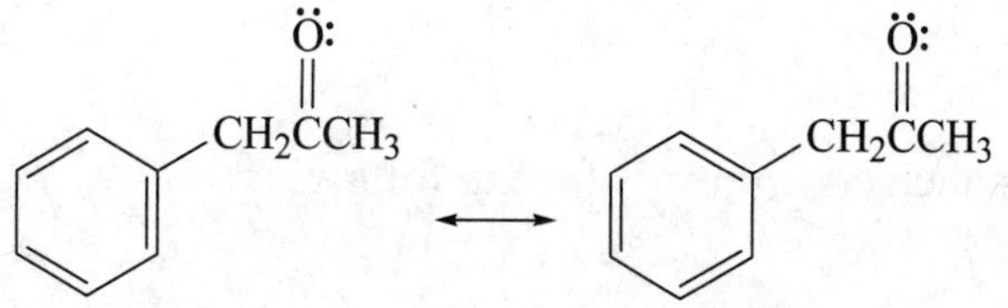
Ö:
CH₂CCH₃
Ö:
CH₂CCH₃

Problem 6. Which of the following have delocalized electrons?

$CH_3CH{=}CHCH_2CH{=}CH_2$ **A**

$CH_3CH{=}CHCH{=}CH_2$ **B**

$CH_3CH{=}CHCH_2\ddot{N}HCH_3$ **C**

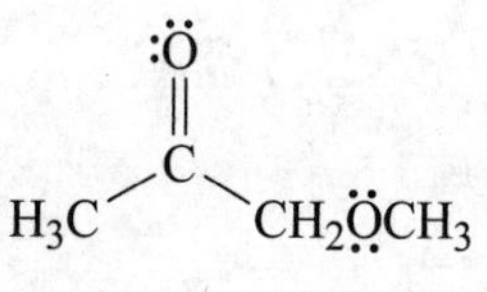
:Ö
C
H₃C
CH₂ÖCH₃
D

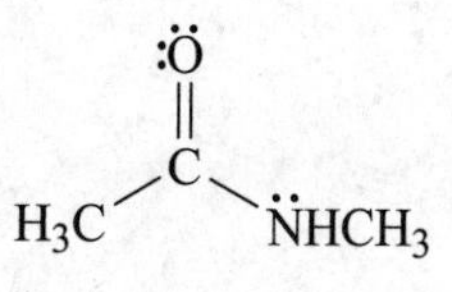
:Ö
C
H₃C
N̈HCH₃
E

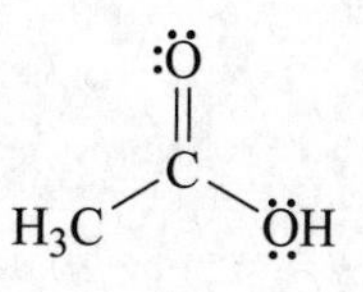
:Ö
C
H₃C
ÖH
F

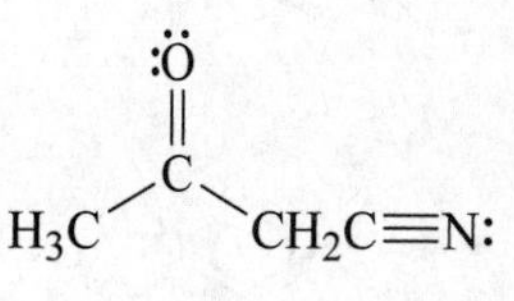
:Ö
C
H₃C
CH₂C≡N:
G

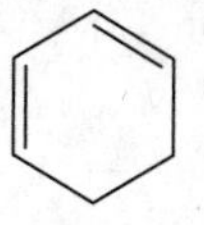
H

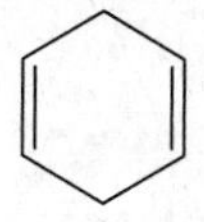
I

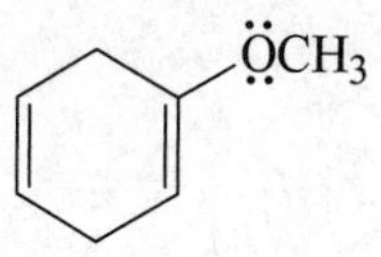
ÖCH₃
J

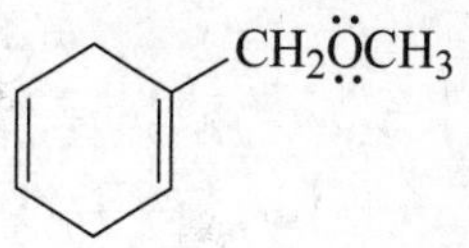
CH₂ÖCH₃
K

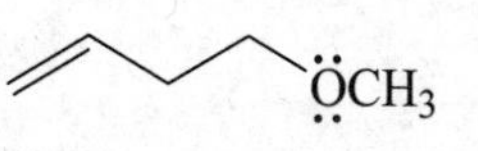
ÖCH₃
L

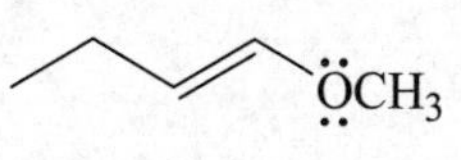
ÖCH₃
M

N

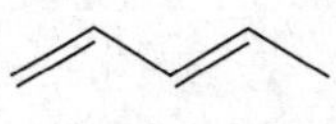
O

Problem 7. Draw the resonance contributors for the compounds in Problem 6 that have delocalized electrons.

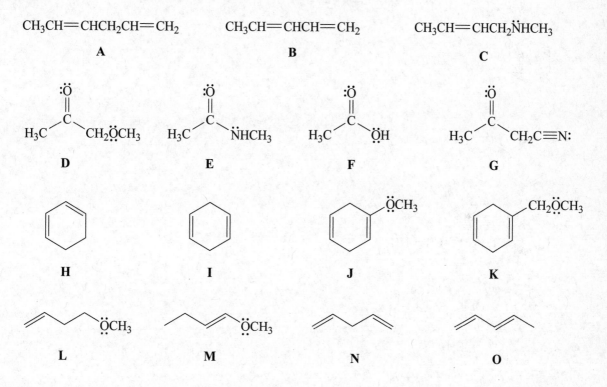

Problem 8. Draw the arrows to show how one resonance contributor leads to the next one.

a. 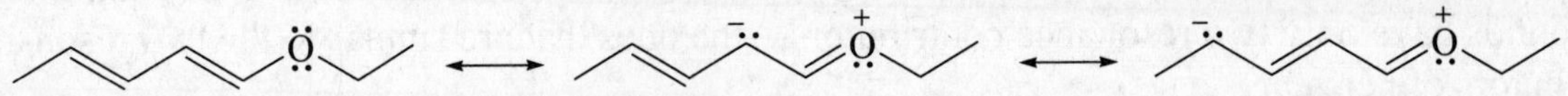

b. 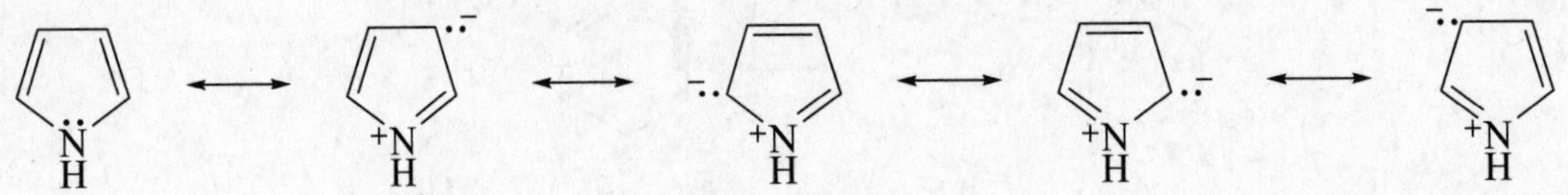

Problem 9. Draw the resonance contributors for the following:

a. 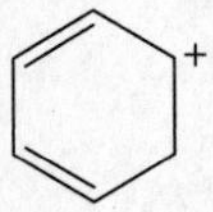

b. 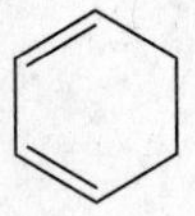

c. 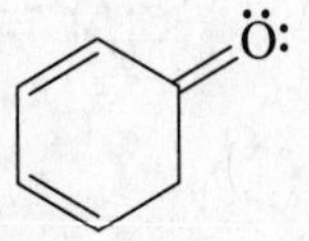

d. $CH_2{=}CH{-}CH{=}CH{-}\ddot{N}H_2$

e. 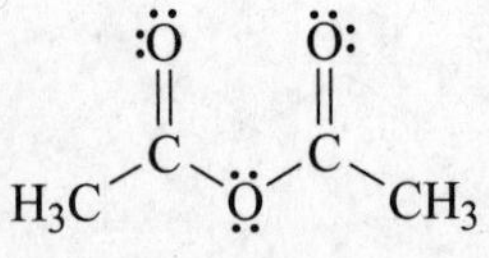

f. 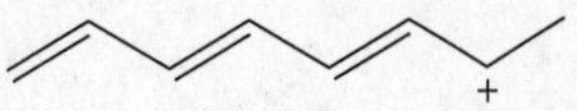

Problem 10. Draw the resonance contributors for the following:

a. 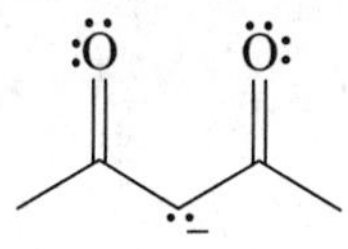

b. 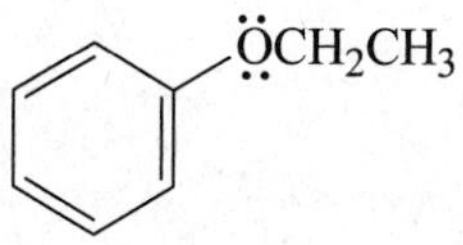

c. 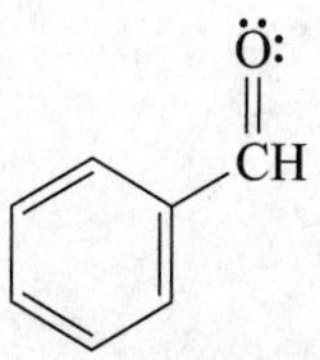

d. 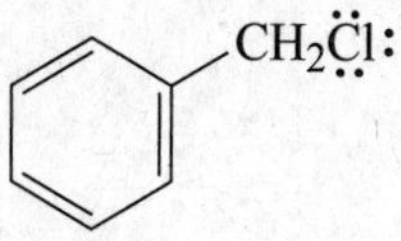

e. 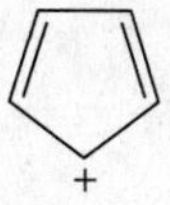

f. 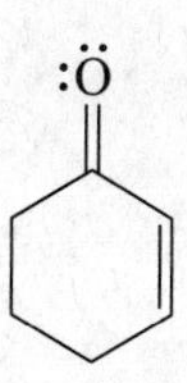

Problem 11. Draw the resonance contributors for the following:

a. 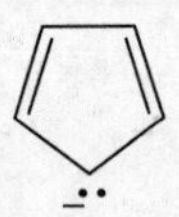

b. 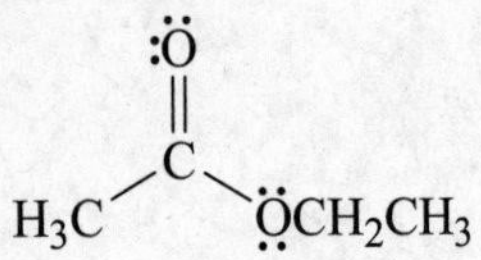

c. 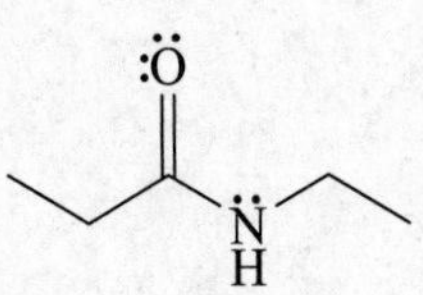

d. 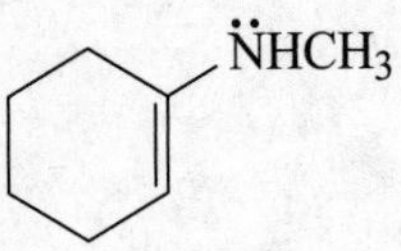

e. 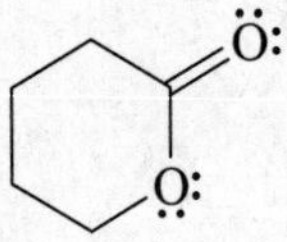

f. 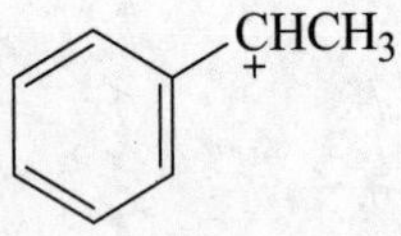

Answers to the Problems

Problem 1.

$$CH_3CH{=}CH{-}CH{=}CH{-}CH{=}CH{-}\overset{+}{C}H_2 \quad \longleftrightarrow \quad CH_3CH{=}CH{-}CH{=}CH{-}\overset{+}{C}H{-}CH{=}CH_2$$

$$CH_3\overset{+}{C}H{-}CH{=}CH{-}CH{=}CH{-}CH{=}CH_2 \quad \longleftrightarrow \quad CH_3CH{=}CH{-}\overset{+}{C}H{-}CH{=}CH{-}CH{=}CH_2$$

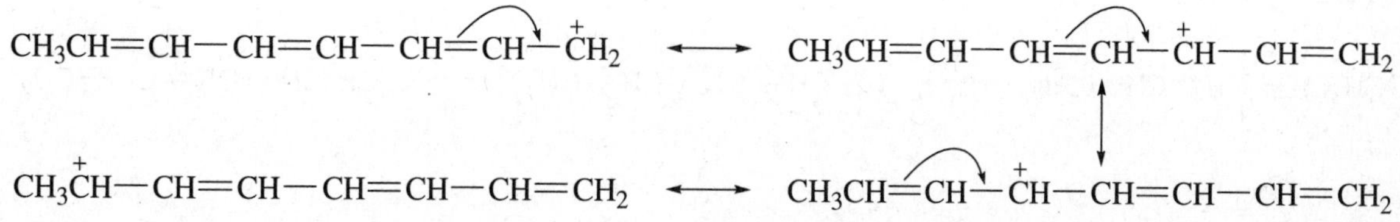

Problem 2.

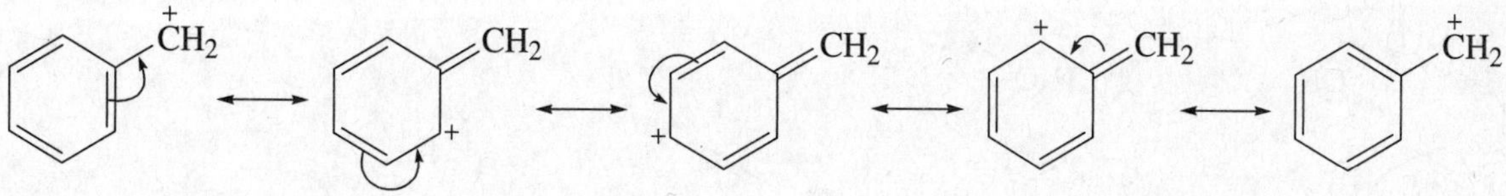

Problem 3.

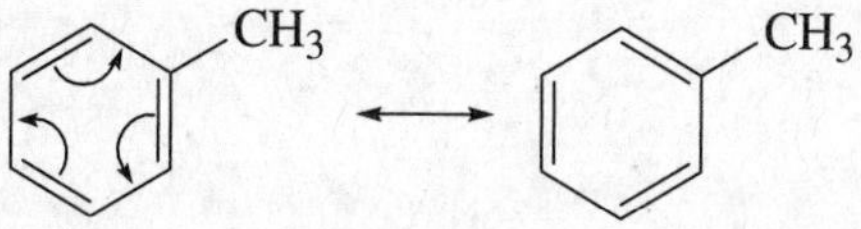

Problem 4.

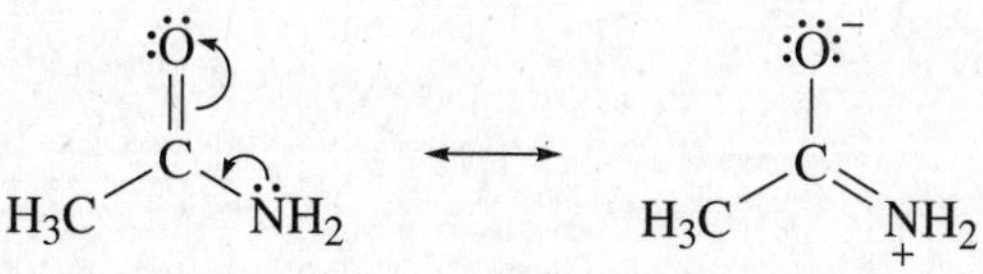

Problem 5.

a. $CH_3CH_2CH{=}CH{-}C{\equiv}N \quad \longleftrightarrow \quad CH_3CH_2\overset{+}{C}H{-}CH{=}C{=}\overset{..}{N}{:}^-$ 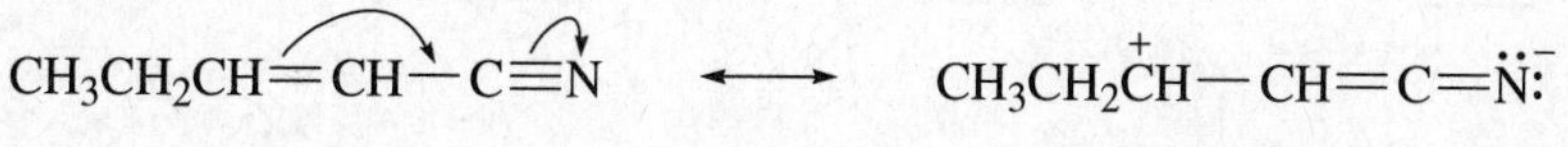

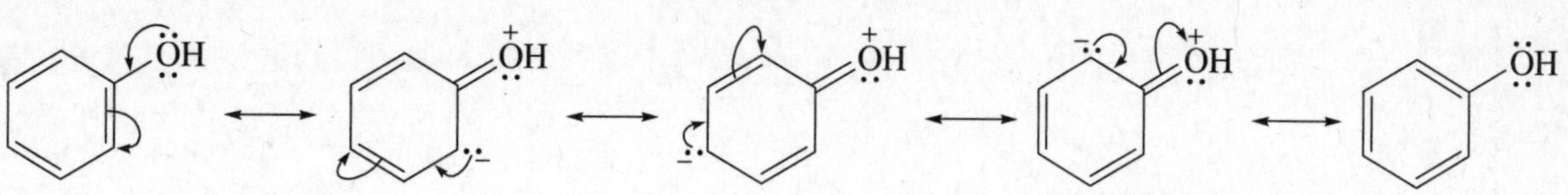

b.

Problem 6.

B, E, F, H, J, M, O

Problem 7.

B 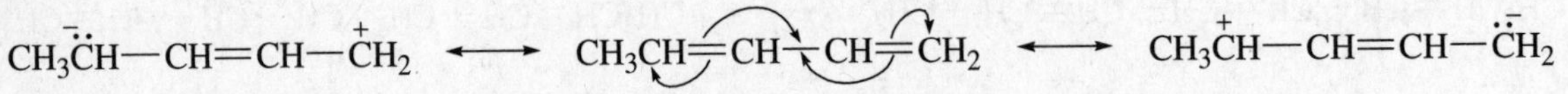

E 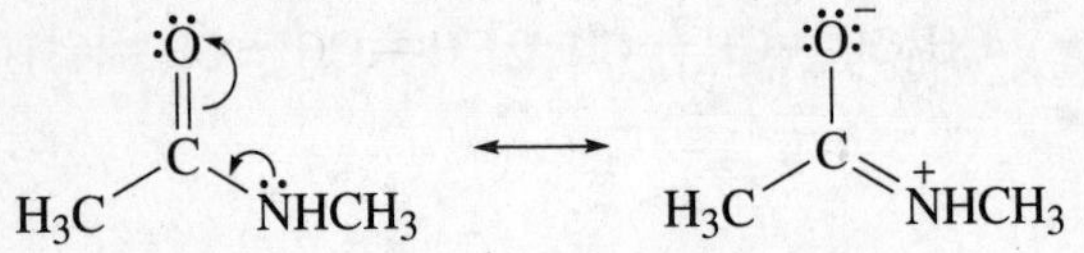

F 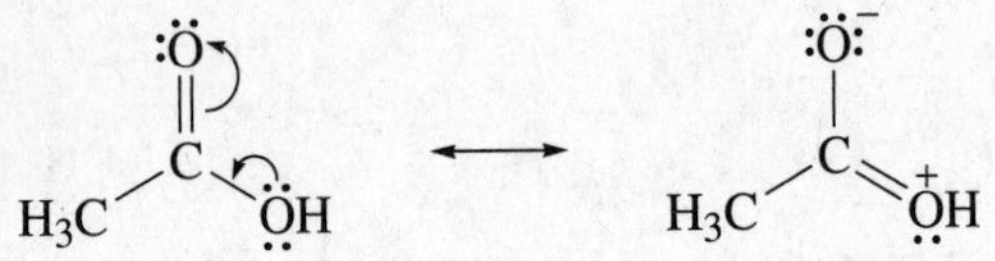

H 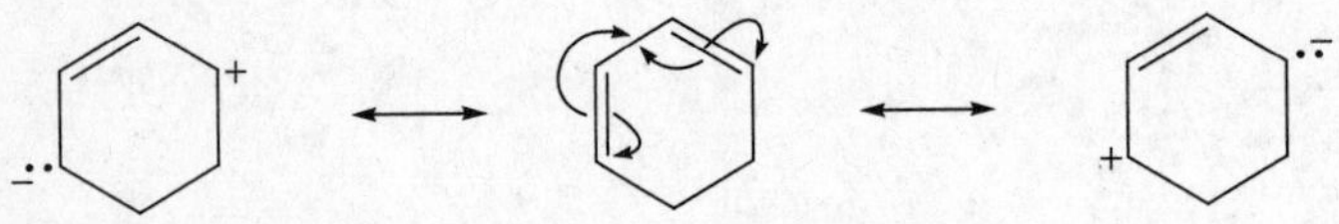

J 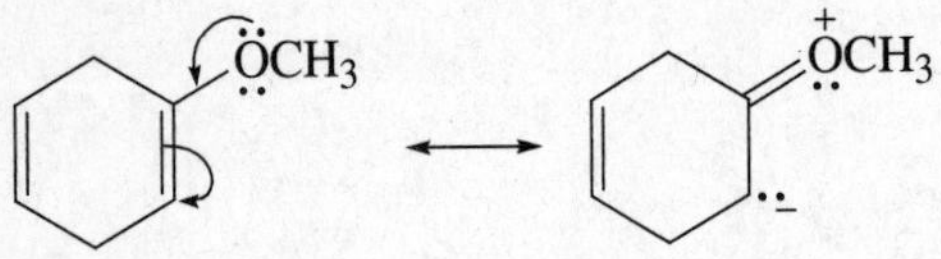

M 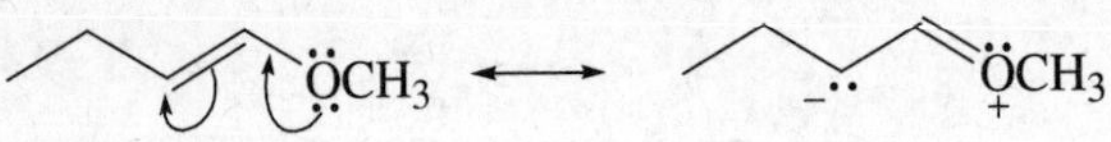

O 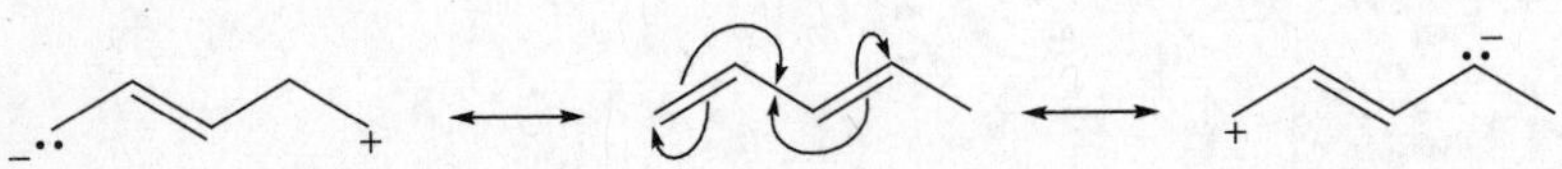

Problem 8.

a. 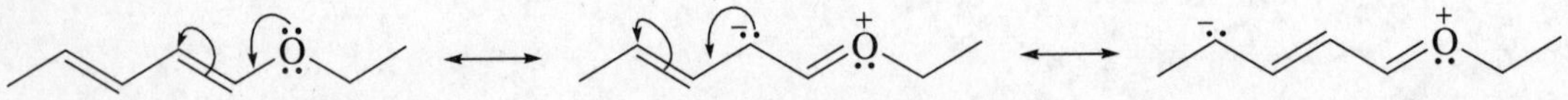

b. 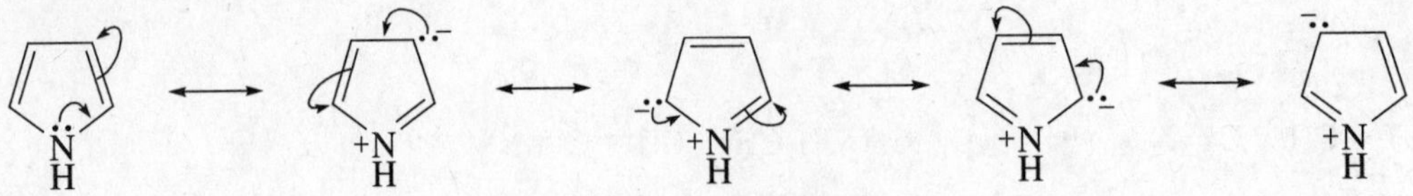

Problem 9.

a. 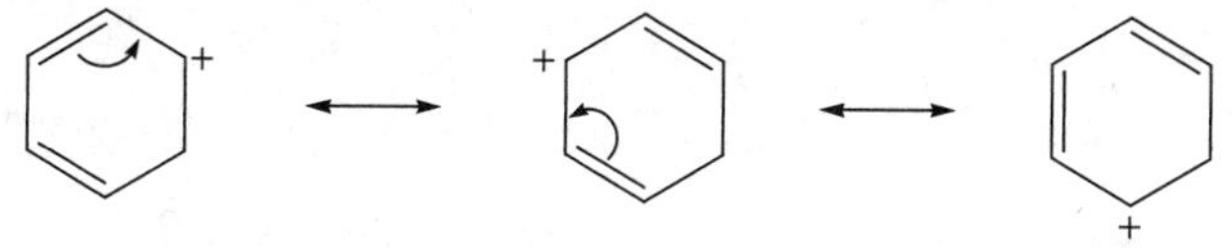

b. Notice in the following example that the electrons can move either clockwise or counterclockwise.

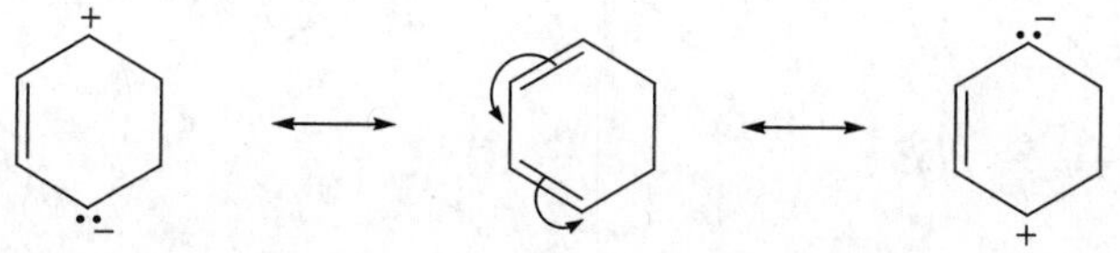

c. 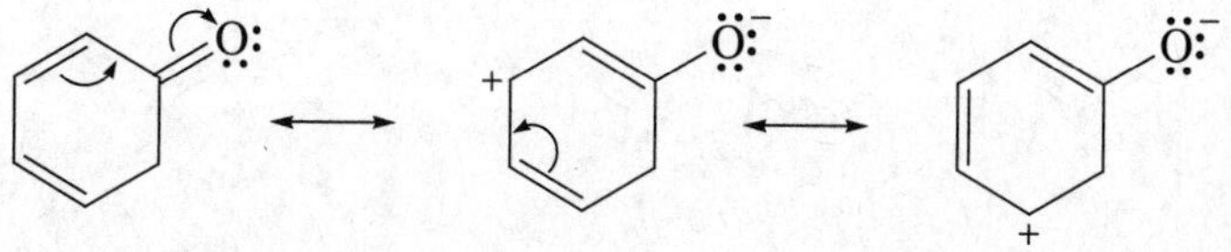

d. 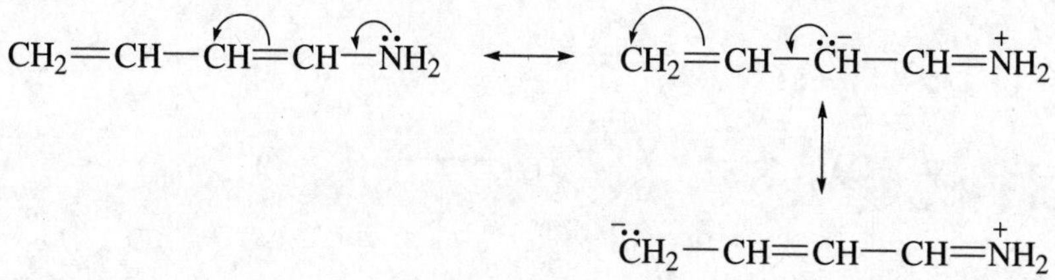
CH2=CH−CH=CH−NH2
CH2=CH−CH−CH=NH2
CH2−CH=CH−CH=NH2

e. Notice in the following example that the electrons can be delocalized onto either of the two sp^2 oxygens.

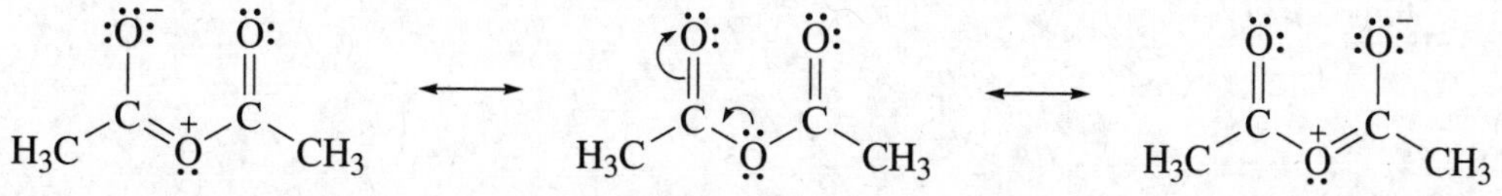

f. 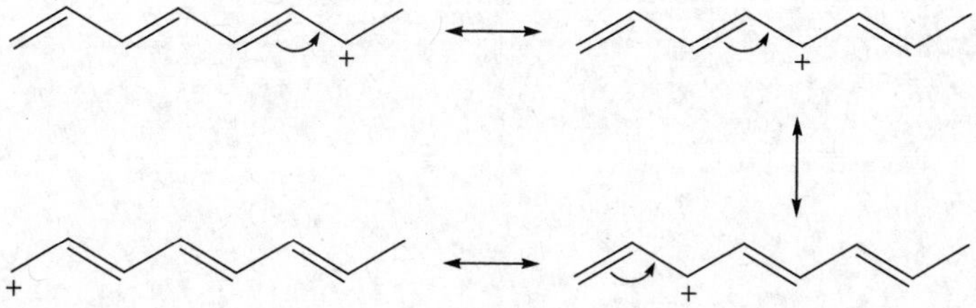

Problem 10.

a.

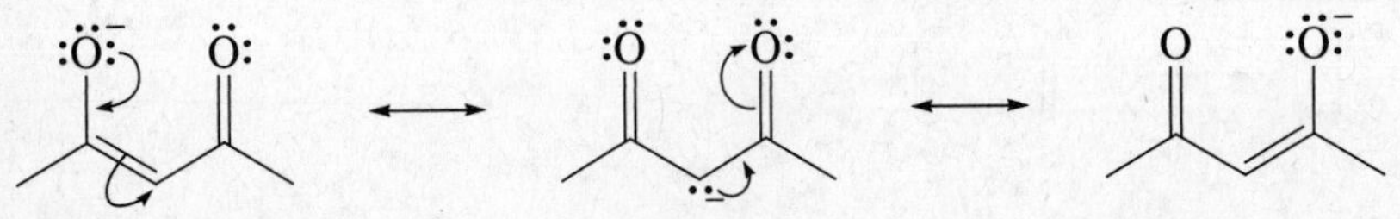

b.

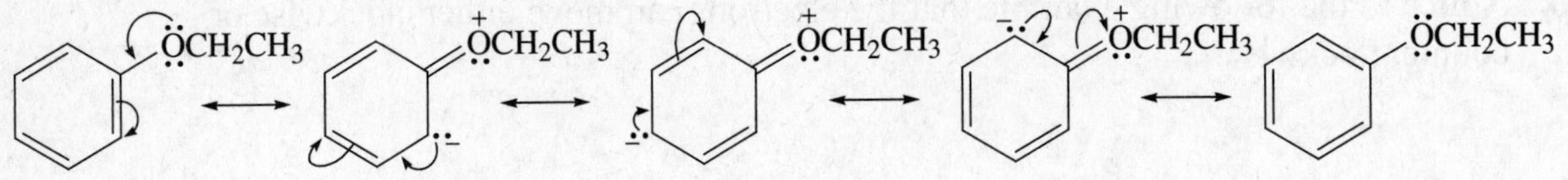
ÖCH₂CH₃

c.

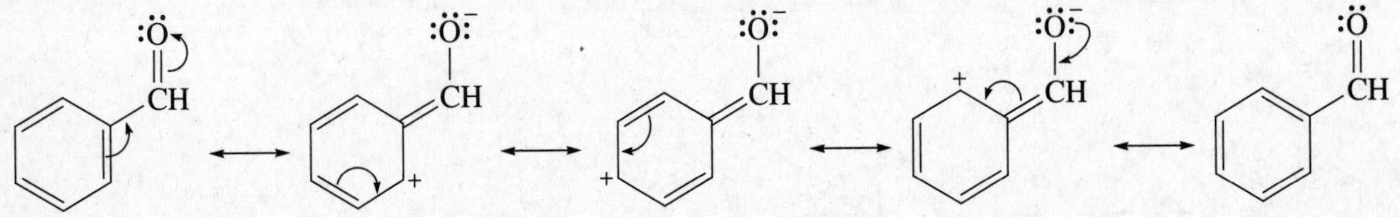
CH

d.

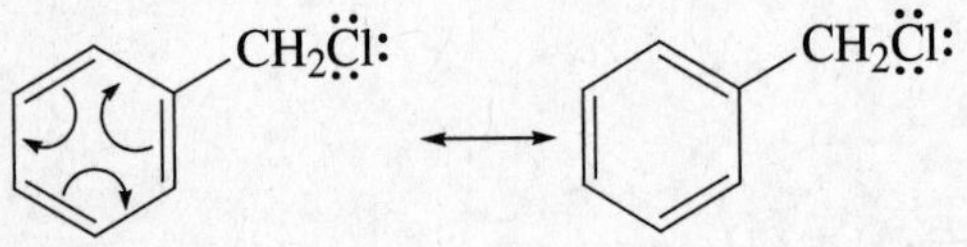
CH₂C̈l:

e.

f.

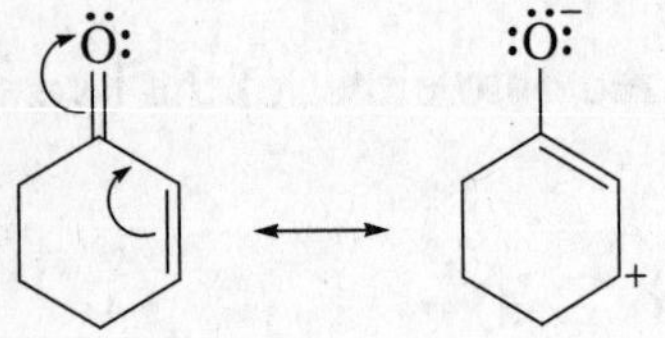

Problem 11.

a.
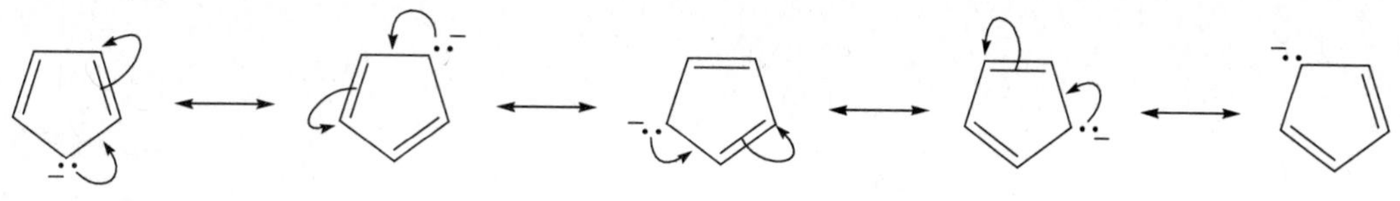

b.
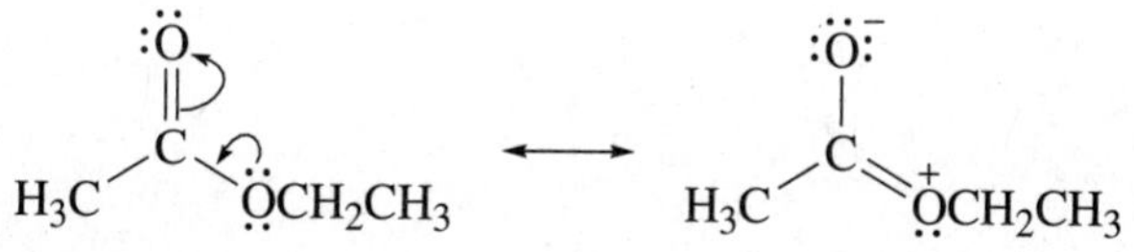
H3C
OCH2CH3
H3C
OCH2CH3

c.
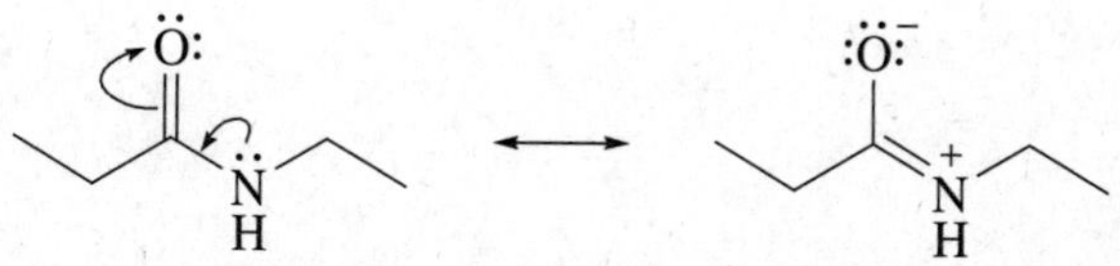
N
H
N
H

d.
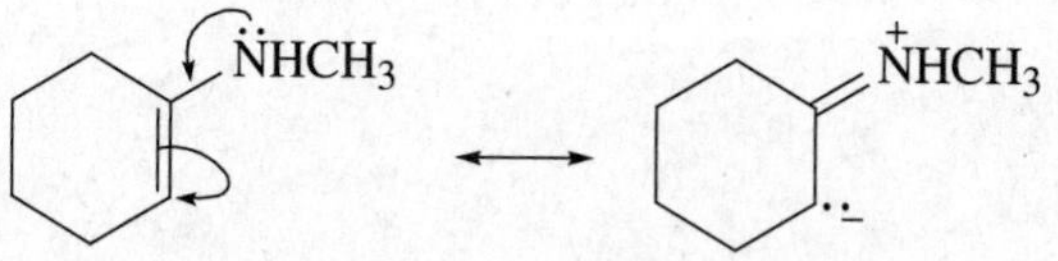
NHCH3
NHCH3

e.
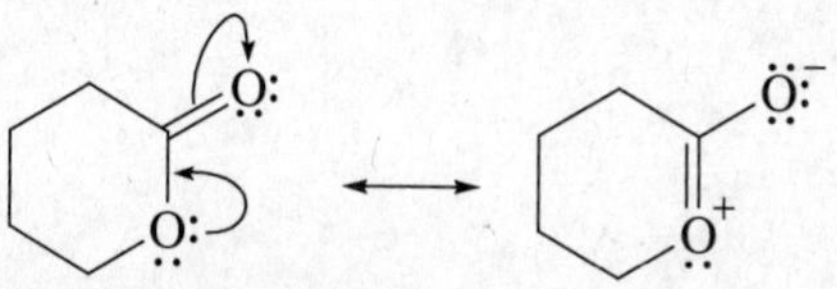

f.
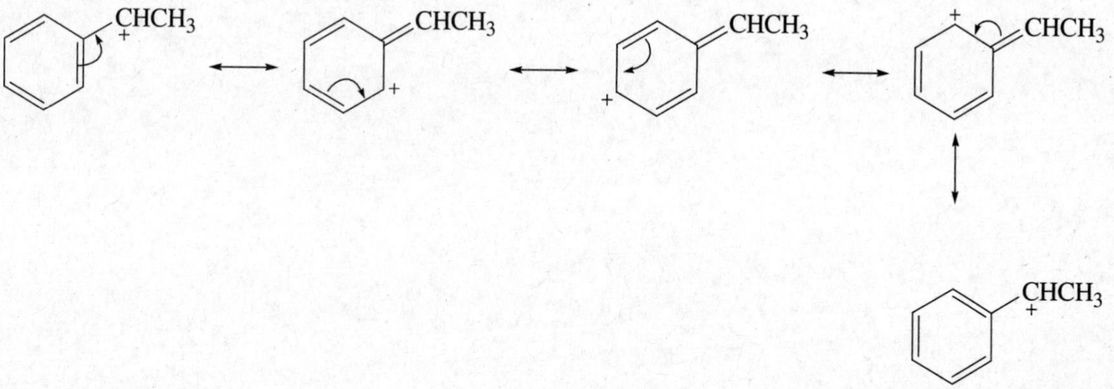
CHCH3
CHCH3
CHCH3
CHCH3
CHCH3

CHAPTER 8

Aromaticity • Reactions of Benzene and Substituted Benzenes

1. **a.** Notice that each resonance contributor has a charge of –1.

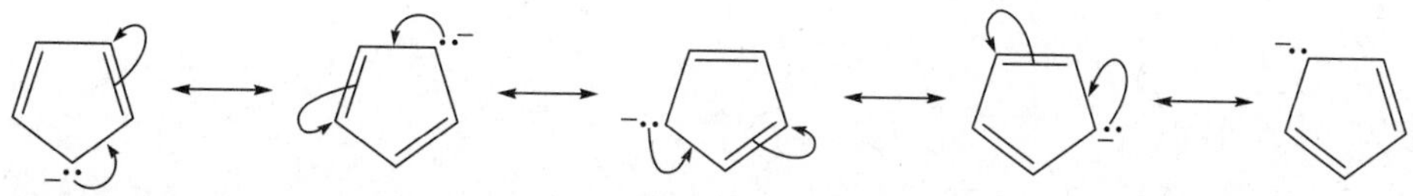

b. five ring atoms

2. The base formed when cyclopentadiene loses a proton is aromatic. The base formed when cyclopentane loses a proton is not aromatic. Because it is aromatic, the cyclopentadienyl anion is more stable than the cyclopentyl anion. The more stable the base, the stronger is its conjugate acid. Therefore, cyclopentadiene is a stronger acid than is cyclopentane.

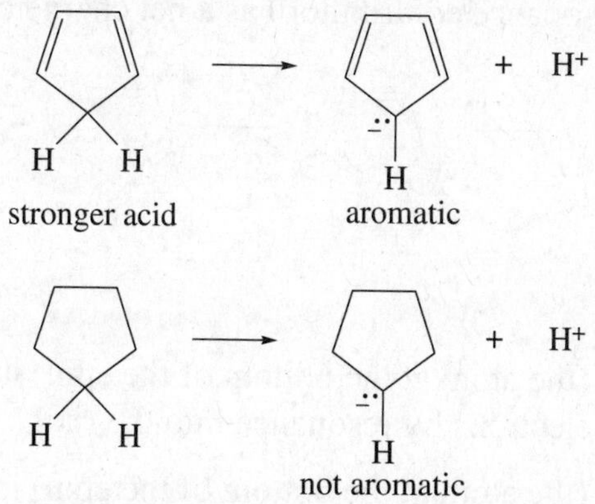
+ H+
H H
H
stronger acid
aromatic
+ H+
H H
H
not aromatic

3. The uncharged compound is not aromatic because one of the carbons is sp^3 hydridized. Therefore, the π electrons cannot form a π cloud.

The cation is aromatic; it has three pairs of π electrons.

The anion is not aromatic; the lone pair is delocalized into the ring, giving the anion four pairs of π electrons.

4. only **b** is aromatic; it is cyclic, planar, every atom in the ring has a *p* orbital, and it has seven pairs of π electrons.

a is not aromatic, because it has two pairs of π electrons and every atom in the ring does not have a *p* orbital

c is not aromatic, because every atom in the ring does not have a *p* orbital

d is not aromatic, because it is not cyclic.

5. **a.** Solved in the text.

b. There are five monobromophenanthrenes.

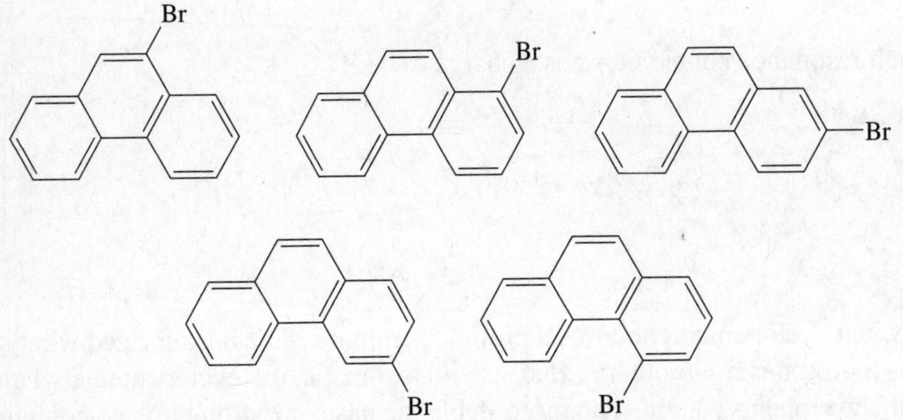

6. **a.** Notice that each resonance contributor has a net charge of 0.

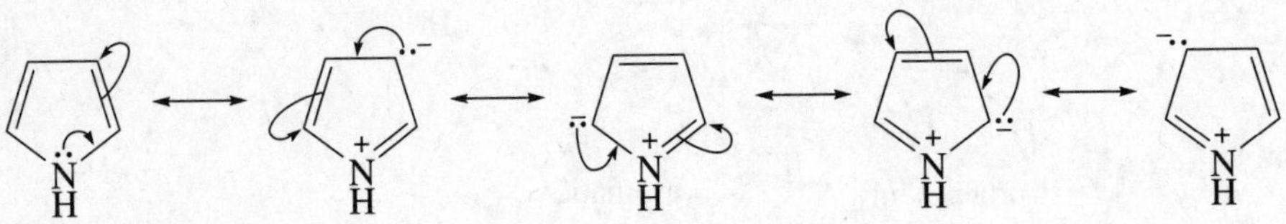

b. four ring atoms

7. **a.** The nitrogen atom (the atom at the bottom of the epm) in pyrrole has a partial positive charge because it donates electrons by resonance into the ring.

b. The nitrogen atom (the atom at the bottom of the epm) in pyridine is the most electronegative atom in the molecule; it withdraws electrons inductively from the ring.

c. The relatively electronegative nitrogen atom in pyridine withdraws electrons inductively from the ring, which causes the center of the epm to be less red.

8. **a.** 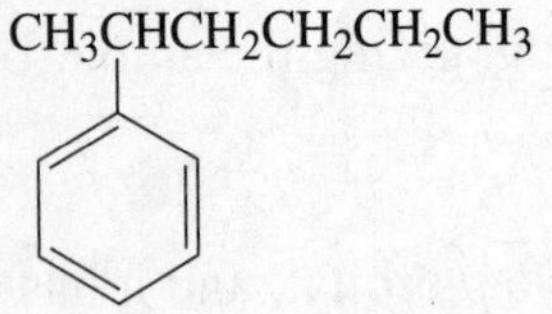 **b.** 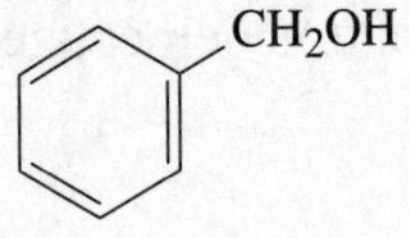 **c.** 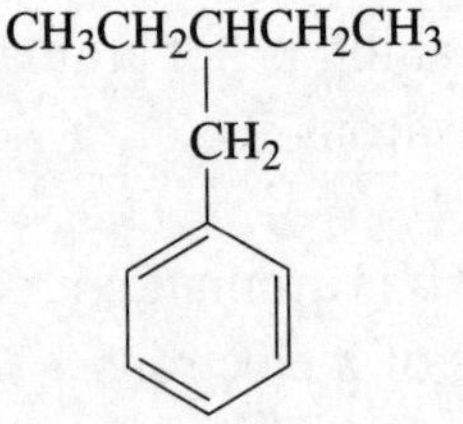

9.

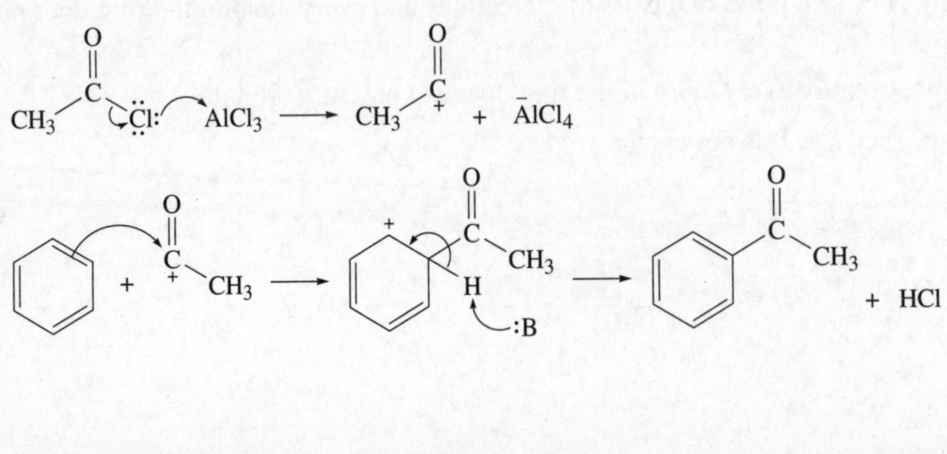

10. 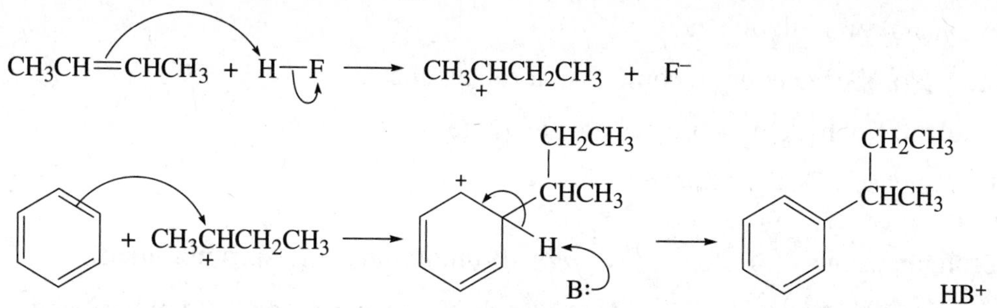

11. **a.** 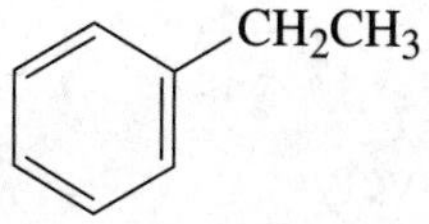 **b.** 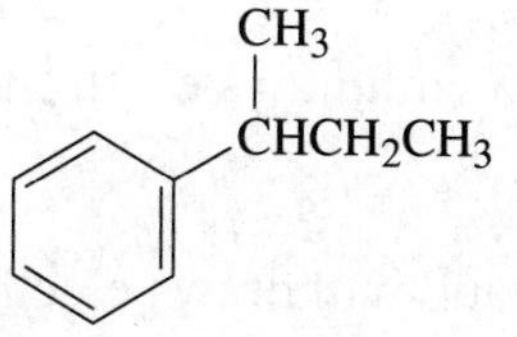 **c.** 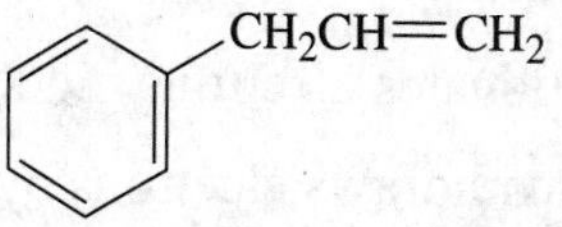

12. **a.** 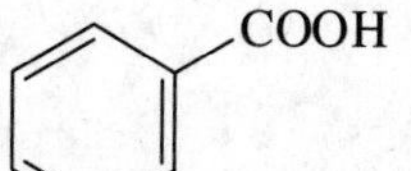 **b.** 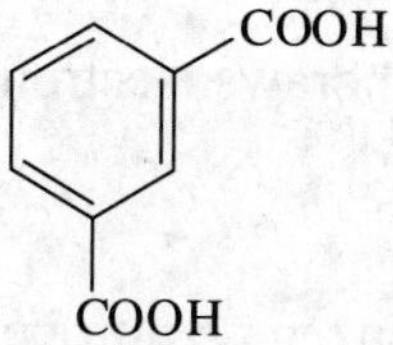

13. **a.** 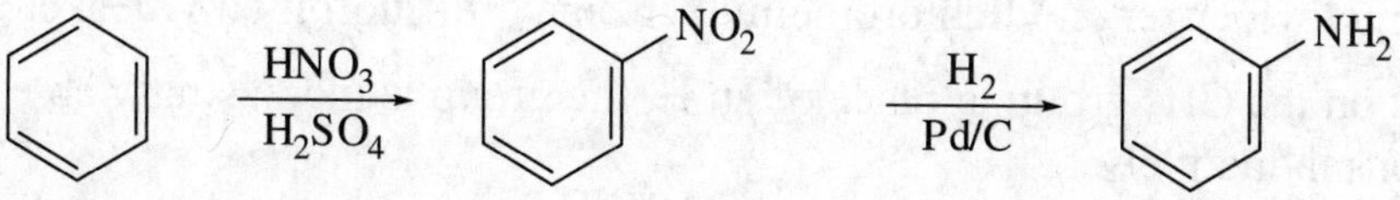

b. 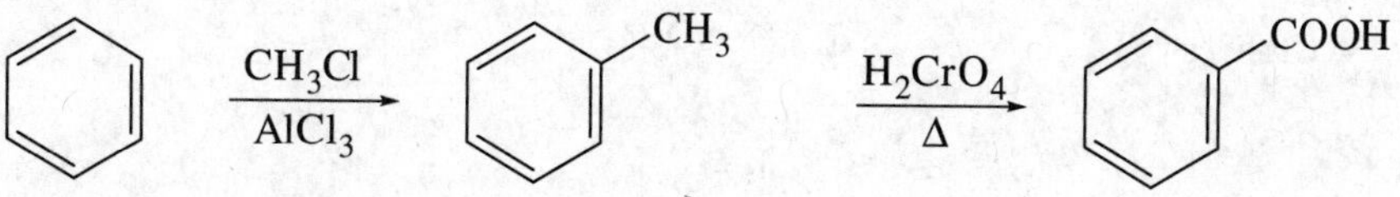

c. 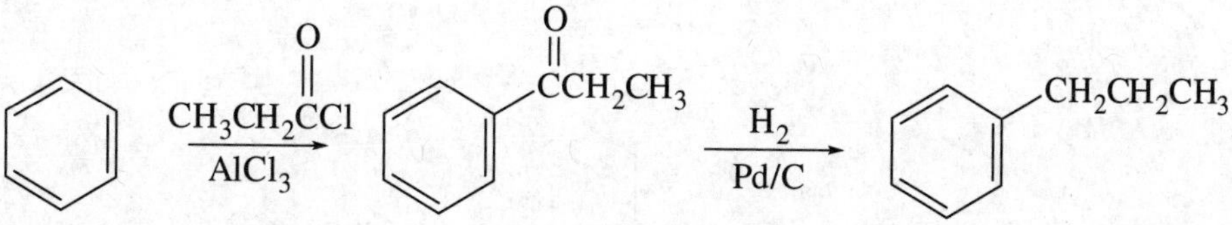

14. **a.** 

b. 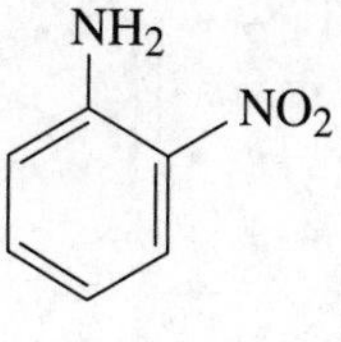

c. 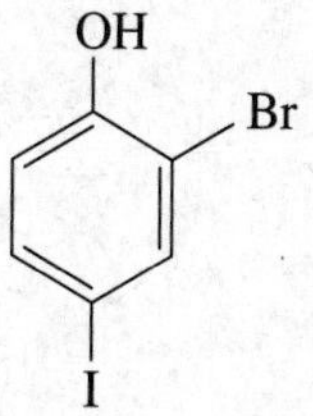

d. 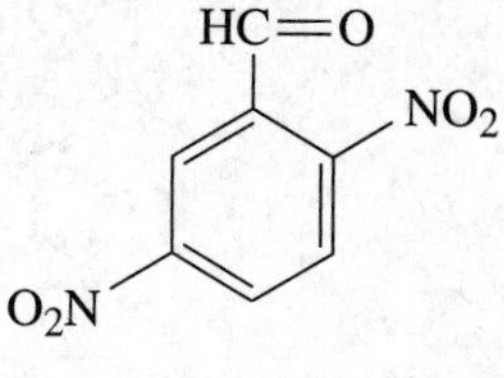

15. **a.** *meta*-ethylphenol or 3-ethylphenol

b. *meta*-bromochlorobenzene or 1-bromo-3-chlorobenzene

c. *para*-bromobenzaldehyde or 4-bromobenzaldehyde

d. *ortho*-ethyltoluene or 2-ethyltoluene

16. **a.** 1,3,5-tribromobenzene

b. *meta*-nitrophenol or 3-nitrophenol

c. *para*-bromotoluene or 4-bromotoluene

d. *ortho*-dichlorobenzene or 1,2-dichlorobenzene

17. **a.** donates electrons by resonance and withdraws electrons inductively

b. donates electrons inductively

c. withdraws electrons by resonance and withdraws electrons inductively

d. donates electrons by resonance and withdraws electrons inductively

e. donates electrons by resonance and withdraws electrons inductively

f. withdraws electrons inductively

18. **a.** phenol > toluene > benzene > bromobenzene > nitrobenzene

b. toluene > chloromethylbenzene > dichloromethylbenzene > difluoromethylbenzene

As each Cl is put on the CH_3 group in place of an H, the group is increasingly able to withdraw electrons inductively.

19. **a.**

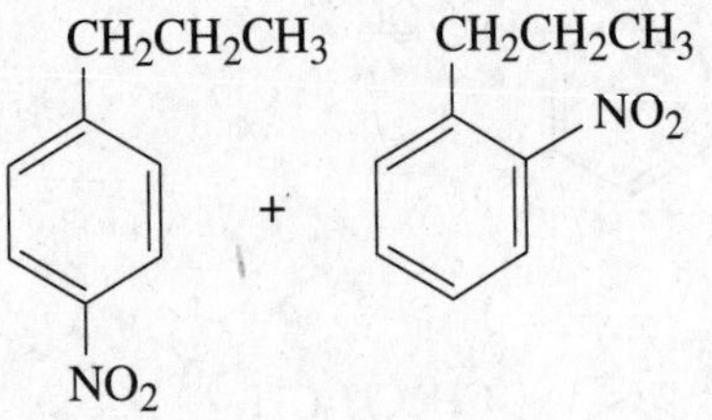
CH2CH2CH3
CH2CH2CH3
NO2
+
NO2

d.

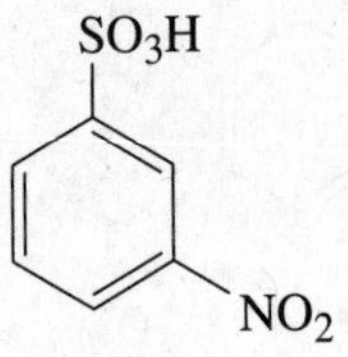
SO3H
NO2

b.

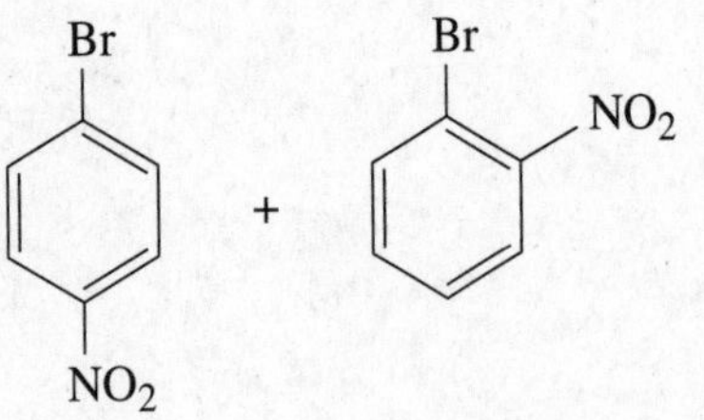
Br
Br
NO2
+
NO2

e.

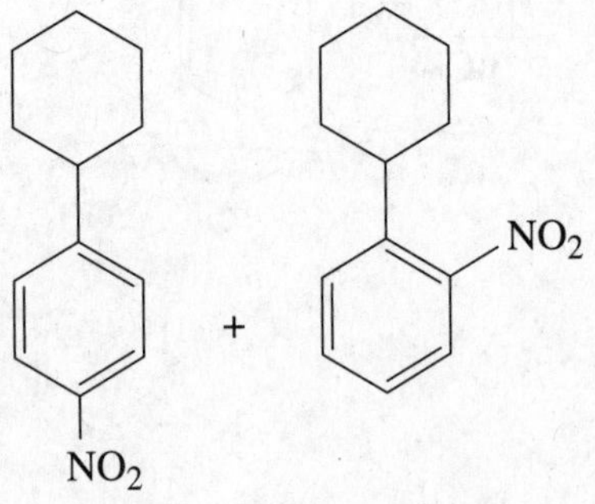
NO2
+
NO2

c.

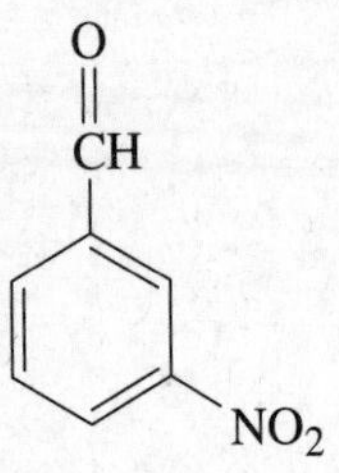
O
CH
NO2

f.

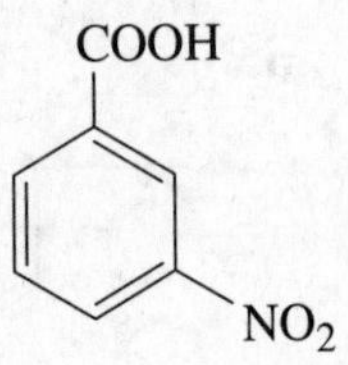
COOH
NO2

20. **a.**

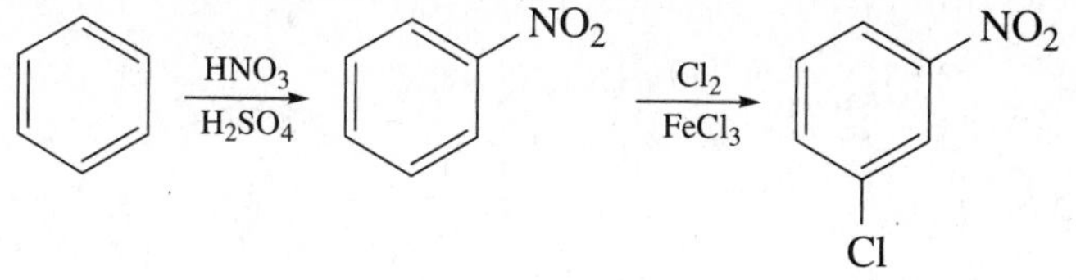

b.

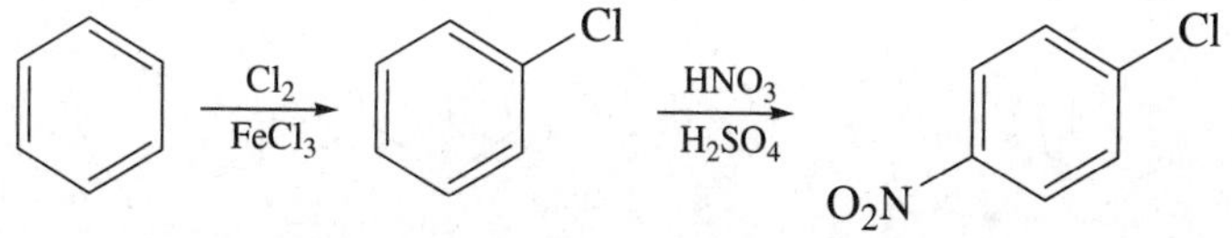

c.

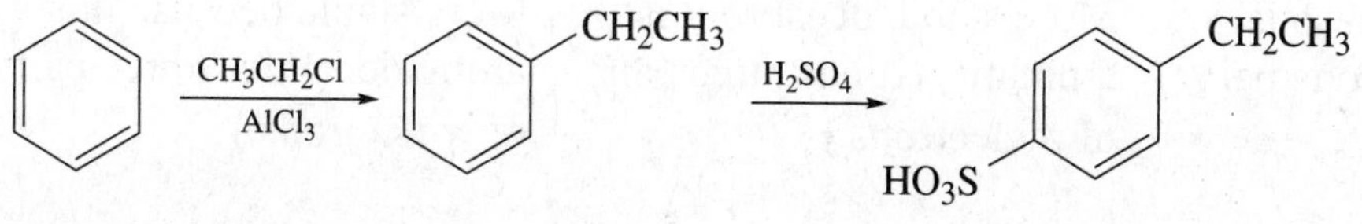

21. **a.** Solved in the text.

b. First, determine which benzene ring is more highly activated. For "**21b**", the ring on the left has a CH_3 substituent that donates electrons inductively; the ring on the right has a substituent that withdraws electrons inductively and by resonance. Therefore, the ring on the left is more highly activated so it is the one that undergoes electrophilic aromatic substitution.

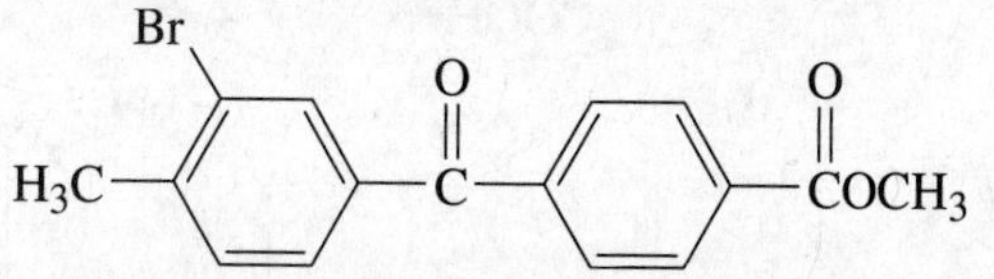

The ring on the right is more highly activated because the oxygen can donate electrons by resonance.

c.

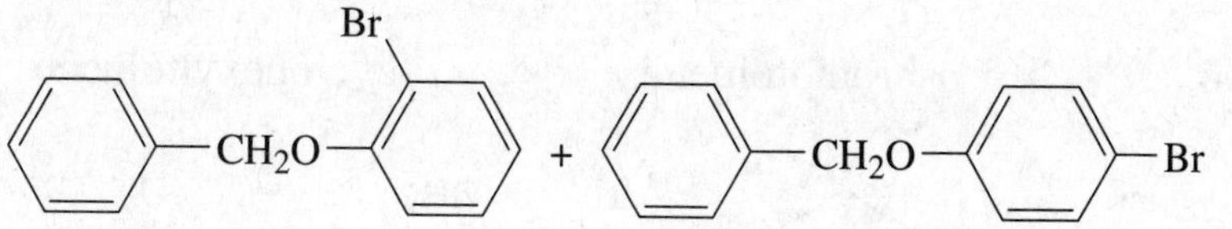

22. **a.** $ClCH_2\overset{O}{\overset{\|}{C}}OH$ **c.** $FCH_2\overset{O}{\overset{\|}{C}}OH$ Fluorine is more electronegative than chlorine

b. $H_3\overset{+}{N}CH_2\overset{O}{\overset{\|}{C}}OH$ **d.** $H\overset{O}{\overset{\|}{C}}OH$ A hydrogen is electron-withdrawing compared with a methyl group

23.

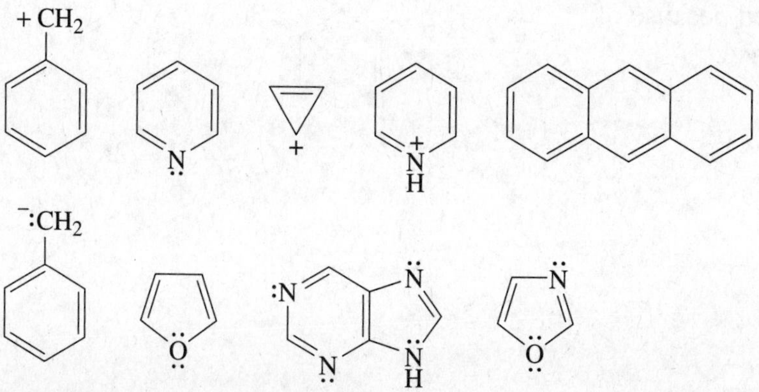

24. **a.** 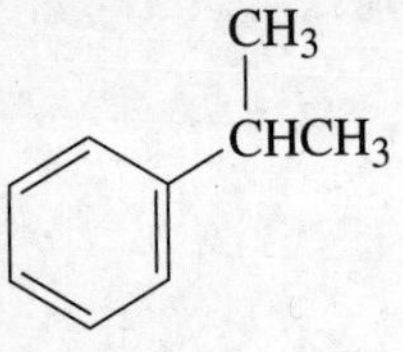 **b.** 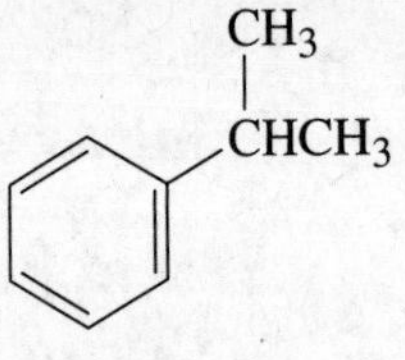 **c.** 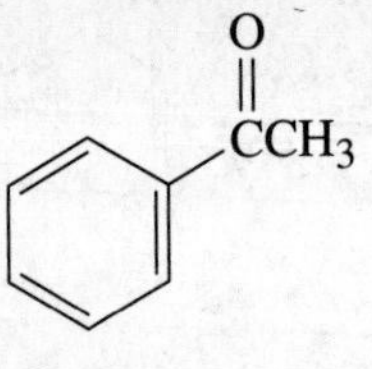

25. **a.** 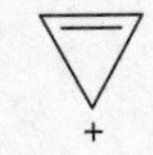

More stable because it is aromatic. (It has one pair of π electrons.)

b. 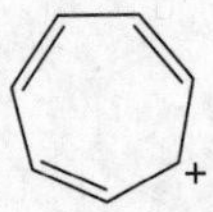

More stable because it is aromatic. (It has three pairs of π electrons.)

c. 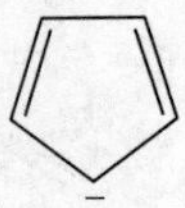

More stable because it is aromatic. (It has three pairs of π electrons.)

26. **a.** 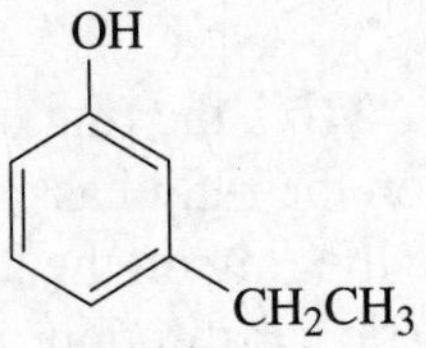 **c.** 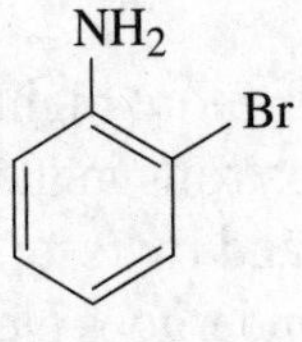 **e.** 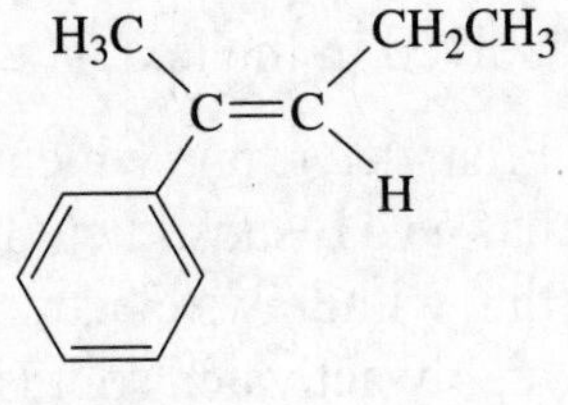

b. 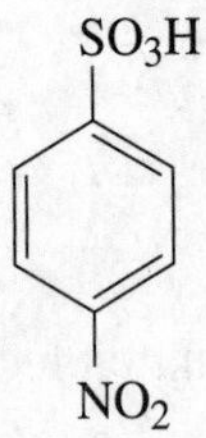 **d.** 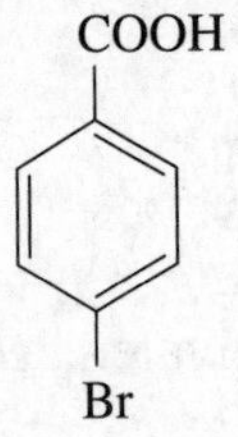 **f.** 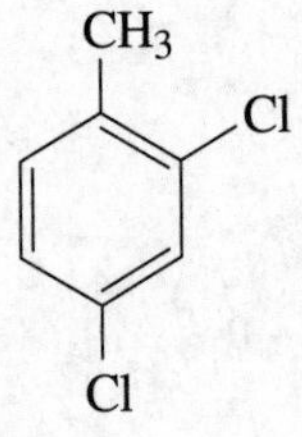

27. **a.** *m*-bromobenzoic acid **b.** *o*-bromotoluene **c.** *p*-cyclohexyltoluene

28. **a.** 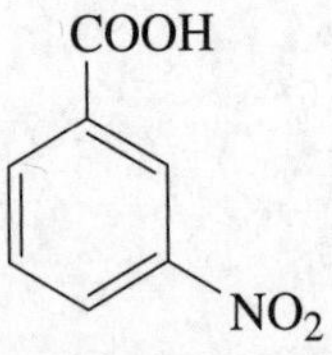 **c.** 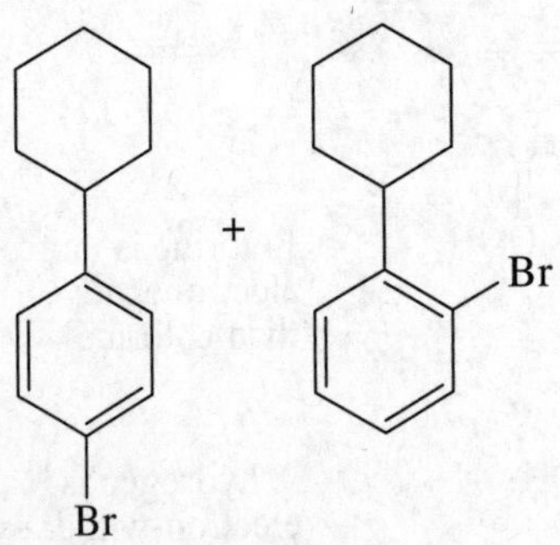 **d.** 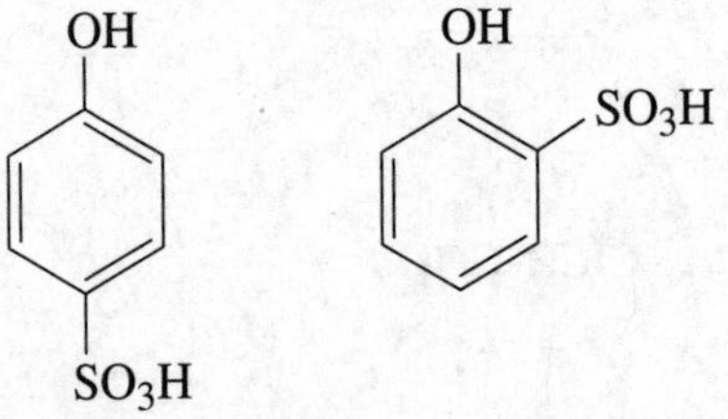

b. 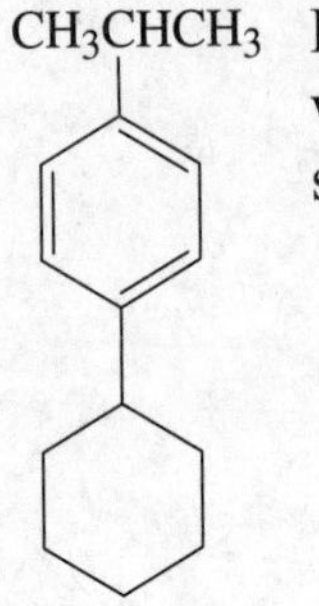 Little of the ortho isomer will be formed because of steric hindrance.

e. 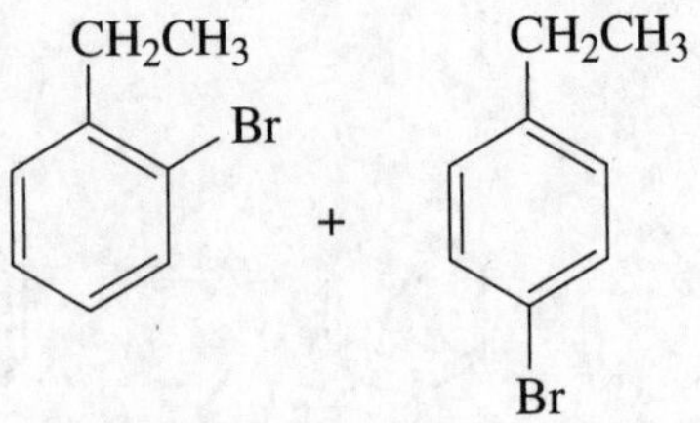

29. The more the substituent can donate electrons into the ring, the more basic will be the oxyanion.

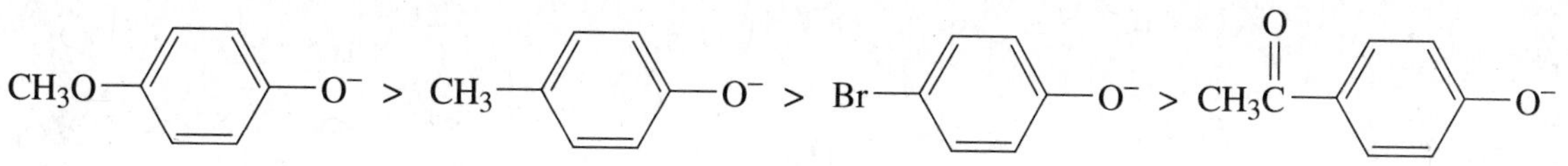

30. a.

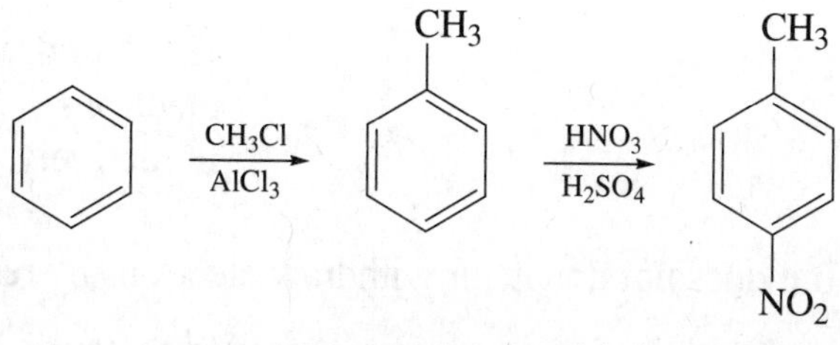

b.

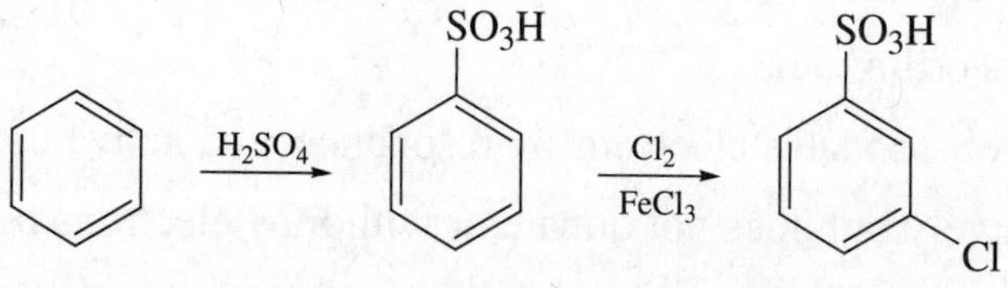

c.

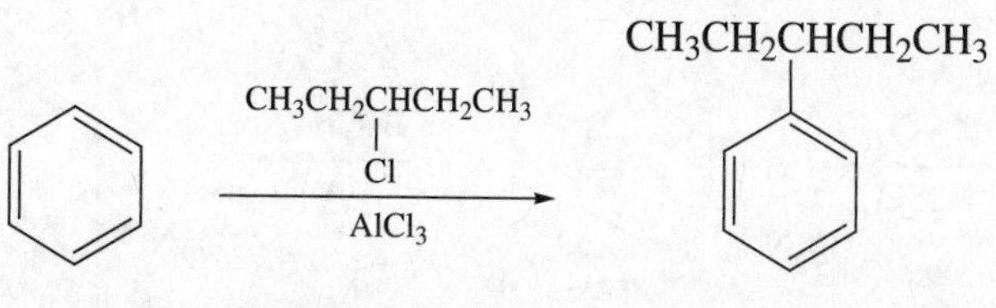

31. a.

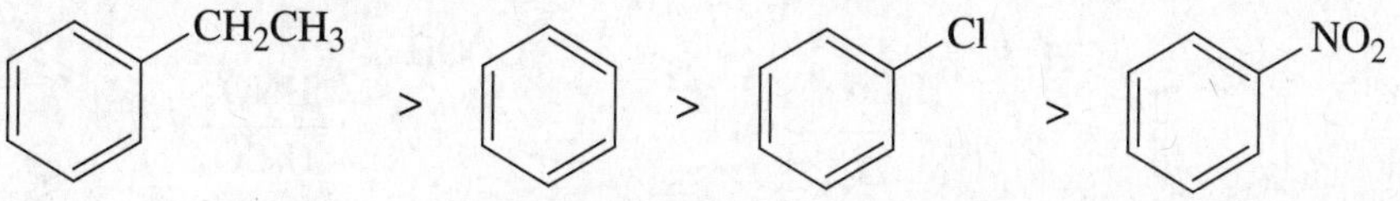

b.

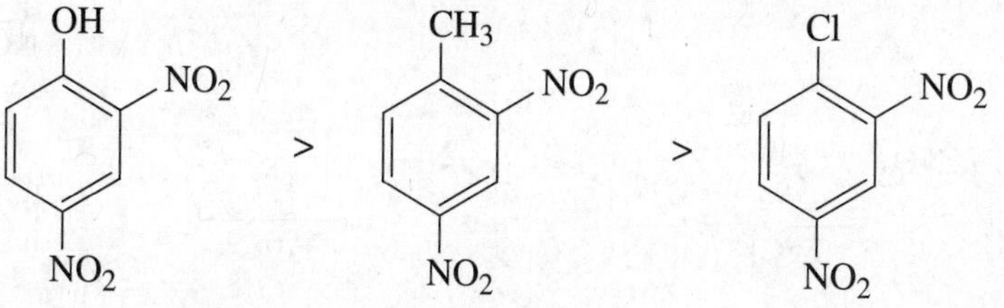

c.

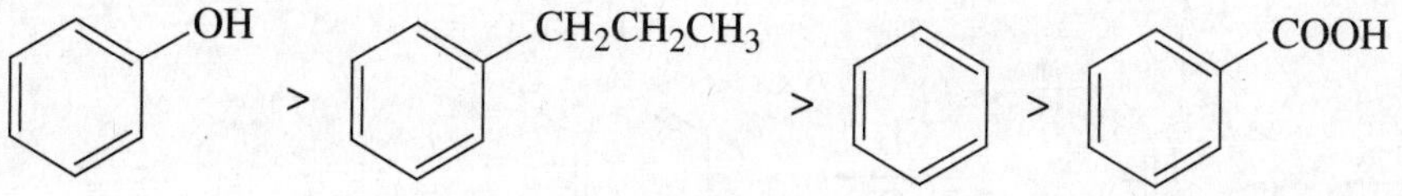

d.

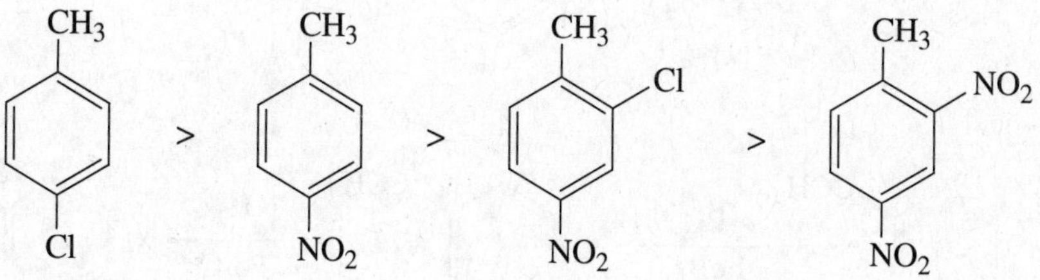

32. **a.** 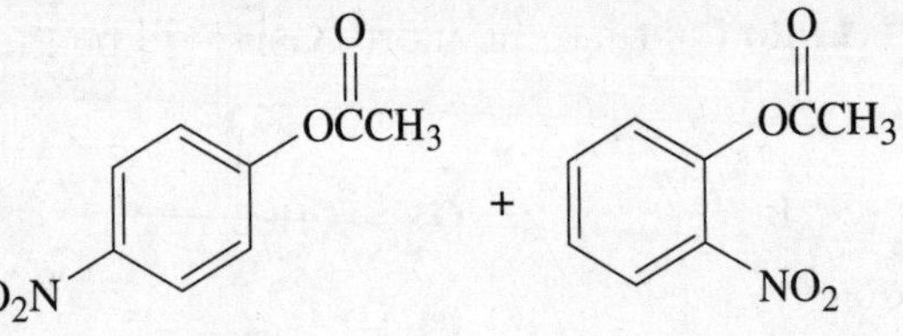
O
OCCH3
O2N
+
O
OCCH3
NO2

b. 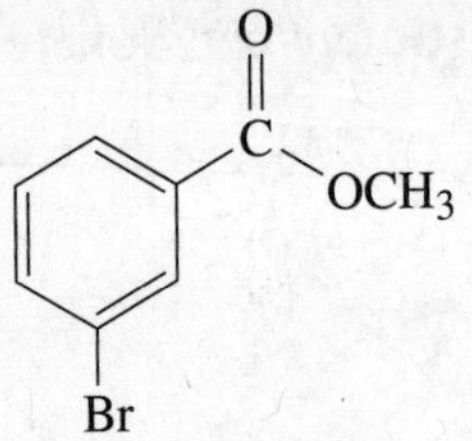
O
C
OCH3
Br

c. 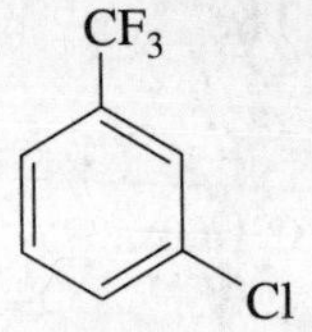
CF3
Cl

CF_3 withdraws electrons inductively.

33. **a.** CH_2CH_3 donates electrons inductively but does not donate or withdraw electrons by resonance.

b. NO_2 withdraws electrons inductively and withdraws electrons by resonance.

c. Br deactivates the ring and directs ortho/para.

d. OH withdraws electrons inductively, donates electrons by resonance, and activates the ring.

e. $^{+}NH_3$ withdraws electrons inductively but does not donate or withdraw electrons by resonance.

34. **a.** 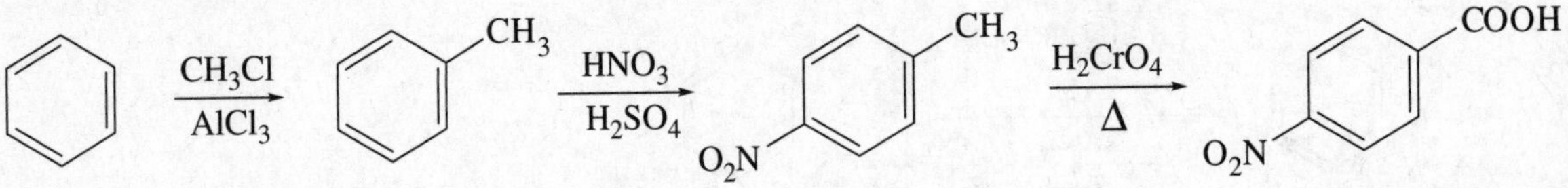
CH3Cl
AlCl3
CH3
HNO3
H2SO4
CH3
O2N
H2CrO4
Δ
COOH
O2N

b. 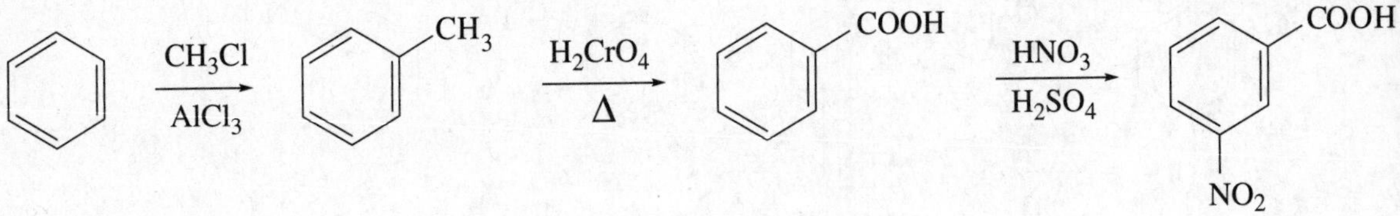
CH3Cl
AlCl3
CH3
H2CrO4
Δ
COOH
HNO3
H2SO4
COOH
NO2

c. 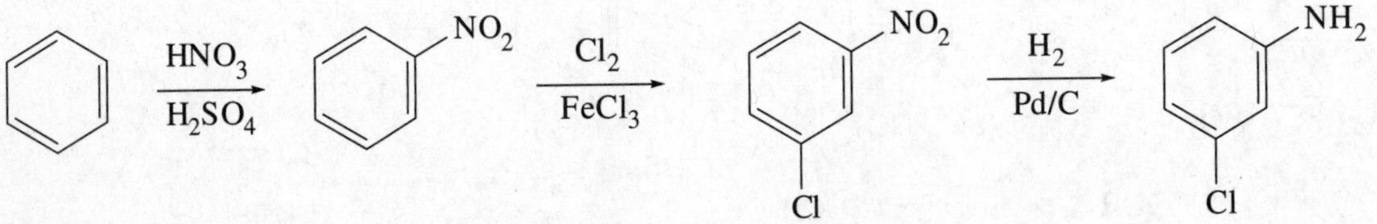
HNO3
H2SO4
NO2
Cl2
FeCl3
NO2
Cl
H2
Pd/C
NH2
Cl

d. 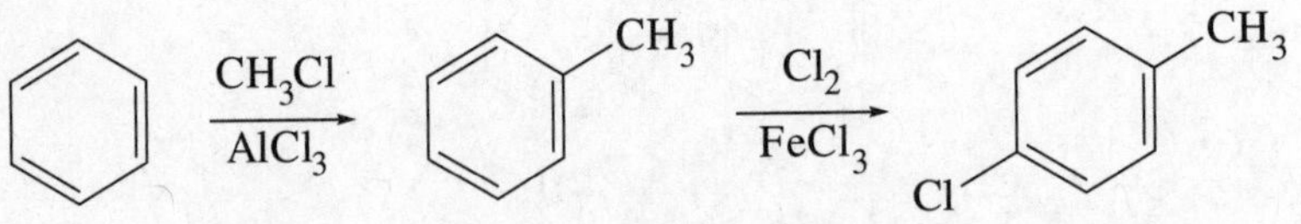
CH3Cl
AlCl3
CH3
Cl2
FeCl3
CH3
Cl

e. 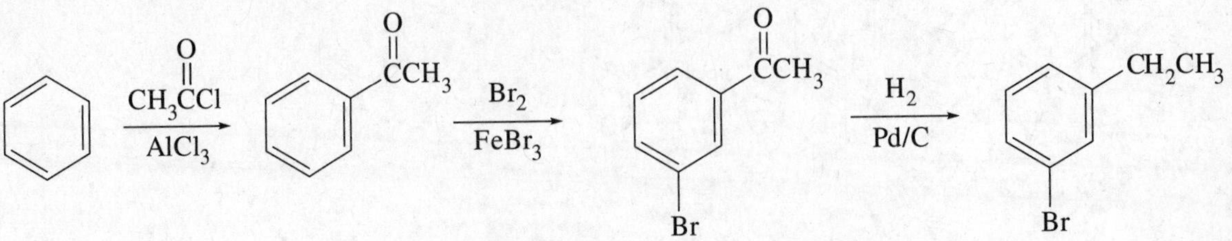
O
CH3CCl
AlCl3
O
CCH3
Br2
FeBr3
O
CCH3
Br
H2
Pd/C
CH2CH3
Br

f. 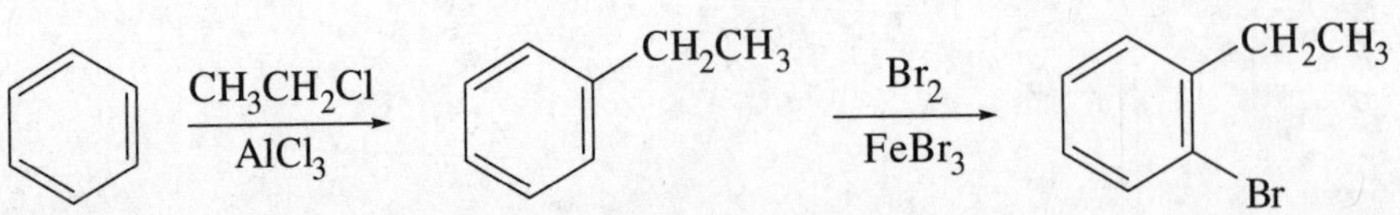
CH3CH2Cl
AlCl3
CH2CH3
Br2
FeBr3
CH2CH3
Br

35. toluene > chloromethylbenzene > dichloromethylbenzene > difluoromethylbenzene

36. a. 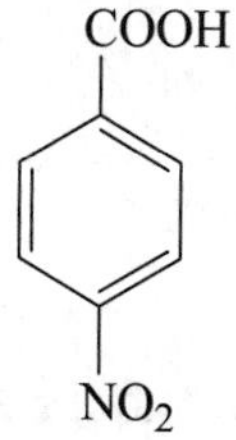

b. 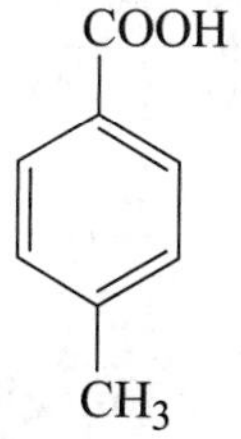

37. a. 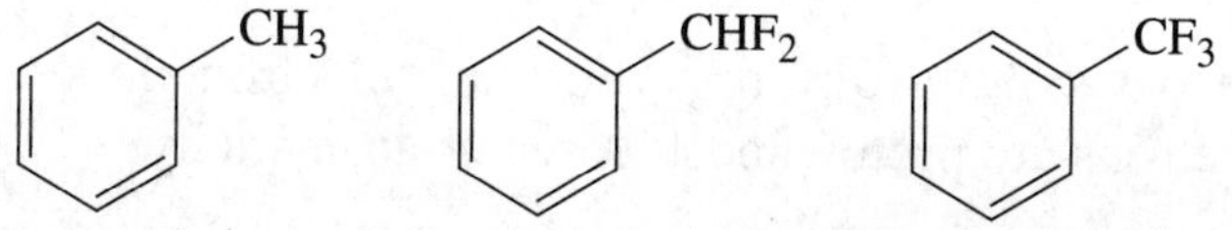

most reactive

least reactive
highest % meta product

b. 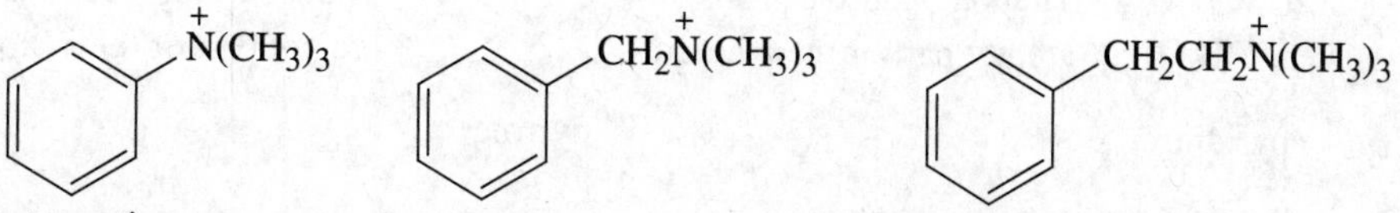

least reactive
highest % meta product

most reactive

c. 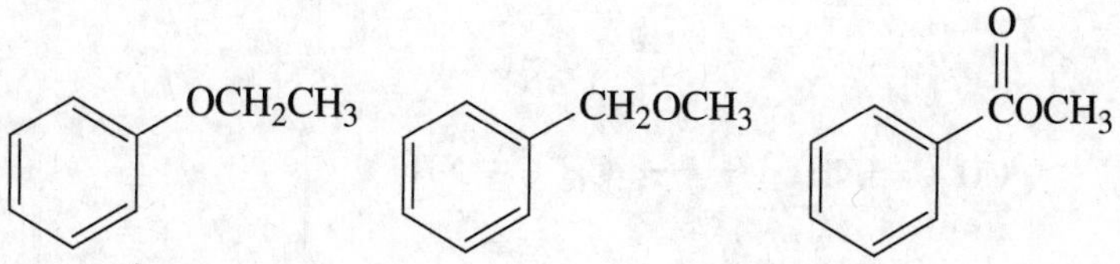

most reactive

least reactive
highest % meta product

38. The compound with the methoxy substituent is more reactive because it forms a more stable carbocation intermediate. The carbocation intermediate is stabilized by resonance electron donation.

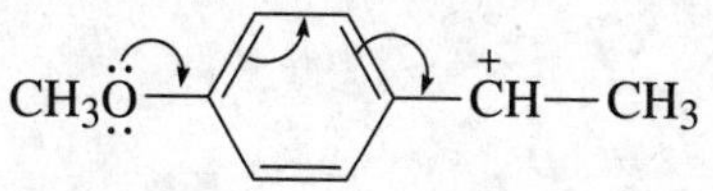

39. 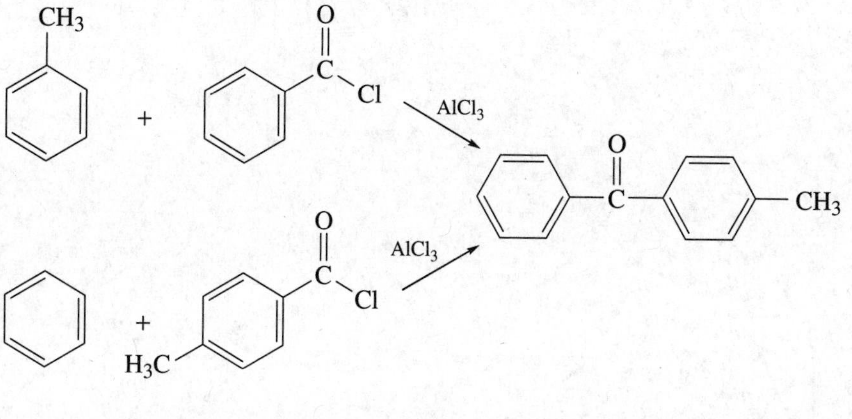

40. **a.** 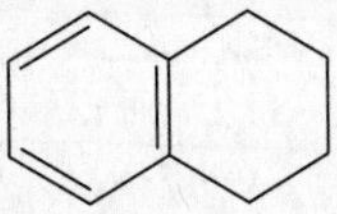 **b.** 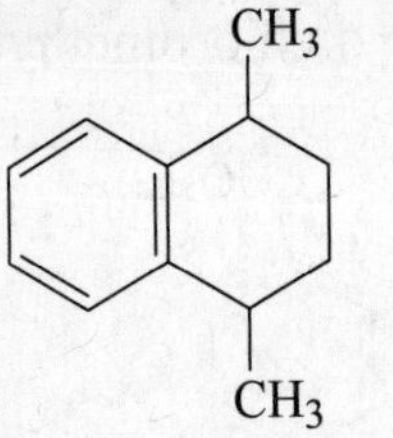
CH3
CH3

41. **a.** Pyridine is a stronger base because its nonbonding electrons are not part of the π cloud. Therefore, when its nonbonding electrons are protonated, it is still aromatic.

In contrast, the nonbonding electrons of pyrrole are part of the π electron cloud. Thus, when its nonbonding electrons are protonated, it loses its aromaticity.

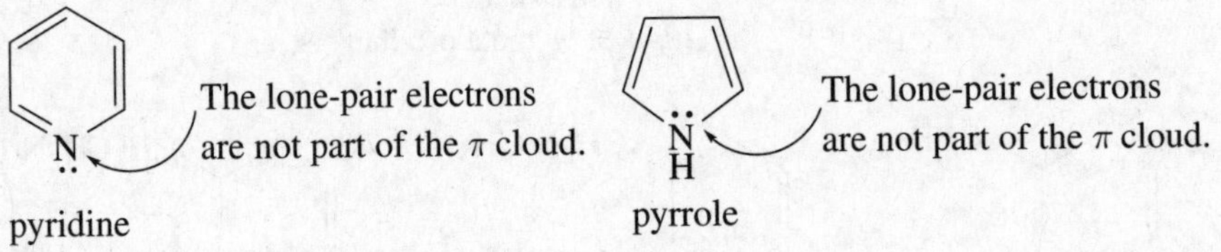
The lone-pair electrons
are not part of the π cloud.
pyridine
The lone-pair electrons
are not part of the π cloud.
pyrrole

b. The compound with the carbon-nitrogen double bond is a stronger base, because it has increased electron density on one of the nitrogens as a result of electron delocalization.

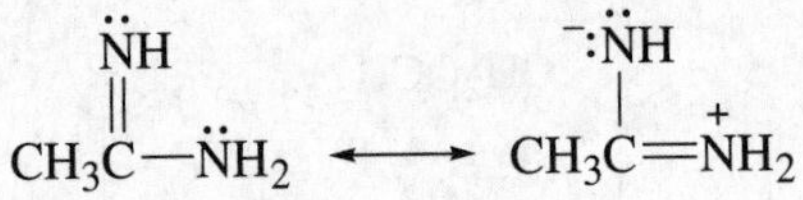

42.

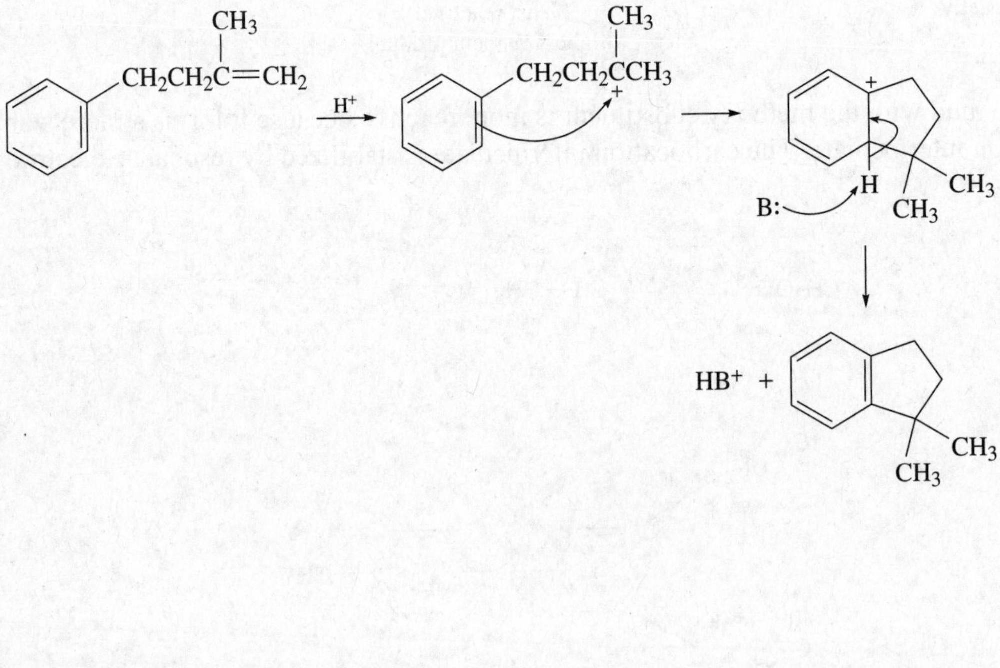
H+
B:
HB+ +

43. **a.**

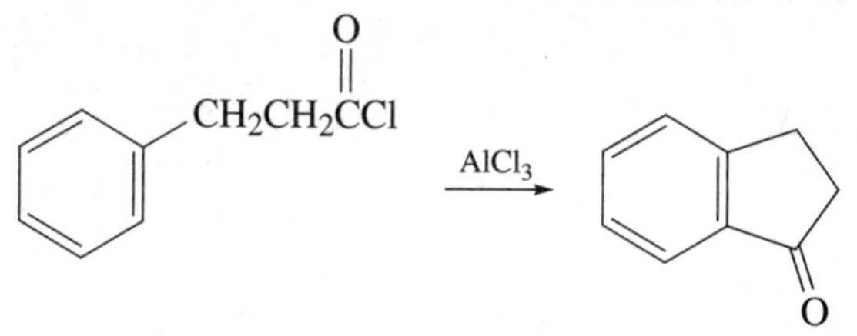
CH₂CH₂CCl
O
AlCl₃
O

b.

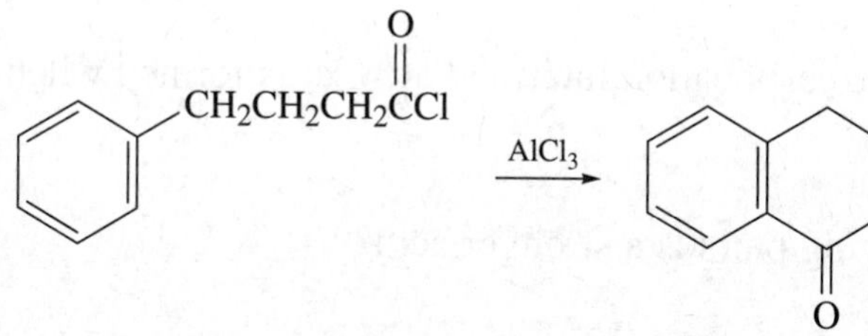
CH₂CH₂CH₂CCl
O
AlCl₃
O

44. Each ring hydrogen is replaced by a D^{+} electrophile. The hydrogens can be replaced in any order.

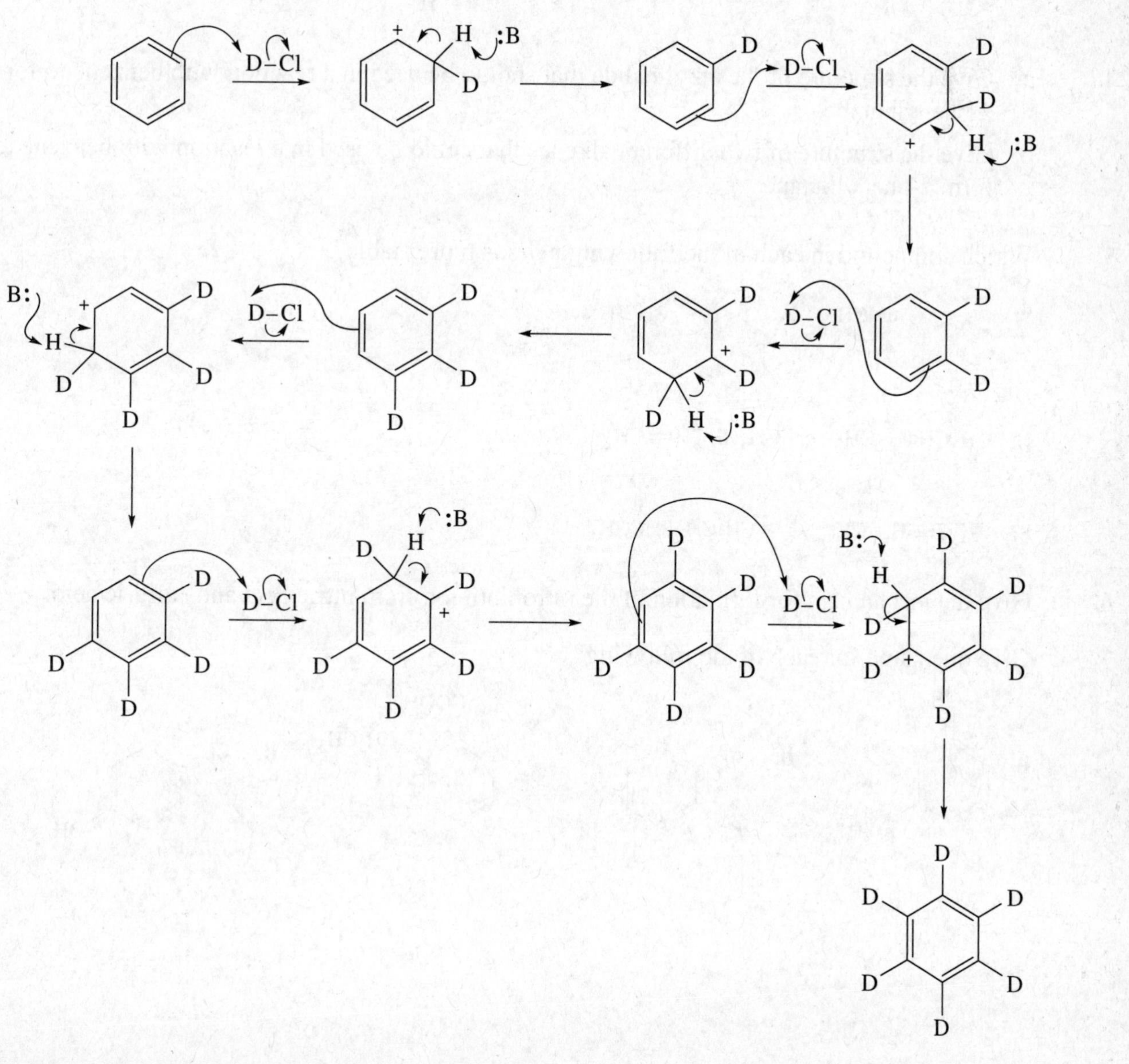

Chapter 8 Practice Test

1. Which are aromatic compounds?

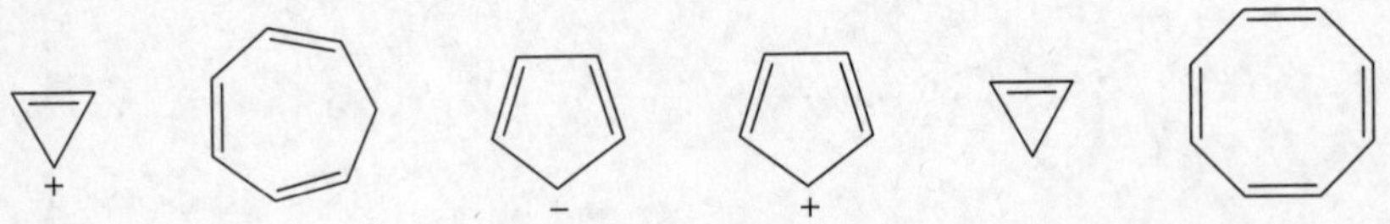

2. Draw the resonance contributors for the carbocation intermediate that is formed when benzene reacts with an electrophile (Y^+).

3. Which compound in each of the following pairs is a stronger acid?

a. 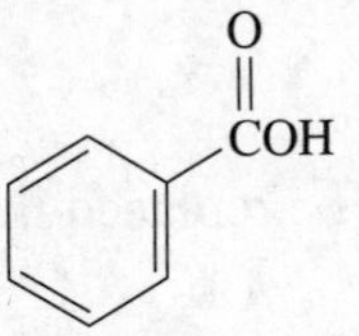 or 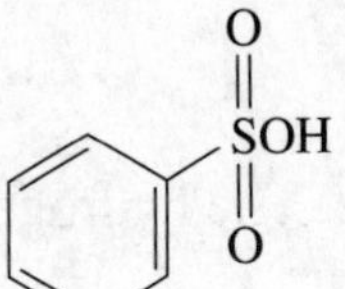

b.  or 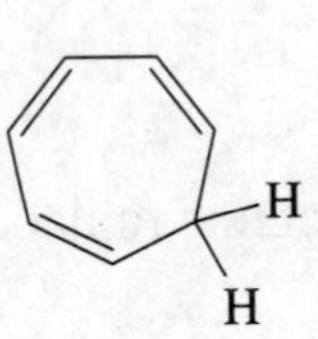

4. **a.** Give the structure of the alkyl halide that should be used in a reaction with benzene to form 2-phenylbutane.

b. Give the structure of two different alkenes that could be used in a reaction with benzene to form 2-phenylbutane.

5. Which compound in each of the following pairs is more stable?

a. 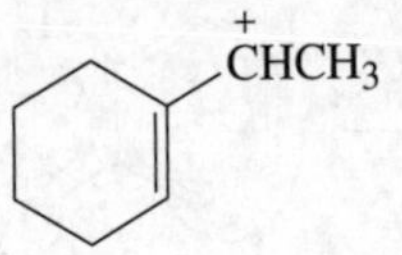 or 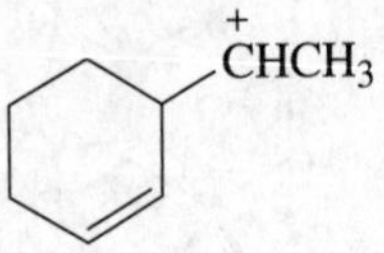

b. $CH_3\overset{+}{C}HCH_2CH_3$ or $CH_3\overset{+}{C}HCH{=}CH_2$

c. $CH_3\overset{-}{C}HCH_2\overset{\overset{O}{\|}}{C}CH_3$ or $CH_3CH_2\overset{-}{C}H\overset{\overset{O}{\|}}{C}CH_3$

6. Give the mechanism for formation of the nitronium ion from nitric acid and sulfuric acid.

7. Give one name for each of the following:

a. 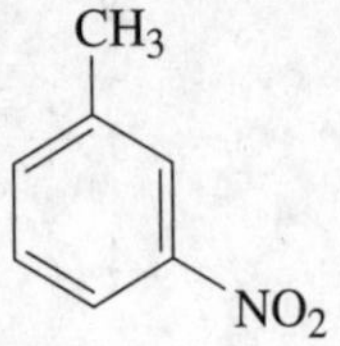

b. 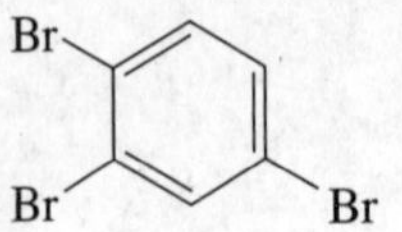

c. 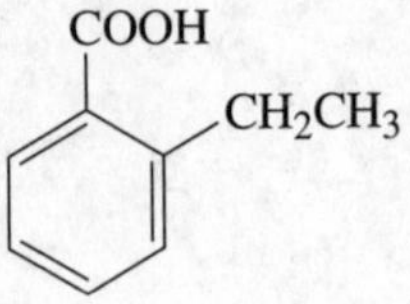

d. 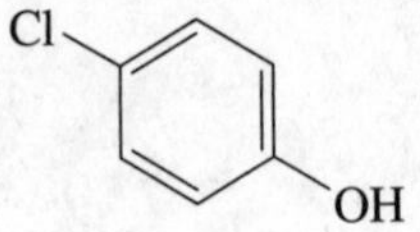

8. Rank the following compounds in order of decreasing reactivity toward $Br_2/FeBr_3$.

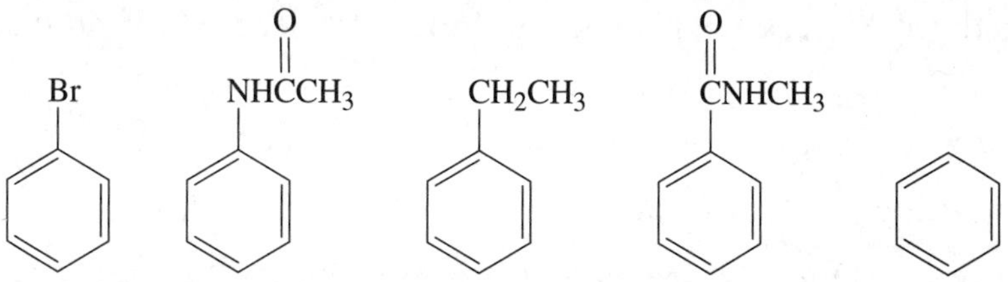
Br
NHCCH3
CH2CH3
CNHCH3

9. For each pair of compounds, indicate the one that is the stronger acid.

a.
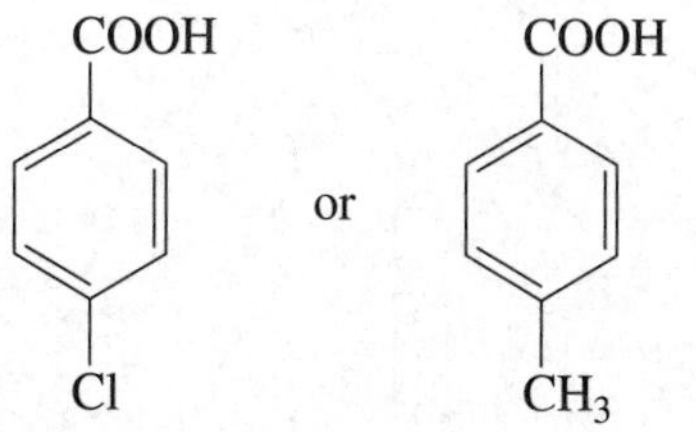
COOH
Cl
or
COOH
CH3

c.
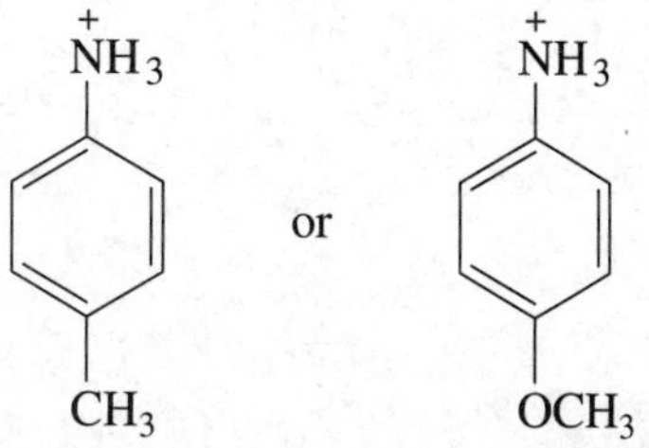
NH3
CH3
or
NH3
OCH3

b.
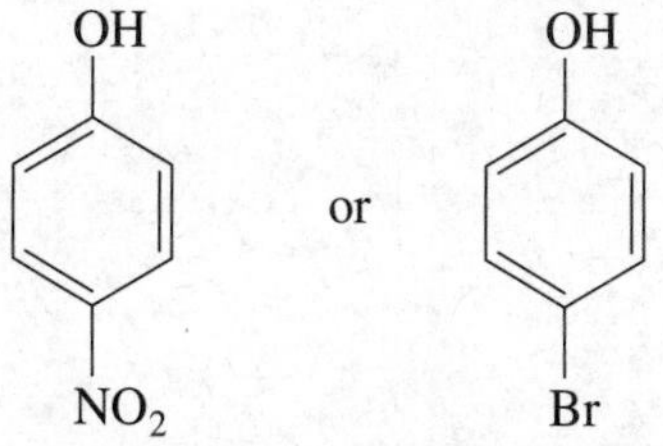
OH
NO2
or
OH
Br

d.
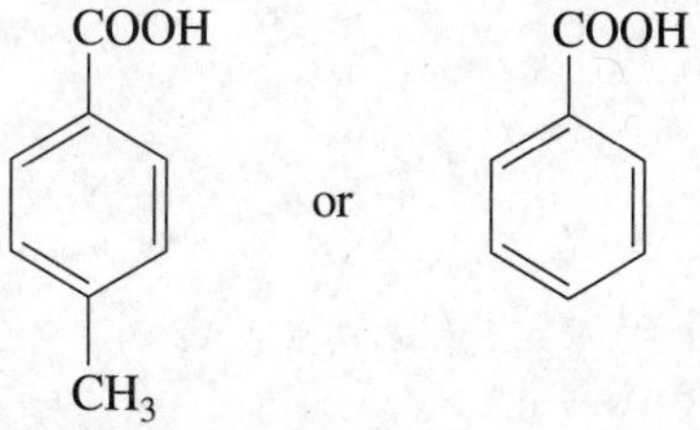
COOH
CH3
or
COOH

10. Which is more reactive in an electrophilic substitution reaction, *para*-bromonitrobenzene or *para*-bromoethylbenzene?

11. Give the major product(s) of each of the following reactions.

a.
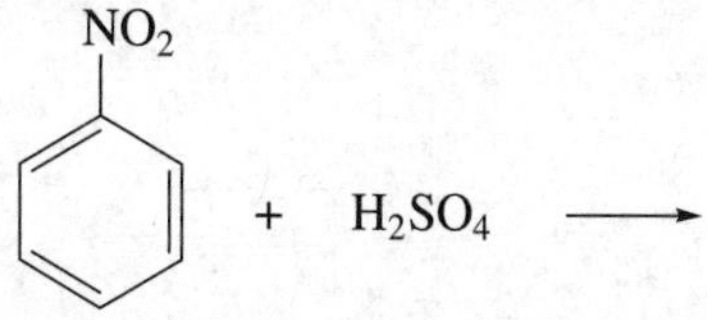
NO2
+ H2SO4

b.
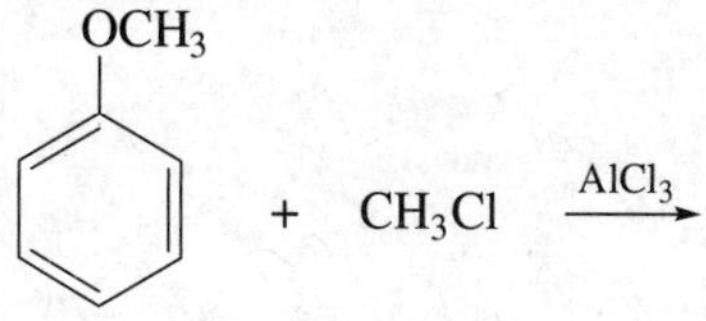
OCH3
+ CH3Cl
AlCl3

c.
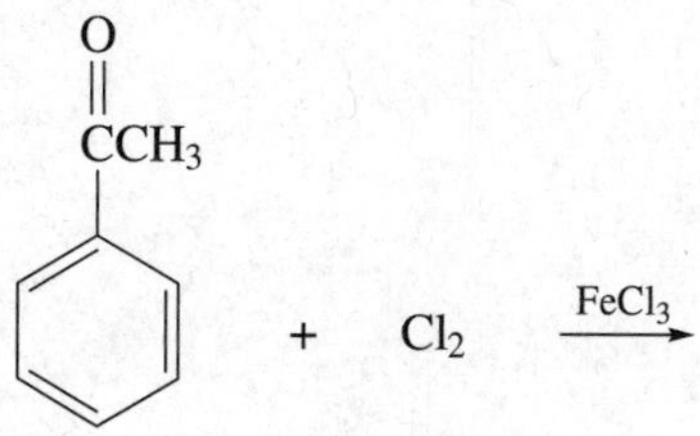
CCH3
+ Cl2
FeCl3

12. Indicate whether each of the following statements is true or false:

a. Benzoic acid is more reactive than benzene towards electrophilic aromatic substitution. T F

b. *para*-Chlorobenzoic acid is more acidic than *para*-methoxybenzoic acid. T F

c. A $CH{=}CH_2$ group is a meta director. T F

d. *para*-Nitroaniline is more basic than *para*-chloroaniline. T F

CHAPTER 9

Substitution and Elimination Reactions of Alkyl Halides

1. **a.** The rate is tripled.

b. The rate is one-half of the original rate.

2. 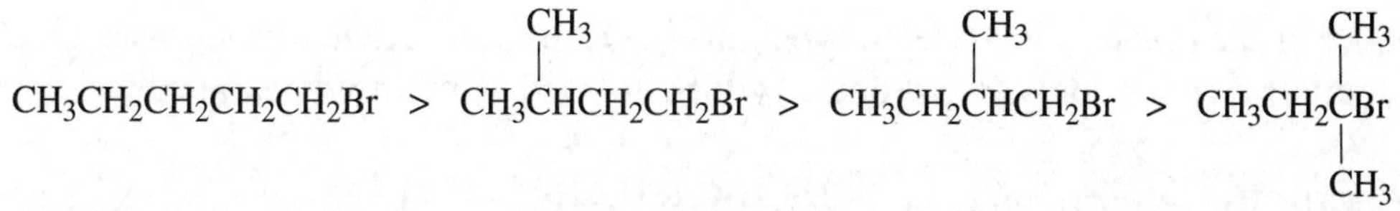

3. **a.** Solved in the text.

b. The S_N2 reaction of (*R*)-2-bromobutane with hydroxide ion will form (*S*)-2-butanol.

c. The S_N2 reaction of (*S*)-3-chlorohexane and with hydroxide ion will form (*R*)-3-hexanol.

d. The S_N2 reaction of 3-iodopentane (it does not have an asymmetric center) with hydroxide ion will form 3-pentanol (it does not have an asymmetric center).

4. Solved in the text.

5. **a.** $CH_3CH_2Br + HO^-$ — HO^- is a stronger nucleophile than H_2O.

b. 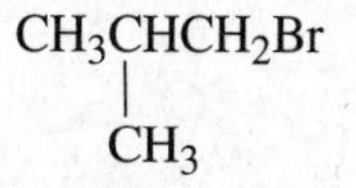 $+ HO^-$ — This alkyl halide has less steric hindrance.

c. $CH_3CH_2Br + I^-$ — Br^- is a weaker base than Cl^-, so Br^- is a better leaving group.

d. $CH_3CH_2CH_2I + HO^-$ — I^- is a weaker base than Br^-, so I^- is a better leaving group.

6. **a.** $CH_3CH_2Br + CH_3CH_2CH_2\ddot{\underset{..}{O}}:^- \longrightarrow CH_3CH_2\ddot{\underset{..}{O}}CH_2CH_2CH_3 + Br^-$

b. $CH_3CH_2Br + CH_3C{\equiv}C:^- \longrightarrow CH_3CH_2C{\equiv}CCH_3 + Br^-$

c. $CH_3CH_2Br + (CH_3)_3\ddot{N} \longrightarrow CH_3CH_2\overset{+}{N}(CH_3)_3 + Br^-$

d. $CH_3CH_2Br + CH_3CH_2\ddot{\underset{..}{S}}:^- \longrightarrow CH_3CH_2\ddot{\underset{..}{S}}CH_2CH_3 + Br^-$

7. 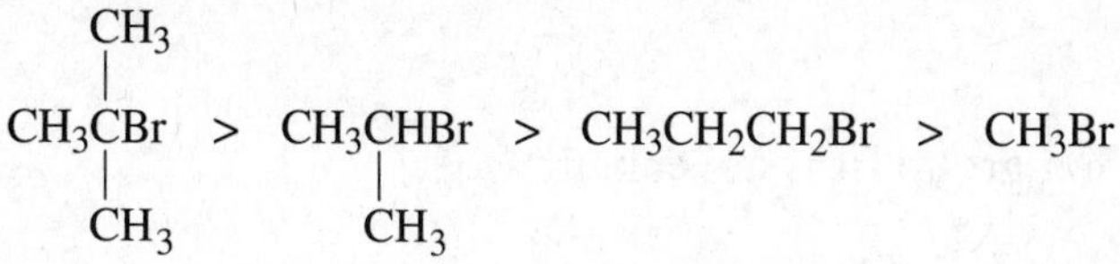

8. **a.** The product of the reaction is 2-methoxypentane. Because the leaving group was attached to an asymmetric center, two products will be formed, (*R*)-2-methoxypentane and (*S*)-2-methoxypentane.

$$CH_3CHCH_2CH_2CH_3 \; + \; CH_3OH \; \rightarrow \; CH_3CHCH_2CH_2CH_3 \; + \; Br^-$$
 | |
 Br OCH_3

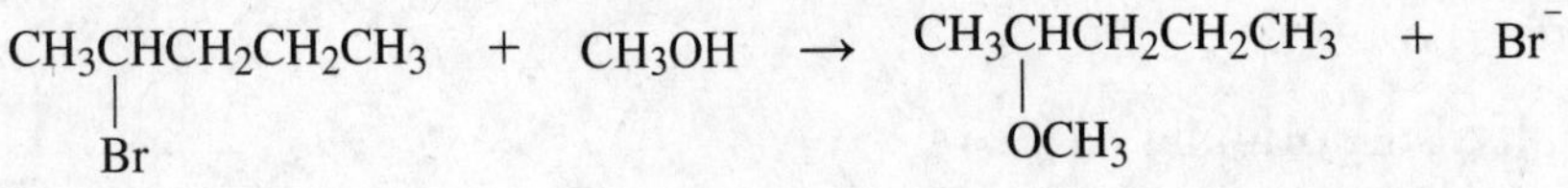

b. The product of the reaction is 3-methoxypentane. Because the leaving group was not attached to an asymmetric center, only one product will be formed, 3-methoxypentane.

$$CH_3CH_2CHCH_2CH_3 \; + \; CH_3OH \; \rightarrow \; CH_3CH_2CHCH_2CH_3 \; + \; Br^-$$
 | |
 Br OCH_3

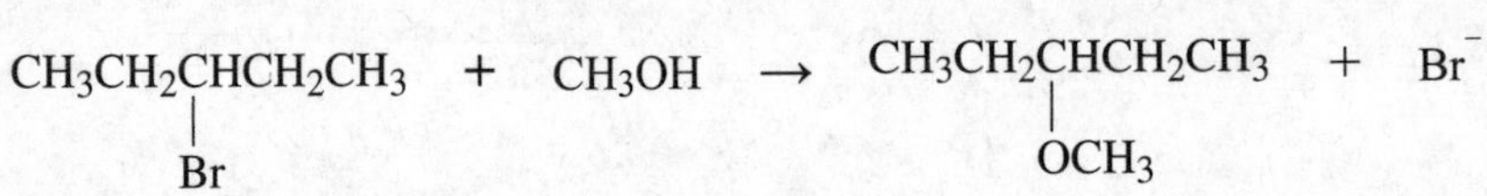

9. 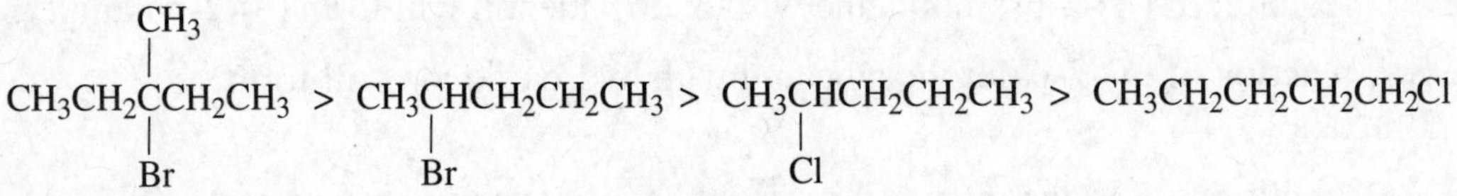

$$\underset{\underset{Br}{|}}{CH_3CH_2\overset{\overset{CH_3}{|}}{C}CH_2CH_3} \; > \; \underset{\underset{Br}{|}}{CH_3CHCH_2CH_2CH_3} \; > \; \underset{\underset{Cl}{|}}{CH_3CHCH_2CH_2CH_3} \; > \; CH_3CH_2CH_2CH_2CH_2Cl$$

10. **a.** $CH_3CH_2CH_2Br$ It has less steric hindrance.

b. 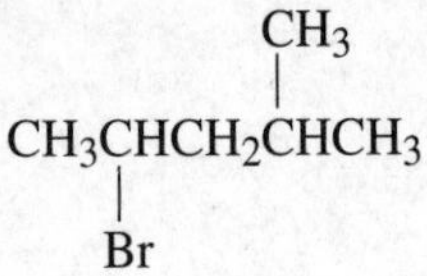 It has less steric hindrance.

c. 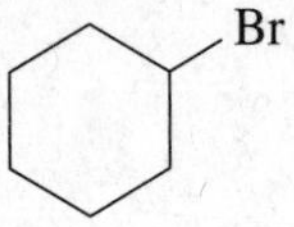 Br^- is a better leaving group (it is a weaker base)

d. 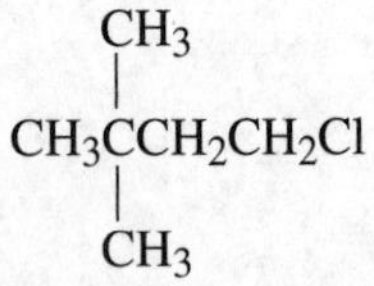 It has less steric hindrance.

11. **a.** 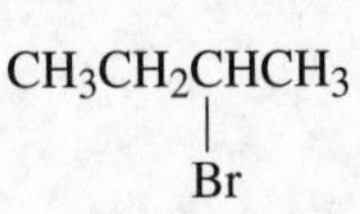 A secondary carbocation is more stable and therefore easier to form.

b. 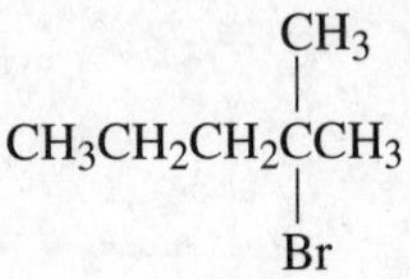 A tertiary carbocation is more stable and therefore easier to form.

c. 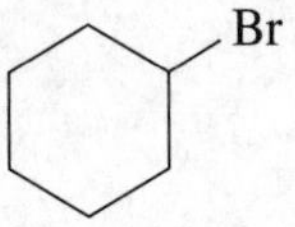 Br^- is a better leaving group (it is a weaker base)

d. Neither can undergo an S_N1 reaction, because primary carbocations are too unstable to be formed.

12. **a.**

1. The S_N2 reaction proceeds with back-side attack.

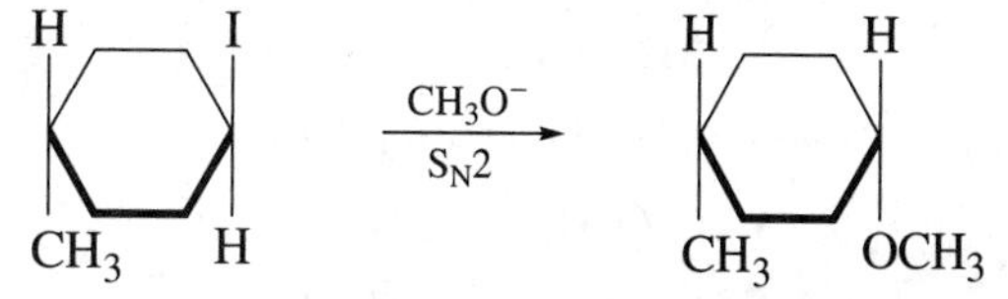

2.

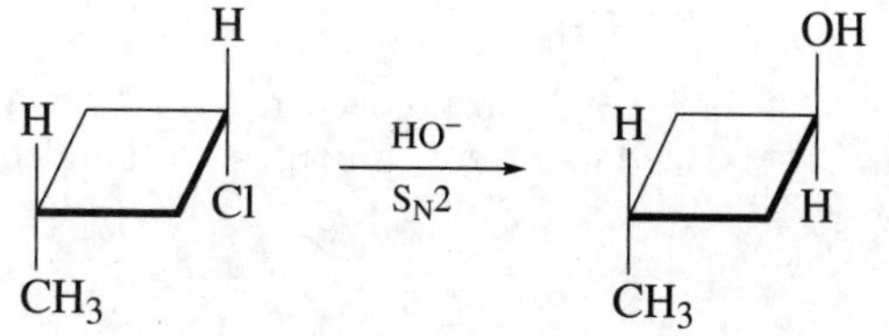

b.

1. The S_N1 reaction proceeds with both back-side and front-side attack.

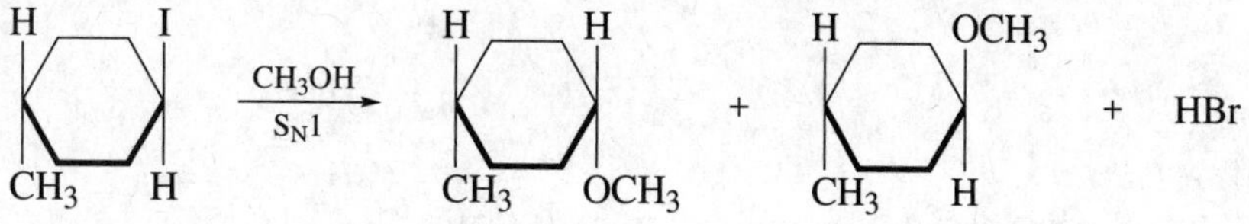

2.

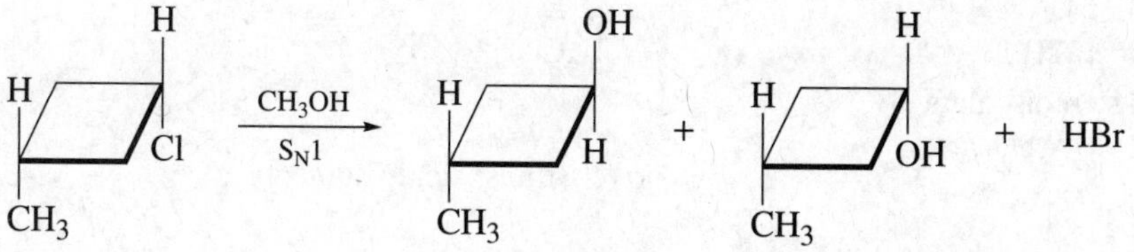

13. **a.** 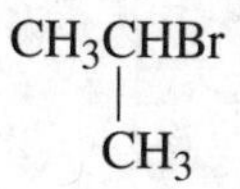 Because Br^- is a better leaving group than Cl^-

b. 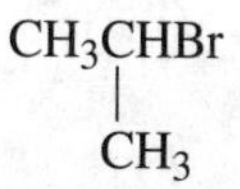 Because Br^- is a better leaving group than Cl^-

14. **a.** 2 is more reactive than 1

b. 2 is more reactive than 1

c. 2 is more reactive than 1

d. 1 is more reactive than 2

15. **a.** Solved in the text.

b. 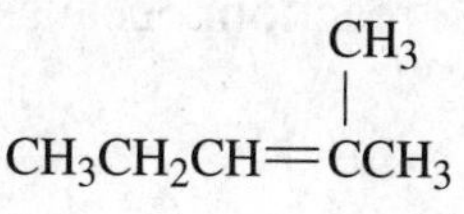

c. 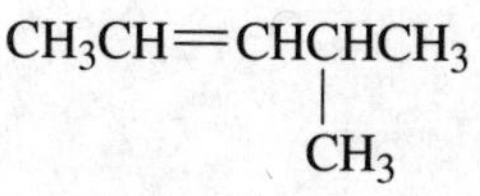

d. 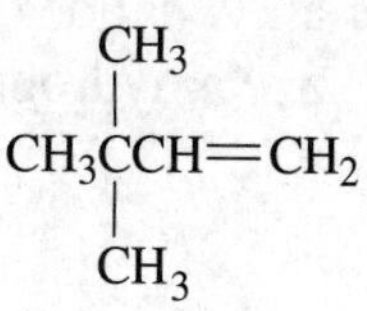

16.

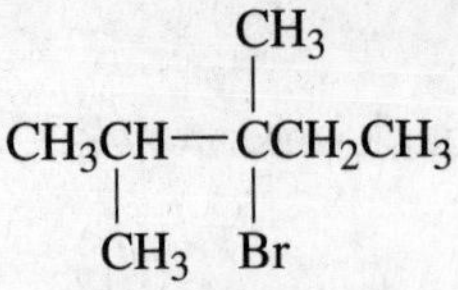

3-bromo-2,3-dimethylpentane

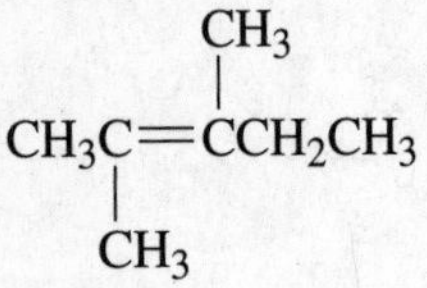 > 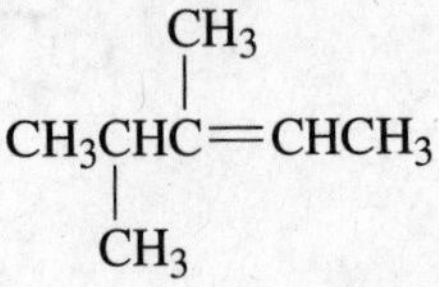 > 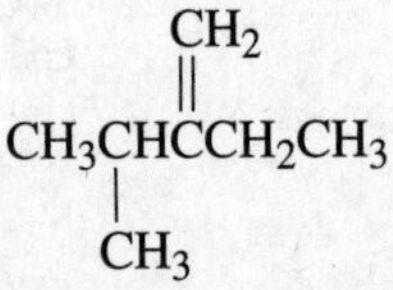

Four alkyl substituents are bonded to the sp^2 carbons.

major product

Three alkyl substituents are bonded to the sp^2 carbons.

Two alkyl substituents are bonded to the sp^2 carbons.

17. Only the major elimination product of **a** and **c** can exist as stereoisomers.

a. Solved in the text.

c. 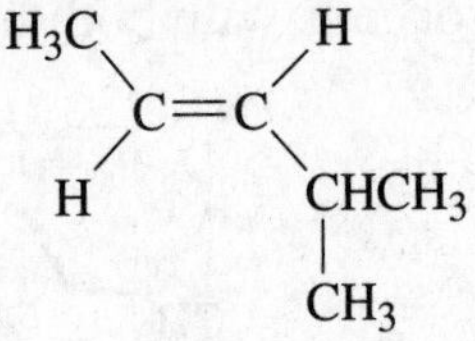

b. 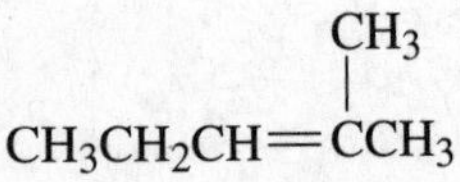

no (cis-trans) stereoisomers

d. 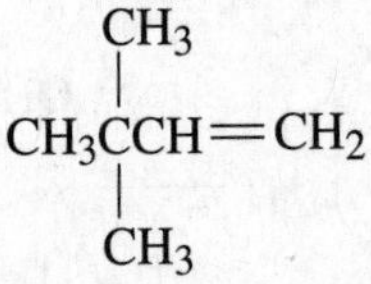

no (cis-trans) stereoisomers

18. **a.** 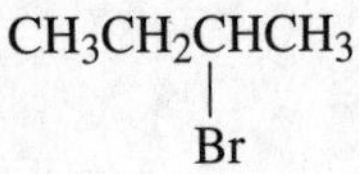 It forms a more stable akene.

b. 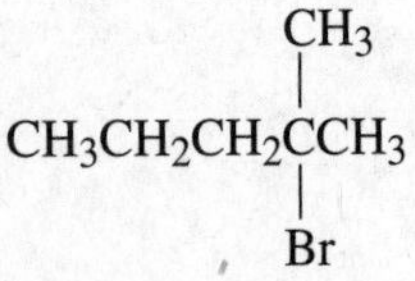 It forms a more stable alkene.

c. 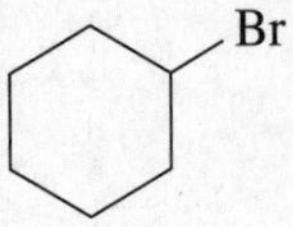 Br^- is a better leaving group (it is a weaker base)

d. 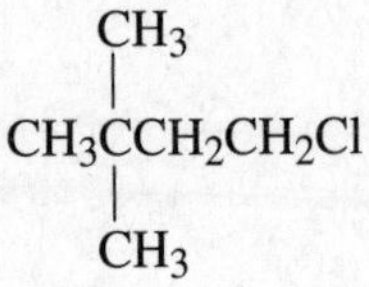 The other alkyl halide cannot undergo an E2 reaction, because it does not have a hydrogen on a β-carbon.

19. **a.** 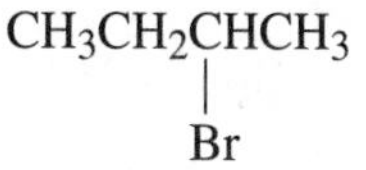

A secondary carbocation is more stable and therefore easier to form.

b. 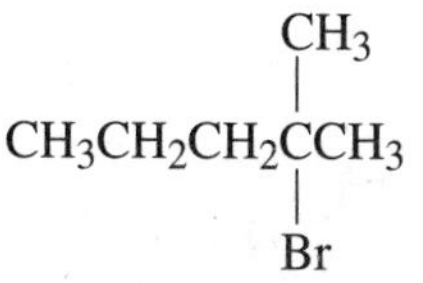

A tertiary carbocation is more stable and therefore easier to form.

c. 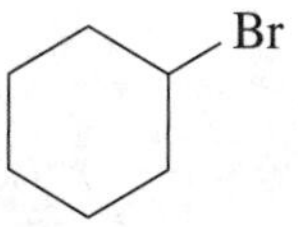

Br^- is a better leaving group (it is a weaker base)

20. **a.** The high concentration of a good nucleophile causes the substitution reaction to be an S_N2 reaction. Therefore, the product of the reaction will have the inverted configuration compared to the configuration of the reactant.

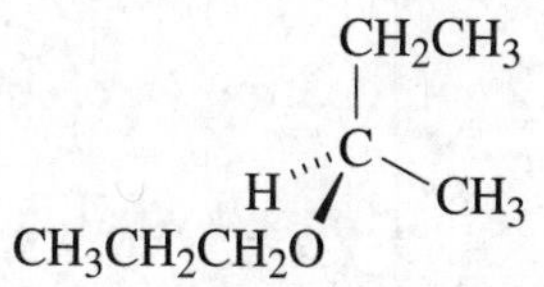

b. The high concentration of a good nucleophile causes the substitution reaction to be an S_N2 reaction. Therefore, the product of the reaction will have the inverted configuration compared to the configuration of the reactant.

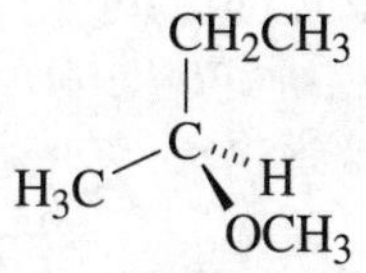

c. The poor nucleophile causes the substitution reaction to be an S_N1 reaction. Therefore, the product of reaction will have both the inverted configuration and the same configuration as that of the reactant.

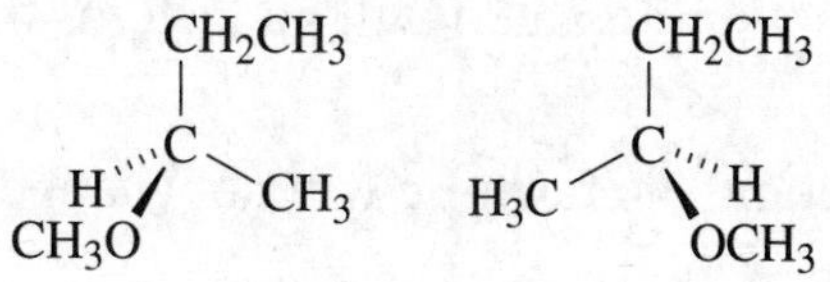

21. **a.** Solved in the text. **b.** (*E*)-2-butene **c.** (*E*)-2-butene

22. **a.** Mainly substitution, because the alkyl halide is primary.

b. The alkyl halide is primary, but substitution has more steric hindrance than the example in "a," so substitution will be slower than usual, allowing both substitution and elimination to occur.

c. Substitution and elimination because the alky halide is secondary.

d. Only elimination because the alkyl halide is tertiary.

23. The reaction conditions (a strong nucleophile) favor $S_N2/E2$ reactions.

a. It is a secondary alkyl halide.

Both a substitution and an elimination product will be formed. Because the alkyl chloride has the *S* configuration, the substitution product will have the *R* configuration.

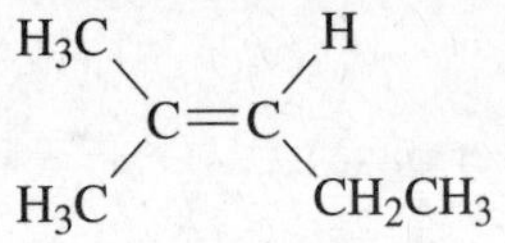

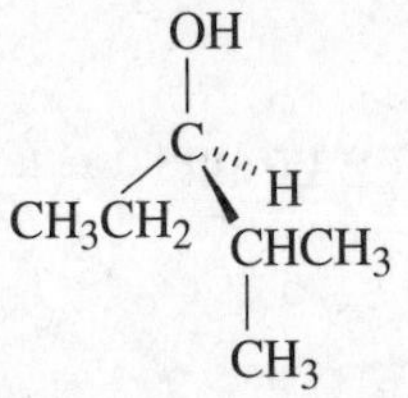

(*R*)-2-methyl-3-pentanol

b. It is a tertiary alkyl halide.

Only an elimination product will be formed

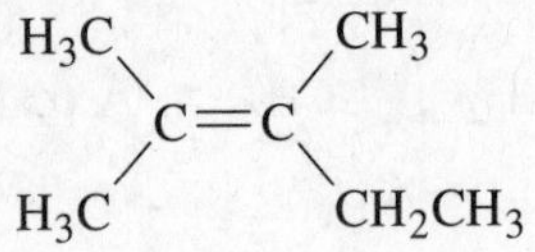

24. **a.** No products, because a primary carbocation is too unstable to be formed.

b. No products, because a primary carbocation is too unstable to be formed.

c. Substitution and elimination because the alky halide is secondary.

d. Substitution and elimination because the alkyl halide is tertiary.

25. Because the rate of an S_N1 reaction is not affected by increasing the concentration of the nucleophile, whereas the rate of an S_N2 reaction is increased when the concentration of the nucleophile is increased, you first have to determine whether the reactions are S_N1 or S_N2 reactions.

a is an S_N2 reaction because the configuration of the product is inverted compared with the reactant.

b is an S_N2 reaction because the reactant is a primary alkyl halide.

c is an S_N1 reaction because the reactant is a tertiary alkyl halide.

Because they are S_N2 reactions, **a** and **b** will go faster if the concentration of the nucleophile is increased.

Because it is an S_N1 reaction, the rate of **c** will not change if the concentration of the nucleophile is increased.

26. **a.** $CH_3Br + HO^- \xrightarrow{DMSO} CH_3OH + Br^-$

Because one of the reactants is charged, the reaction will take place more rapidly in an aprotic polar solvent.

b. $CH_3Br + NH_3 \longrightarrow CH_3\overset{+}{N}H_3 + Br^-$

NH_3 is a better nucleophile than H_2O.

c. $CH_3Br + NH_3 \xrightarrow{EtOH} CH_3\overset{+}{N}H_3 + Br^-$

Because the reactants are not charged, the reaction will take place more rapidly in a protic polar solvent.

27.

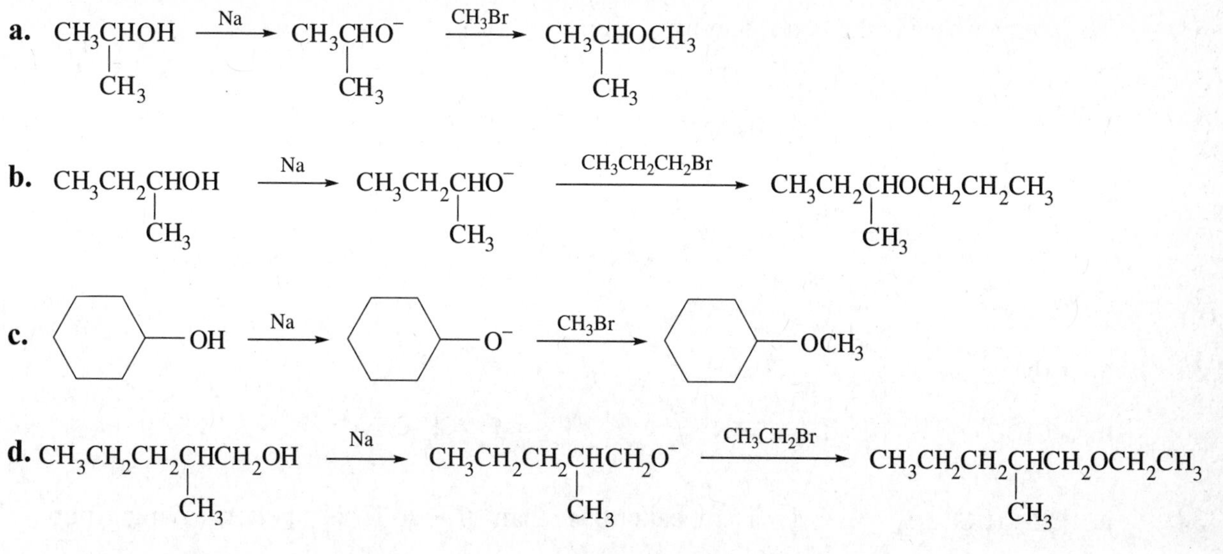

28.

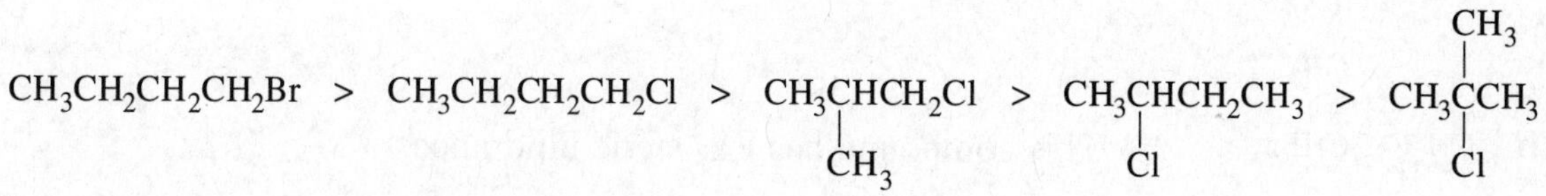

29. **a.** $CH_3Br + CH_3O \longrightarrow CH_3OCH_3 + Br^-$

CH_3O^- is a better nucleophile than CH_3OH.

b. $CH_3I + NH_3 \longrightarrow CH_3\overset{+}{N}H_3 + I^-$

I^- is a better leaving group than Cl^-.

c. $CH_3Cl + CH_3NH_2 \longrightarrow CH_3\overset{+}{N}H_2CH_3 + Cl^-$

CH_3NH_2 is a better nucleophile than CH_3OH.

30. **a.** CH_3OH **d.** CH_3SH

b. CH_3NH_2 **e.** $CH_3OCH_2CH_3$

c. CH_3SH **f.** $CH_3\overset{+}{N}H_2CH_3$

(Notice that the product in "c" is not protonated because its pK_a is ~ −7; the product in "f" is protonated because its pK_a is ~11).

31. The stronger base is the better nucleophile.

a. HO^- **c.** $CH_3CH_2O^-$

b. $^-NH_2$ **d.**

32. The weaker base is the better leaving group.

a. H_2O

b. NH_3

c. $CH_3\overset{\overset{\displaystyle O}{\|}}{C}O^-$

d. 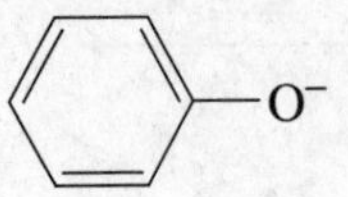

33. **a.** HO^-

b. CH_3O^-

c. $CH_3CH_2S^-$

d. $^-C{\equiv}N$

e. $CH_3\overset{\overset{\displaystyle O}{\|}}{C}O^-$

f. $CH_3C{\equiv}C^-$

34. **a.** $CH_3CH_2\underset{\underset{\displaystyle I}{|}}{C}HCH_3$ — I^- is a weaker base than Br^-, so I^- is a better leaving group.

b. $CH_3CH_2\overset{\overset{\displaystyle CH_3}{|}}{C}HBr$ — This compound has less steric hindrance.

c. $CH_3CH_2\overset{\overset{\displaystyle CH_3}{|}}{C}HCH_2Br$ — A primary carbon is less sterically hindered than a secondary carbon.

d. 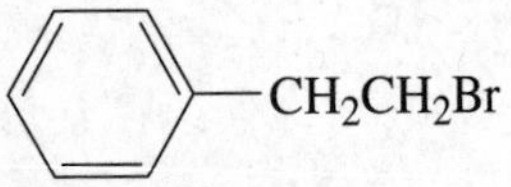 — A primary carbon is less sterically hindered than a secondary carbon.

35. **a.** $CH_3CH_2\underset{\underset{\displaystyle I}{|}}{C}HCH_3$ — I^- is a weaker base than Br^-, so I^- is a better leaving group.

b. The two compounds are equally reactive.

c. $CH_3CH_2CH_2\overset{\overset{\displaystyle CH_3}{|}}{C}HBr$ — A secondary carbocation is more stable than a primary carbocation.

d. 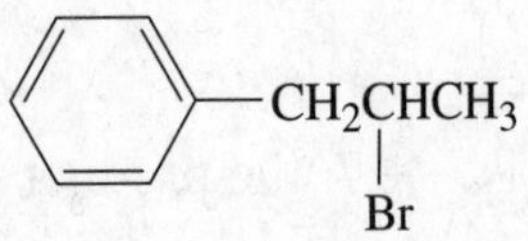 — A secondary carbocation is more stable than a primary carbocation.

36. **a.** An S_N2 reaction forms the isomer with the inverted configuration.

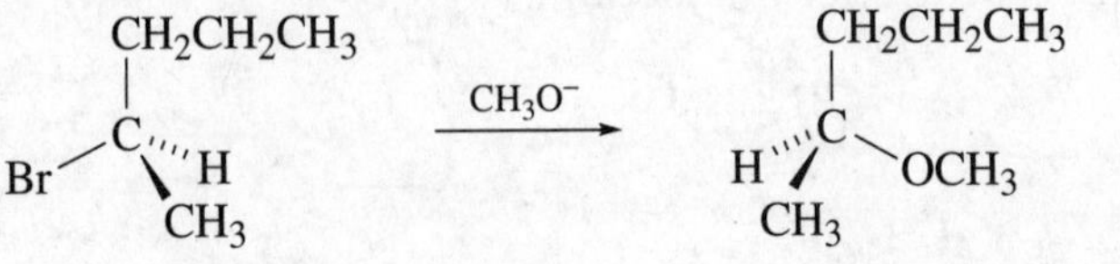

(*R*)-2-bromopentane (*S*)-2-methoxypentane

b. An S_N1 reaction forms a pair of enantiomers.

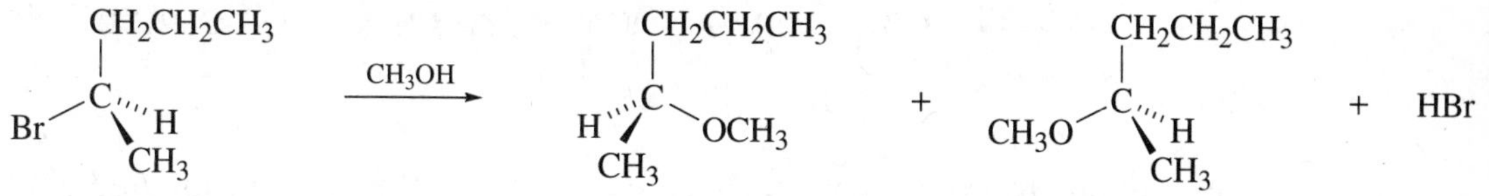
CH2CH2CH3
Br
H
CH3
CH3OH
CH2CH2CH3
H
OCH3
CH3
+
CH2CH2CH3
CH3O
H
CH3
+ HBr

(*R*)-2-bromopentane (*S*)-2-methoxypentane (*R*)-2-methoxypentane

c. An S_N2 reaction forms the isomer with the inverted configuration (that is, the one that results from back-side attack.).

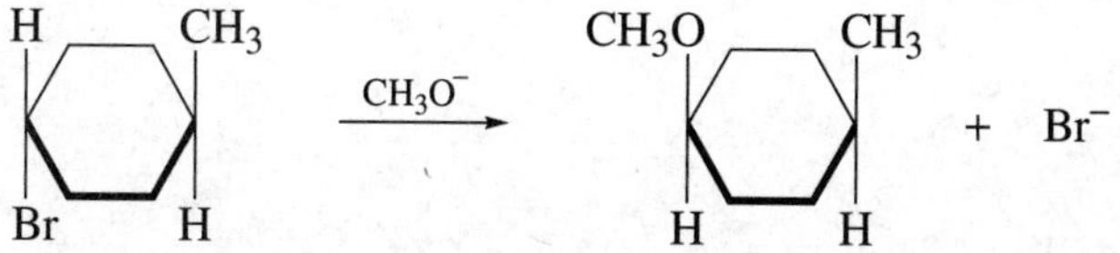
H
CH3
Br
H
CH3O⁻
CH3O
CH3
H
H
+ Br⁻

d. An S_N1 reaction produces products from both back-side and front-side attack.

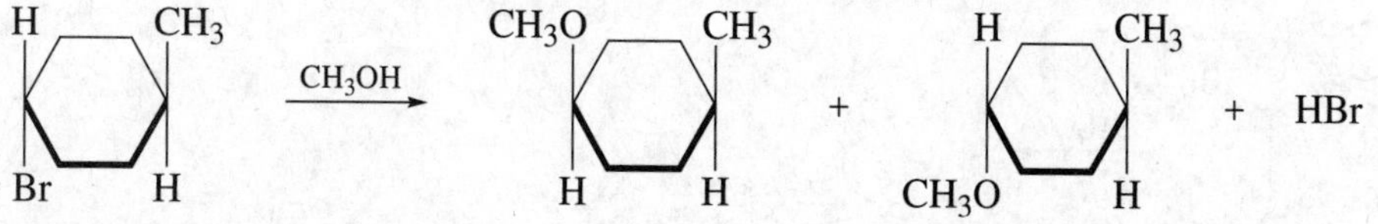
H
CH3
Br
H
CH3OH
CH3O
CH3
H
H
+
H
CH3
CH3O
H
+ HBr

e. 3-Bromo-3-methylpentane does not have an asymmetric center, so there are no stereoisomers.

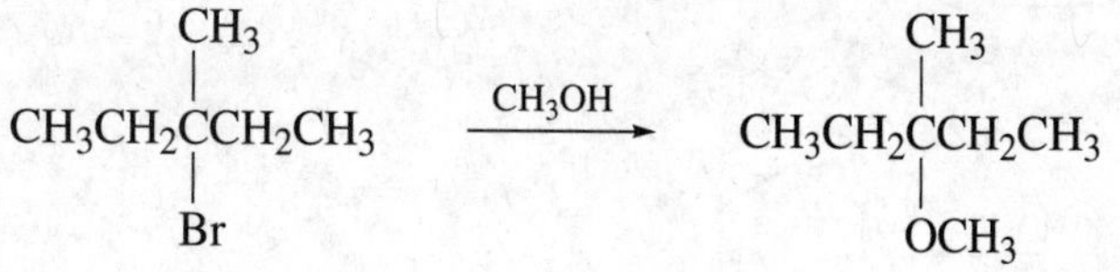
CH3
CH3CH2CCH2CH3
Br
CH3OH
CH3
CH3CH2CCH2CH3
OCH3

37. **a.** 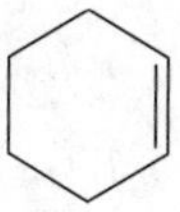 **b.** 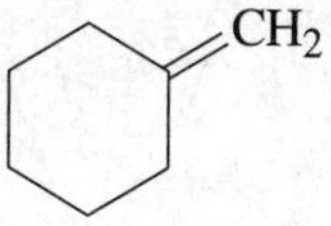
CH2
c. 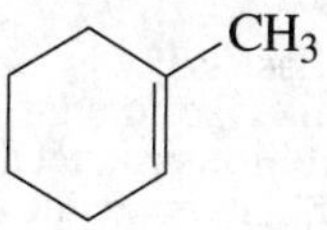
CH3

38. **a.** 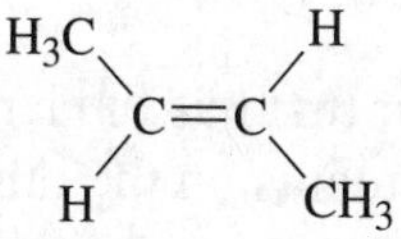
H3C
H
C=C
H
CH3
b. 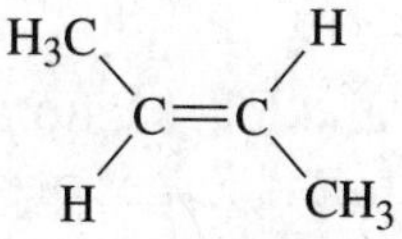
H3C
H
C=C
H
CH3
c. 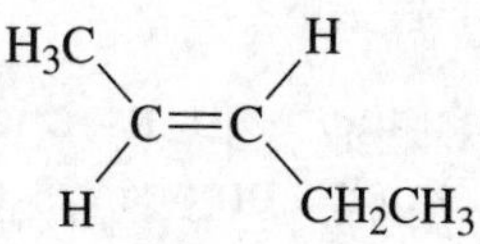
H3C
H
C=C
H
CH2CH3

39. **a.** 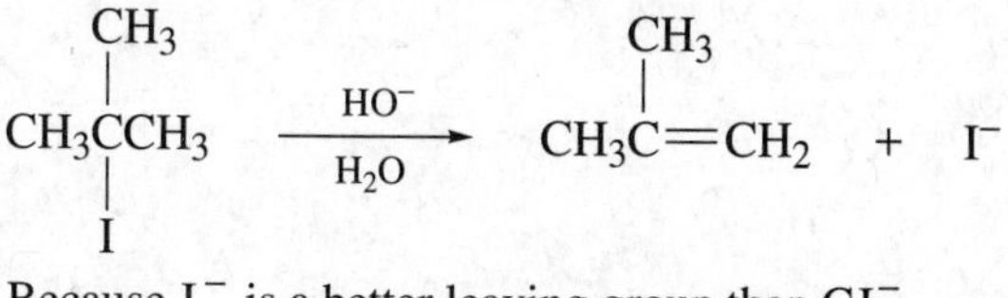
CH3
CH3CCH3
I
HO⁻
H2O
CH3
CH3C=CH2 + I⁻

Because I^- is a better leaving group than CI^-.

b. 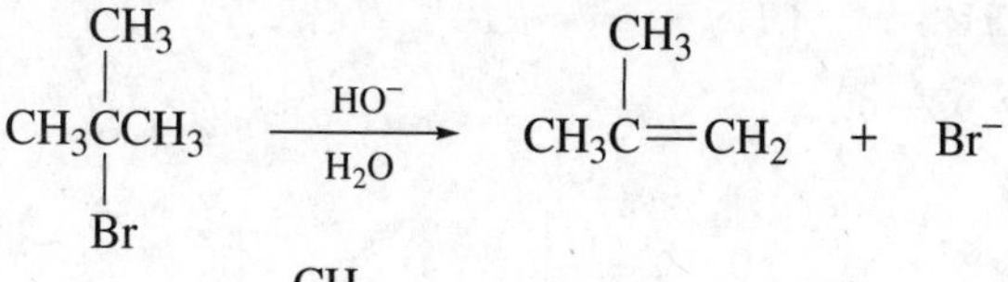
CH3
CH3CCH3
Br
HO⁻
H2O
CH3
CH3C=CH2 + Br⁻

Because $CH_3\overset{CH_3}{\overset{|}{C}}{=}CH_2$ is more stable than $CH_3CH{=}CH_2$, and being more stable, it is easier to form.

40.   **a.** 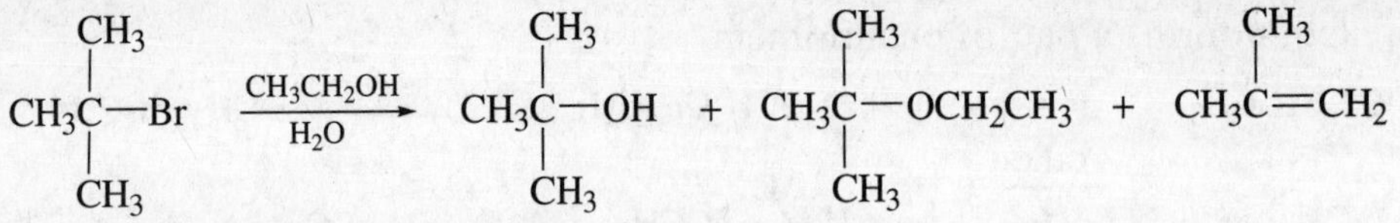

$$CH_3C(CH_3)_2-Br \xrightarrow[H_2O]{CH_3CH_2OH} CH_3C(CH_3)_2-OH + CH_3C(CH_3)_2-OCH_2CH_3 + CH_3C(CH_3)=CH_2$$

b.   Both compounds form the same carbocation. Because the substitution and elimination products are formed from the carbocation, both compounds form the same products.

41.   **a.** 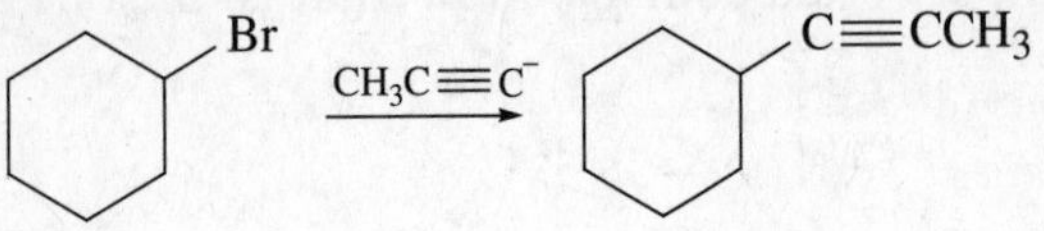

b. 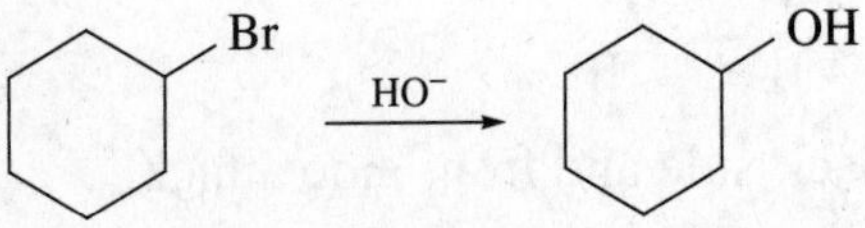

c. 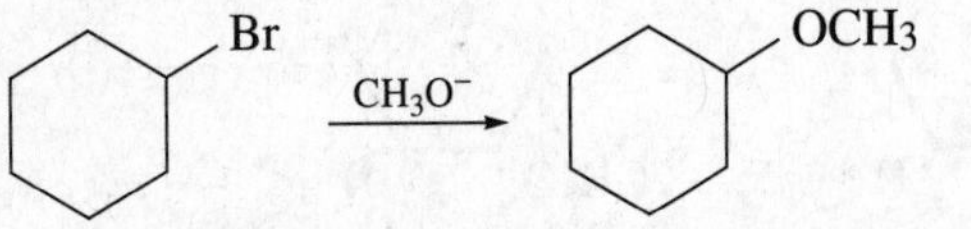

d. 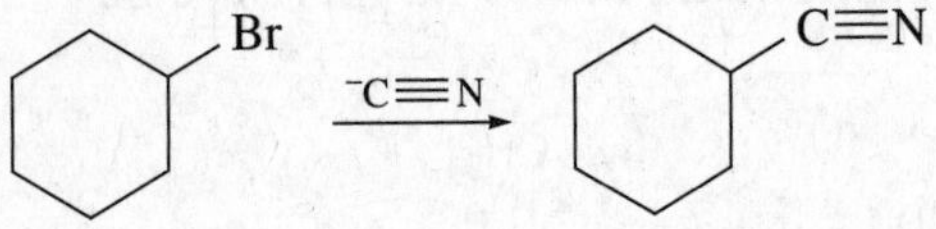

e. 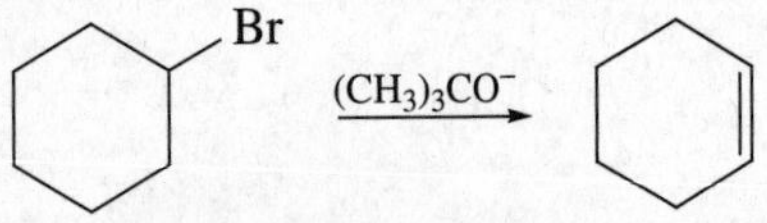

A bulky base is used
to minimize the amount
of substitution product and,
therefore, maximize the amount
of elimination product.

42.   Notice that the major product in both E2 and E1 reactions is the isomer that has the larger of the two substituents on one sp^2 carbon trans to the larger of the two substituents on the other sp^2 carbon.

a. 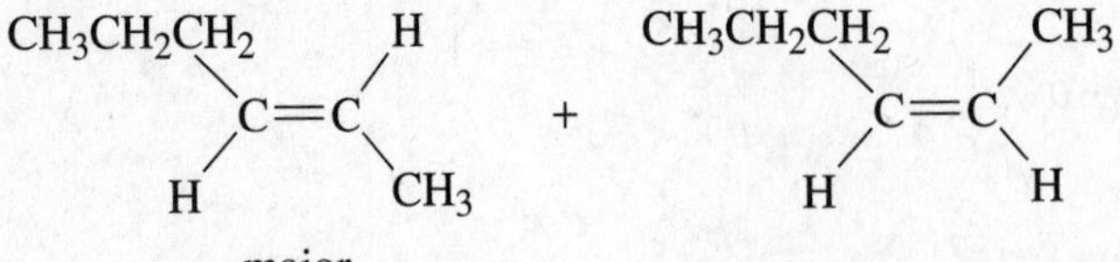

b. 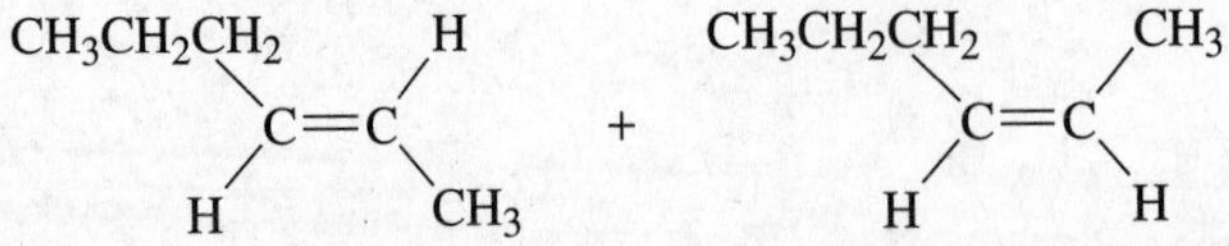

c. 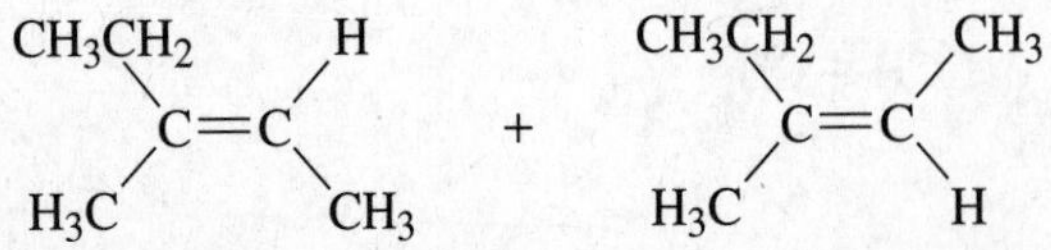

d. 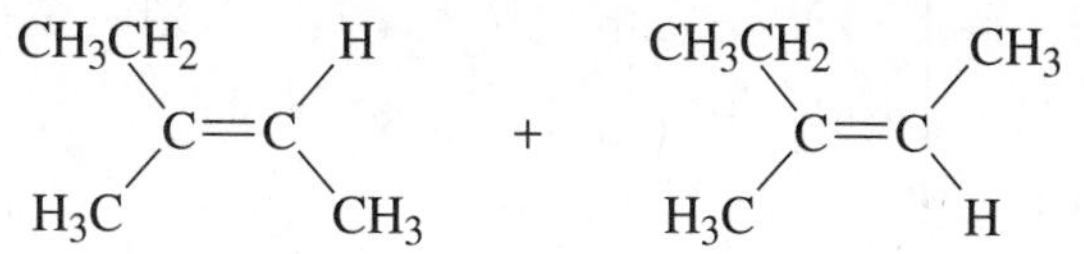

major

43. **a.** The reaction with quinuclidine was faster because quinuclidine is less sterically hindered as a result of the substituents on the nitrogen being pulled back into a ring structure.

b. The reaction with quinuclidine had the larger rate constant because the faster the reaction, the larger the rate constant.

44. **a.** 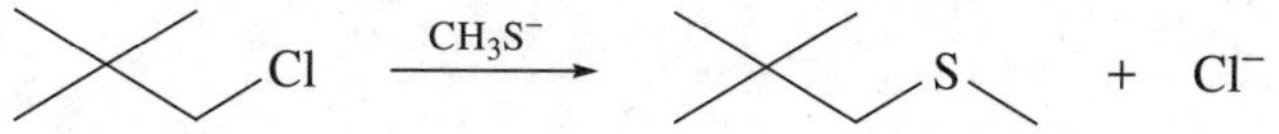

The nucleophile is less sterically hindered.

b. 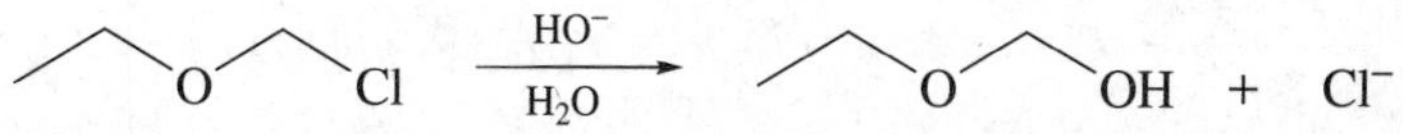

The electron - withdrawing oxygen increases the electrophilicity of the carbon (makes it more positive), so the nucleophile attacks it more readily.

45. **a.** 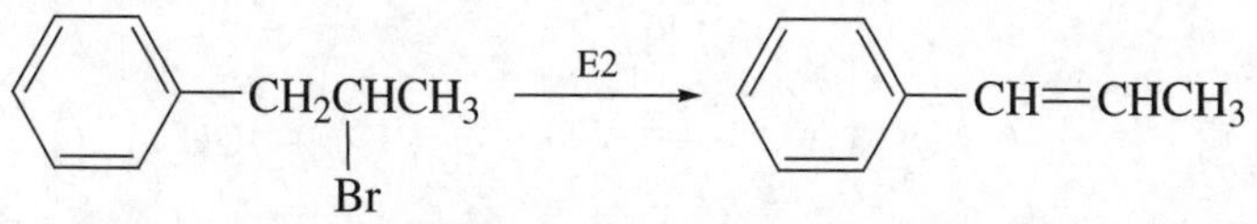

more reactive

The double bond that is formed is conjugated with the benzene ring. A conjugated double bond is more stable than the isolated double bond that would be formed by the other alkyl halide; the more stable the double bond, the easier it is to form.

b. 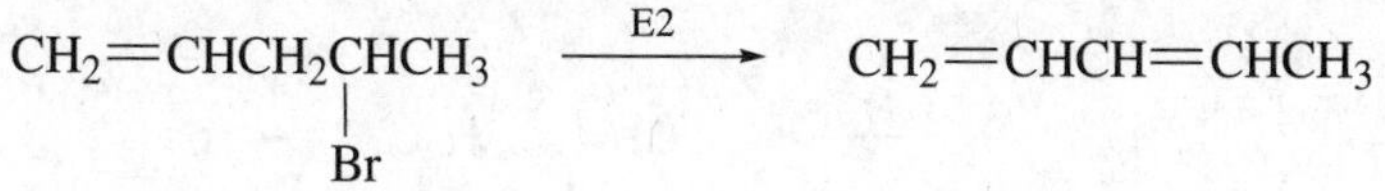

more reactive

This compound is more reactive, because the new double bond is conjugated with the other double bond.

46. Methoxide ion will be a better nucleophile in DMSO because DMSO, unlike methanol, does not have a hydrogen bonded to an oxygen that can surround methoxide ion and decrease its nucleophilicity.

47.

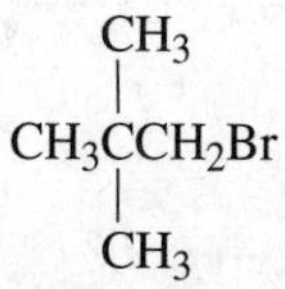

1-bromo-2,2-dimethylpropane

a. The bulky *tert*-butyl substituent blocks the backside of the carbon bonded to the bromine to nucleophilic attack, making an S_N2 reaction difficult. An S_N1 reaction is difficult because the carbocation formed when the bromide ion departs would be an unstable primary carbocation.

b. It cannot undergo an E2 reaction, because the β-carbon is not bonded to a hydrogen. It cannot undergo an E1 reaction, because that would require formation of a primary carbocation.

48. The stereoisomer obtained in greatest yield is the one with the larger of the two substituents on one sp^2 carbon trans to the larger of the two substituents on the other sp^2 carbon.

a. 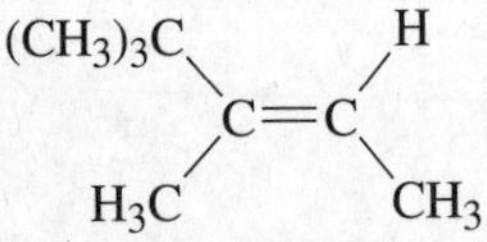

b. 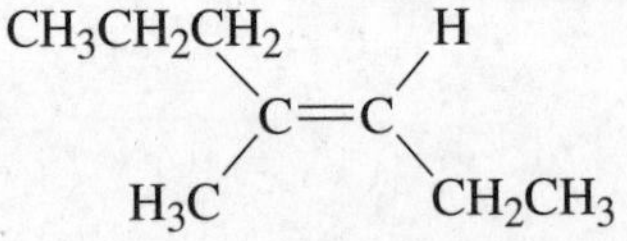

49. The second reaction will give a better yield of cyclopentyl methyl ether because methyl bromide can form only the desired substitution product. In contrast, cyclopentyl bromide, the alkyl halide in the first reaction, forms an elimination product in addition to the desired substitution product.

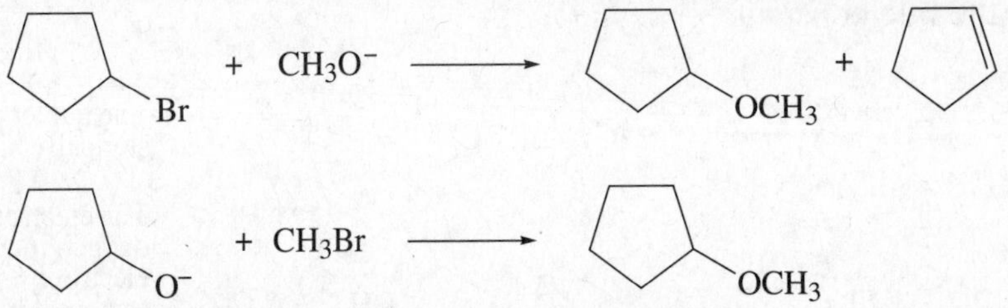

50. The predominant product is the elimination product because tertiary alkyl halides react with nucleophiles under $S_N2/E2$ conditions to form an elimination product and little substitution product.

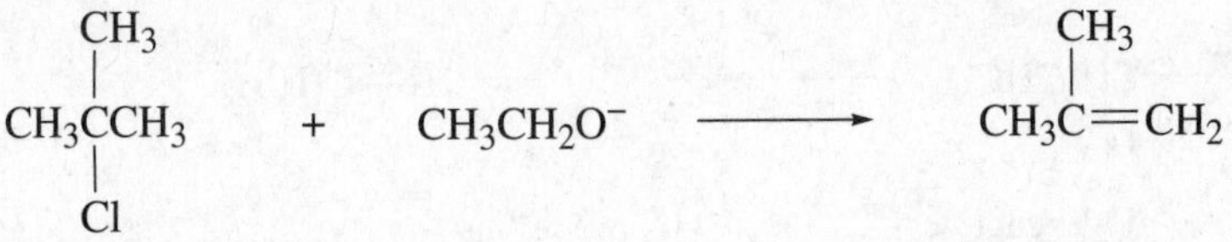

2-chloro-2-methylpropane   ethoxide ion   predominant product

Rather than a tertiary alkyl halide and a primary alkoxide ion, he should have used a primary alkyl halide and a tertiary alkoxide ion.

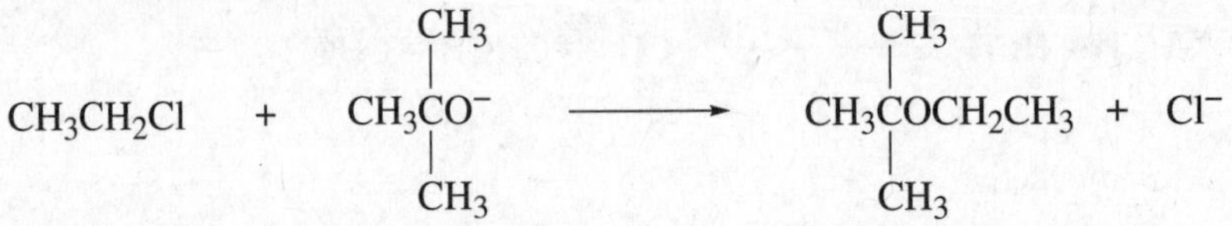

ethyl chloride   *tert*-butoxide ion   2-ethoxy-2-methylpropane

51. The allylic carbocation formed by 3-bromocyclohexene is stabilized by electron delocalization. Thus it is more stable than the carbocation formed by bromocyclohexane. Therefore, 2-bromocyclohexene is more reactive in an E1 reaction, because the more stable the carbocation, the easier it can be formed.

52.

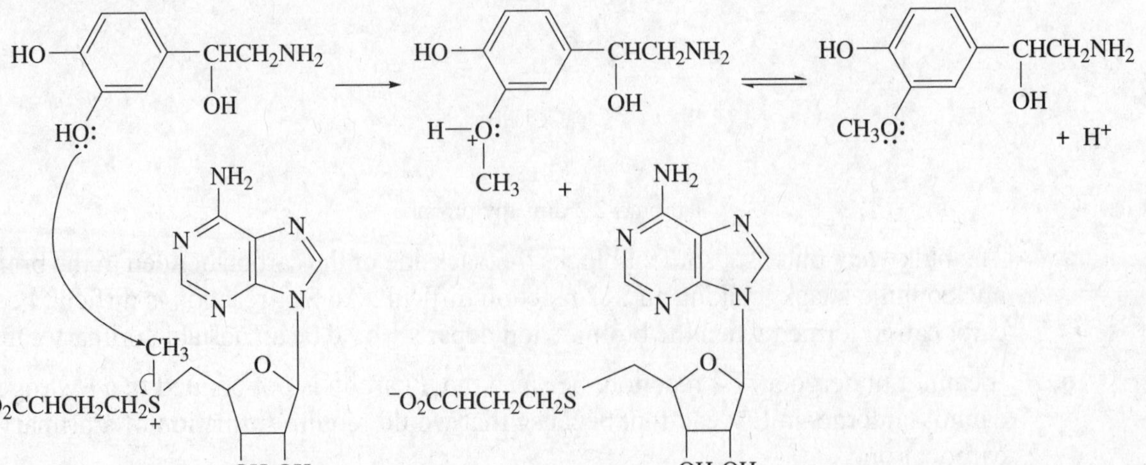

53. **a.**

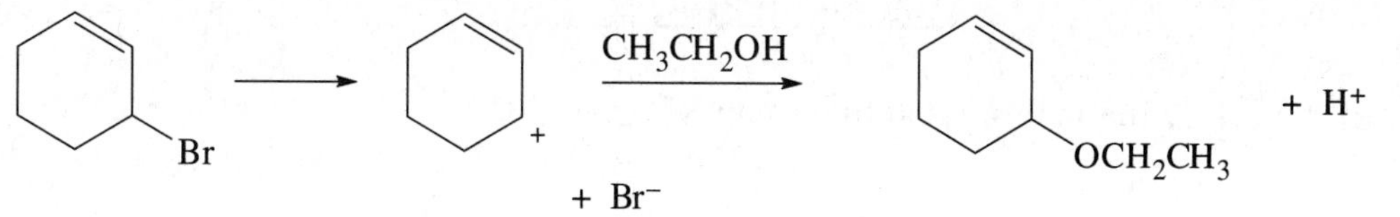

b.

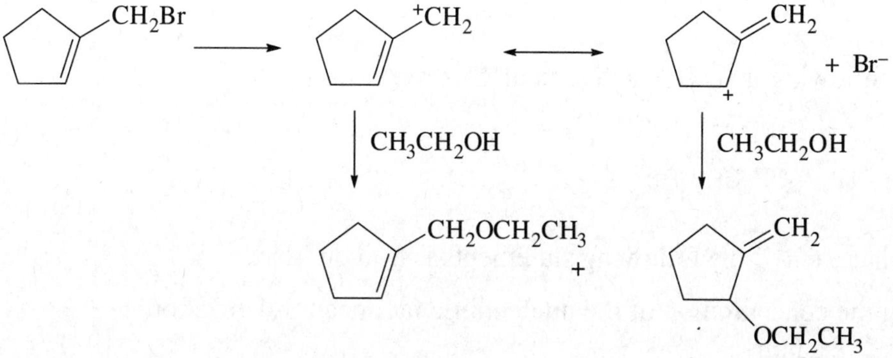

c.

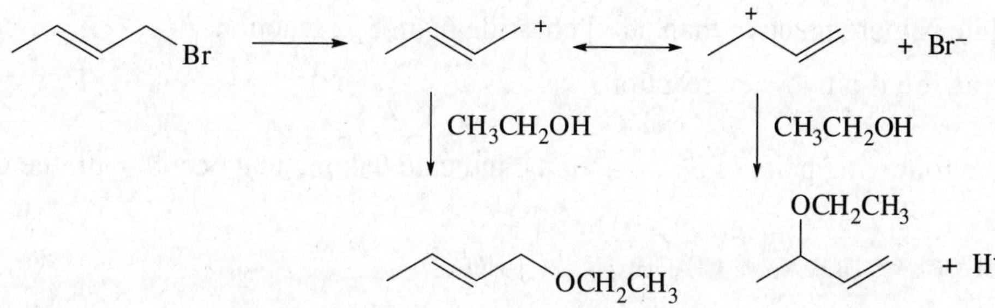

54. **a.** $CH_3CH_2CH_2CH_2Br \xrightarrow{NH_3} CH_3CH_2CH_2CH_2\overset{+}{N}H_3 \xrightarrow{HO^-} CH_3CH_2CH_2CH_2NH_2$

b.

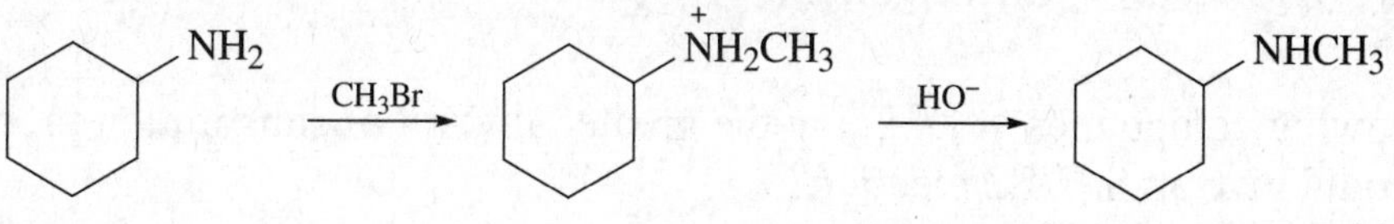

c.

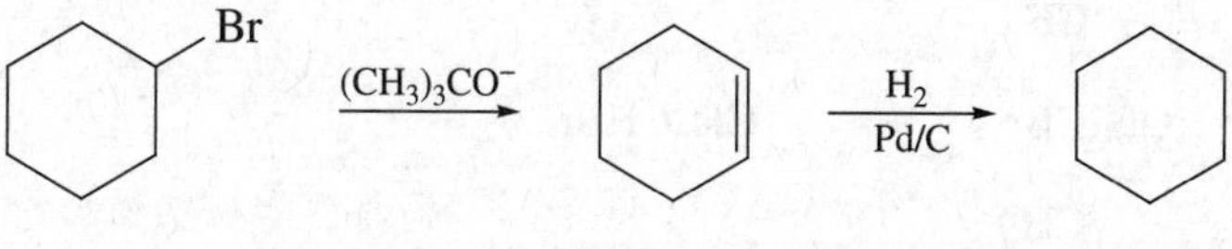

55. **a.**

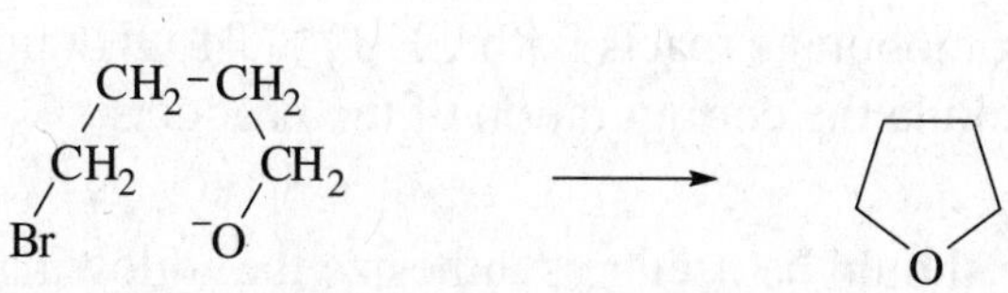

b.

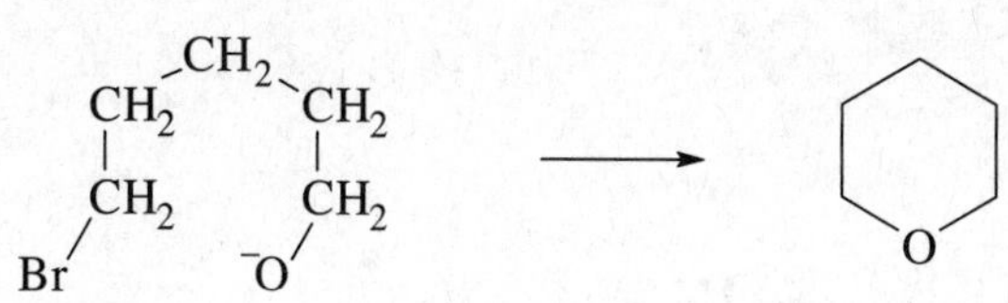

Chapter 9 Practice Test

1. Which of the following is more reactive in an S_N1 reaction?

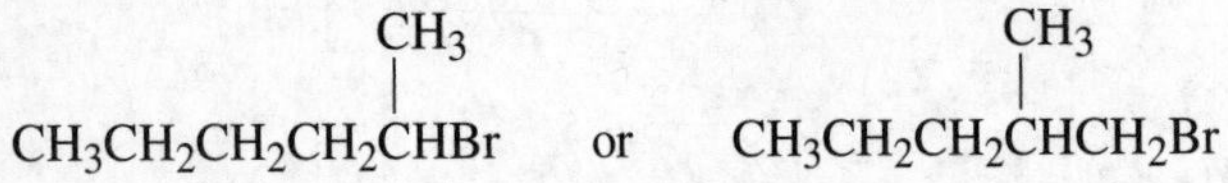

2. Which of the following is more reactive in an S_N2 reaction?

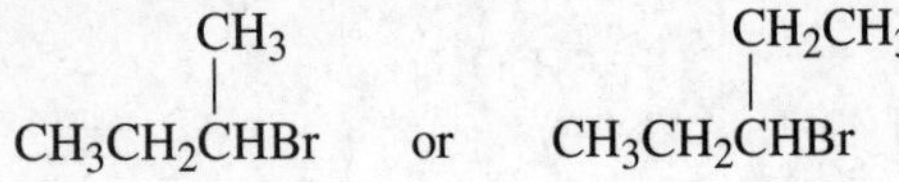

3. Indicate whether each of the following statements is true or false:

a. Increasing the concentration of the nucleophile favors an S_N1 reaction over an S_N2 reaction. T F

b. Ethyl iodide is more reactive than ethyl chloride in an S_N2 reaction. T F

c. An S_N2 reaction is a two-step reaction. T F

4. For each of the following pairs of S_N2 reactions, indicate the one that occurs with the greater rate constant:

a. 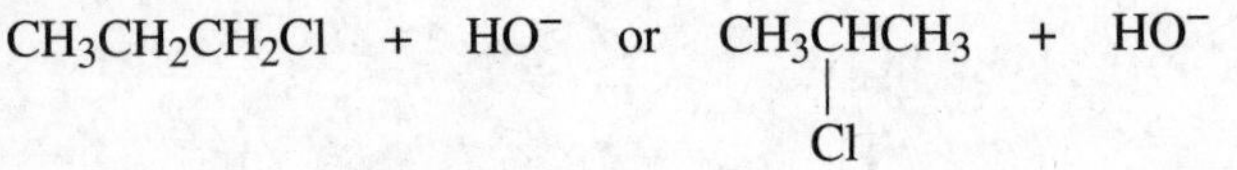

b. $CH_3CH_2CH_2Cl + HO^-$ or $CH_3CH_2CH_2I + HO^-$

c. $CH_3CH_2CH_2Br + HO^-$ or $CH_3CH_2CH_2Br + H_2O$

5. Which of the following compounds would give the greater amount of substitution product under conditions that would give an S_N2/E2 reaction?

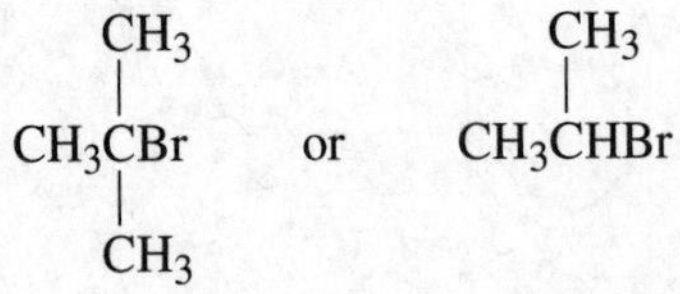

6. What products are obtained when (*R*)-2-bromobutane reacts with CH_3O^-/CH_3OH under conditions that favor S_N1/E1 reactions? Include the configuration of the products.

7. What alkoxide ion and what alkyl bromide should be used to synthesize the following ethers?

a. 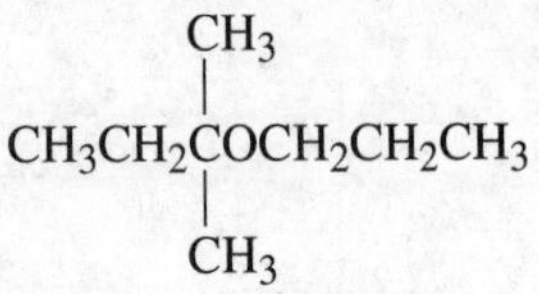

b. 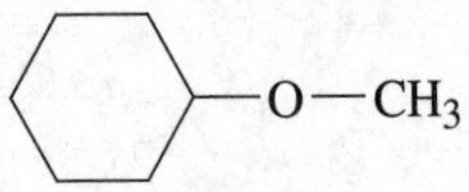

8. For each of the following pairs of E2 reactions, indicate the one that occurs with the greater rate constant:

a. $CH_3CH_2CH_2Cl$ + HO^- or 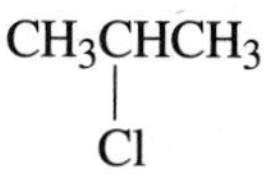 + HO^-

b. $CH_3CH_2CH_2Cl$ + HO^- or $CH_3CH_2CH_2I$ + HO^-

c. $CH_3CH_2CH_2Br$ + HO^- or $CH_3CH_2CH_2Br$ + H_2O

9. Give the major product of an E2 reaction of each of the following compounds with hydroxide ion:

a. 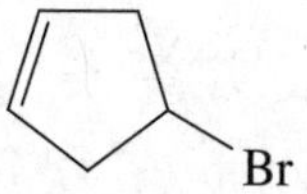
Br

b. 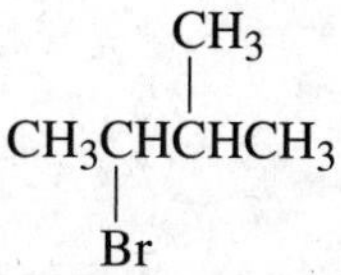
CH3
CH3CHCHCH3
Br

CHAPTER 10

Reactions of Alcohols, Amines, Ethers, and Epoxides

1. CH_3OH
common = methyl alcohol
systematic = methanol

CH_3CH_2OH
common = ethyl alcohol
systematic = ethanol

$CH_3CH_2CH_2OH$
common = propyl alcohol or *n*-propyl alcohol
systematic = 1-propanol

$CH_3CH_2CH_2CH_2OH$
common = butyl alcohol or *n*-butyl alcohol
systematic = 1-butanol

$CH_3CH_2CH_2CH_2CH_2OH$
common = pentyl alcohol or *n*-pentyl alcohol
systematic = 1-pentanol

$CH_3CH_2CH_2CH_2CH_2CH_2OH$
common = hexyl alcohol or *n*-hexyl alcohol
systematic = 1-hexanol

2. **a.** 1-pentanol
primary

b. 4-methylcyclohexanol
secondary

c. 5-chloro-2-methyl-2-pentanol
tertiary

d. 2-ethyl-1-pentanol
primary

e. 5-methyl-3-hexanol
secondary

f. 2,6-dimethyl-4-octanol
secondary

3.

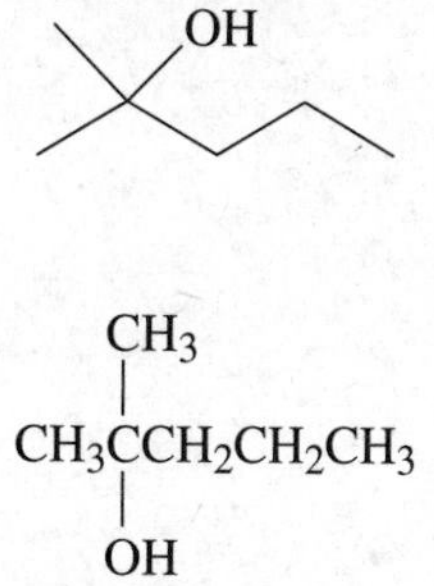

2-methyl-2-pentanol

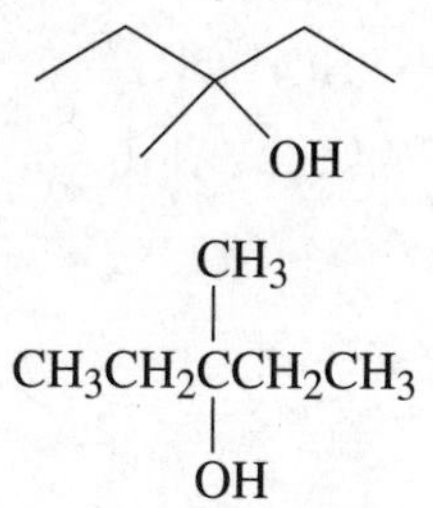

3-methyl-3-pentanol

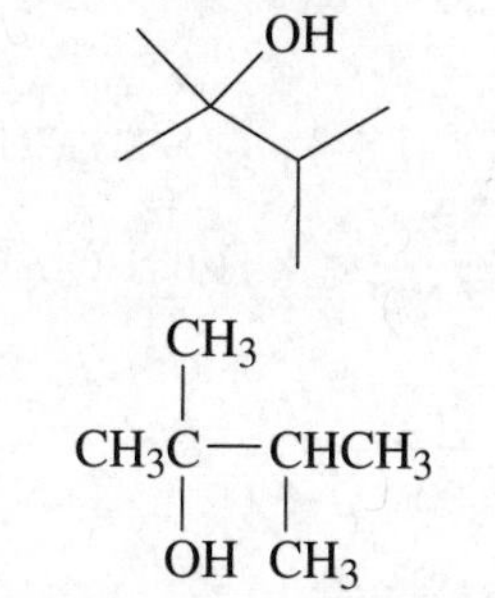

2,3-dimethyl-2-butanol

4. NH_3 and CH_3NH_2 are nucleophiles because of the lone pair electrons on the nitrogen atom. Therefore, when the lone pair is protonated, these compounds are no longer nucleophiles.

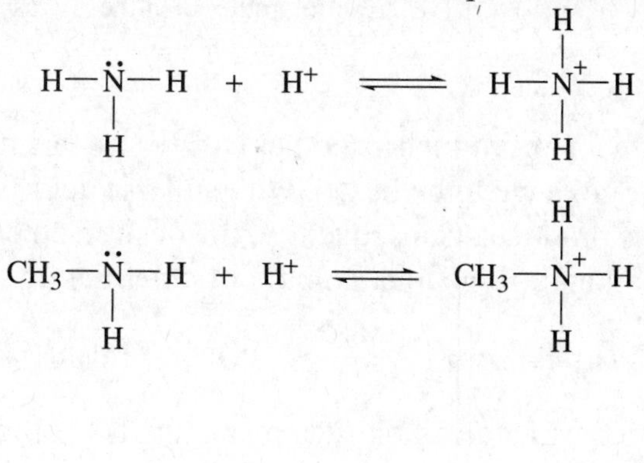

5. **a.** Solved in the text.

b. The conjugate acid of the leaving group of $CH_3OH_2^+$ is H_3O^+; its pK_a is = –1.7. The conjugate acid of the leaving group of CH_3OH is H_2O; its pK_a is = 15.5.

Because H_3O^+ is a much stronger acid than H_2O, H_2O is a weaker base than HO^- and, therefore, a better leaving group. Thus, $CH_3OH_2^+$ is more reactive than CH_3OH.

6. **a.** $$CH_3CH_2CH(OH)CH_3 + HBr \xrightarrow{\Delta} CH_3CH_2CH(Br)CH_3 + H_2O$$

b.

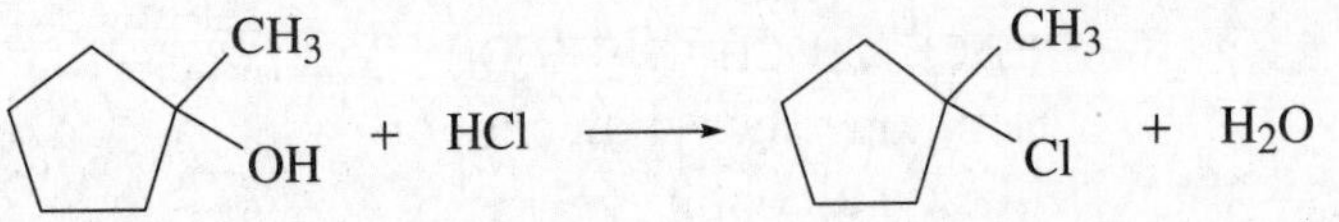

CH3
OH
+ HCl
CH3
Cl
+ H2O

7. **a.** $$CH_3CH_2CH_2CH_2OH \xrightarrow[2.\ CH_3O^-]{1.\ HBr\ +\ \Delta} CH_3CH_2CH_2CH_2OCH_3$$

b. $$CH_3CH_2CH_2CH_2OH \xrightarrow[2.\ CH_3CH_2C(=O)O^-]{1.\ HBr\ +\ \Delta} CH_3CH_2CH_2CH_2OC(=O)CH_2CH_3$$

c. $$CH_3CH_2CH_2CH_2OH \xrightarrow[2.\ CH_3CH_2NH_2]{1.\ HBr\ +\ \Delta} CH_3CH_2CH_2CH_2\overset{+}{N}H_2CH_2CH_3 \xrightarrow{HO^-} CH_3CH_2CH_2CH_2NHCH_2CH_3$$

d. $$CH_3CH_2CH_2CH_2OH \xrightarrow[2.\ ^-C{\equiv}N]{1.\ HBr\ +\ \Delta} CH_3CH_2CH_2CH_2C{\equiv}N$$

8.

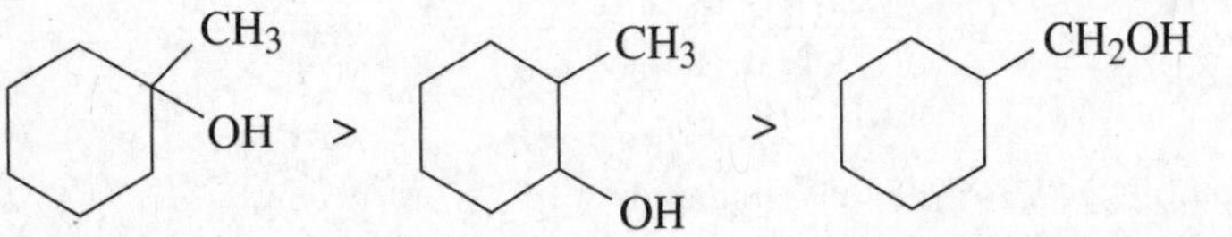

CH3
OH
>
CH3
OH
>
CH2OH

forms a tertiary carbocation

forms a secondary carbocation

it cannot form a primary carbocation; therefore, the dehydration will be an E2 reaction rather than the E1 reaction the others will undergo

9. **a.** In order to synthesize an unsymmetrical ether (ROR') by this method, two different alcohols (ROH and R'OH) would have to be heated with sulfuric acid. Therefore, three different ethers would be obtained as products. Consequently, the desired ether would account for considerably less than half of the total amount of ether that is synthesized.

$$ROH + R'OH \xrightarrow[\Delta]{H_2SO_4} ROR + ROR' + R'OR'$$

b. It could be synthesized using an alkoxide ion and an alkyl bromide.

$$CH_3CH_2CH_2O^- \xrightarrow{CH_3CH_2Br} CH_3CH_2CH_2OCH_2CH_3 + Br^-$$

10. **a.** 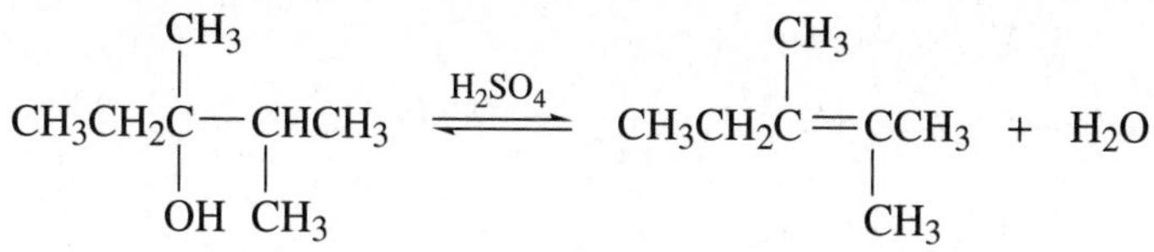

The hydrogen is removed from the β-carbon that is bonded to the fewest hydrogens.

b. 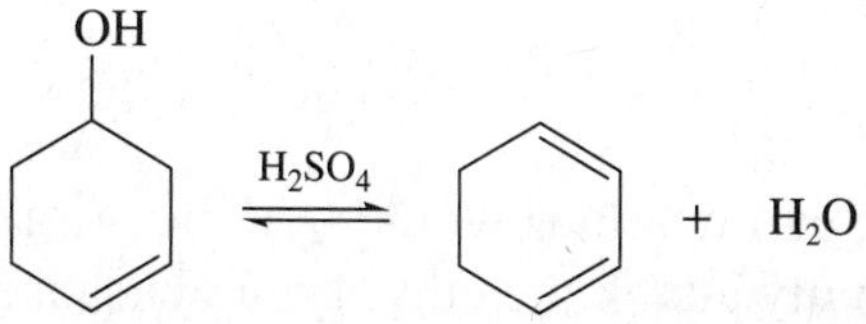

The two β-carbons are bonded to the same number of hydrogens. A hydrogen is removed from the β-carbon that results in formation of a conjugated double bond, since it is more stable than an isolated double bond.

11. **a.** 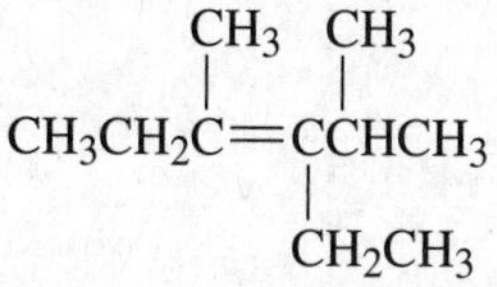 **b.** 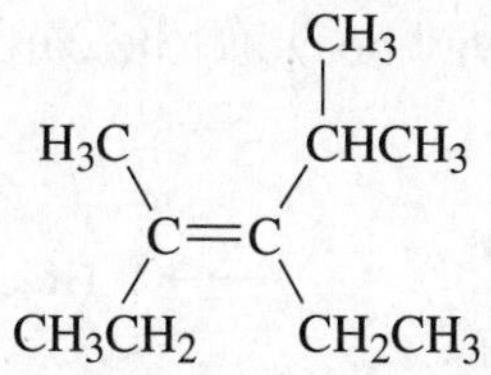

The larger group on one sp^2 carbon and the larger group on the other sp^2 carbon are on opposite sides of the double bond.

12. **a.** 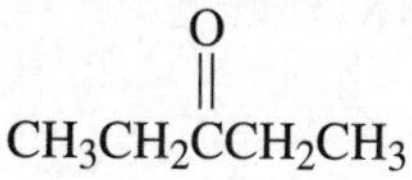 **b.** 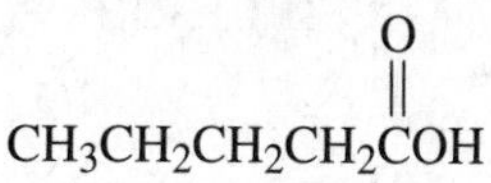 **c.** 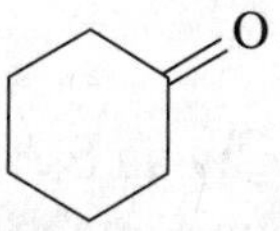 **d.** 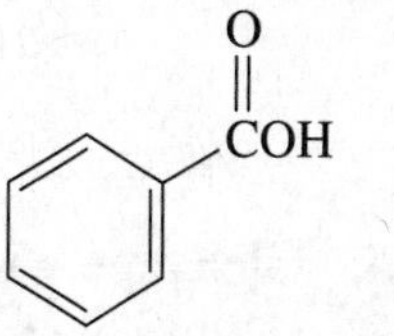

13. **a.** 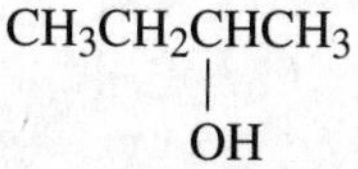 **b.** 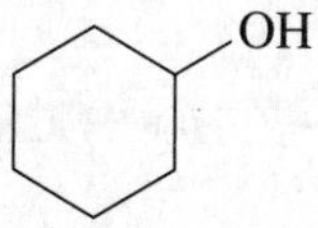 **c.** $CH_3CH_2CH_2CH_2OH$

14. Protonated amino groups cannot be displaced by a strongly basic nucleophile such as HO^- because HO^- would react immediately with the acidic hydrogen of the $^+NH_3$ group and thereby be converted to water, a very poor nucleophile. Water is too poor a nucleophile to displace a protonated amino group.

$$CH_3\overset{+}{N}H_3 + HO^- \rightleftharpoons CH_3NH_2 + H_2O$$

15. **a.** **1.** methoxyethane
2. ethoxyethane
3. 4-methoxyoctane
4. 1-propoxybutane

b. no

c. **1.** ethyl methyl ether
2. diethyl ether
3. no common name
4. butyl propyl ether

16. Solved in the text.

17. **a.** Cleavage occurs by an S_N2 pathway because the carbocations would both be primary and, therefore, too unstable to form; I^- will preferentially attack the ethyl group because it has less steric hindrance.

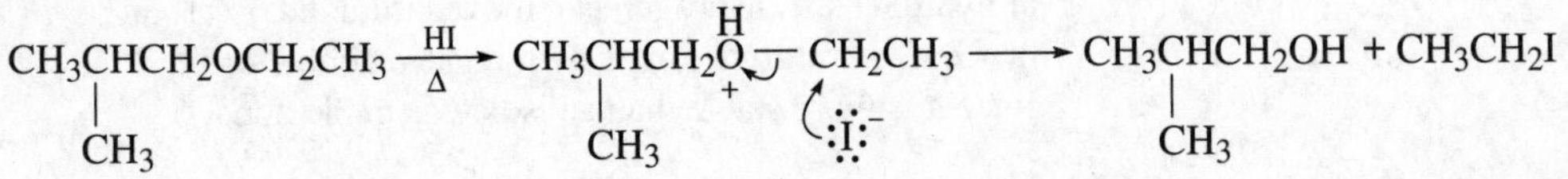

CH3CHCH2OCH2CH3
CH3
HI
Δ
CH3CHCH2O—CH2CH3
H
+
CH3
:I:⁻
CH3CHCH2OH + CH3CH2I
CH3

b. Cleavage occurs by an S_N2 pathway because the carbocations would both be primary and, therefore, too unstable to form.

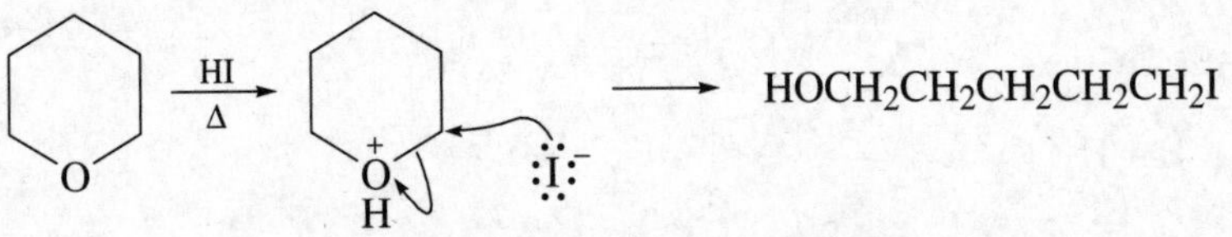

O
HI
Δ
+
O
H
:I:⁻
HOCH2CH2CH2CH2CH2I

c. Cleavage occurs by an S_N1 pathway because the tertiary carbocation that is formed is relatively stable; I^- will attach to the *tert*-butyl carbocation.

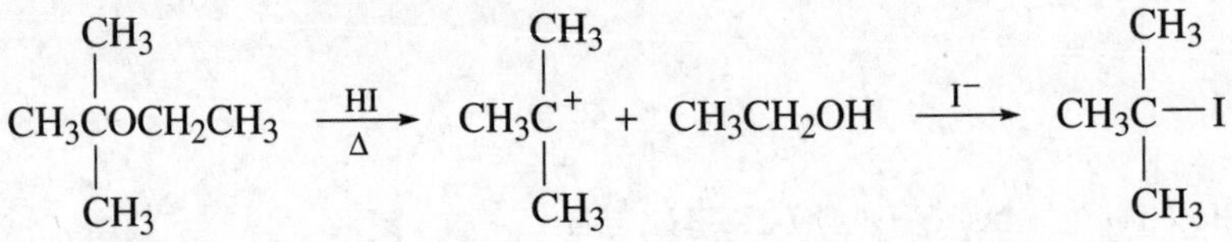

CH3
CH3COCH2CH3
CH3
HI
Δ
CH3
CH3C+ + CH3CH2OH
CH3
I⁻
CH3
CH3C—I
CH3

d. Cleavage will occur by an S_N1 pathway because the tertiary carbocation that is formed is relatively stable; I^- will attach to the tertiary carbocation.

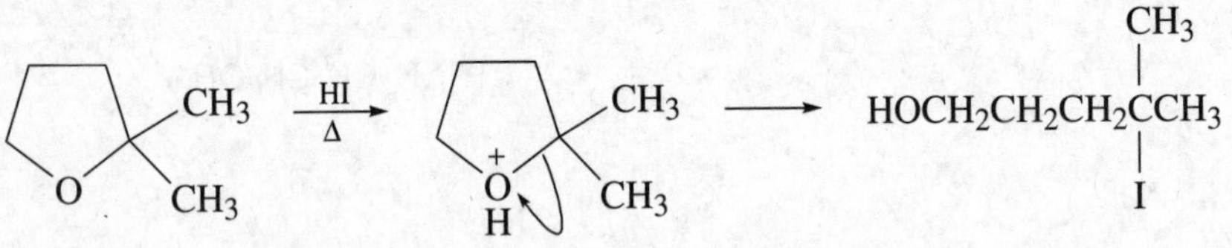

CH3
O
CH3
HI
Δ
CH3
+
O
H
CH3
CH3
HOCH2CH2CH2CCH3
I

18. **a.** 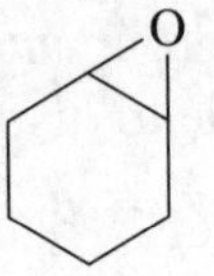

O

b. 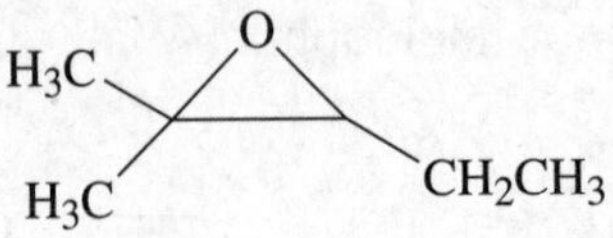

O
H3C
H3C
CH2CH3

19. a. 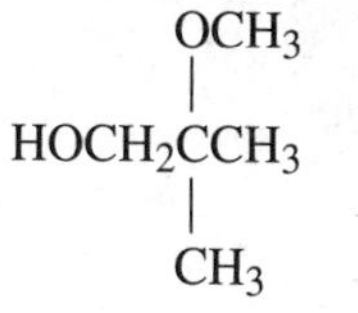

c. 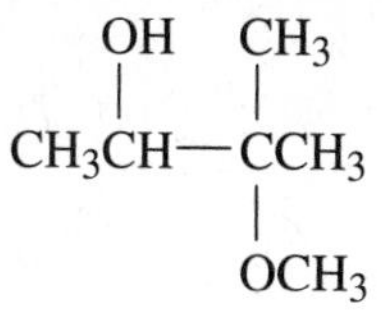

b. 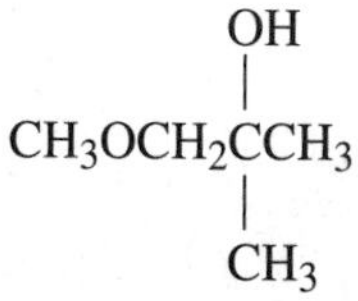

d. 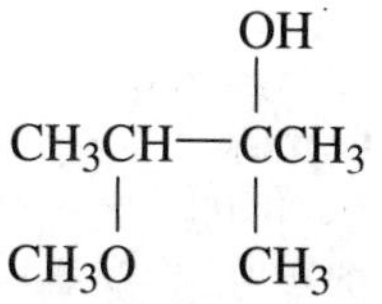

20. The reactivity of tetrahydrofuran is more similar to a noncyclic ether because the five-membered ring does not have the strain that makes the epoxide so reactive.

21. The less stable the carbocation that is formed when the three-membered ring of the arene oxide opens, the more likely it is that the compound will be carcinogenic. This is because the less stable the carbocation, the less likely it will be formed. Thus, the more likely the compound will exist in solution long enough to be attacked by a nucleophile (the carcinogenic pathway.)

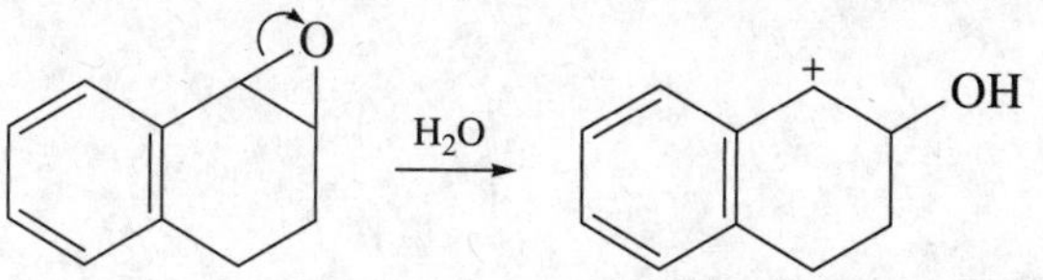
H2O
OH

This one is more apt to be carcinogenic

The carbocation can be stabilized by electron delocalization only if the aromaticity of the benzene ring is destroyed.

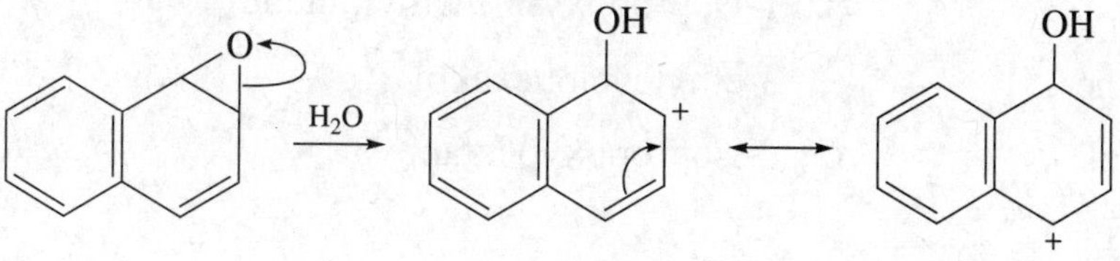
O
H2O
OH
OH

This carbocation can be stabilized by electron delocalization without destroying the aromaticity of the benzene ring.

22. a. 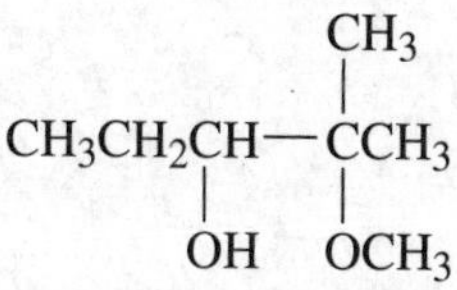

d. 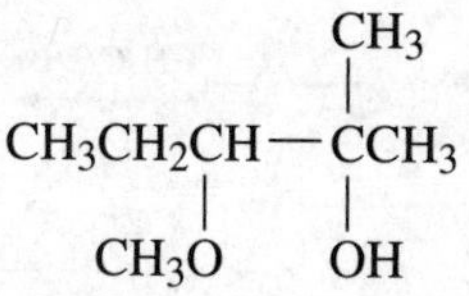

b. 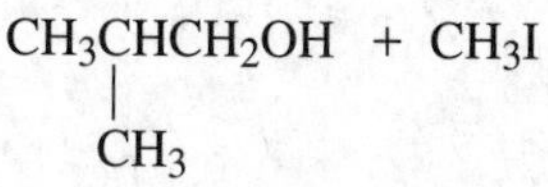

e. 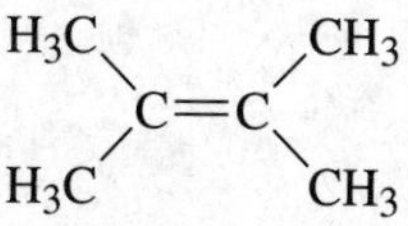

c. 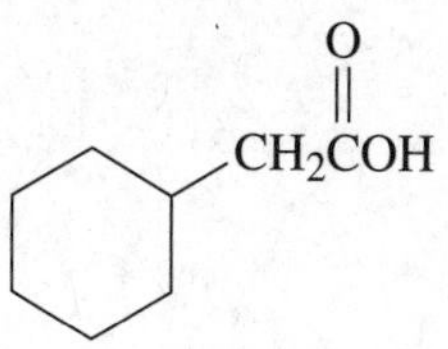

f. 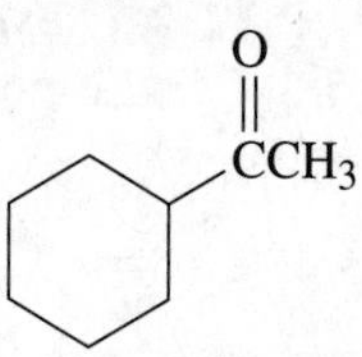

23. **a.** isopropyl propyl ether
1-isopropoxypropane or 2-propoxypropane

b. butyl ethyl ether
1-ethoxybutane

c. *sec*-butyl methyl ether
2-methoxybutane

d. diisopropyl ether
2-isopropoxypropane

24. **a.** 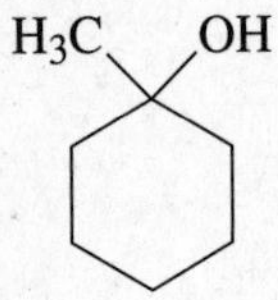

The rate-limiting step in all the dehydration reactions in this question is carbocation formation.
A tertiary carbocation is more stable than a secondary carbocation and therefore, is easier to form.

b. 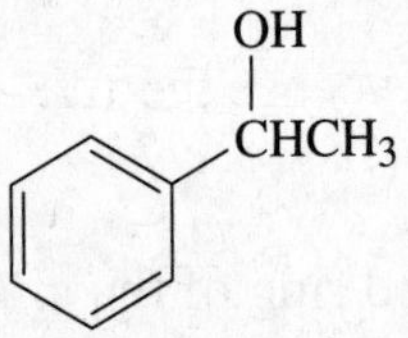

A secondary benzylic carbocation is more stable than a primary carbocation.

c. $CH_3C(CH_3)(OH)CH_2CH_3$ A tertiary carbocation is more stable than a secondary carbocation.

d. 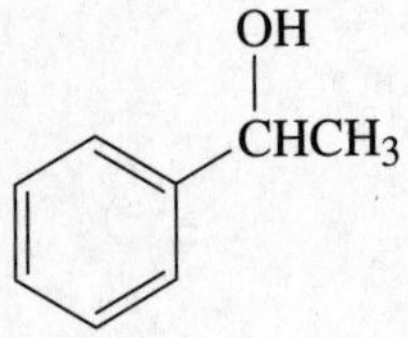

A secondary benzylic carbocation is more stable than a secondary carbocation.

25. **a.** 3-ethoxyheptane

b. methoxycyclohexane

c. 4-methyl-1-pentanol

d. 1-isopropoxy-3-methylbutane

e. 3-ethylcyclohexanol

f. 2-isopropoxypentane

26. **a.** 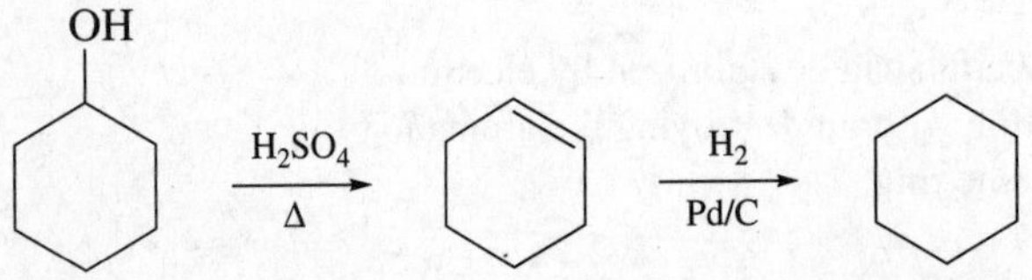

b. $CH_3CH_2C{\equiv}CH \xrightarrow{^-NH_2} CH_3CH_2C{\equiv}C^- \xrightarrow{CH_3CH_2Br} CH_3CH_2C{\equiv}CCH_2CH_3$

c. $CH_3CH_2C{\equiv}CH \xrightarrow{^-NH_2} CH_3CH_2C{\equiv}C^- \xrightarrow[2.\ H^+]{1.\ H_2C\overset{O}{—}CH_2} CH_3CH_2C{\equiv}CCH_2CH_2OH$

27. **a.** $CH_3CH(CH_3)OCH(CH_3)CH_3$

b. $CH_2{=}CHCH_2OCH{=}CH_2$

c. $CH_3CH_2CH(CH_3)OCH_2CH(CH_3)CH_3$

d. 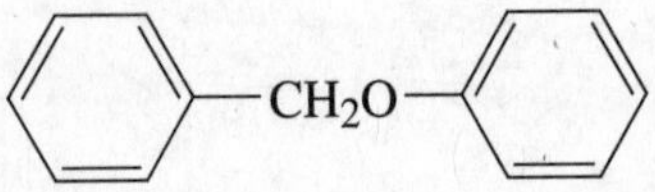

28. "c" is the only one with an asymmetric center, so it the only one of the ethers in Problem 27 that can exist as stereoisomers.

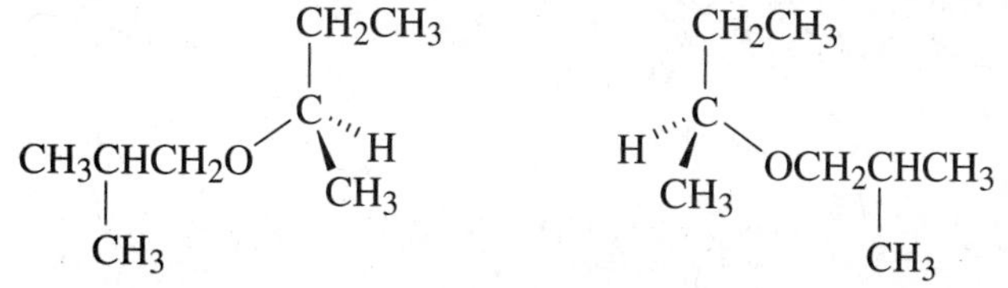

29. **a.**

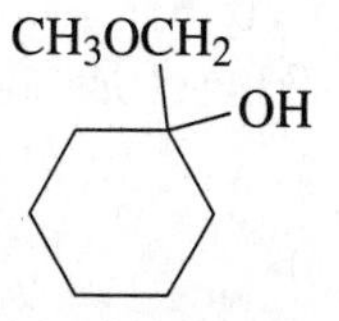

b.

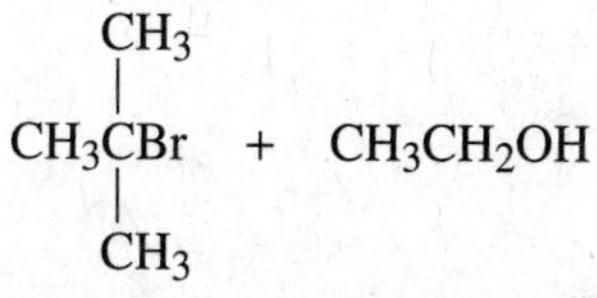

c.

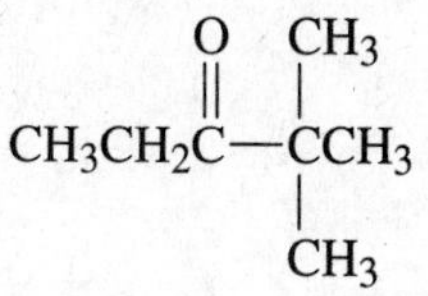

d.

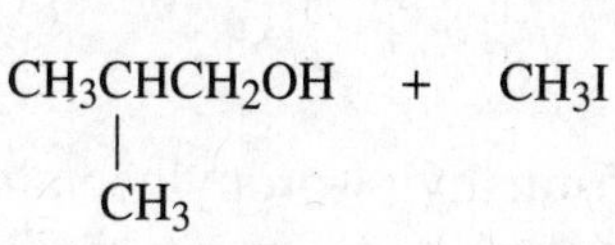

e.

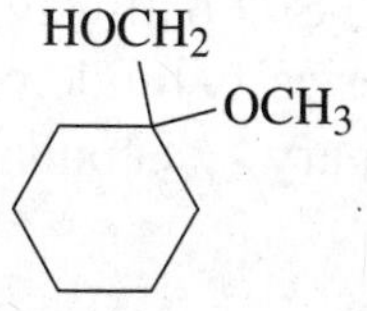

f.

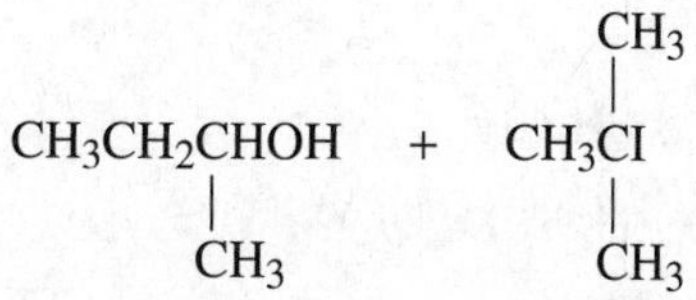

g.

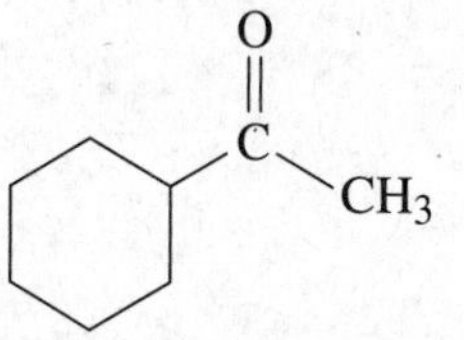

h.

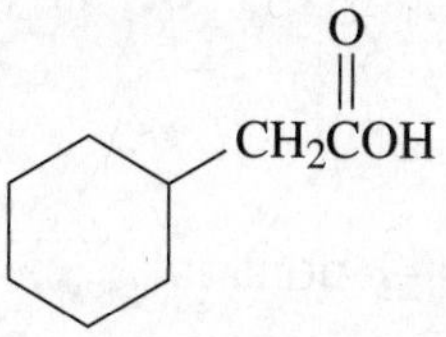

30. **a.**

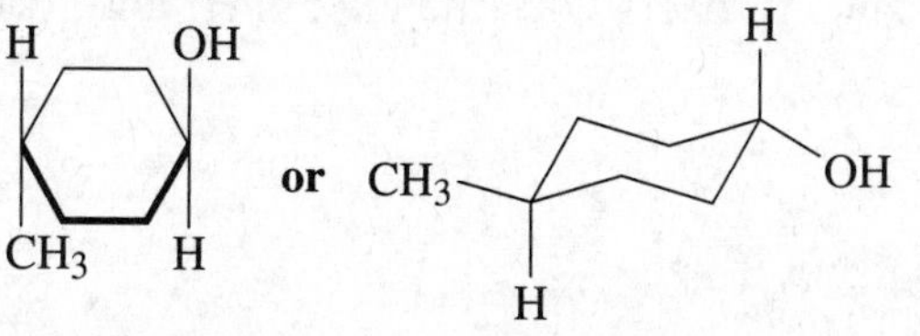

b. $CH_3CH_2OCH_2CH_2CH_2OH$

31. **a.**

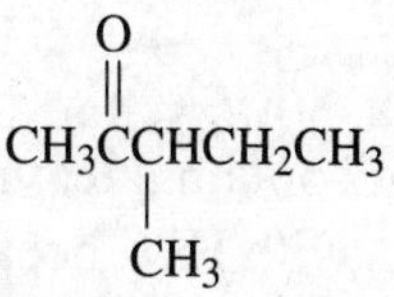

b.

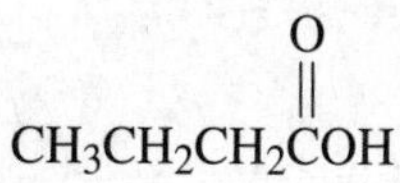

c.

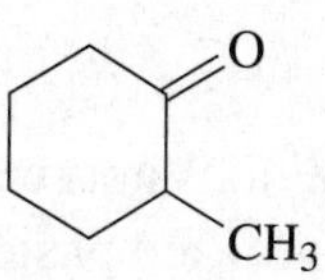

32. **a.**

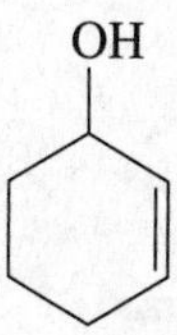

A secondary allylic carbocation is more stable than a secondary carbocation.

b.

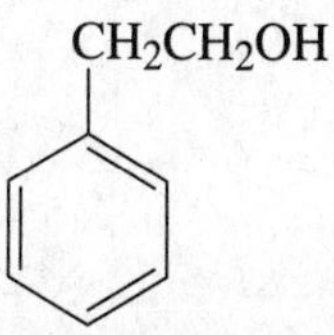

The other alcohol cannot undergo dehydration because its β-carbon is not bonded to a hydrogen.

33. 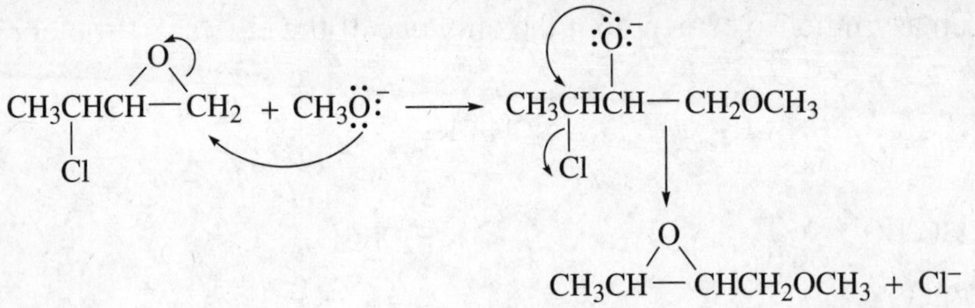

34. If secondary alcohols reacted by an S_N2 pathway, they would be less reactive than primary alcohols because they are more sterically hindered than primary alcohols. Therefore, the relative reactivity would be: tertiary > primary > secondary.

35. 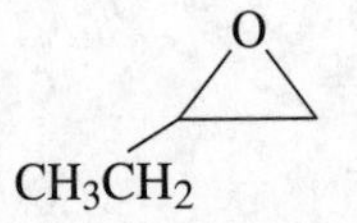

1,2-epoxybutane

a. 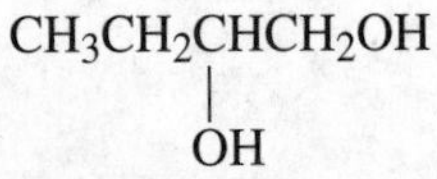

0.1 M HCl is a dilute solution
of HCl in water

b. 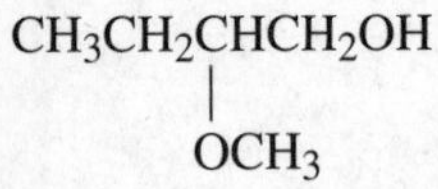

c. 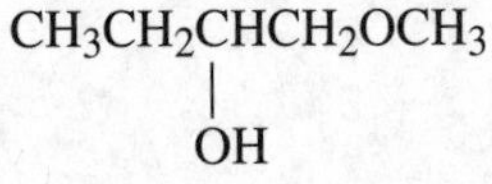

d. 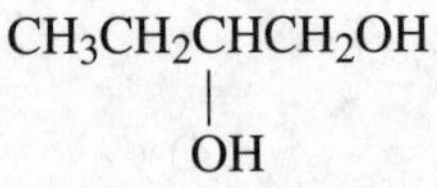

36. **a.** 1-propanol **b.** 4-propyl-1-nonanol **c.** 1-methoxy-5-methyl-3-propylheptane

37. Ethyl alcohol is not obtained as a product, because it reacts with the excess HI and forms ethyl iodide.

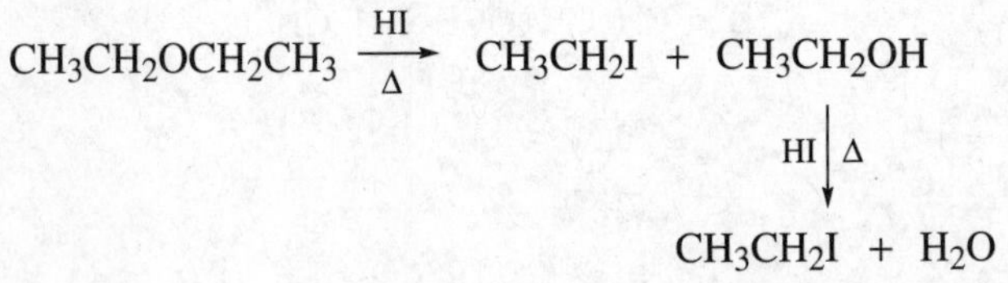

38. Cyclopropane does not react with HO^-, because cyclopropane does not contain a leaving group; a carbanion is far too basic to be a leaving group. Ethylene oxide reacts with HO^- because ethylene oxide contains an RO^- leaving group.

39. **a.** 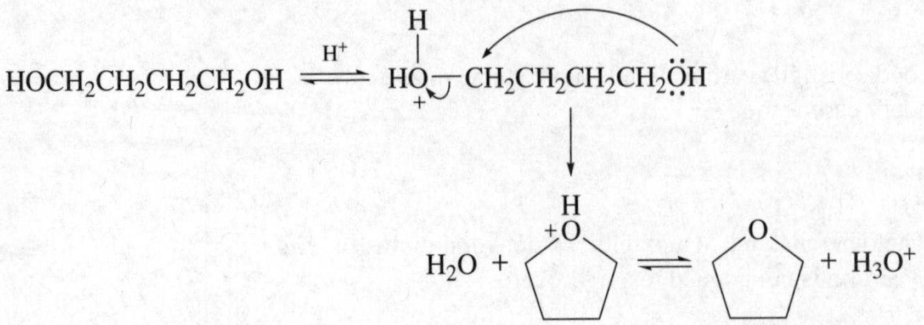

b.

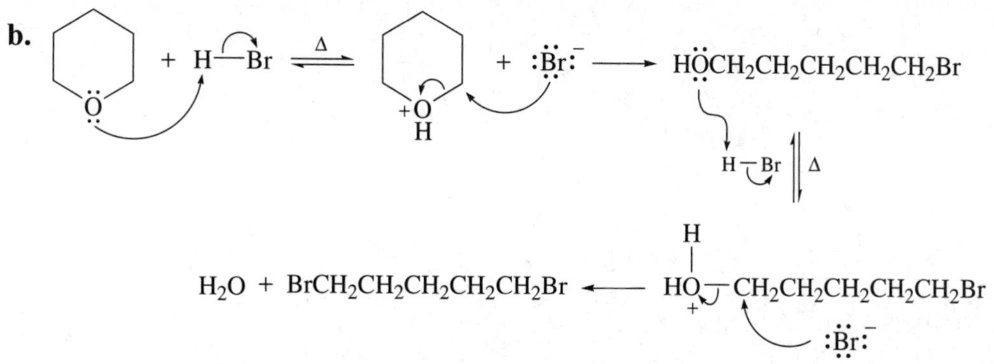

40.

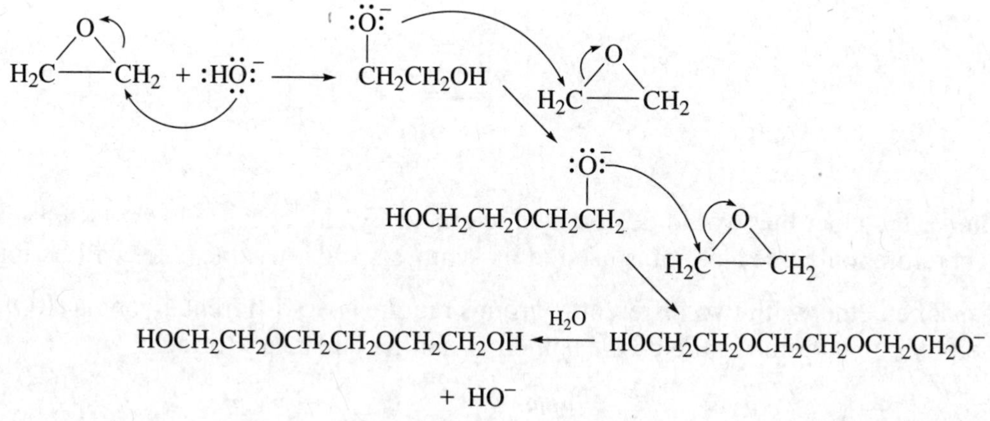

41. 1-Butanol dehydrates in an E2 reaction to form 1-butene. In the acidic solution, 1-butene is protonated to form a secondary carbocation, which loses a proton from the β-carbon bonded to the fewest hydrogens resulting in the formation of 2-butene.

$$\underset{}{CH_3CH_2CH_2CH_2OH} \underset{\Delta}{\overset{H_2SO_4}{\rightleftharpoons}} \underset{\text{1-butene}}{CH_3CH_2CH{=}CH_2} \overset{H^+}{\rightleftharpoons} CH_3CH_2\overset{+}{C}HCH_3 \rightleftharpoons \underset{\text{2-butene}}{CH_3CH{=}CHCH_3} + H^+$$

42. 2-hexene and 3-hexene

$$CH_3CH_2CH_2CH_2CH_2CH_2OH \underset{\Delta}{\overset{H_2SO_4}{\longrightarrow}} CH_3CH_2CH_2CH_2CH{=}CH_2$$

$$CH_3CH_2CH_2CH_2CH{=}CH_2 \overset{H^+}{\rightleftharpoons} CH_3CH_2CH_2CH_2\underset{+}{C}HCH_3$$

$$CH_3CH_2CH_2CH_2\underset{+}{C}HCH_3 \rightleftharpoons \underset{\text{2-hexene} + H^+}{CH_3CH_2CH_2CH{=}CHCH_3}$$

$$\overset{H^+}{\rightleftharpoons} CH_3CH_2CH_2\underset{+}{C}HCH_2CH_3 \rightleftharpoons \underset{\substack{\text{3-hexene} \\ \text{(cis and trans)} + H^+}}{CH_3CH_2CH{=}CHCH_2CH_3} + \underset{\substack{\text{2-hexene} \\ \text{(cis and trans)}}}{CH_3CH_2CH_2CH{=}CHCH_3}$$

43.

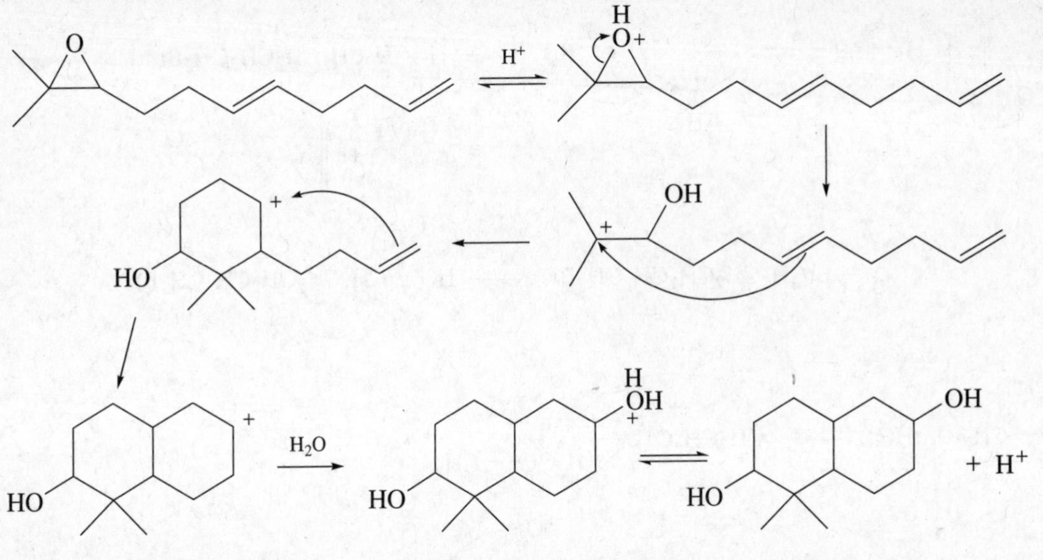

44. **Diethyl ether** is the ether that would be obtained in greatest yield, because its two R groups are the same. Therefore, only one alcohol is used in its synthesis and only one ether will be formed.

The synthesis of an ether with two different R groups requires two different alcohols (ROH and R'OH). Therefore, in addition to the desired ether, two other ethers are formed.

$$\text{ROH} + \text{R'OH} \xrightarrow{H^+} \text{ROR} + \text{ROR'} + \text{R'OR'}$$

45. When (*S*)-2-butanol is heated, it forms a planar carbocation. Water has equal access to the front and the back of the carbocation, so equal amounts (a racemic mixture) of (*S*)-2-butanol and (*R*)-2-butanol are formed.

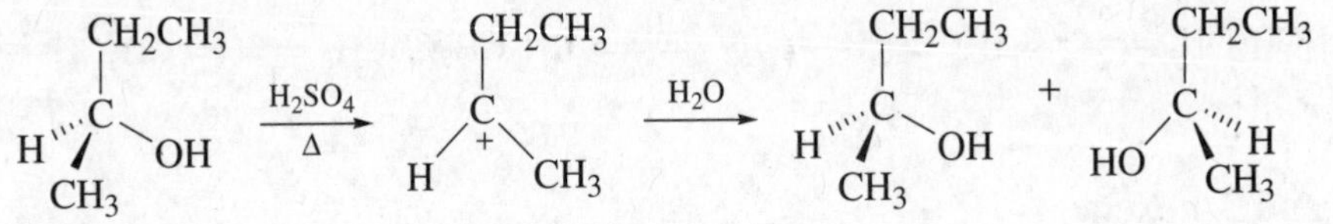

(*S*)-2-butanol a planar carbocation

46. The amine attacks the epxoide on the backside of the oxygen leaving group. Therefore, the two substituents on the ring will be trans to one another. Because the amine can attack two different ring carbons, two stereoisomers will be formed, the *R*,*R*-isomer and the *S*,*S*-isomer.

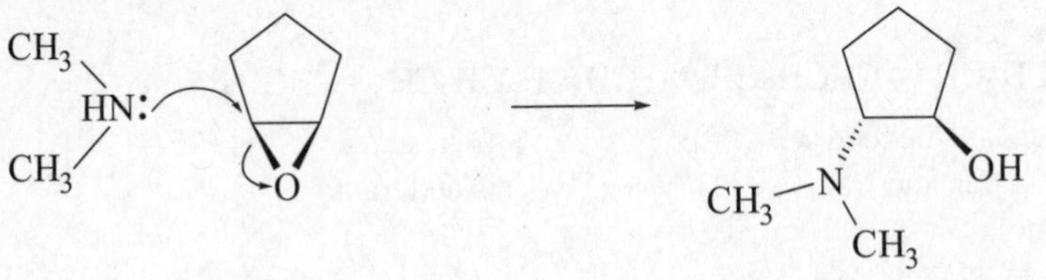

R,*R*-isomer

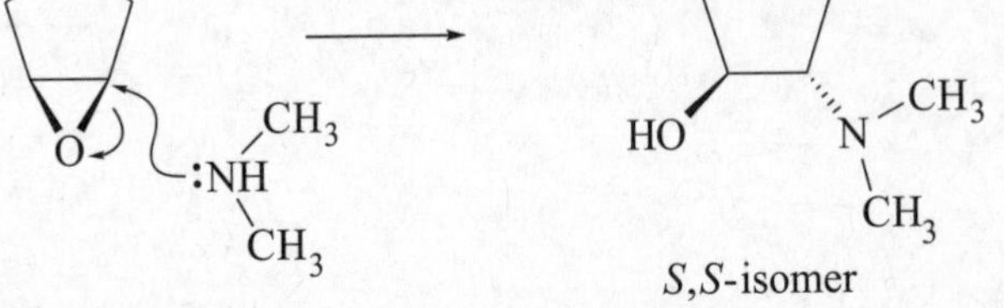

S,*S*-isomer

47. The two different carbocations formed by naphthalene oxide differ in stability.

One carbocation is more stable than the other because it can be stabilized by electron delocalization without disrupting the aromaticity of the adjacent ring. The more stable carbocation leads to the major product.

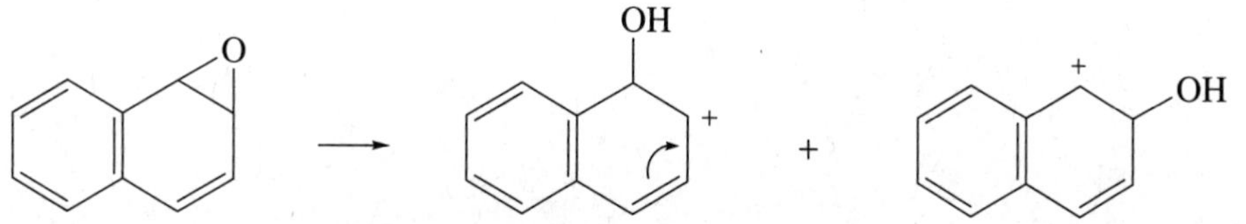

more stable carbocation because it can be stabilized without destroying the aromaticity of the benzene ring

cannot be stabillized without destroying the aromaticity of the benzene ring

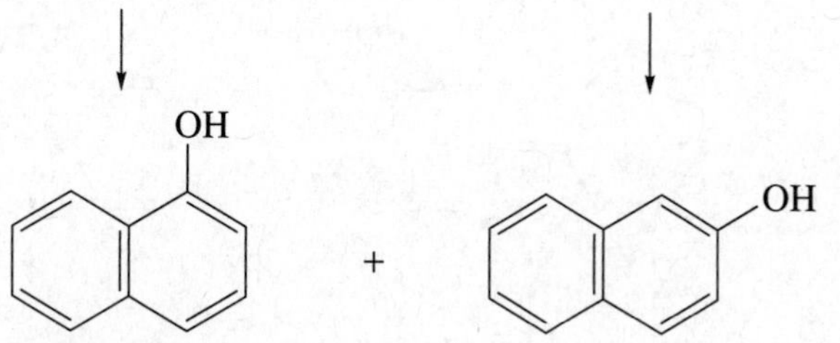

1-naphthol
major product

2-naphthol

48. **a.** Note that a bond joining two rings cannot be epoxidized.

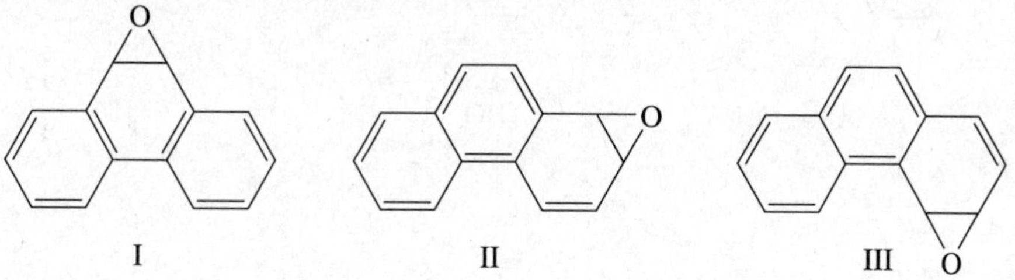

I II III

b. The epoxide ring in phenanthrenes II and III can open in two different directions to give two different carbocations and, therefore, two different phenols.

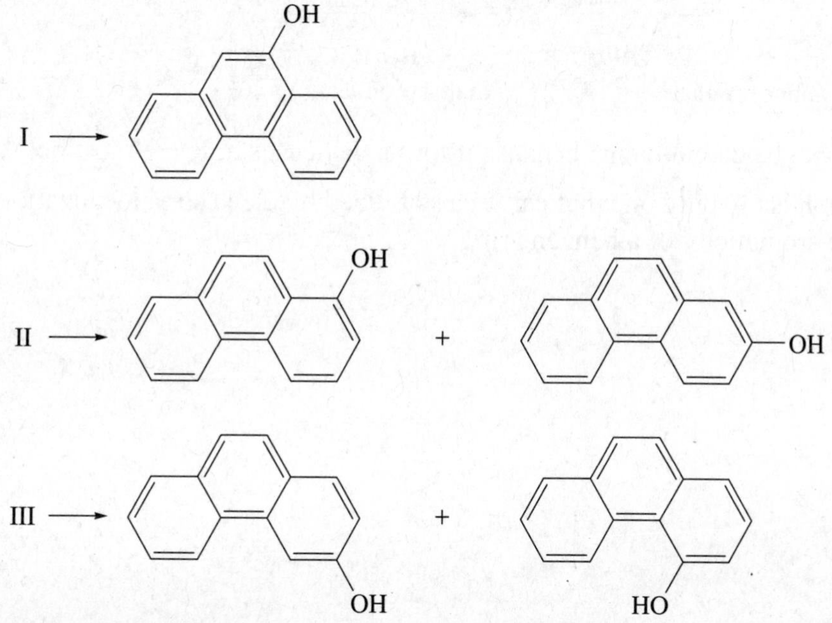

c. The two different carbocations formed by phenanthrenes II and III differ in stability.

One carbocation is more stable than the other because it can be stabilized by electron delocalization without disrupting the aromaticity of the adjacent ring. The more stable carbocation leads to the major product.

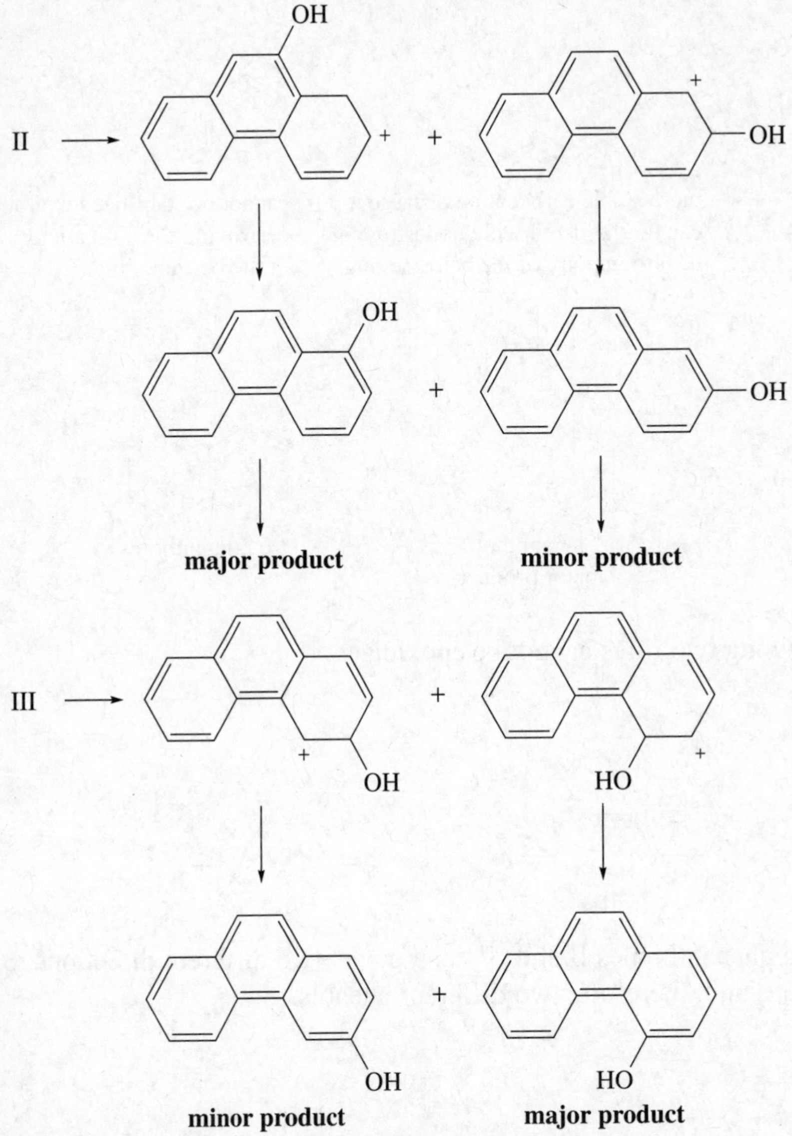
II
III
major product
minor product
minor product
major product

d. I is most likely to be carcinogenic because it forms the least stable carbocation.

This carbocation is the only one that can't be stabilized by electron delocalization without destroying the aromaticity of a benzene ring.

49. **a.**

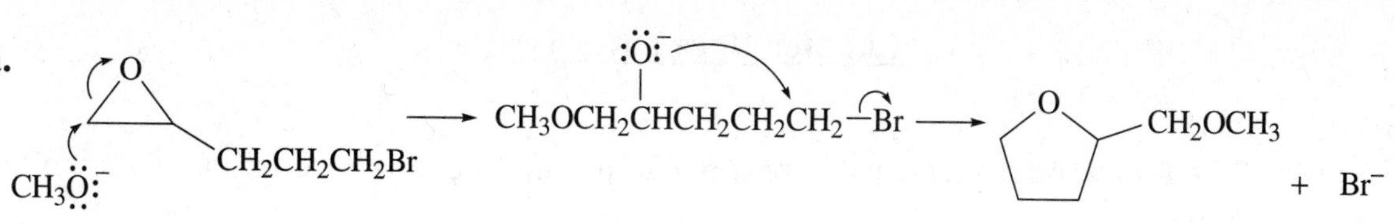

b.

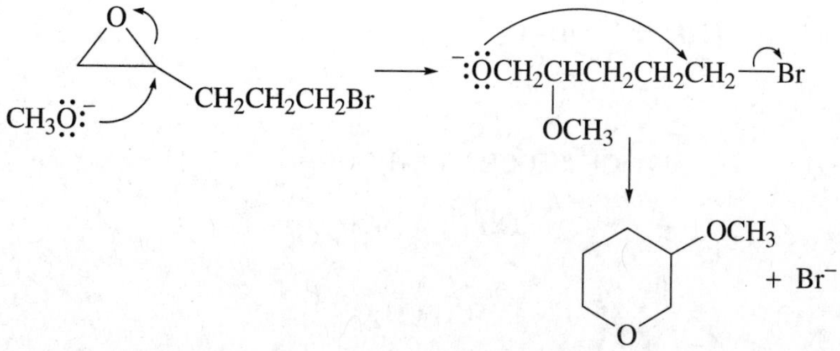

c. The six-membered ring is formed by attack on the more sterically hindered carbon of the epoxide. Attack on the less sterically hindered carbon, which forms the five-membered ring, is preferred.

50. **a.**

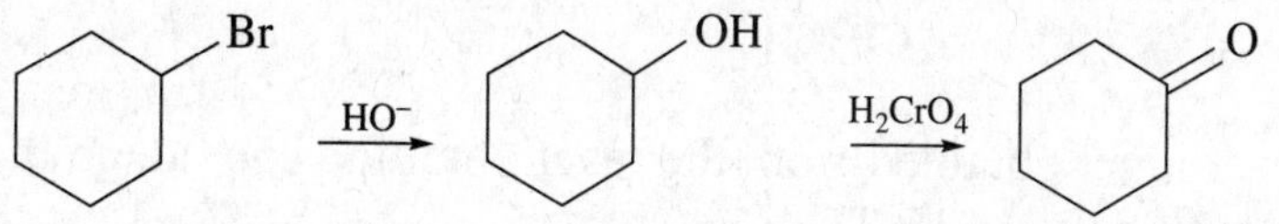

b.

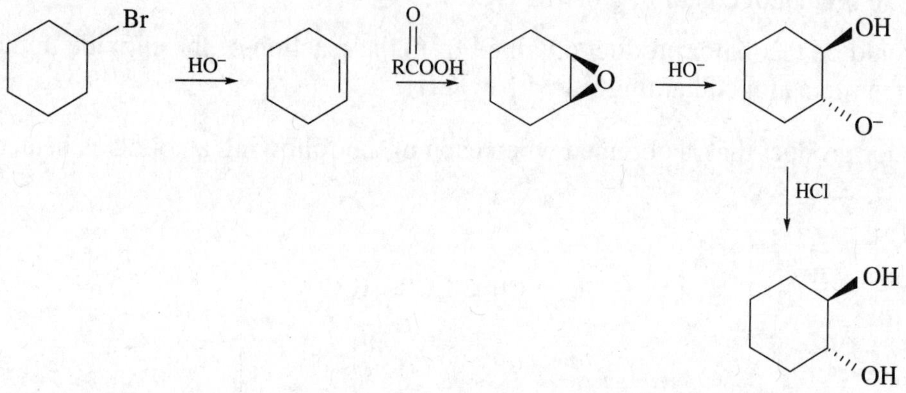

Chapter 10 Practice Test

1. Which of the following reagents is the best one to use in order to convert methyl alcohol into methyl bromide?

Br^- HBr Br_2 NaBr Br^+

2. Name the following compounds:

a. 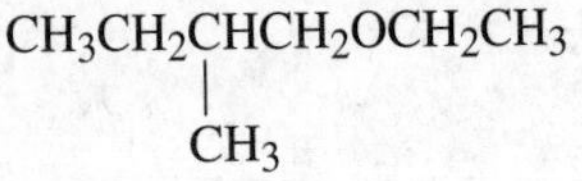

b. 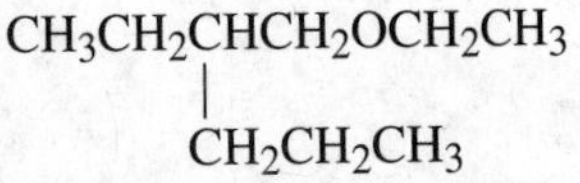

c. 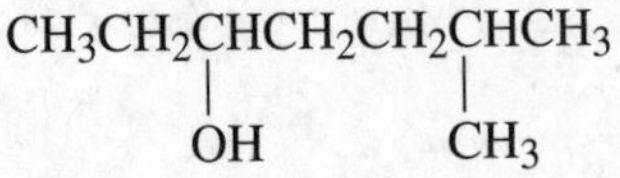

d. 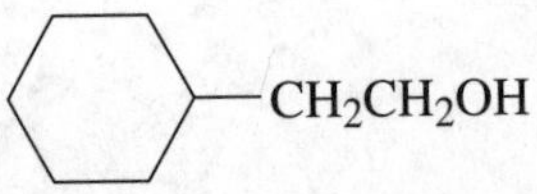

3. Give two names for the following compound:

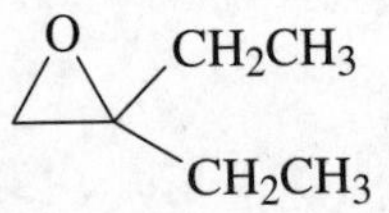

4. a. What would be the major product obtained from the reaction of the epoxide in the above problem in methanol containing 0.1 M HCl?

b. What would be the major product obtained from the reaction of the epoxide in the above problem in methanol containing 0.1 M $NaOCH_3$?

5. Give the major product that is obtained when each of the following alcohols is heated in the presence of H_2SO_4.

a. 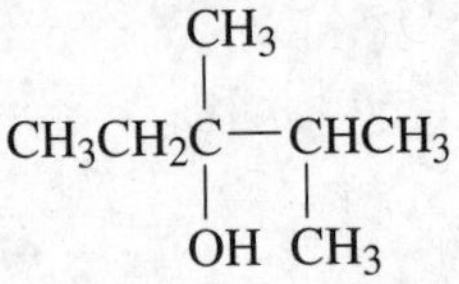

b. 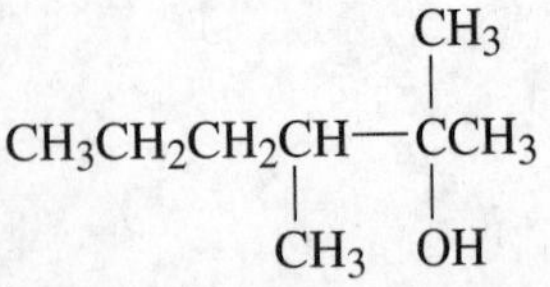

c. 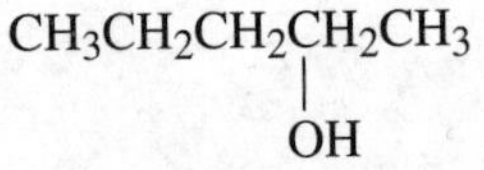

d. 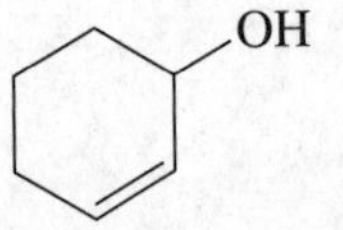

6. Indicate whether the following statements are true or false:

a. Tertiary alcohols are easier to dehydrate than secondary alcohols. T F

b. 1-Methylcyclohexanol reacts more rapidly than 2-methylcyclohexanol with HBr. T F

c. 1-Butanol forms a ketone when it is oxidized by chromic acid. T F

7. What products would be obtained from heating the following ethers with one equivalent of HI?

a. 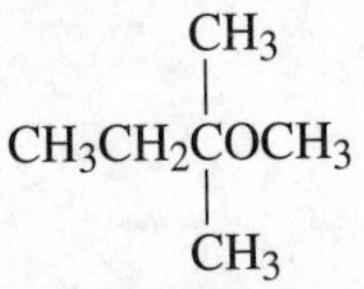

b. 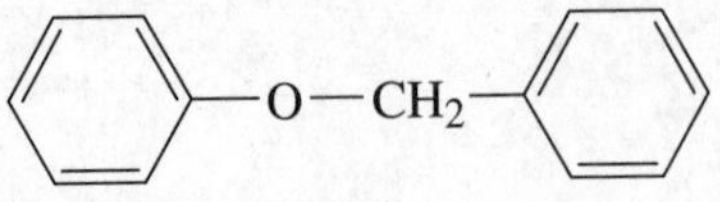

CHAPTER 11

Carbonyl Compounds I: Nucleophilic Acyl Substitution

1. **a.** propanamide
propionamide
b. isobutyl butanoate
isobutyl butyrate
c. potassium butanoate
potassium butyrate
d. pentanoyl chloride
valeryl chloride
e. *N,N*-dimethylhexanamide
f. *N*-ethyl-2-methylbutanamide

2. **a.** 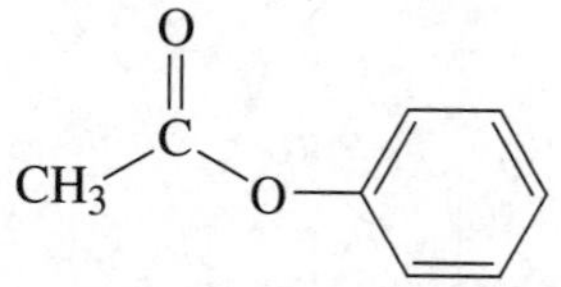

b. 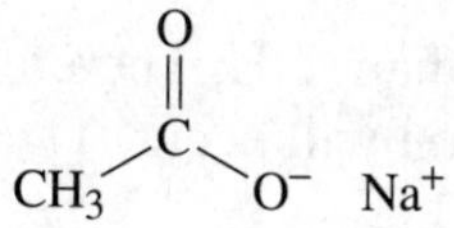

c. 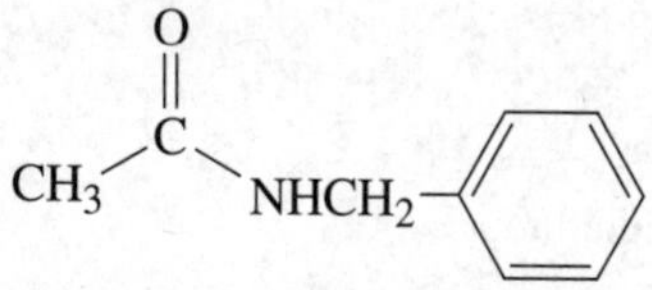

d. 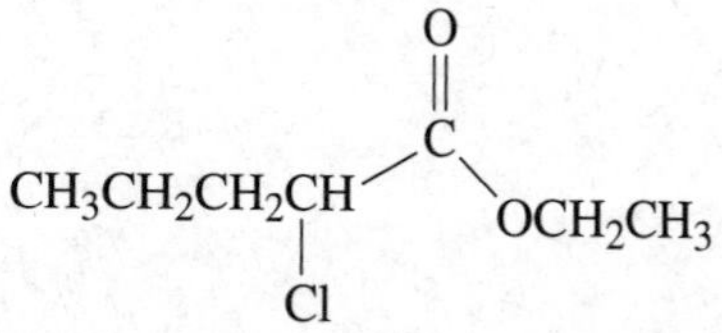

e. 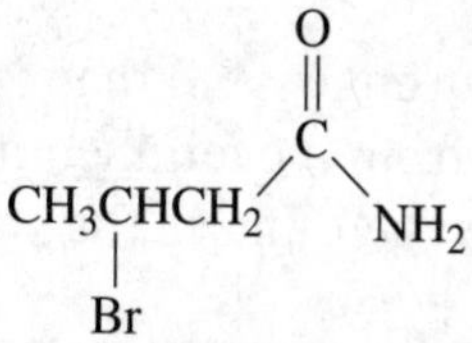

f. 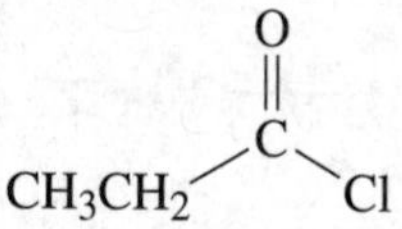

3. The carbon-oxygen bond in an alcohol is longer because, as a result of electron delocalization, the carbon-oxygen single bond in a carboxylic acid has some double-bond character.

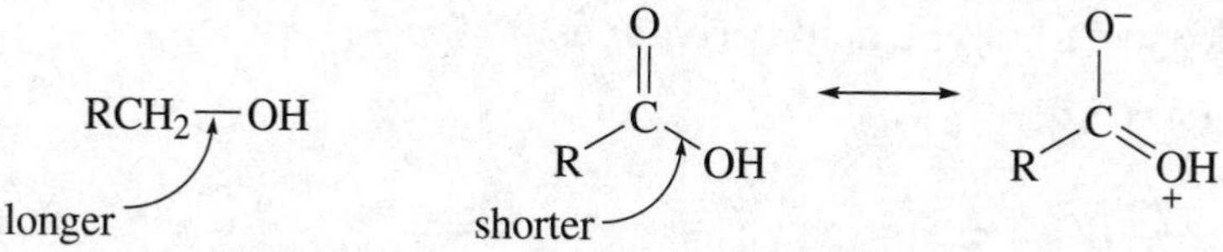
RCH2—OH
longer
shorter

4. The bond between oxygen and the methyl group is the longest because it is a pure single bond, whereas the other two carbon-oxygen bonds have some double-bond character.
The bond between carbon and the carbonyl oxygen is the shortest because it has the most double bond character.

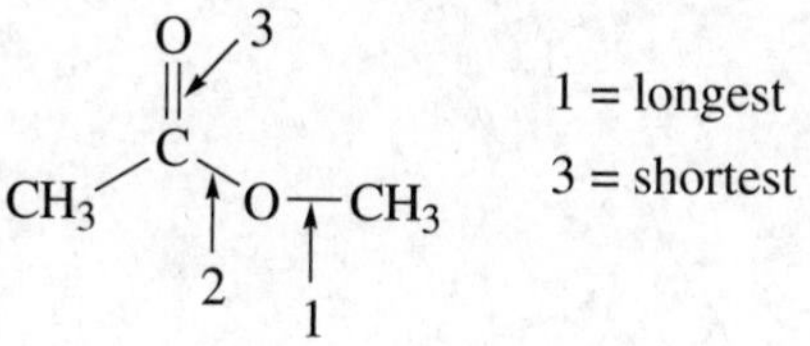
1 = longest
3 = shortest

5. **a.** Because HCl is a stronger acid than H_2O, Cl^- is a weaker base than HO^-. Therefore, Cl^- will be eliminated from the tetrahedral intermediate, so the product of the reaction will be acetic acid.

However, since the solution is basic, acetic acid will be in its basic form as a result of losing a proton.

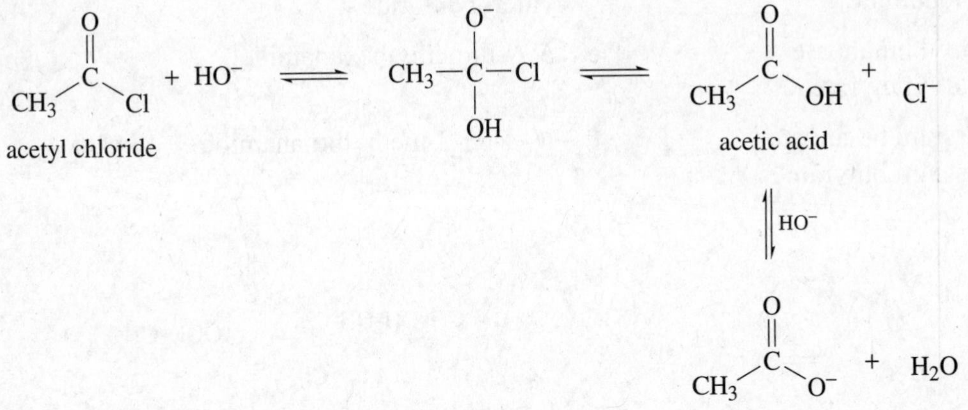
acetyl chloride
acetic acid

b. Because H_2O is a stronger acid than NH_3, HO^- is a weaker base than ^-NH, Therefore, HO^- will be eliminated from the tetrahedral intermediate, so no reaction will occur. The product of the reaction will be the same as the starting material.

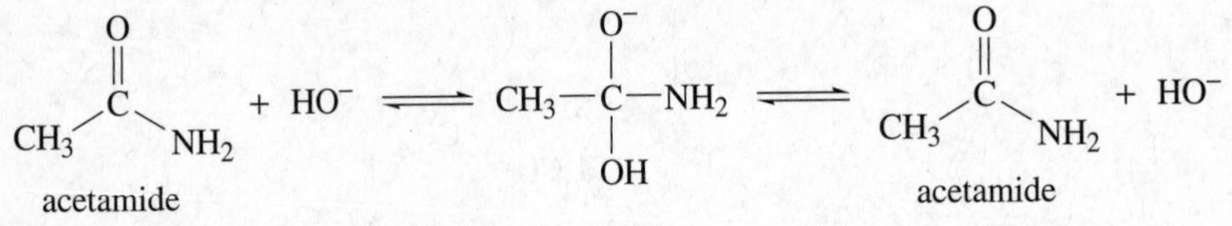
acetamide
acetamide

6. **a.** a new carboxylic acid derivative

b. no reaction

c. a mixture of two carboxylic acid derivatives

7. **a.** 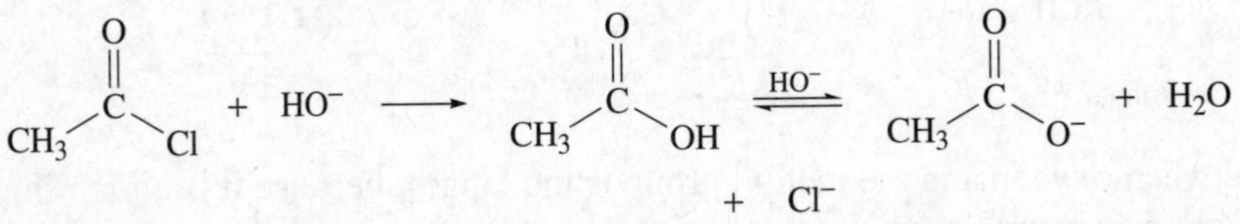

b. 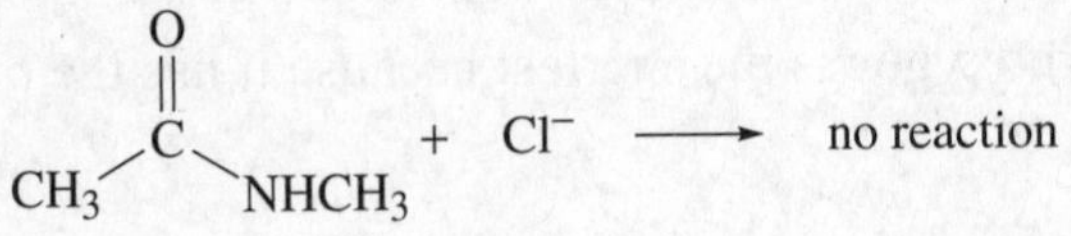
no reaction

8. Solved in the text.

9.   **a.** 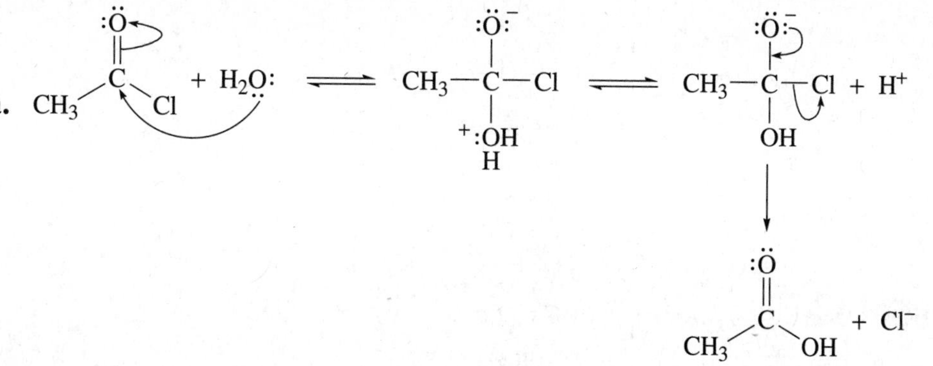

$$CH_3\overset{\overset{\displaystyle :\ddot{O}}{\|}}{C}Cl \;+\; H_2\ddot{O}: \;\rightleftharpoons\; CH_3-\overset{\overset{\displaystyle :\ddot{O}:^-}{|}}{\underset{\underset{\displaystyle H}{\overset{+}{:}OH}}{C}}-Cl \;\rightleftharpoons\; CH_3-\overset{\overset{\displaystyle :\ddot{O}:^-}{|}}{\underset{\underset{\displaystyle OH}{}}{C}}-Cl \;+\; H^+$$

$$\downarrow$$

$$CH_3\overset{\overset{\displaystyle :\ddot{O}}{\|}}{C}OH \;+\; Cl^-$$

b. 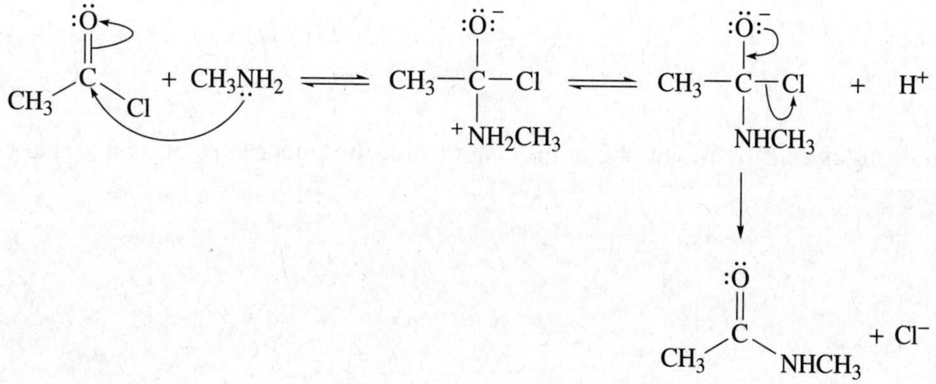

$$CH_3\overset{\overset{\displaystyle :\ddot{O}}{\|}}{C}Cl \;+\; CH_3\ddot{N}H_2 \;\rightleftharpoons\; CH_3-\overset{\overset{\displaystyle :\ddot{O}:^-}{|}}{\underset{\underset{\displaystyle +NH_2CH_3}{}}{C}}-Cl \;\rightleftharpoons\; CH_3-\overset{\overset{\displaystyle :\ddot{O}:^-}{|}}{\underset{\underset{\displaystyle NHCH_3}{}}{C}}-Cl \;+\; H^+$$

$$\downarrow$$

$$CH_3\overset{\overset{\displaystyle :\ddot{O}}{\|}}{C}NHCH_3 \;+\; Cl^-$$

10.   **a.** $CH_3CH_2CH_2OH$

   b. $CH_3CH_2NH_2$

   c. $(CH_3)_2NH$

   d. 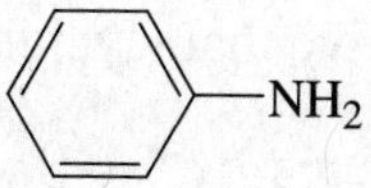

   e. H_2O

   f. 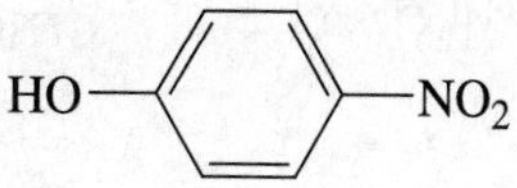

11.   **a.** 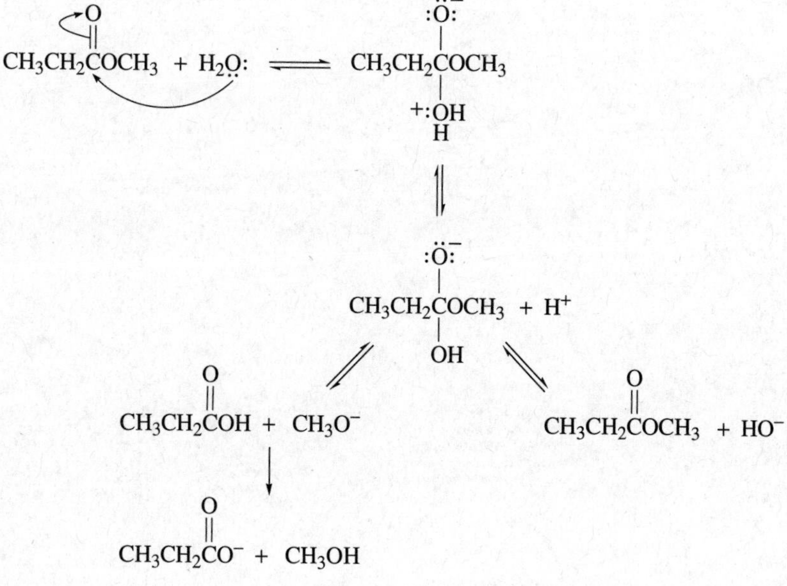

$$CH_3CH_2\overset{\overset{\displaystyle O}{\|}}{C}OCH_3 \;+\; H_2\ddot{O}: \;\rightleftharpoons\; CH_3CH_2\overset{\overset{\displaystyle :\ddot{O}:^-}{|}}{\underset{\underset{\displaystyle H}{\overset{+}{:}OH}}{C}}OCH_3$$

$$\updownarrow$$

$$CH_3CH_2\overset{\overset{\displaystyle :\ddot{O}:^-}{|}}{\underset{\underset{\displaystyle OH}{}}{C}}OCH_3 \;+\; H^+$$

$$\swarrow \qquad\qquad \searrow$$

$$CH_3CH_2\overset{\overset{\displaystyle O}{\|}}{C}OH \;+\; CH_3O^- \qquad\qquad CH_3CH_2\overset{\overset{\displaystyle O}{\|}}{C}OCH_3 \;+\; HO^-$$

$$\downarrow$$

$$CH_3CH_2\overset{\overset{\displaystyle O}{\|}}{C}O^- \;+\; CH_3OH$$

b.

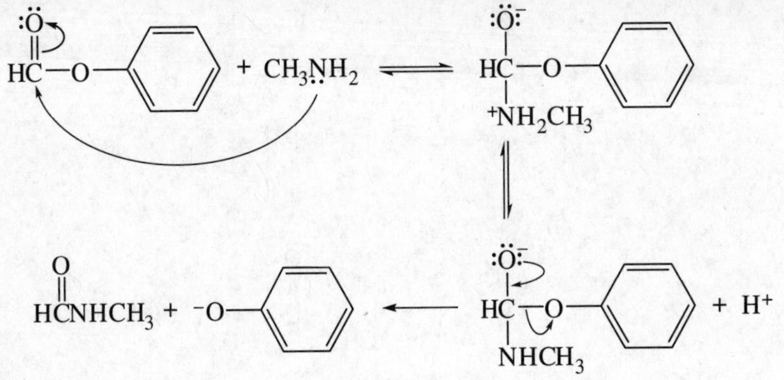

12. Solved in the text.

13. Phenol is a stronger acid than cyclohexanol. Therefore, the phenolate ion is a weaker base and a better leaving group

pK_a = 10 pK_a = 16

Thus, phenyl acetate is more reactive than cyclohexyl acetate toward hydrolysis.

14. **a.**

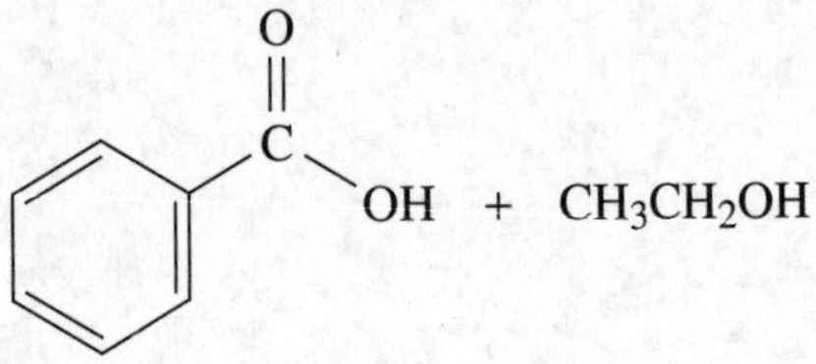

b.

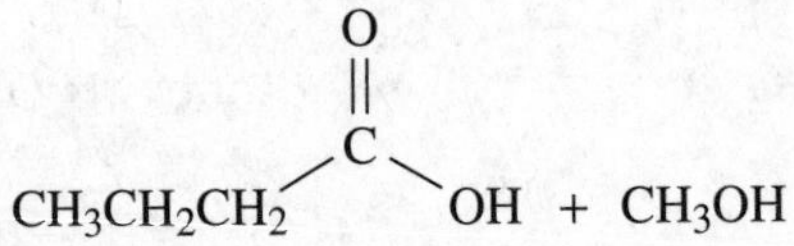

15.

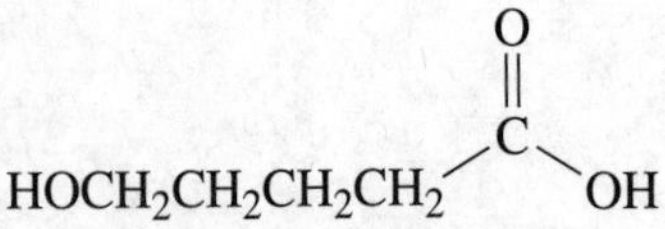

16.

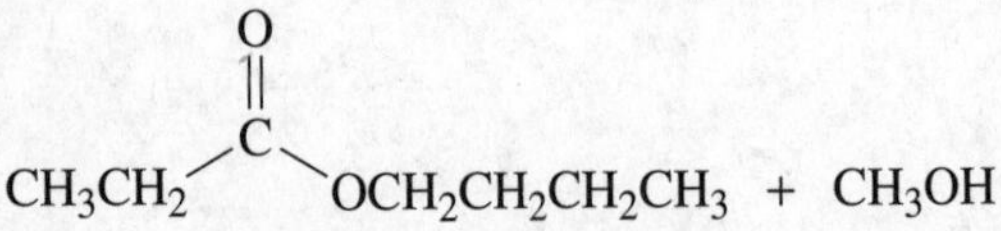

17.

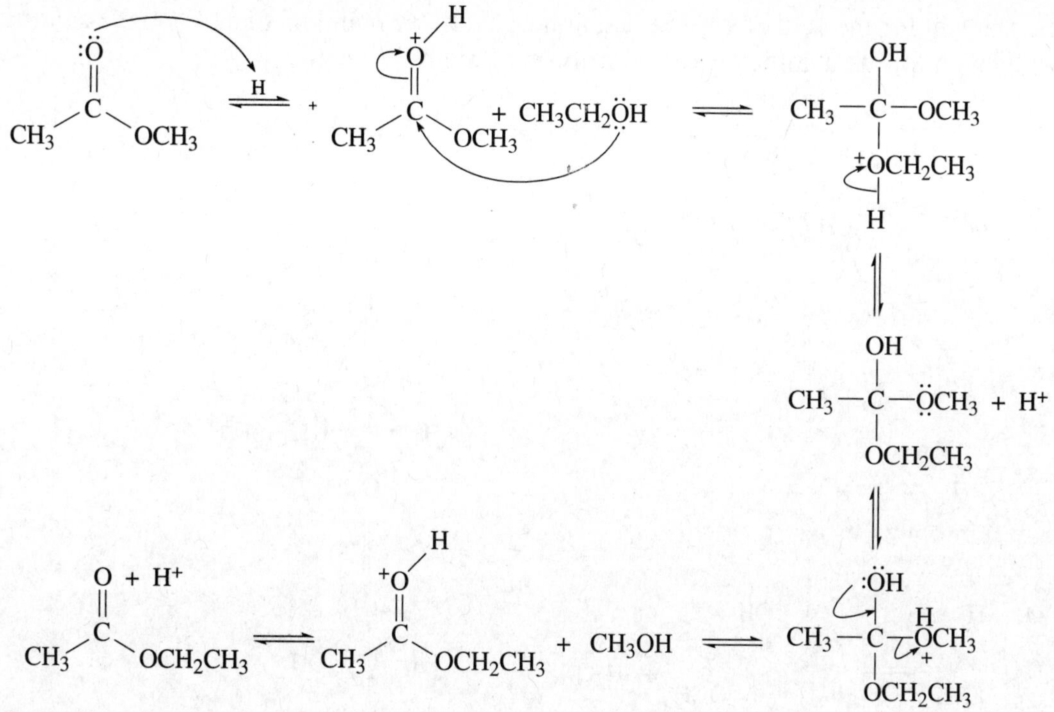

18. Solved in the text

19. **a.**

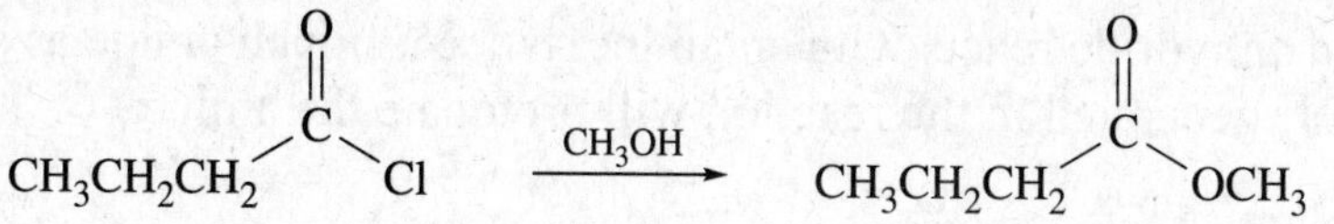

methyl butyrate

b.

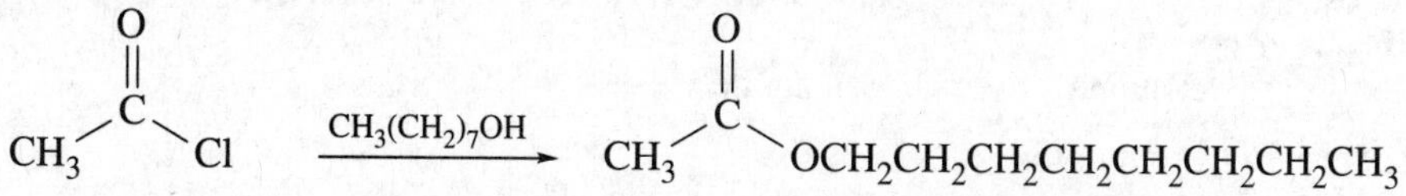

octyl acetate

20. The mechanism for the acid-catalyzed reaction of acetic acid and methanol is the exact reverse of the mechanism for the acid-catalyzed hydrolysis of methyl acetate.

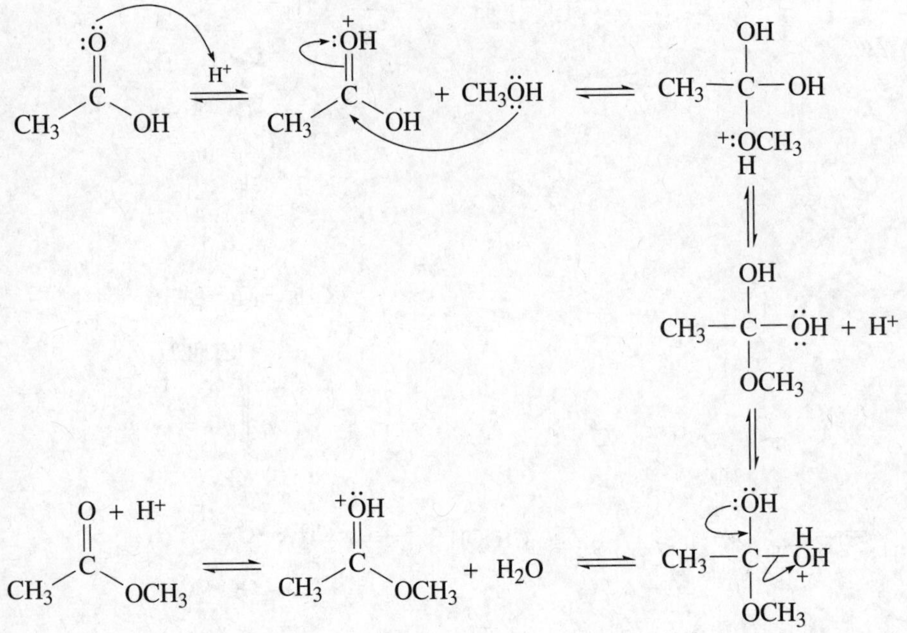

21. **a.** Solved in the text.

b. Notice that when an acid anhydride reacts with an amine, twice as much amine as anhydride is required because the H^+ generated in the reaction will protonate the amine.

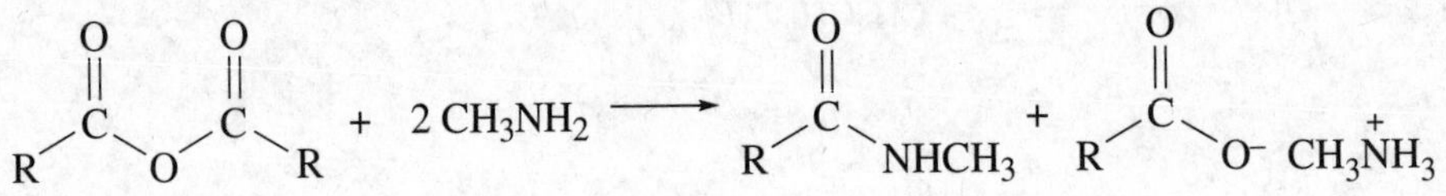

an anhydride   an amine   an amide

22. **a.** 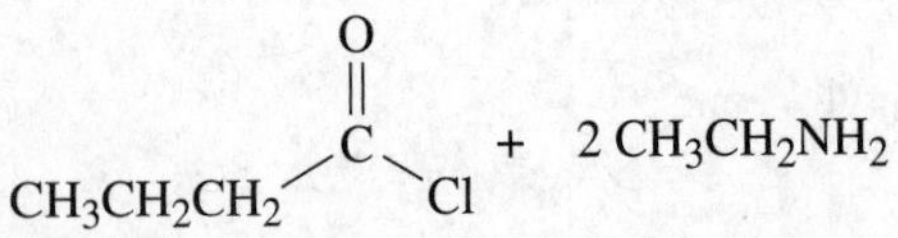

b. 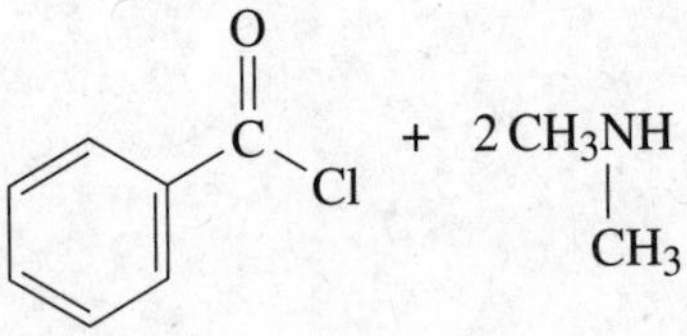

23. #2 and #4 will form amides
#1 will form a salt
#3 no reaction

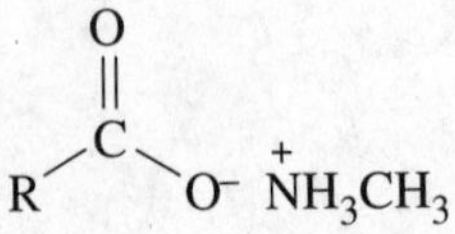

a salt

24. **a.** 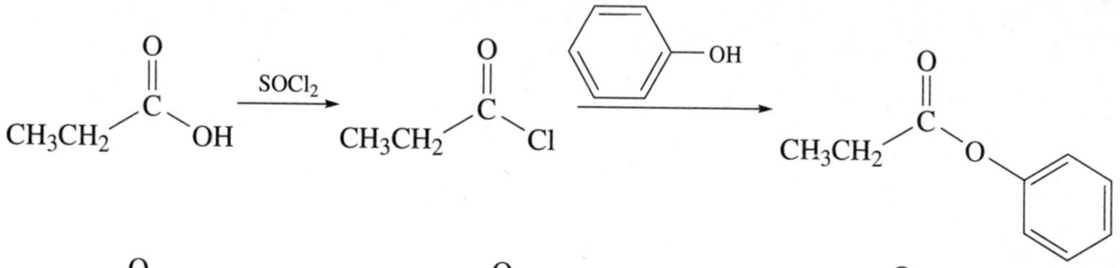

b. 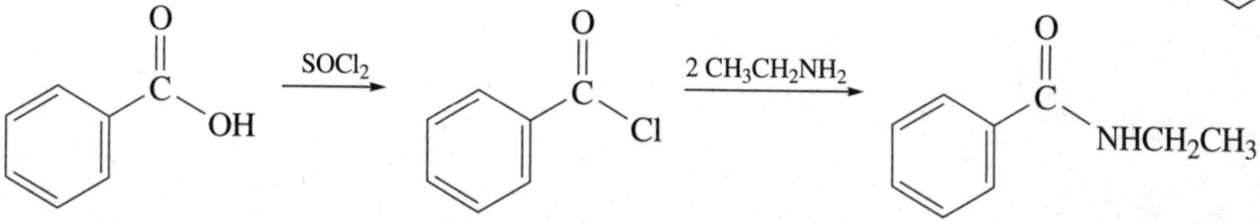

25. **a.** $CH_3CH_2CH_2Br$

b. 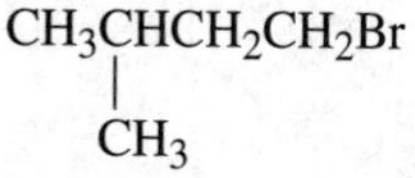

26. **a.** 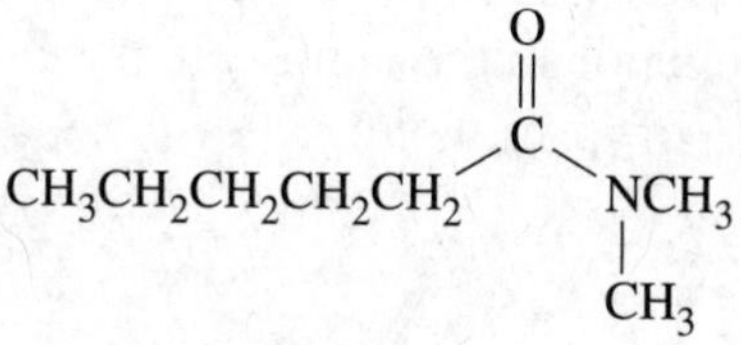

d. 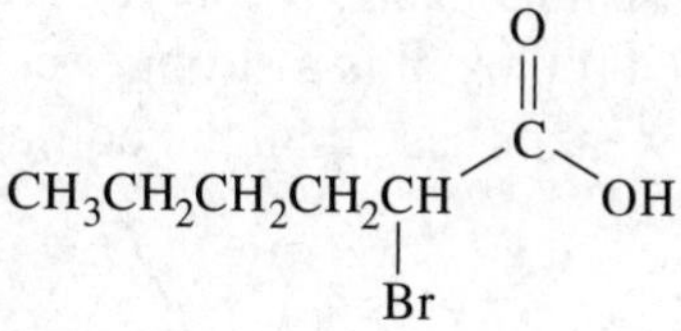

b. 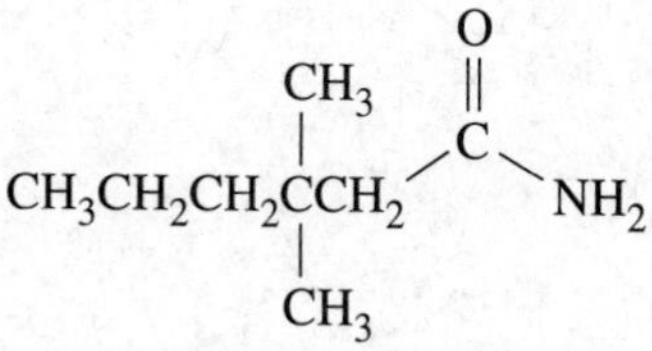

e. 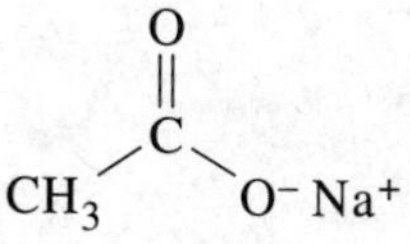

c. 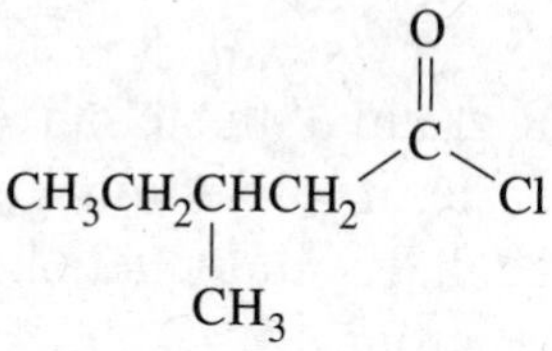

f. 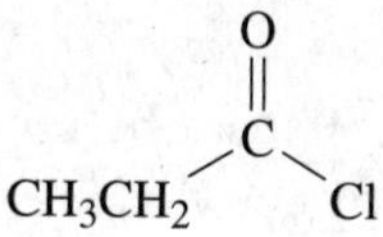

27. **a.** 5-ethylheptanoic acid

b. propyl propanoate
propyl propionate

c. cyclohexyl butanoate
cyclohexyl butyrate

d. *N,N*-dimethylbutanamide
N,N-dimethylbutyramide

e. pentanoyl chloride
valeryl chloride

f. methyl 3-methylpentanoate
methyl *β*-methylvalerate

g. *N*-methyl-3-butenamide

h. (*S*)-3-methylpentanoic acid
(*S*)-*β*-methylvaleric acid

i. pentanenitrile
butyl cyanide

28. a. 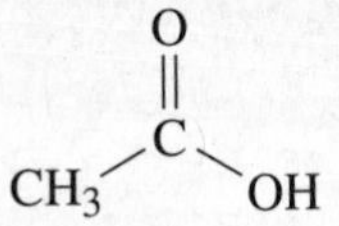

d. 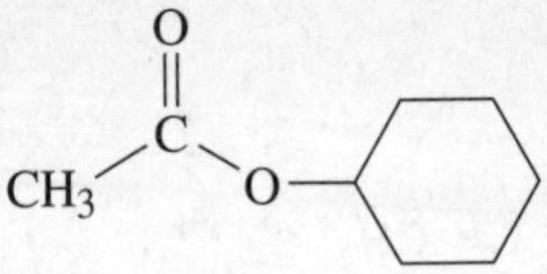

b. 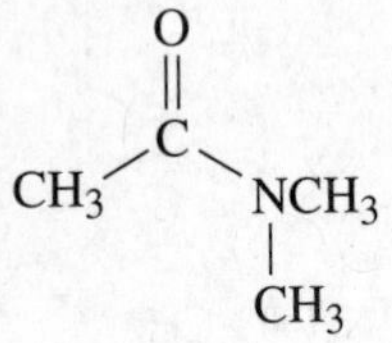

e. 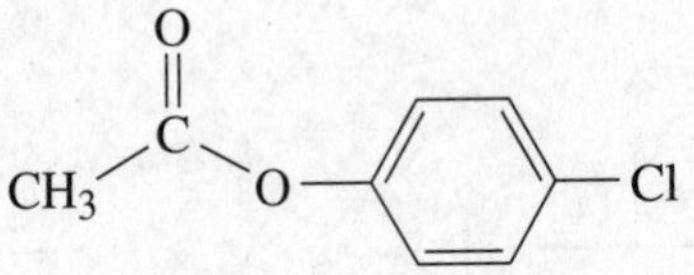

c. 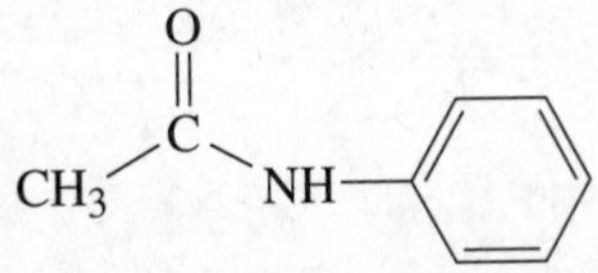

f. 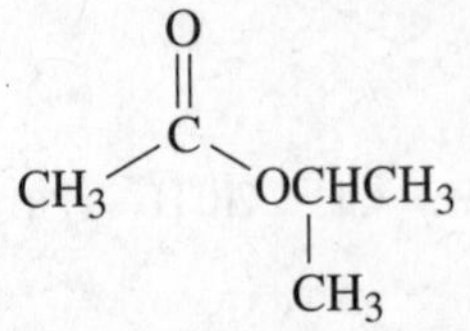

29. a. The weaker the base attached to the acyl group, the easier it is to form the tetrahedral intermediate. (*para*-Chlorophenol is a stronger acid than phenol so the conjugate base of *para*-chlorophenol is a weaker base than the conjugate base of phenol, etc.)

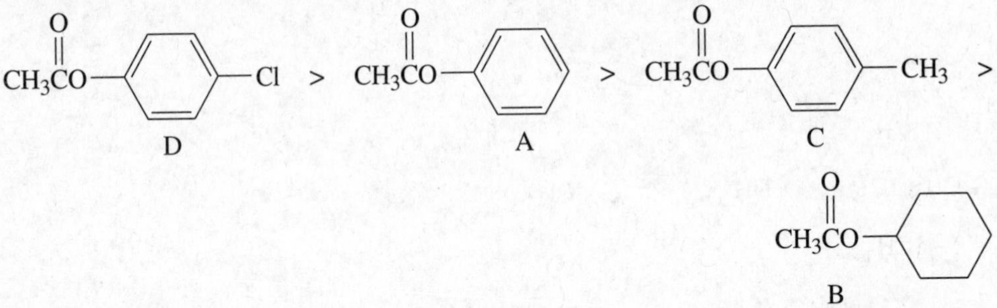
D
A
C
B

b. The tetrahedral intermediate collapses by eliminating the $^{-}$OR group of the tetrahedral intermediate. The weaker the basicity of the $^{-}$OR group, the easier it is to eliminate it. Thus, the rate of both formation of the tetrahedral intermediate and collapse of the tetrahedral intermediate is decreased by increasing the basicity of the $^{-}$OR group.

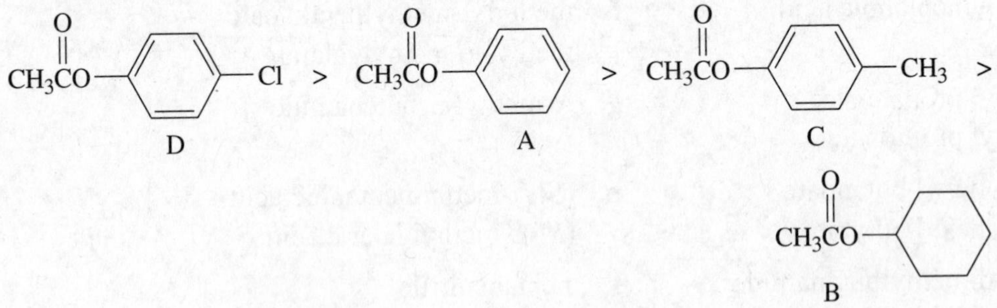
D
A
C
B

30. a. 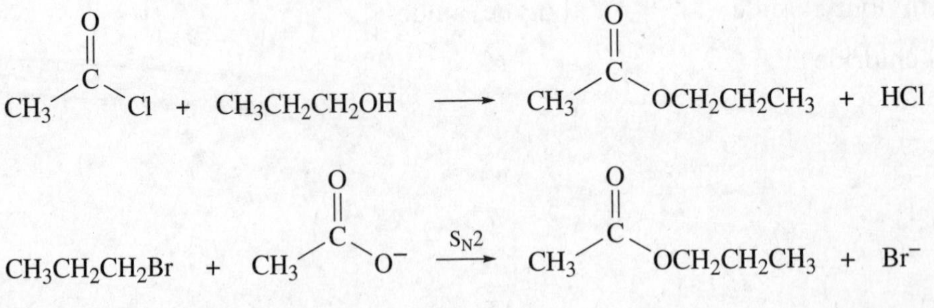
SN2

b.

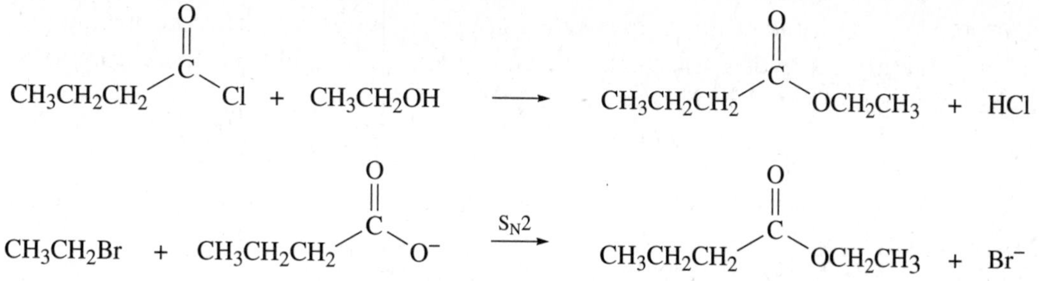

c.

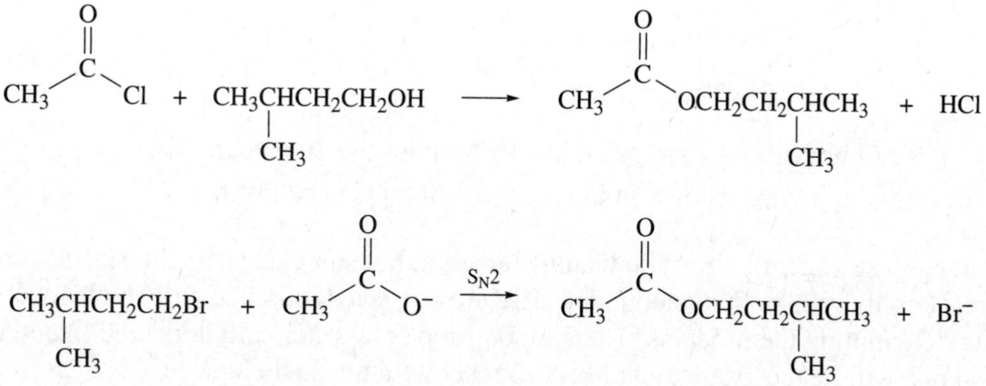

d.

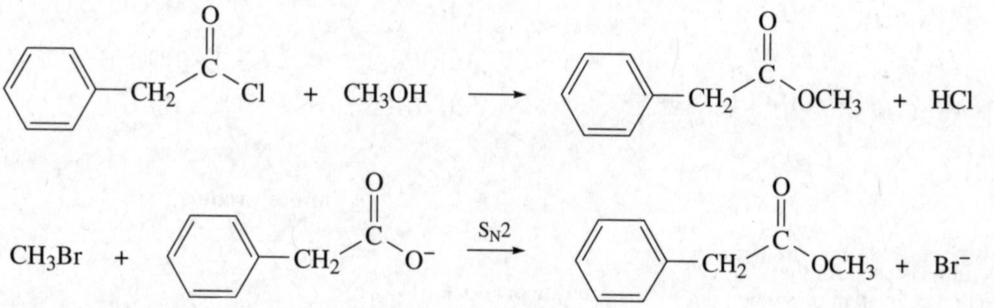

31. The carboxylic acid will have a higher boiling point because it can form hydrogen bonds. The ester cannot form hydrogen bonds.

32. The reaction of methylamine with propionyl chloride generates a proton that will protonate unreacted amine, thereby destroying its nucleophilicity. If two equivalents of CH_3NH_2 are used, one equivalent will remain unprotonated and be able to react with propionyl chloride to form about one equivalent of *N*-methylpropanamide.

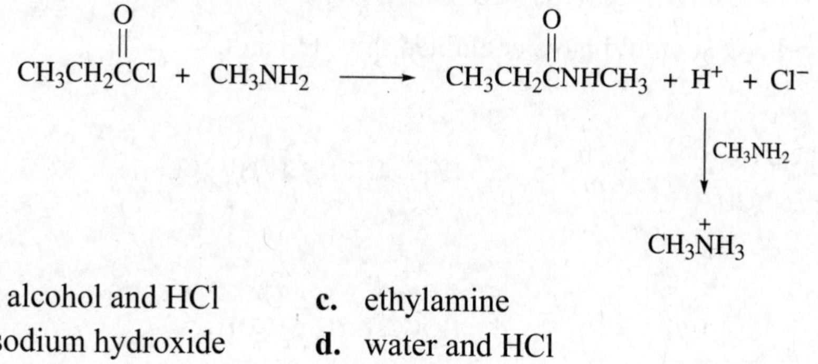

33. **a.** isopropyl alcohol and HCl **c.** ethylamine

b. aqueous sodium hydroxide **d.** water and HCl

34. Aspartame has an amide group and an ester group that will be hydrolyzed in an aqueous solution of HCl. Because the hydrolysis is carried out in an acidic solution, the carboxylic acid groups and the amino groups will be in their acidic forms.

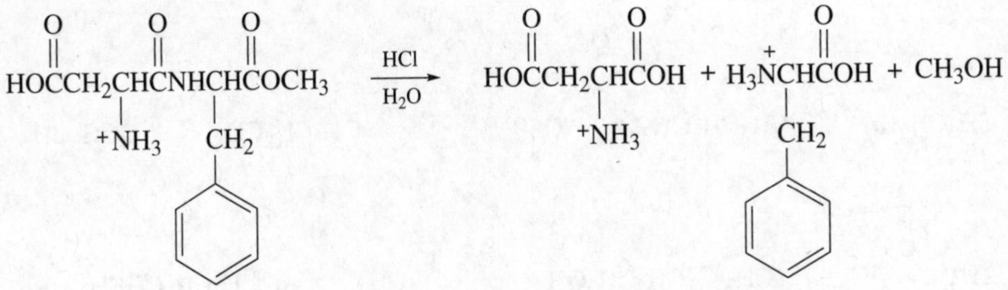

35. **a.** 1, 2, 4, 6 will not form the indicated products under the given conditions.
b. 6 will form the product shown in the presence of an acid catalyst.

36. The amine is a stronger nucleophile than the alcohol, but since the acyl chloride is very reactive, it can react easily with both nucleophiles. Therefore, steric hindrance will be the most important factor in determining the products. The amino group is less sterically hindered than the alcohol group so that will be the group most likely to react with the acyl chloride.

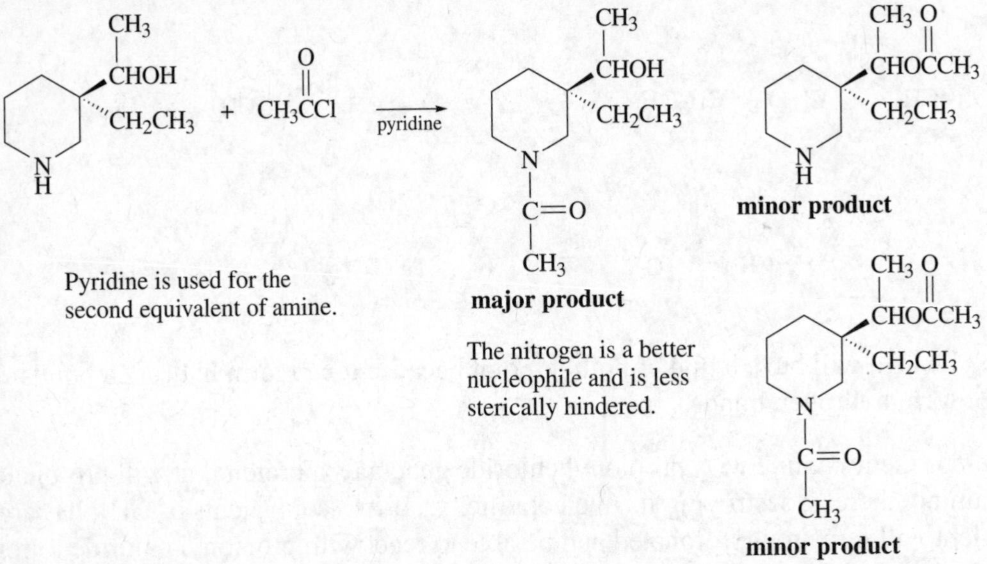
pyridine
Pyridine is used for the second equivalent of amine.
major product
The nitrogen is a better nucleophile and is less sterically hindered.
minor product
minor product

37. **a.** The alcohol (CH_3CH_2OH) contained the ^{18}O label.
b. The carboxylic acid would have contained the ^{18}O label.

38. **a.**

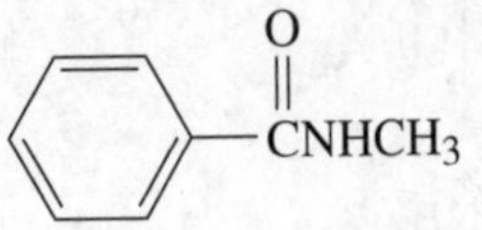

b. $CH_3\overset{O}{\overset{\|}{C}}F$

c. $HO\overset{O}{\overset{\|}{C}}CH_2CH_2CH_2\overset{+}{N}H_3\ Cl^-$

d. $HO\overset{O}{\overset{\|}{C}}CH_2CH_2CH_2OH$

39. In the tetrahedral intermediate, the potential leaving groups will be a chloride ion and an alkoxide ion; the chloride ion is the one that will be eliminated. Therefore, it is easier for the intermediate to collapse back to starting materials, reforming the ester, than to eliminate the alkoxide ion and form an acyl chloride. This means that "a" represents the reaction coordinate diagram for the reaction.

40. The alcohol component of phenyl acetate is a stronger acid ($pK_a = 10.0$) than is the alcohol component of methyl acetate ($pK_a = 15.5$). Therefore, the phenoxide ion is a weaker base than the methoxide ion, causing phenyl acetate to be the more reactive ester.

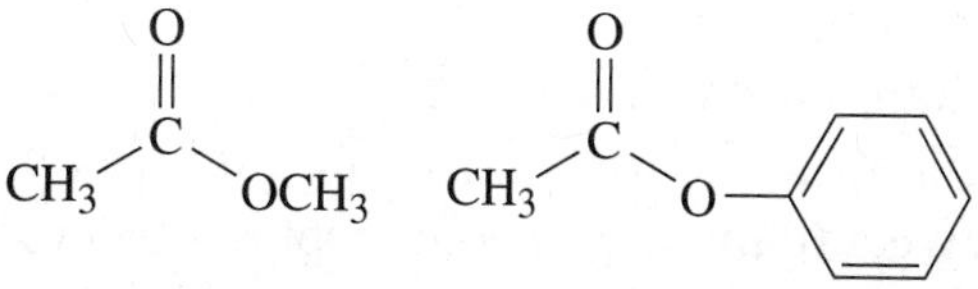

methyl acetate phenyl acetate

CH_3OH 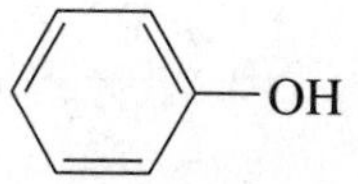

$pK_a = 15.5$ $pK_a = 10.0$

41. The relative reactivities of the amides depend on the basicities of their leaving groups: the weaker the base, the more reactive the amide.

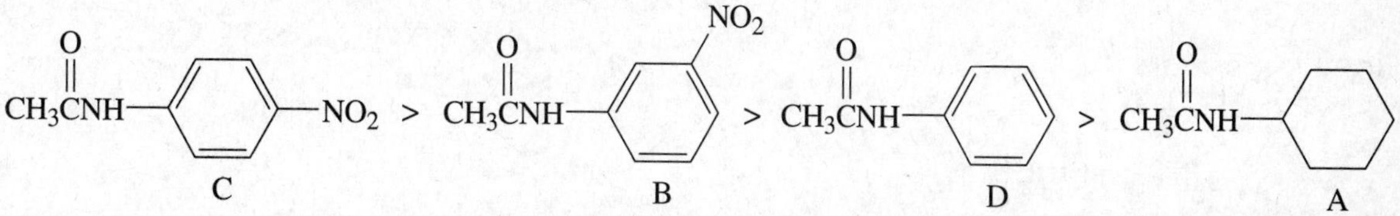
C
B
D
A

A *para*-nitro-substituted aniline is less basic than a *meta*-substituted aniline because when the nitro group is in the para position, electrons can be delocalized onto the nitro group.

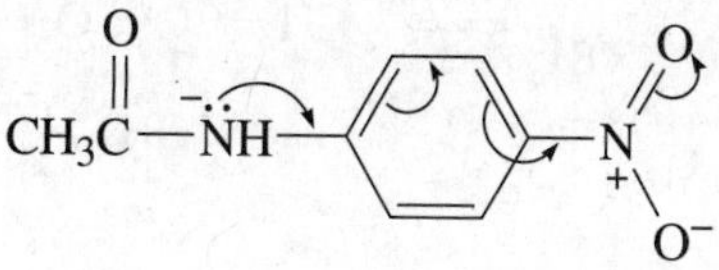

42. **a.** $HO\overset{O}{\overset{\|}{C}}CH_2CH_2\overset{O}{\overset{\|}{C}}OH$ **b.** $CH_3\overset{O}{\overset{\|}{C}}OCH_3 + HOCH_2\overset{O}{\overset{\|}{C}}CH_3$

43. If the amine is tertiary, the nitrogen atom in the amide cannot get rid of its positive charge by losing a proton. An amide with a positively charged amino group is very reactive because the positively charged amino group is a weak base and, therefore, an excellent leaving group. Water will immediately react with the amide, and because the positively charged amine is a better leaving group than the OH group, the amine will be expelled and the product will be a carboxylic acid.

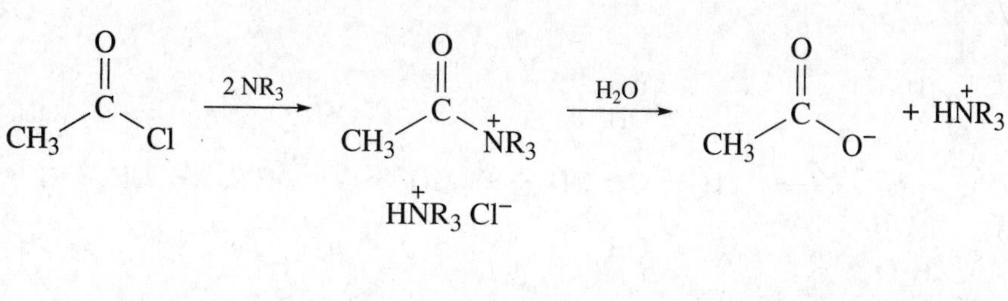
2 NR3
H2O

44. The acid-catalyzed hydrolysis of acetamide forms acetic acid and ammonium ion. It is an *irreversible reaction*, because the pK_a of acetic acid is less than the pK_a of the ammonium ion. Therefore, it is impossible to have the carboxylic acid in its reactive acidic form and ammonia in its reactive basic form.

For example, if the solution is sufficiently acidic to have the carboxylic acid is in its acidic form, ammonia will also be in its acidic form so it will not be a nucleophile. If the pH of the solution is sufficiently basic to have ammonia in its nucleophilic basic form, the carboxylic acid will also be its basic form; a negatively charged carboxylic acid is not attacked by nucleophiles.

$$CH_3COH\ (\text{C=O}) + \overset{+}{N}H_4 \longrightarrow \text{no reaction}$$

The ammonium ion is not a nucleophile.

$$CH_3CO^-\ (\text{C=O}) + NH_3 \longrightarrow \text{no reaction}$$

The carboxylate ion is not attached by nucleophiles.

45. **a.** 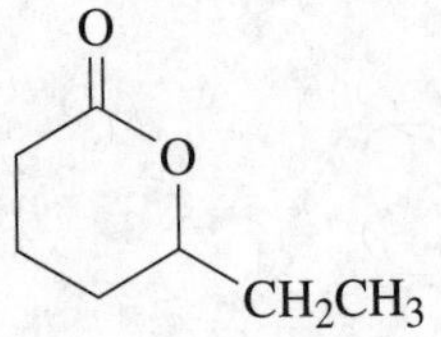

b. 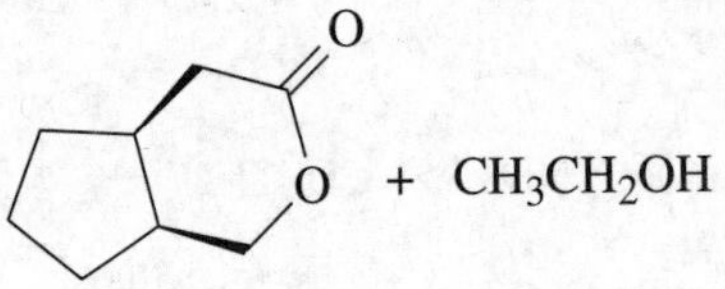

46. **a.** 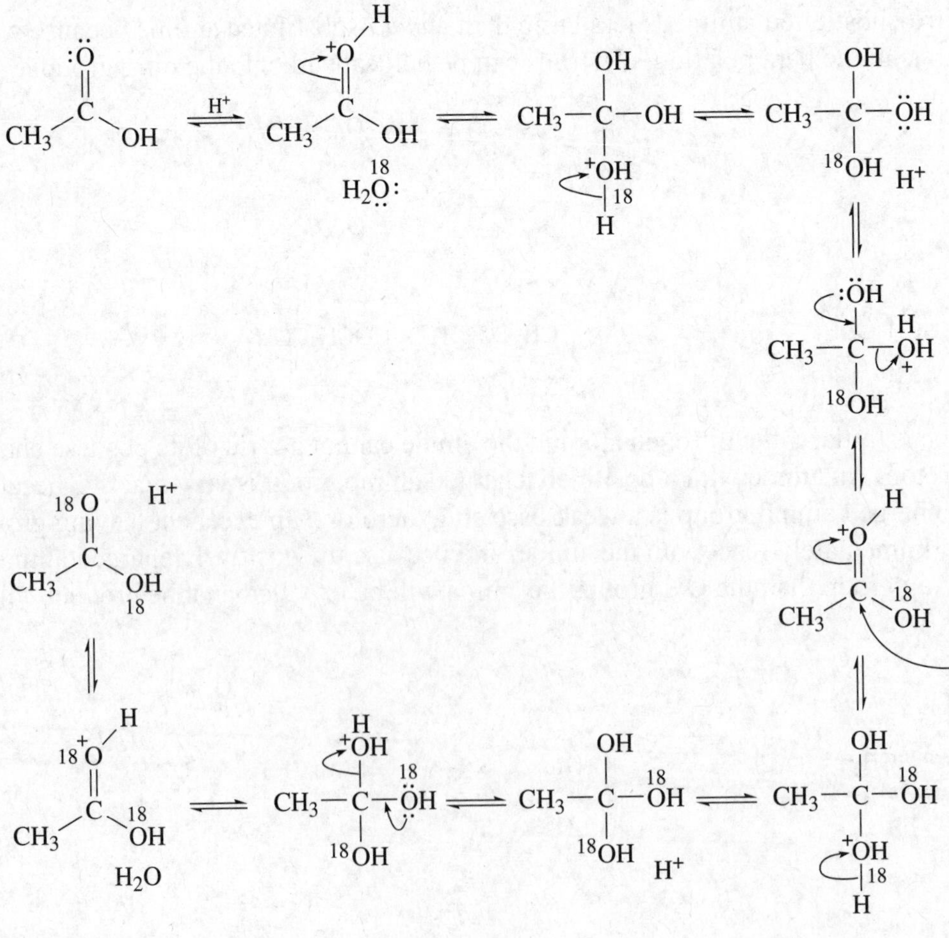

b. Only one isotopically labeled oxygen can be incorporated into the ester because the bond between the methyl group and the labeled oxygen does not break. That bond would have to break in order for the carbonyl oxygen to become labeled.

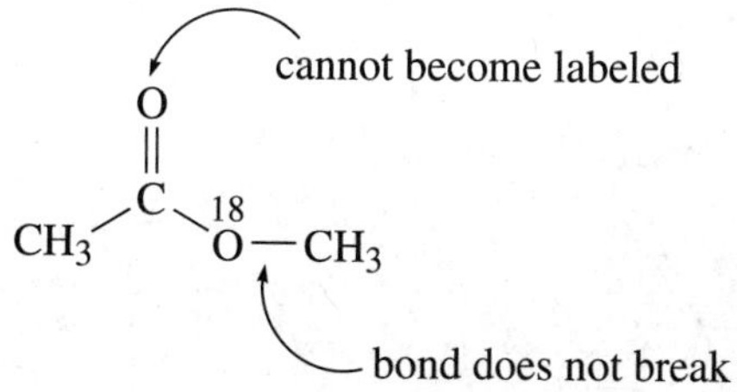
cannot become labeled
O
C
CH3
18
O
CH3
bond does not break

Chapter 11 Practice Test

1. Circle the compound in each pair that is more reactive toward nucleophilic acyl substitution.

a. $CH_3\overset{O}{\overset{\|}{C}}OCH_3$ or $CH_3\overset{O}{\overset{\|}{C}}NHCH_3$

b. $CH_3\overset{O}{\overset{\|}{C}}OCH_3$ or 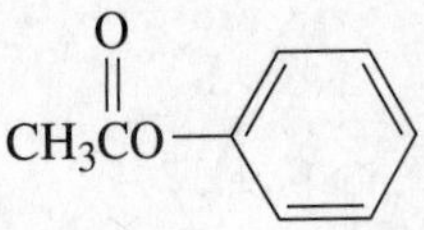

c. 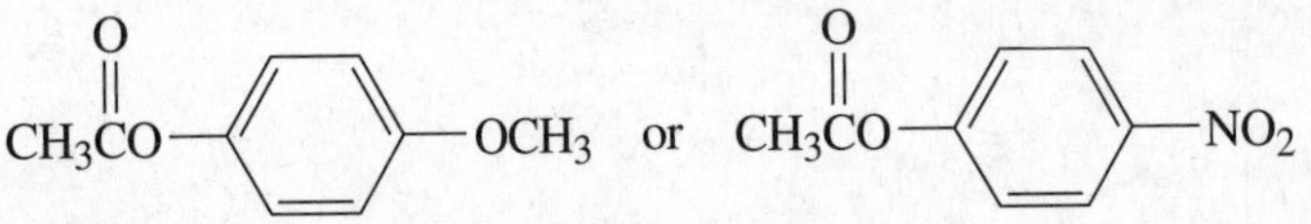

d. $CH_3\overset{O}{\overset{\|}{C}}OCH_3$ or $CH_3\overset{O}{\overset{\|}{C}}Cl$

2. Give the systematic name for each of the following:

a. $CH_3CH_2CH_2CH_2\overset{O}{\overset{\|}{C}}NHCH_2CH_3$

b. $CH_3CH_2\underset{CH_3}{\underset{|}{C}}HCH_2\overset{O}{\overset{\|}{C}}OH$

c. 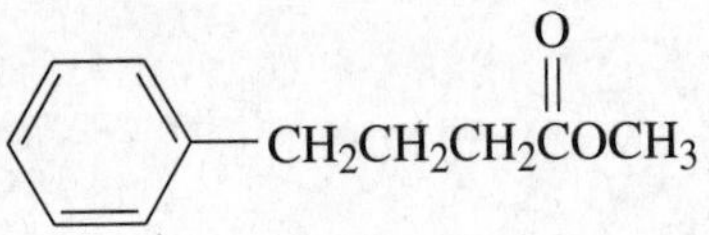

3. Give an example of each of the following:

a. an aminolysis reaction

b. a hydrolysis reaction

c. a transesterification reaction

4. What carbonyl compound would be obtained from collapse of each of the following tetrahedral intermediates?

a. $CH_3\text{—}\overset{OH}{\underset{OH}{C}}\text{—}\overset{+}{N}H_3$

b. $CH_3\text{—}\overset{OH}{\underset{+OH\ H}{C}}\text{—OCH}_3$

c. $CH_3\text{—}\overset{O^-}{\underset{OH}{C}}\text{—NH}_2$

d. $CH_3\text{—}\overset{OH}{\underset{OH}{C}}\text{—}\overset{+}{O}CH_3\ H$

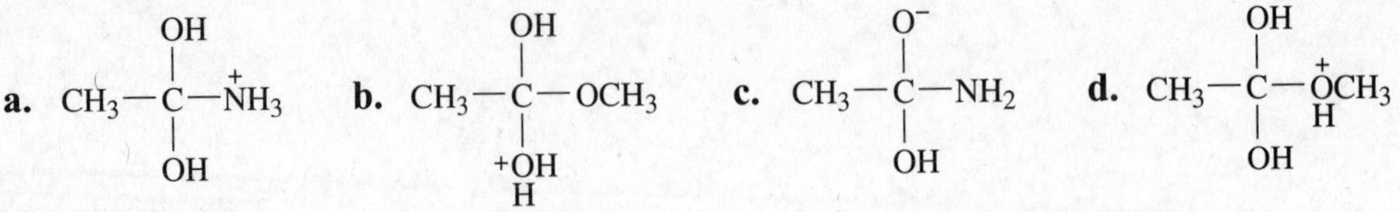

5. Give the product of each of the following reactions.

a. $CH_3CH_2C{\equiv}N + H_2O \xrightarrow[\Delta]{HCl}$

b. $CH_3CH_2\overset{O}{\overset{\|}{C}}OCH_2CH_2CH_3 + \underset{\text{excess}}{H_2O} \xrightarrow{HCl}$

c. $CH_3CH_2\overset{O}{\overset{\|}{C}}Cl + 2\ CH_3CH_2NH_2 \longrightarrow$

d. 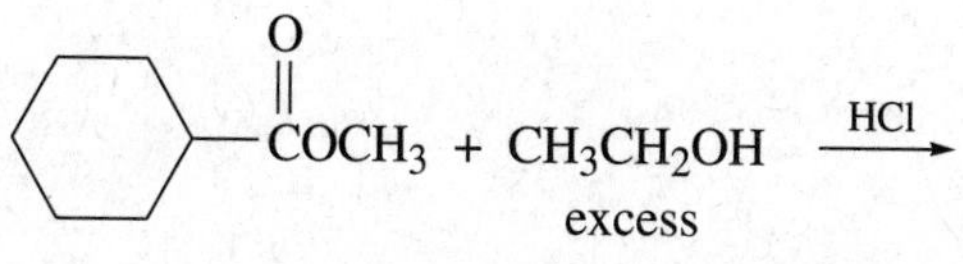
COCH3 + CH3CH2OH
HCl
excess

e. $CH_3CH_2CH_2\overset{O}{\overset{\|}{C}}OH \xrightarrow[\text{2. } CH_3CH_2CH_2OH]{\text{1. } SOCl_2}$

f. $CH_3CH_2\overset{O}{\overset{\|}{C}}NHCH_3 + H_2O \xrightarrow[\Delta]{HCl}$

g. 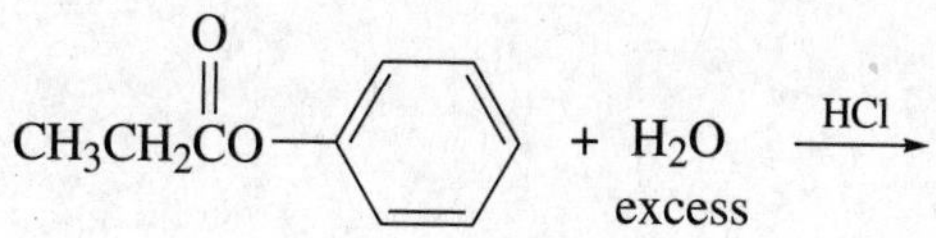
CH3CH2CO
+ H2O
HCl
excess

CHAPTER 12
Carbonyl Compounds II:
Reactions of Aldehydes and Ketones • More Reactions of Carboxylic Acid Derivatives

1. If the ketone functional group were anywhere else in these compounds, they would not be ketones and, therefore, would not have the "one" suffix.

2. **a.** 4-heptanone **b.** 4-phenylbutanal

3. **a.** 3-methylpentanal and *β*-methylvaleraldehyde

b. 2-methyl-4-heptanone and isobutyl propyl ketone

c. 4-ethylhexanal and *γ*-ethylcaproaldehyde

d. 6-methyl-3-heptanone and ethyl isopentyl ketone

4. **a.** 2-Heptanone is more reactive because it has less steric hindrance. There is little difference in the amount of steric hindrance provided at the carbonyl carbon (the site of nucleophilic attack) by a pentyl and a propyl group because they differ at a point somewhat removed from the site of nucleophilic attack. However, the difference in size between a methyl group and a propyl group is significant at the site of nucleophilic attack.

$$CH_3\overset{\overset{O}{\|}}{C}CH_2CH_2CH_2CH_2CH_3 \qquad CH_3CH_2CH_2\overset{\overset{O}{\|}}{C}CH_2CH_2CH_3$$

2-heptanone 4-heptanone

b. 5-Methyl-3-hexanone is more reactive because the methyl substituent is farther away from the carbonyl carbon.

$$CH_3CH_2\overset{\overset{O}{\|}}{C}CH_2\underset{\underset{CH_3}{|}}{C}HCH_3 \qquad CH_3CH_2\overset{\overset{O}{\|}}{C}\underset{\underset{CH_3}{|}}{C}HCH_2CH_3$$

5-methyl-3-hexanone 4-methyl-3-hexanone

5. Notice that a Grignard reagent will react with an H bonded to an O or an N, forming an alkane.

a. $CH_3CH_2MgBr + H_2O \longrightarrow CH_3CH_3 + Mg^{2+}\ Br^-\ HO^-$

b. $CH_3CH_2MgBr + CH_3OH \longrightarrow CH_3CH_3 + Mg^{2+}\ Br^-\ CH_3O^-$

c. $CH_3CH_2MgBr + CH_3NH_2 \longrightarrow CH_3CH_3 + Mg^{2+}\ Br^-\ CH_3\bar{N}H$

6. **a.** $CH_3CH_2CH_2CH_2\overset{\overset{OH}{|}}{C}HCH_3$ **b.** $CH_3CH_2CH_2\underset{\underset{CH_3}{|}}{\overset{\overset{OH}{|}}{C}}CH_3$ **c.**

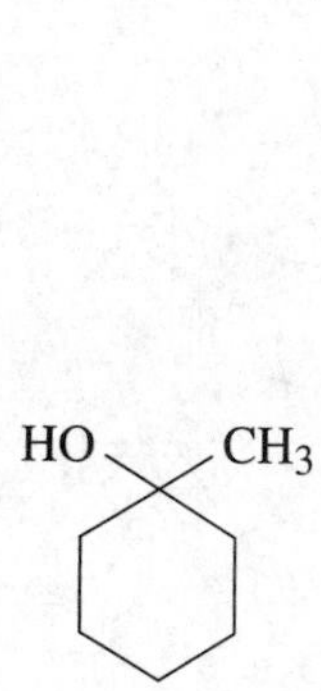
HO
CH3

7. $CH_3\overset{\overset{O}{\|}}{C}CH_2CH_3 + CH_3CH_2CH_2MgBr$ and $CH_3CH_2\overset{\overset{O}{\|}}{C}CH_2CH_2CH_3 + CH_3MgBr$

8. **a.** Solved in the text.

b. **1.** Solved in the text

2. $CH_3C(=O)OR + 2\ CH_3MgBr$

4. $CH_3C(=O)OR + 2\ CH_3CH_2MgBr$

6. 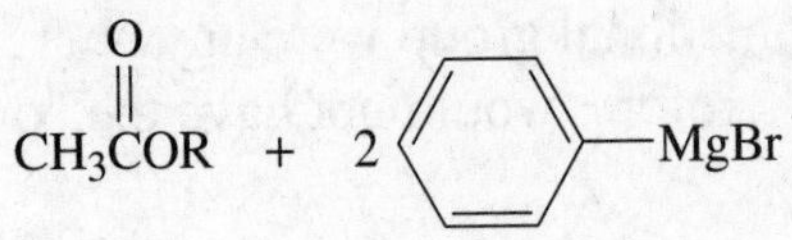
O
CH3COR + 2
MgBr

9. The reaction of methyl formate with excess Grignard reagent will form a secondary alcohol with two identical substituents, since the two substituents come from the Grignard reagent.

Therefore, only the following two alcohols can be prepared that way.

$CH_3CH(OH)CH_3$ $\quad$ $CH_3CH_2CH(OH)CH_2CH_3$

10. **A** will not undergo a nucleophilic addition reaction with a Grignard reagent because the Grignard reagent will react with the H on the N.

C will not undergo a nucleophilic addition reaction with a Grignard reagent because the Grignard reagent will react with the H on the O.

11. **a.** $CH_3CH(CH_3)CH_2OH$

b. 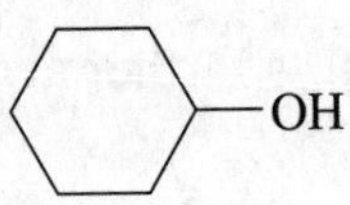
OH

c. 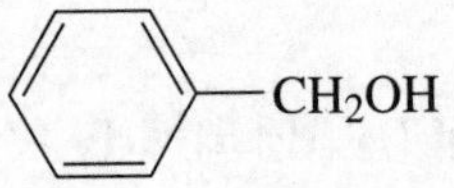
CH2OH

d. 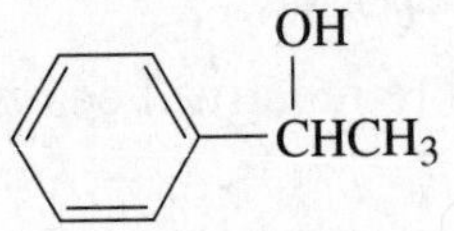
OH
CHCH3

12. **a.** 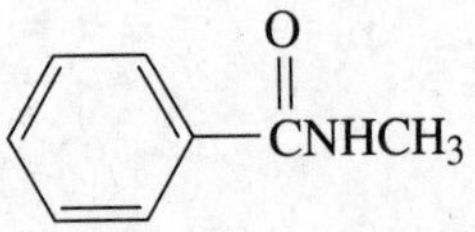
O
CNHCH3

b. $CH_3C(=O)NH_2$

c. $CH_3C(=O)NHCH_2CH_3$

d. $CH_3C(=O)N(CH_2CH_3)_2$

13. **a.** 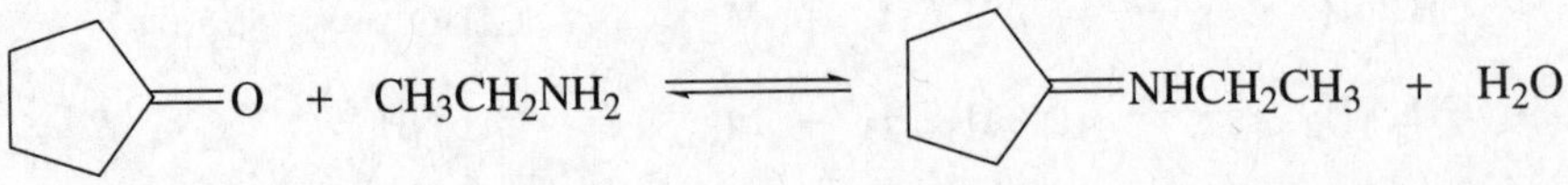
O + CH3CH2NH2
NHCH2CH3 + H2O

b. 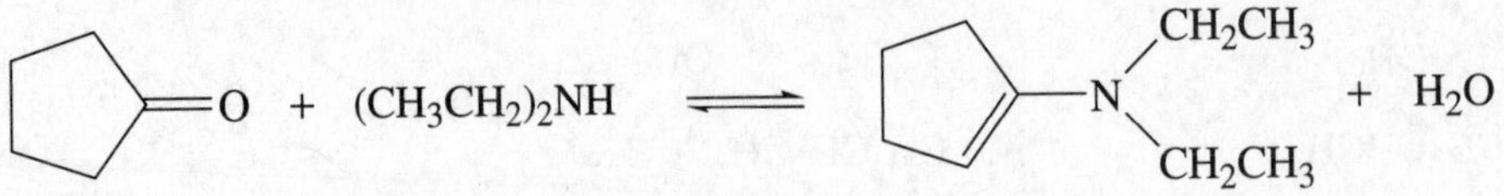
O + (CH3CH2)2NH
CH2CH3
N
CH2CH3
+ H2O

c. 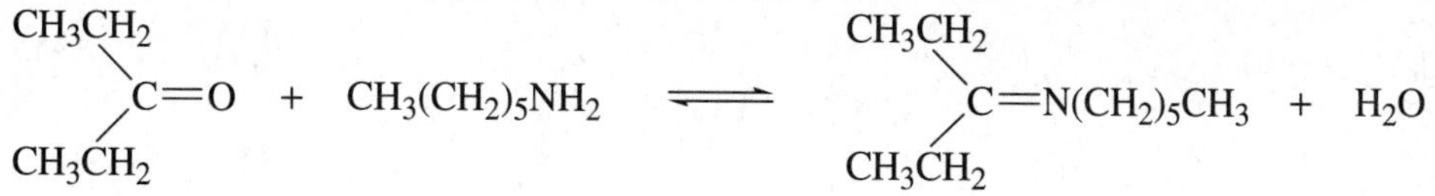

d. 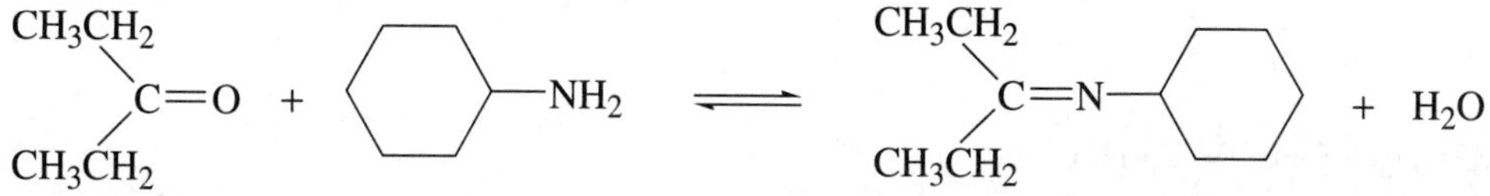

14. Electron-withdrawing groups decrease the stability (concentration) of the aldehyde and increase the stability (concentration) of the hydrate. Therefore, the three electron-withdrawing chlorines cause trichloroacetaldehyde to have a large equilibrium constant for hydrate formation.

$$K_{eq} = \frac{[\text{hydrate}]}{[\text{aldehyde}]\,[H_2O]}$$

15. Because an electron-withdrawing substituent decreases the stability of a ketone and increases the stability of a hydrate, the compound with the electron-withdrawing *para*-nitro substituents has the largest equilibrium constant for addition of water and, therefore, forms the most hydrate.

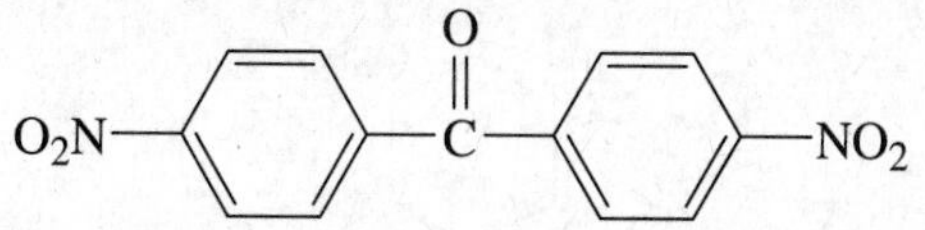

16.

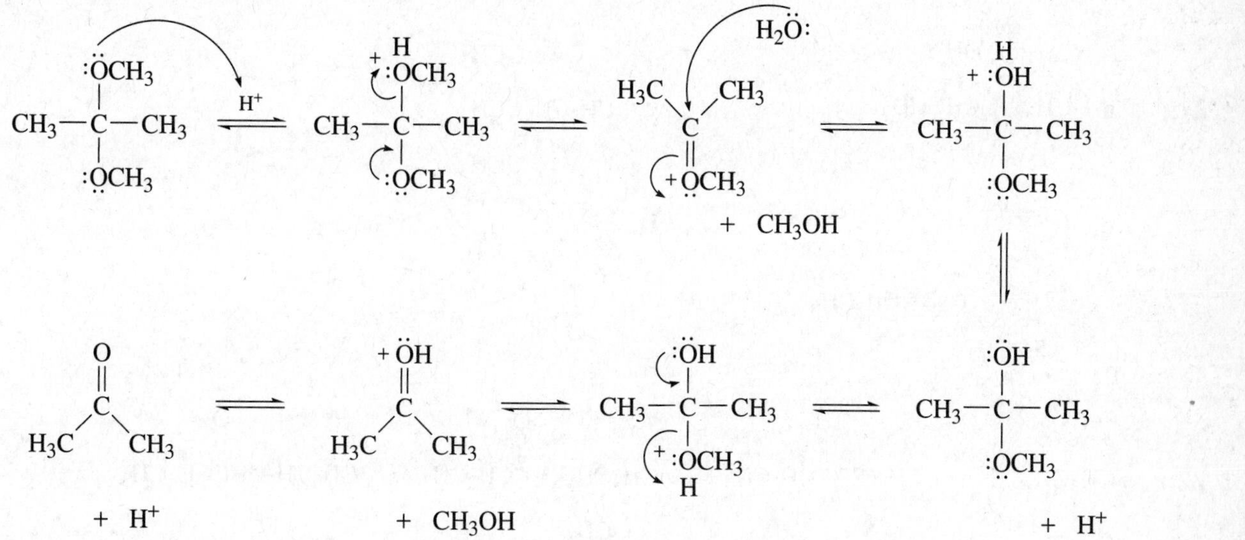

17. **a.** hemiacetals: 7

b. acetals: 2, 3

c. hemiketals: 1, 8

d. ketals: 5

e. hydrates: 4, 6

18. **a.** 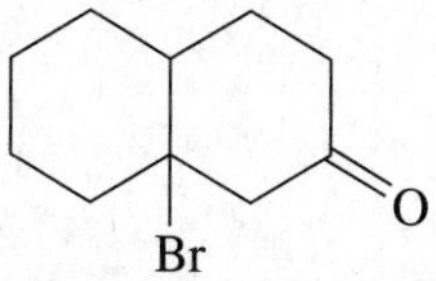

b. 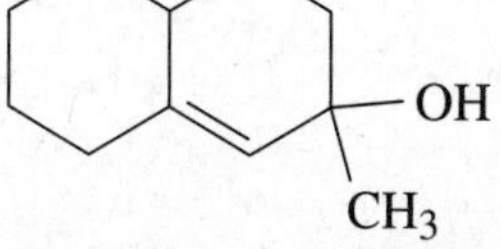

19. **a.** $CH_3CH(CH_3)CHO$ — O above CH; CH_3CHCH with CH_3 below

d. $CH_3CH_2CCHCH_2CH_2CH_3$ (O above C; Br below CH)

b. $CH_3CH_2CH_2CCH_2CH_2CH_2CH_3$ (O above C)

e. 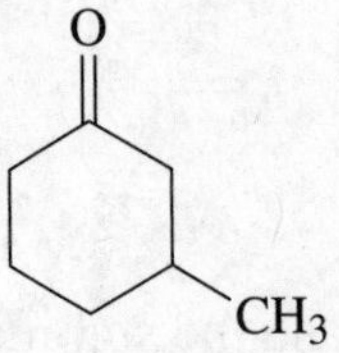

c. $CH_3CH_2CHCH_2CH_2CH$ (O above terminal CH; Br below CH)

f. $CH_3CCH_2CCH_3$ (O above each C)

20. **a.** $CH_3CH_2CHOCH_2CH_3$ (OCH_2CH_3 above CH)

c. $CH_3CH_2CH_2CH_2OH + CH_3CH_2OH$

b. $CH_3CH_2CHCH_3$ (OH above CH)

d. 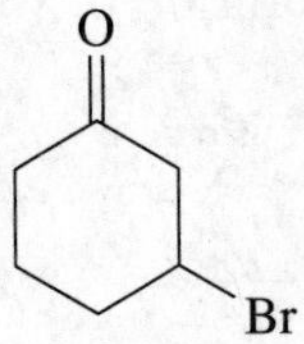

21. **a.** $CH_3CH_2CHOCH_3$ (OH above CH)

c. $CH_3CH_2CCH_3$ (OCH_3 above and below C)

b. 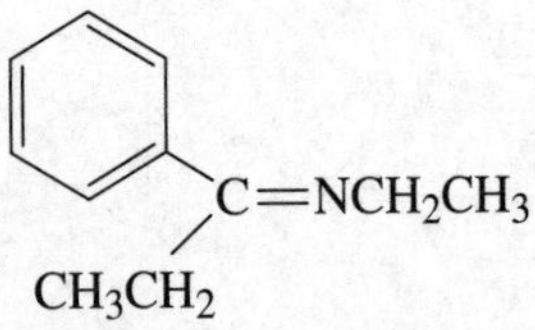

d. 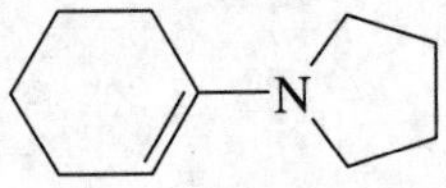

22. CH_3CH_2CH (O above CH) > $CH_3CH_2CCH_2CH_3$ (O above C) > $CH_3CHCH_2CCH_2CH_3$ (O above C; CH_3 below CH) > $CH_3CH_2CHCCH_2CH_3$ (O above C; CH_3 below CH) >

$CH_3CH_2CHCCHCH_2CH_3$ (O above C; CH_3 CH_3 below each CH) > $CH_3CH_2CHC{-}CCH_2CH_3$ (O above C; CH_3 above and below the second C; CH_3 below CH)

23.

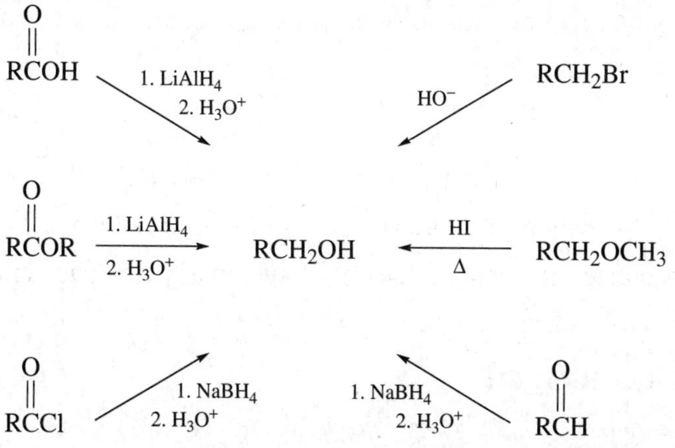
RCOH
1. LiAlH4
2. H3O+
HO−
RCH2Br
RCOR
1. LiAlH4
2. H3O+
RCH2OH
HI
Δ
RCH2OCH3
RCCl
1. NaBH4
2. H3O+
1. NaBH4
2. H3O+
RCH

24.

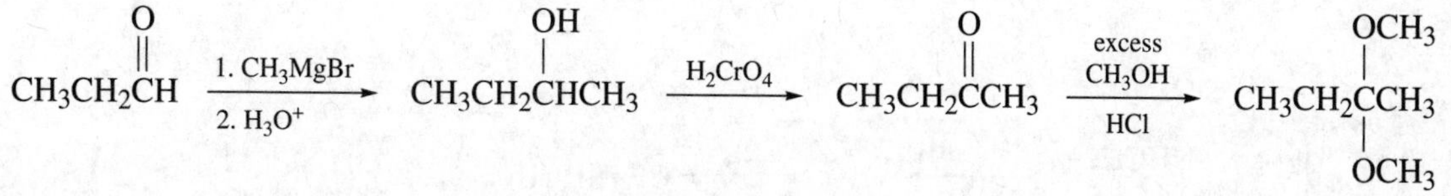
CH3CH2CH
1. CH3MgBr
2. H3O+
OH
CH3CH2CHCH3
H2CrO4
CH3CH2CCH3
excess
CH3OH
HCl
OCH3
CH3CH2CCH3
OCH3

25. **a.**

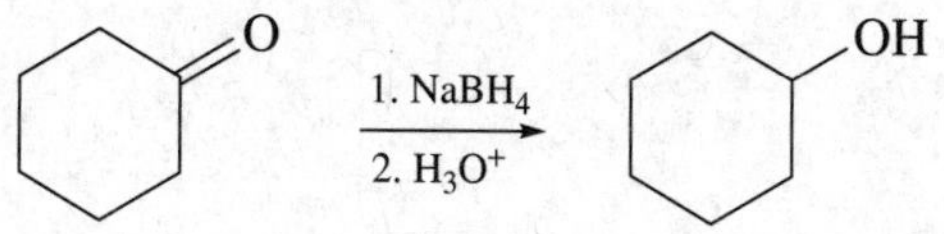
O
1. NaBH4
2. H3O+
OH

b.

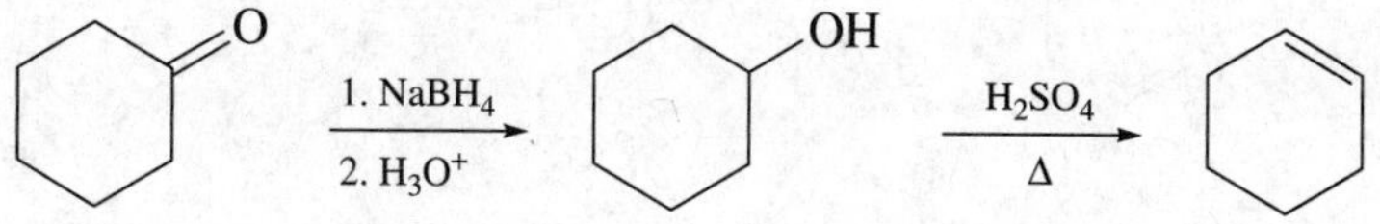
O
1. NaBH4
2. H3O+
OH
H2SO4
Δ

c.

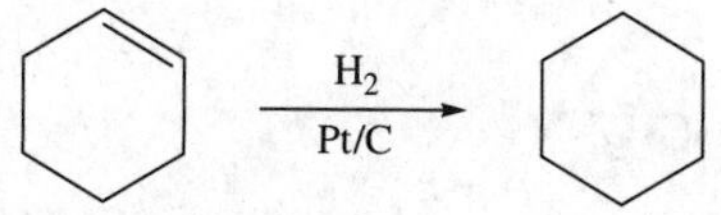
H2
Pt/C

product of **b**

d.

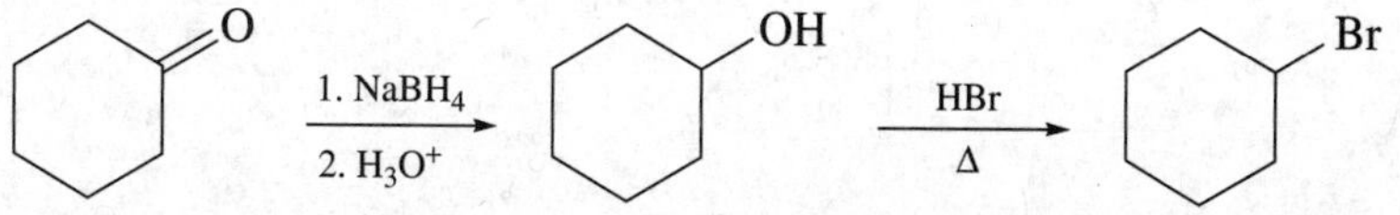
O
1. NaBH4
2. H3O+
OH
HBr
Δ
Br

e.

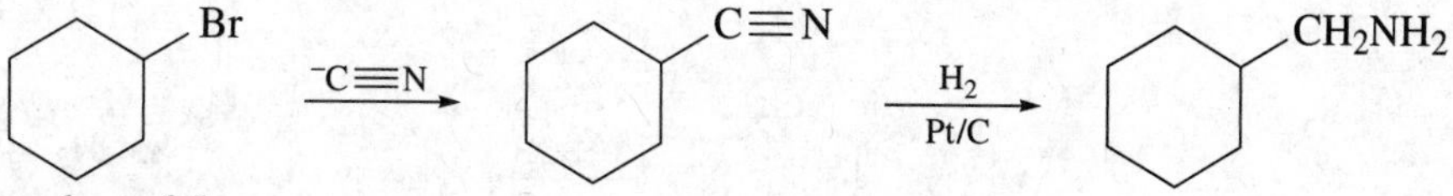
Br
−C≡N
C≡N
H2
Pt/C
CH2NH2

product of **d**

f.

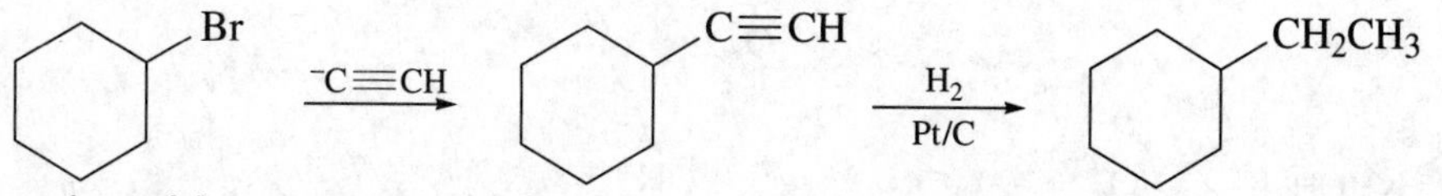
Br
−C≡CH
C≡CH
H2
Pt/C
CH2CH3

product of **d**

or

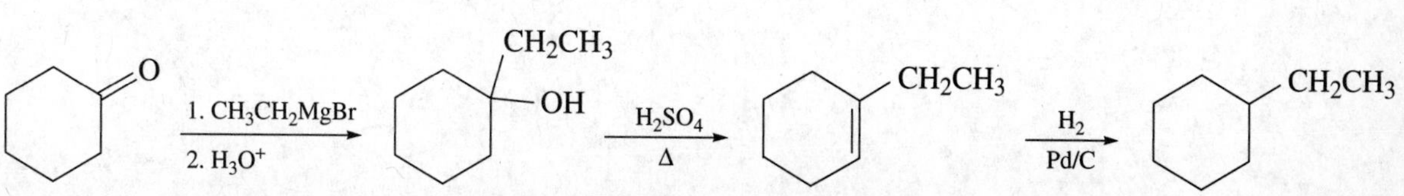
O
1. CH3CH2MgBr
2. H3O+
CH2CH3
OH
H2SO4
Δ
CH2CH3
H2
Pd/C
CH2CH3

26. **a.** Two isomers are obtained because the reaction creates an asymmetric center in the product.

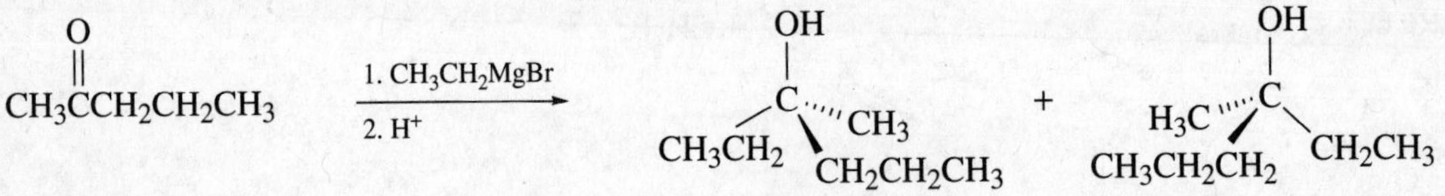

(*R*)-3-methyl-3-hexanol (*S*)-3-methyl-3-hexanol

b. Only one compound is obtained because the product does not have an asymmetric center.

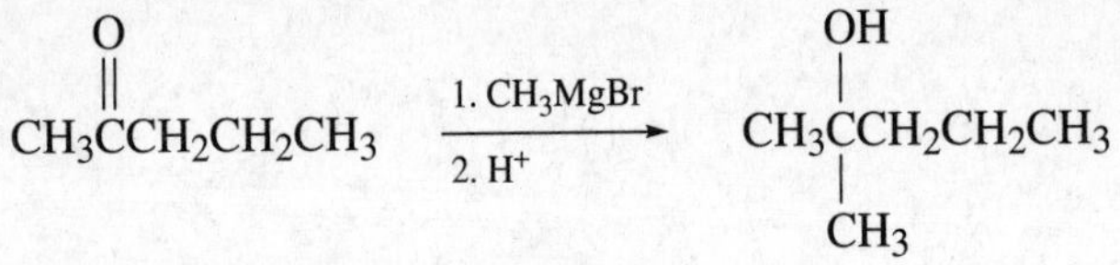

27. **a.**

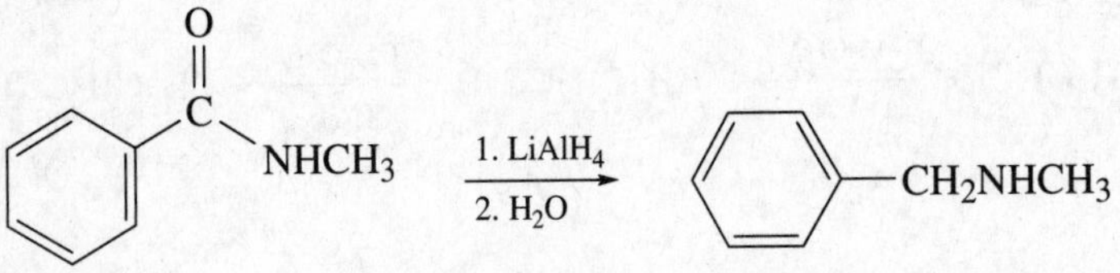

b.

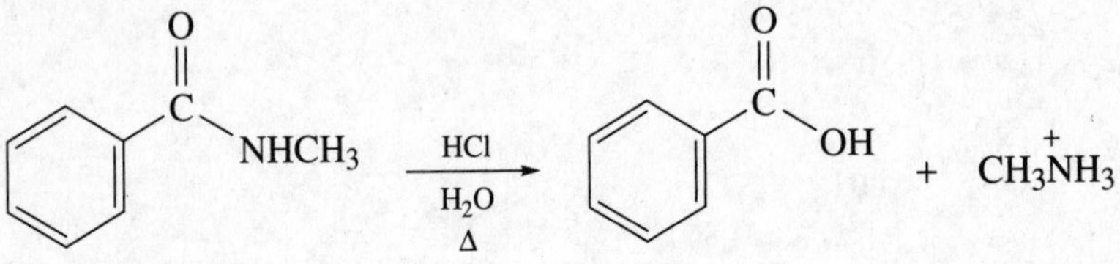

c.

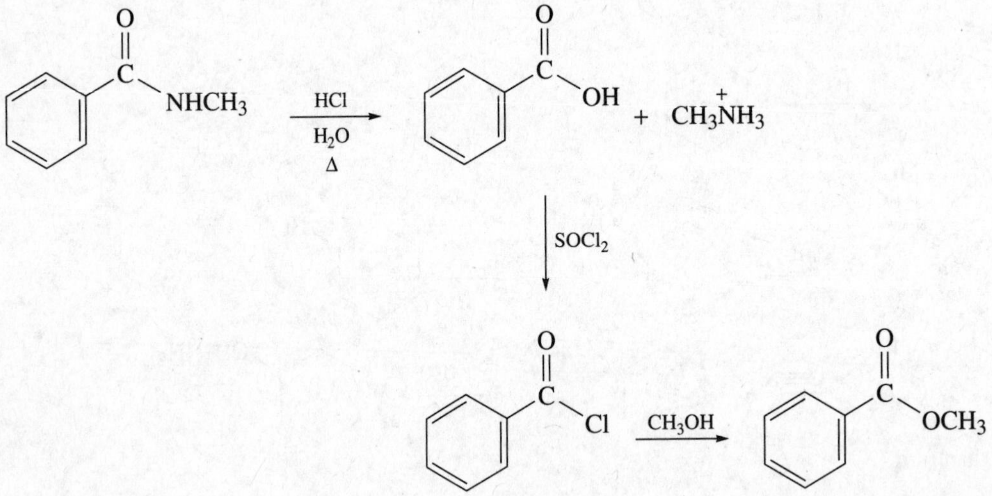

d.

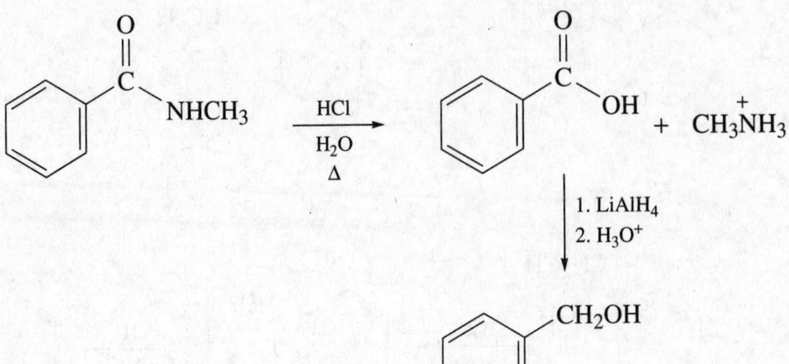

28.

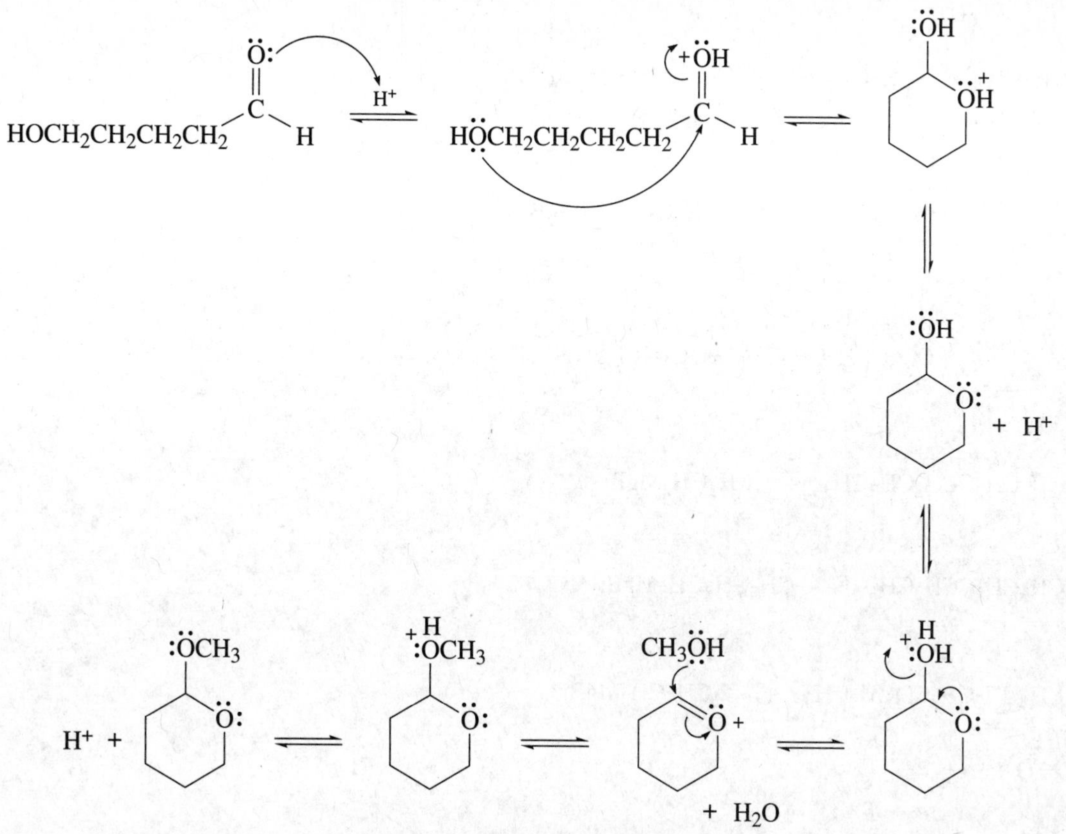

29. The greater the electron-withdrawing ability of the para substituent, the greater the K_{eq} for hydrate formation.

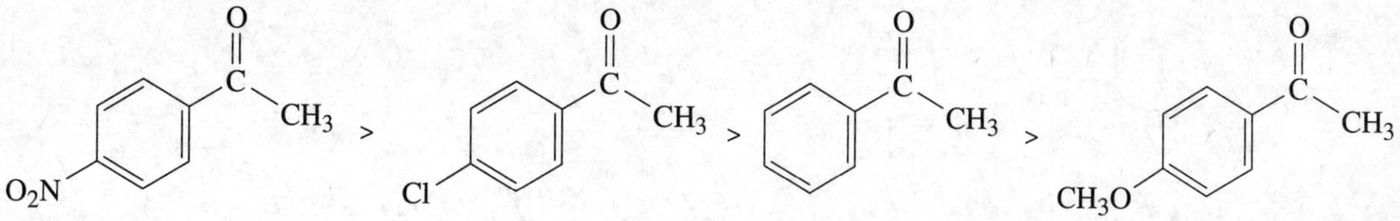

30.

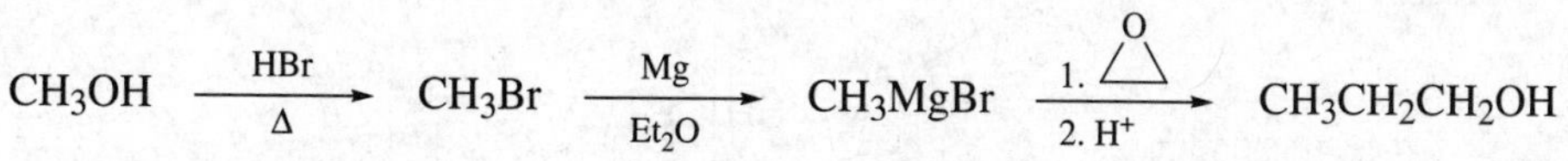

31. **a.**

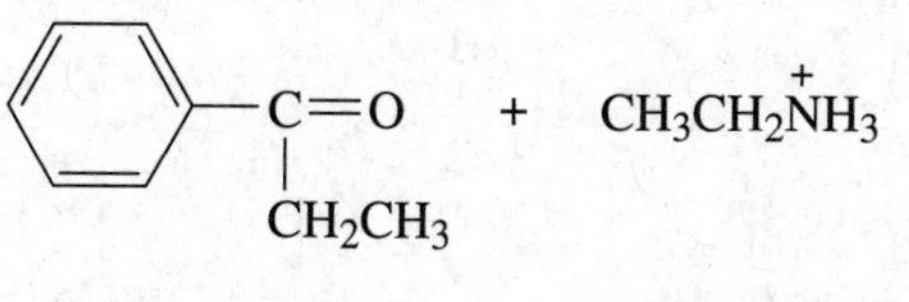

c.

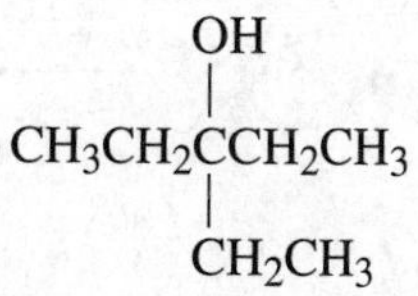

b.

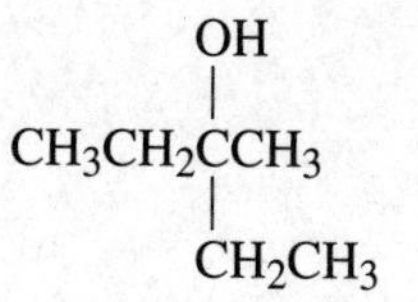

d.

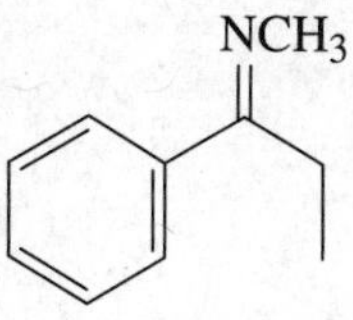

32. **a.**

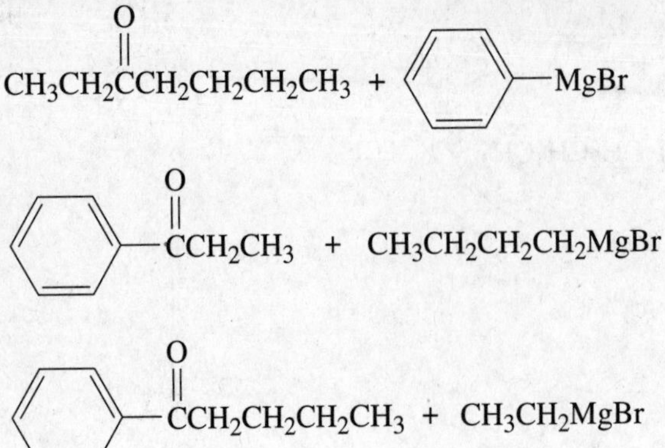

b.

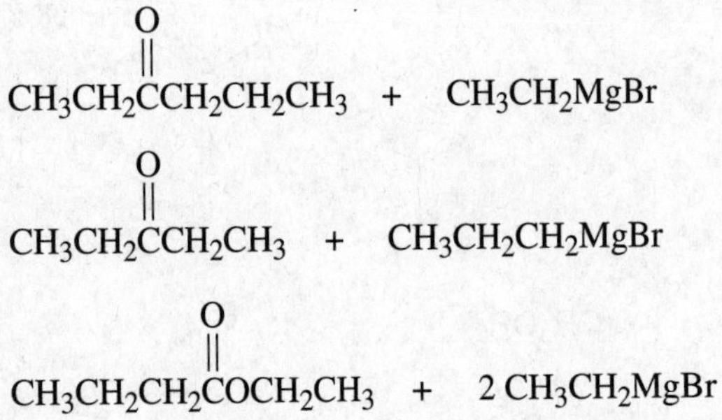

33. **a.** **b.** **c.**

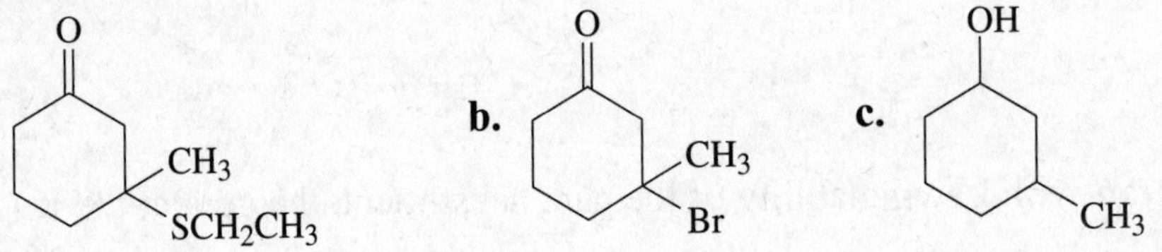

34. **a.**

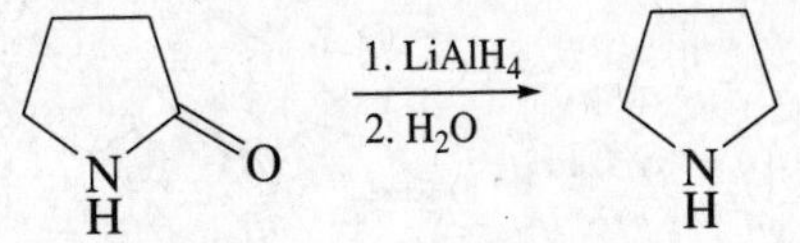

b.

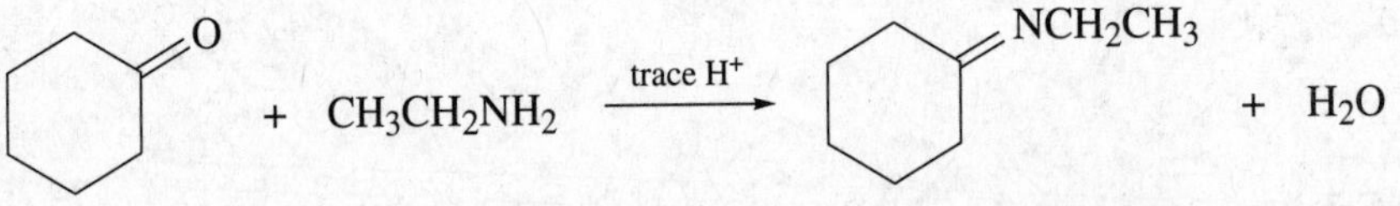

c.

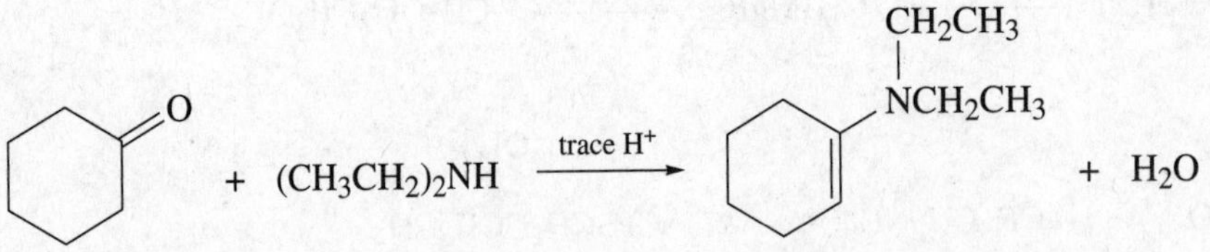

d.

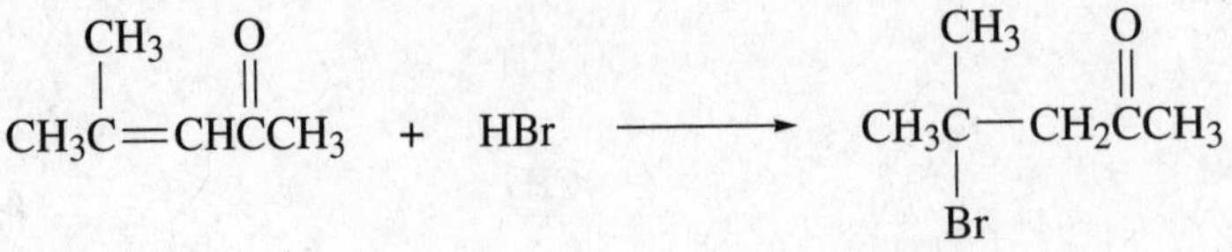

35. **a.** excess CH_3MgBr followed by H_3O^+

b. $LiAlH_4$ followed by H_2O

36. **a.**

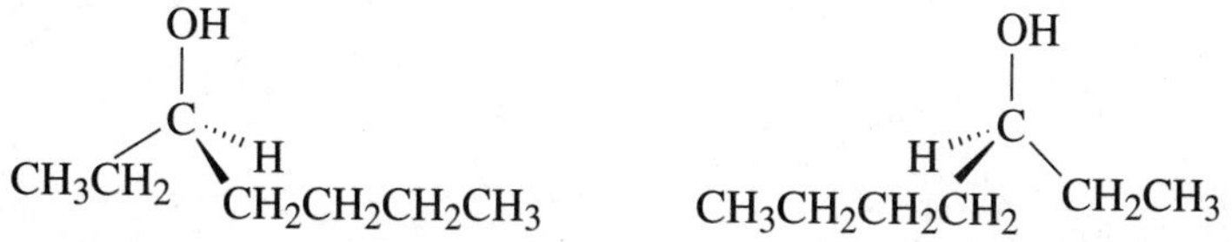

b.

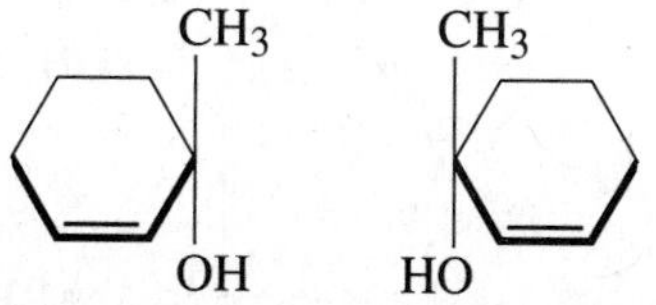

37.

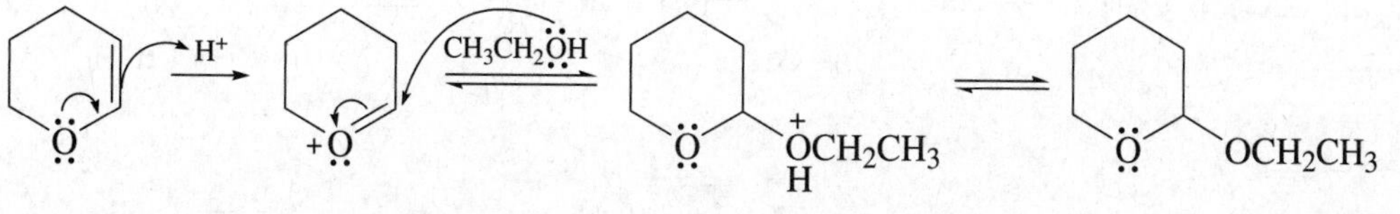

38. **a.** $CH_3CH_2CH_2CH_2Br \xrightarrow{^-C\equiv N} CH_3CH_2CH_2CH_2C\equiv N \xrightarrow[\Delta]{H^+,\ H_2O} CH_3CH_2CH_2CH_2COH$ (with C=O)

b. $CH_3CH_2CH_2CH_2Br \xrightarrow{^-C\equiv N} CH_3CH_2CH_2CH_2C\equiv N \xrightarrow[Pd/C]{H_2} CH_3CH_2CH_2CH_2CH_2NH_2$

39. Methyl formate and excess Grignard reagent will form a secondary alcohol because, unlike other esters that have an alkyl or aryl group on the carbonyl carbon that cause them to form tertiary alcohols, methyl formate has a hydrogen.

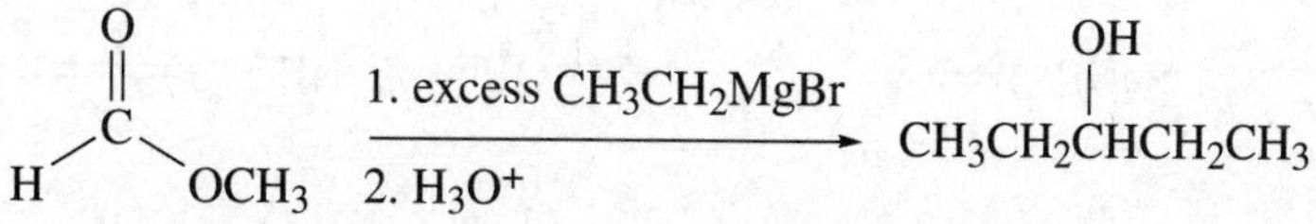

40.

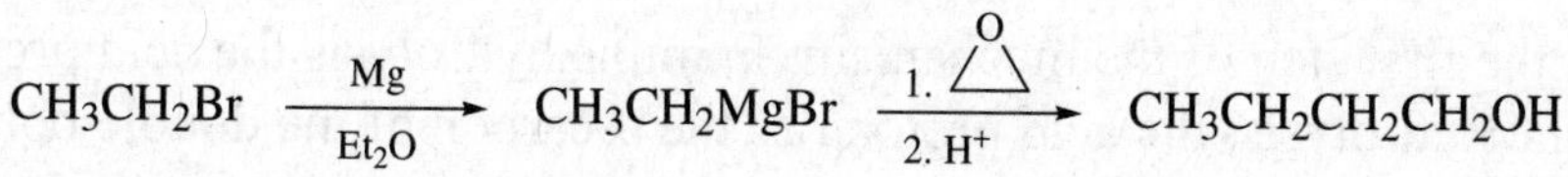

41.

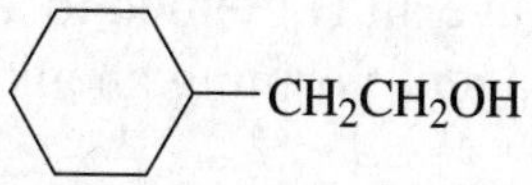

42. a. **1.**

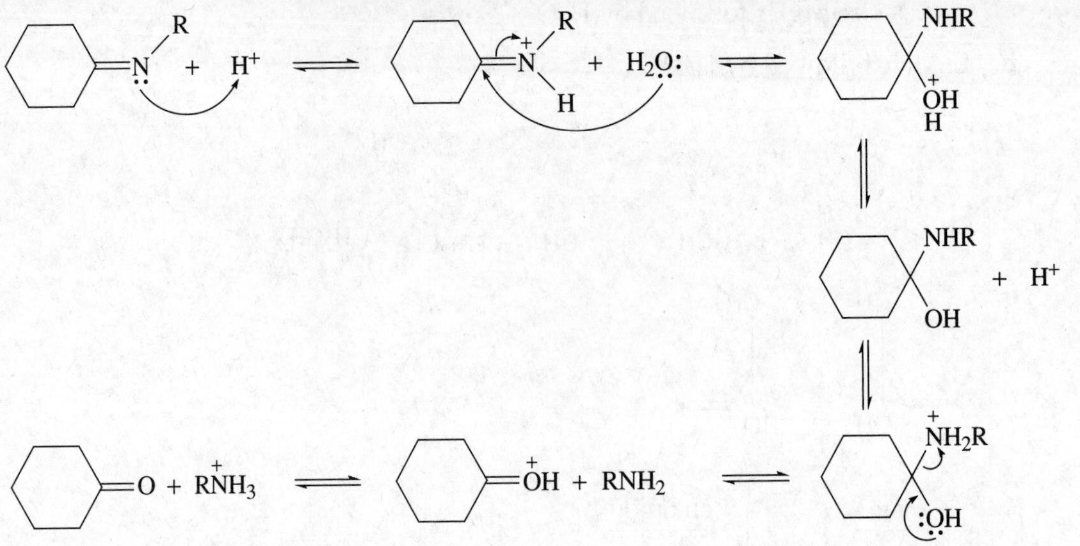

2.

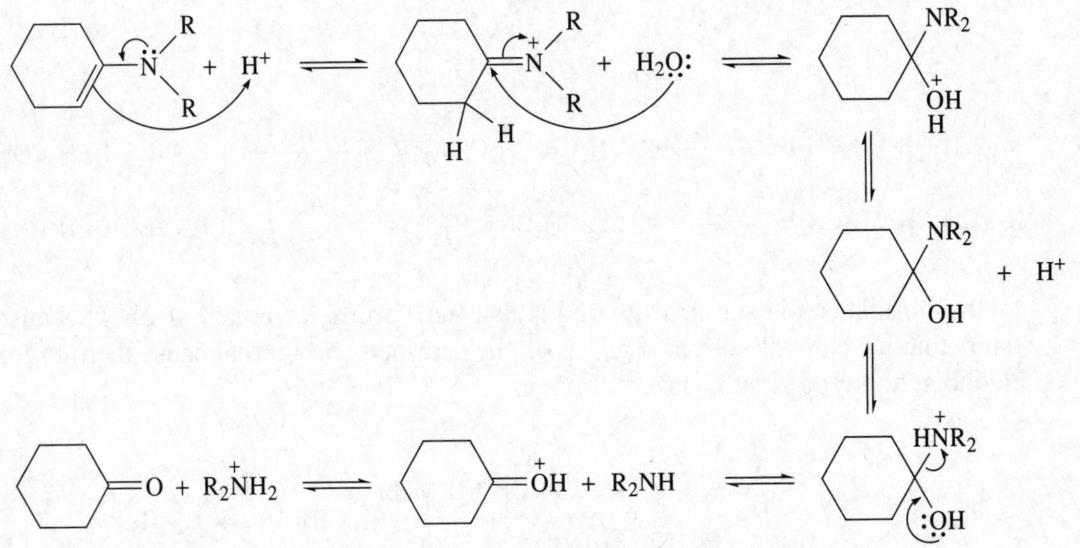

b. The only difference is the first step of the mechanism: in imine hydrolysis the acid protonates the nitrogen; in enamine hydrolysis, the acid protonates the β-carbon of the double bond.

43. Only "c" can be used to form a Grignard reagent. Both "a" and "b" have an H bonded to an oxygen that would immediately react as an acid with the Grignard reagent, forming an alkane.

44. **a.** The carboxyl group on the left is more acidic because an electron-withdrawing oxygen is bonded to the carbon immediately adjacent to the COOH group.

b. The data show that the amount of hydrate decreases with increasing pH until about pH = 6 and that increasing the pH beyond 6 has no effect on the amount of hydrate.

A hydrate is stabilized by electron-withdrawing groups. A COOH group is electron withdrawing but a COO^- group is less so. In acidic solutions, where both carboxylic acid groups are in their acidic (COOH) forms, the compound exists as essentially all hydrate. As the pH of the solution increases and the COOH groups become COO^- groups, the amount of hydrate decreases. Above pH = 6, where both carboxyl groups are in their basic (COO^-) forms, there is only a small amount of hydrate.

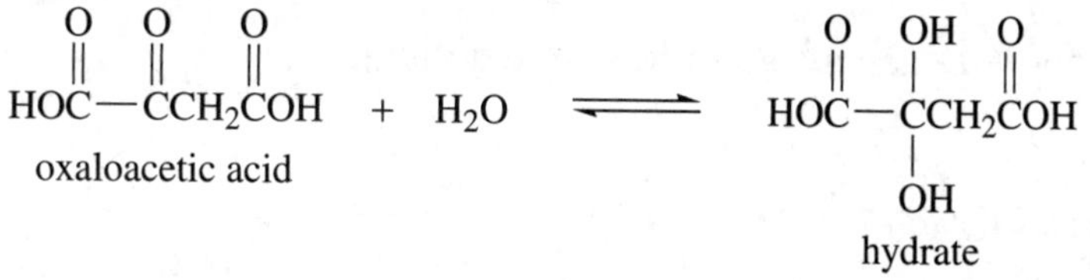
oxaloacetic acid
hydrate

45.

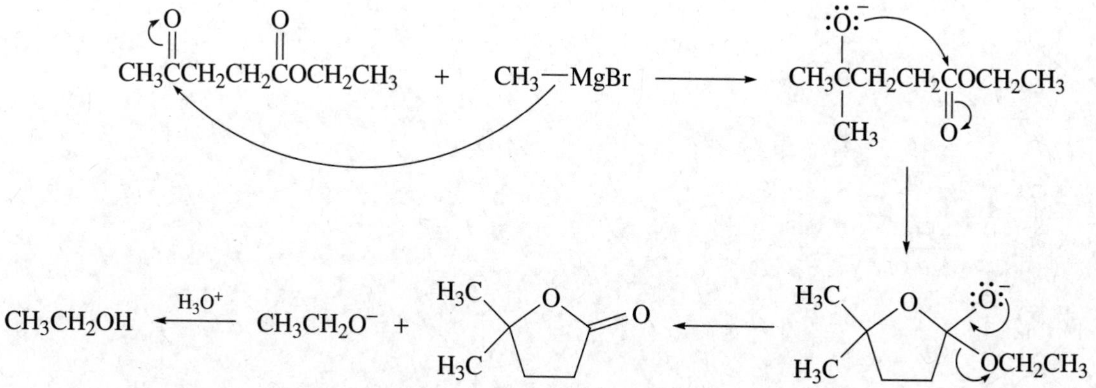

Chapter 12 Practice Test

1. Give the product of each of the following reactions:

a. 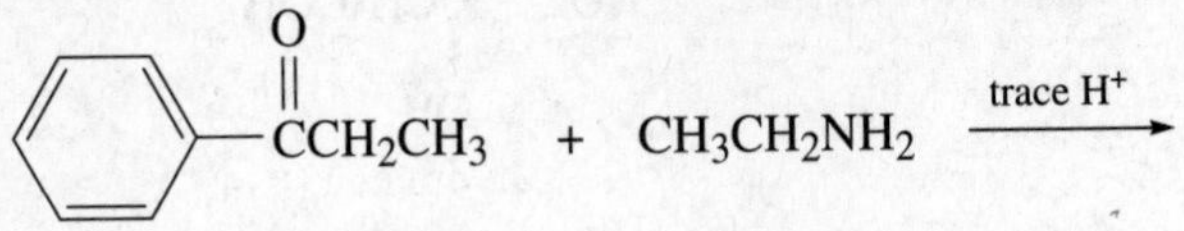

b. 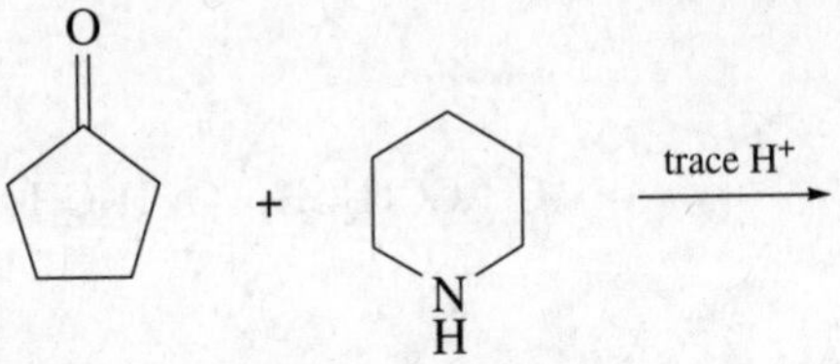

c. $H\overset{O}{\overset{\|}{C}}H \xrightarrow[2.\ H_3O^+]{1.\ CH_3CH_2CH_2MgBr}$

d. 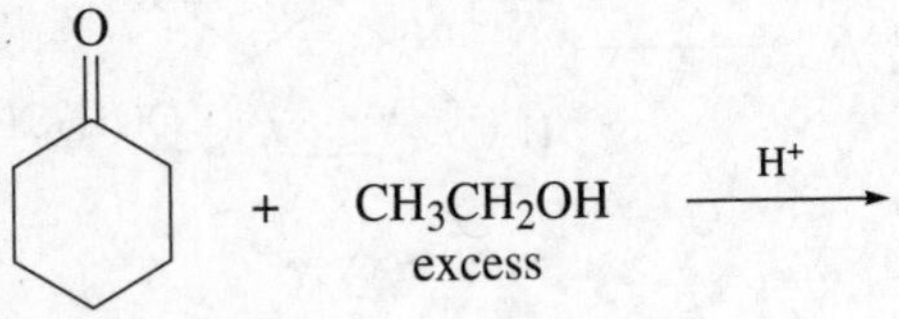

e. $CH_3CH_2\overset{O}{\overset{\|}{C}}CH_2CH_3 \xrightarrow[2.\ H_3O^+]{1.\ CH_3MgBr}$

f. 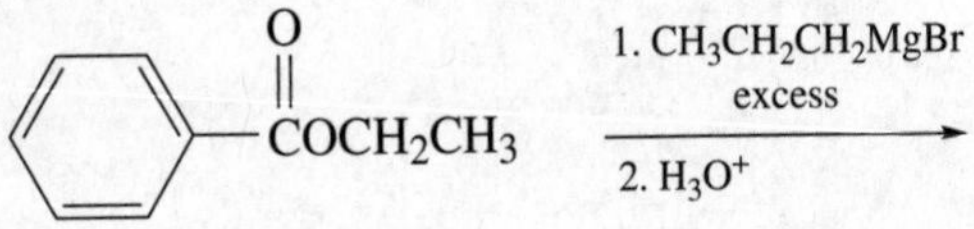

2. Which of the following alcohols cannot be prepared by the reaction of an ester with excess Grignard reagent?

$$\underset{\overset{\displaystyle |}{CH_3}}{CH_3CH_2\overset{\overset{\displaystyle OH}{|}}{C}CH_2CH_2CH_3} \qquad \underset{\overset{\displaystyle |}{CH_3}}{CH_3\overset{\overset{\displaystyle OH}{|}}{C}CH_2CH_3} \qquad \underset{\overset{\displaystyle |}{CH_3}}{CH_3\overset{\overset{\displaystyle OH}{|}}{C}CH_3}$$

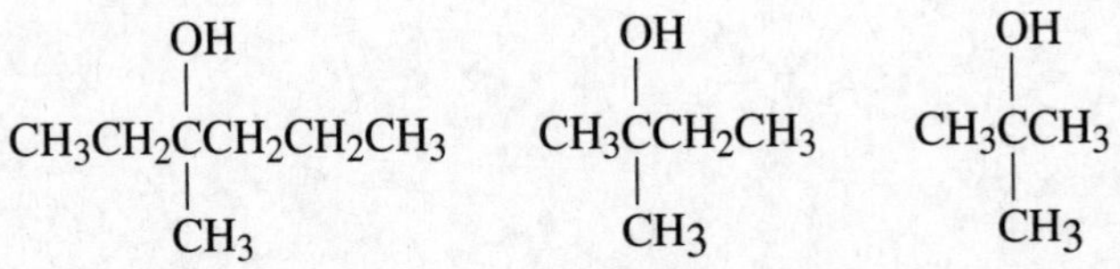

3. Which of the following ketones would form the greatest amount of hydrate in an aqueous solution?

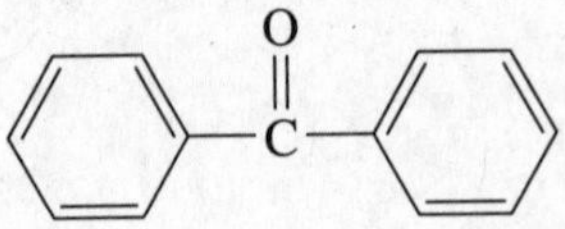

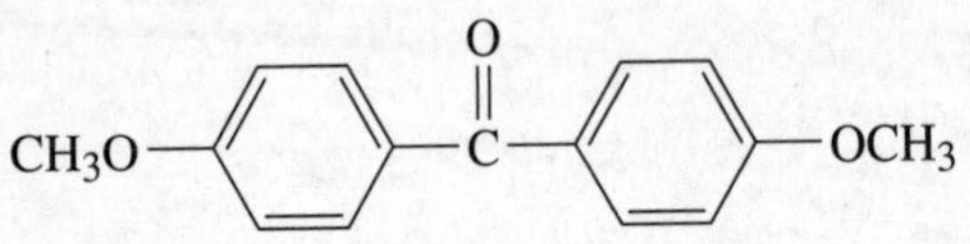

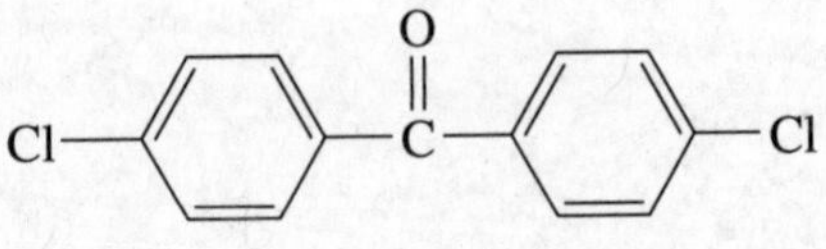

4. Give an example of each the following:

a. an enamine

b. an acetal

c. an imine

d. a hemiacetal

5. Which is more reactive toward nucleophilic addition?

a. butanal or 2-pentanone

b. 4-heptanone or 3-pentanone

6. Indicate how the following compound could be prepared using the given starting material.

$$CH_3CH_2CH_2Br \longrightarrow CH_3CH_2CH_2\overset{\overset{\displaystyle O}{\|}}{C}OCH_2CH_3$$

CHAPTER 13

Carbonyl Compounds III: Reactions at the α-Carbon

1.

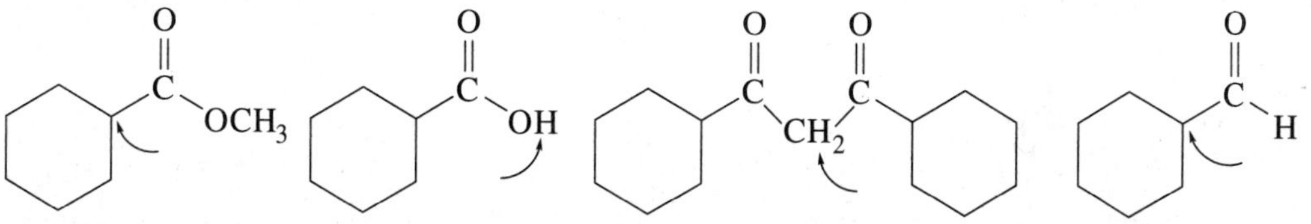

2. **a.** The ketone is a stronger acid than the ester because the electrons left behind when a proton is removed from the α-carbon of the ketone are more readily delocalized onto the carbonyl oxygen atom. When the ester loses a proton the electrons have to compete with the lone pair on the alkoxy group oxygen for delocalization onto the carbonyl oxygen.

b. Because it is the weaker acid, the ester has the greater pK_a value.

3. 2,4-Pentanedione is a stronger acid because the electrons left behind when a proton is removed can be readily delocalized onto two carbonyl oxygen atoms. The electrons left behind when a proton is removed from ethyl 3-oxobutyrate can be readily delocalized onto one carbonyl oxygen but there is competition for delocalization onto the second carbonyl oxygen by the ethoxy group.

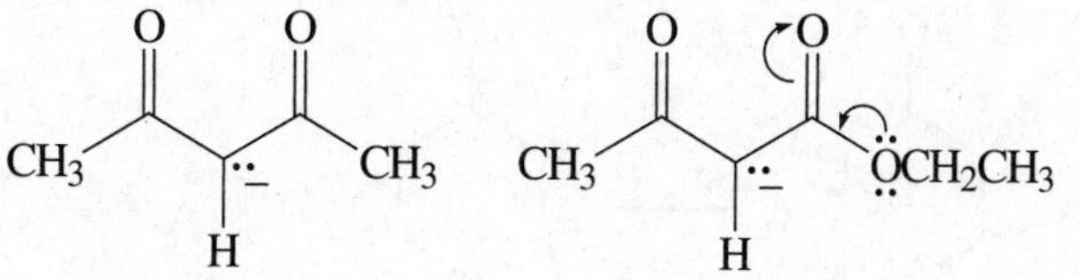

2,4-pentanedione   ethyl 3-oxobutyrate

4. A proton cannot be removed from the α-carbon of *N*-methylethanamide or ethanamide because these compounds have a hydrogen bonded to the nitrogen and this hydrogen is more acidic than the one attached to the α-carbon.

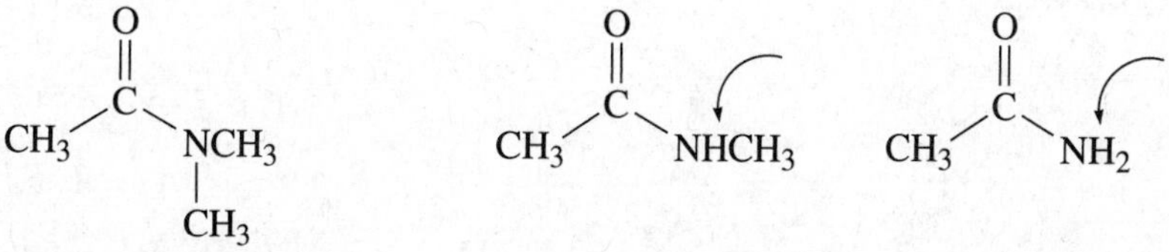

N,N-dimethylethanamide   *N*-methylethanamide   ethanamide

The contributing resonance structures show why the hydrogen attached to the nitrogen is more acidic (the nitrogen has a partial positive charge) than the hydrogen attached to the α-carbon.

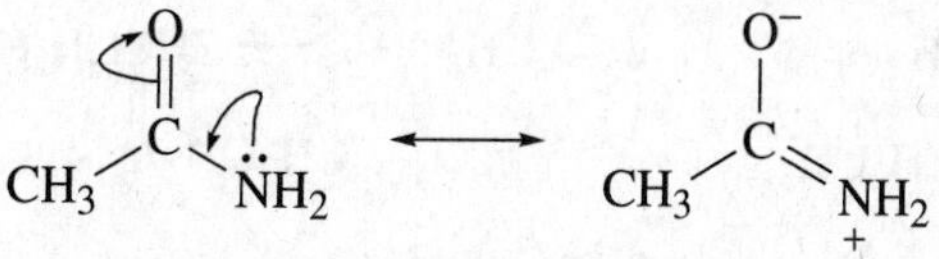

5. Acyl chlorides are much more reactive than ketones or esters, so it is easier for hydroxide ion to attack the reactive carbonyl group than to remove a hydrogen from an α-carbon.

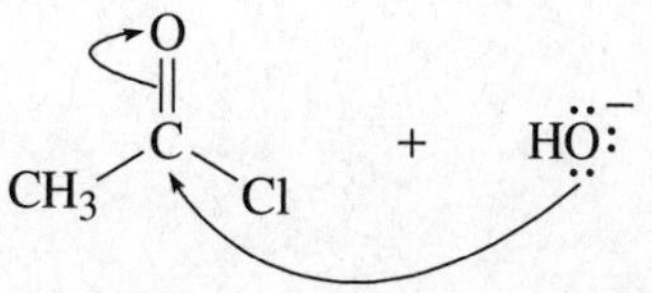

6. **a.** 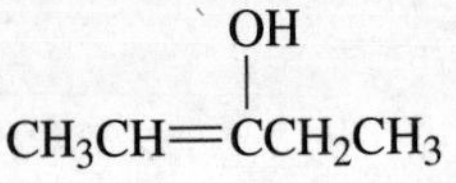 **b.** 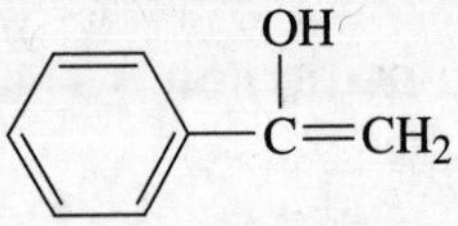 **c.** 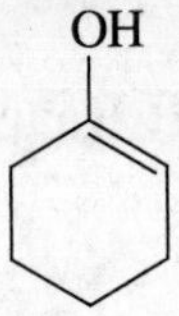

7. 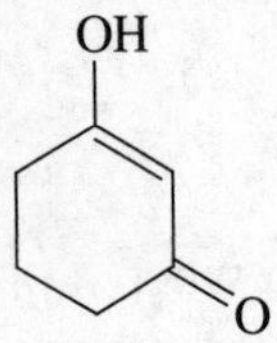 and 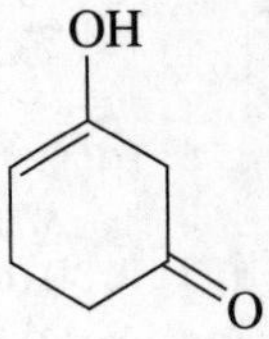

more stable
because the double
bonds are conjugated

8. **a.** 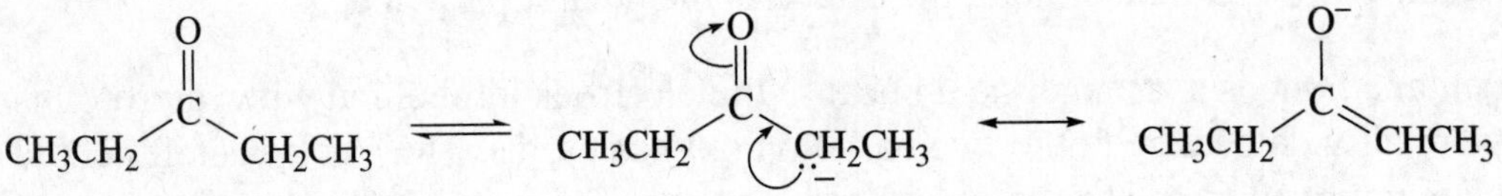

3-pentanone + H^+

b. 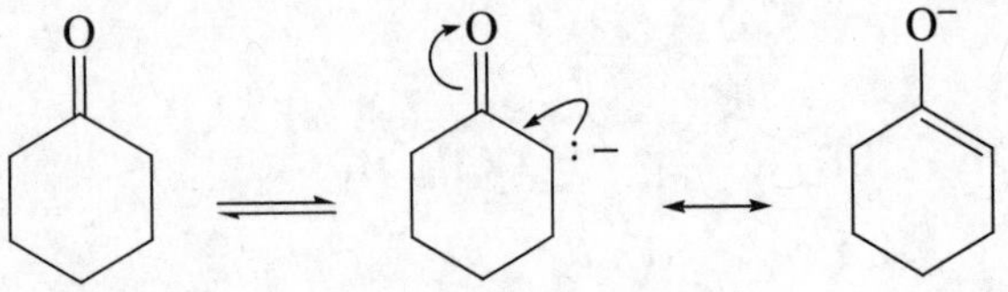

cyclohexanone + H^+

9. **a.** 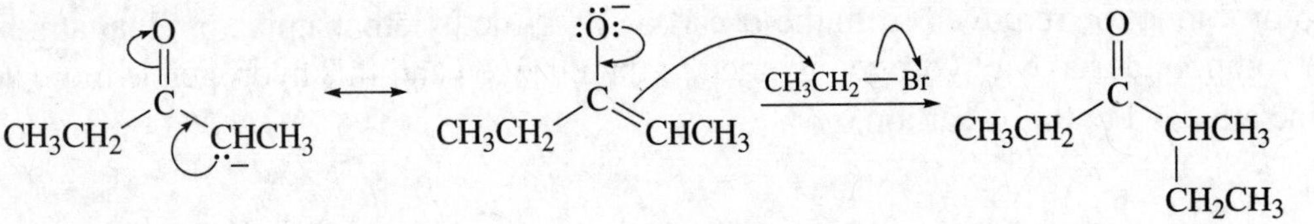

b. 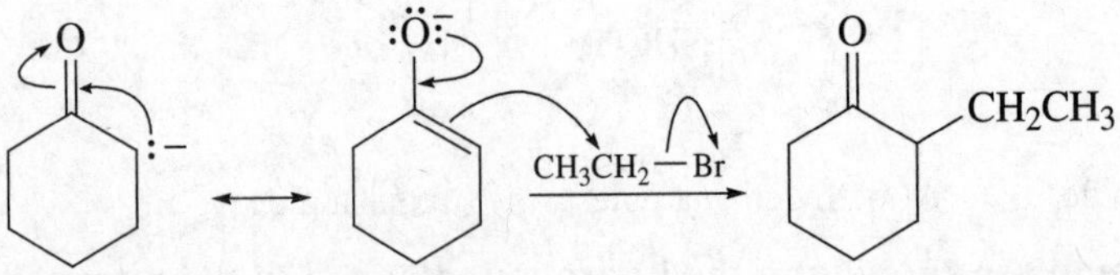

10. **a.** 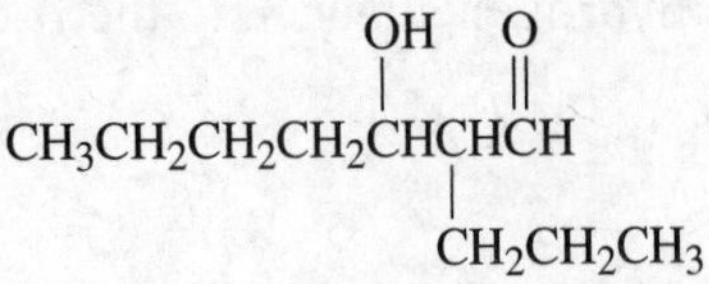 **c.** 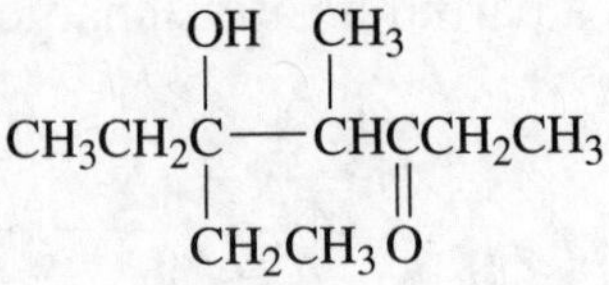

b. 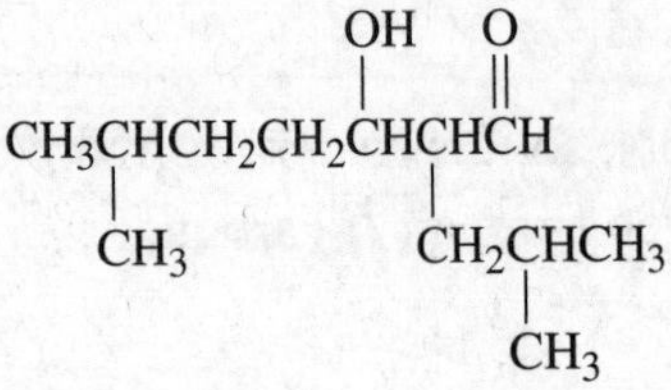 **d.** 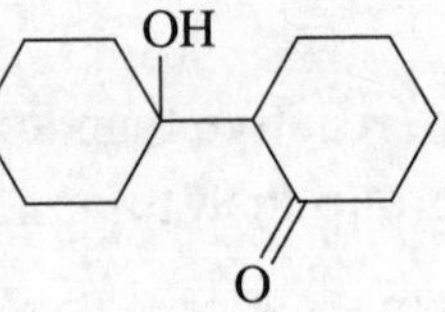

11. 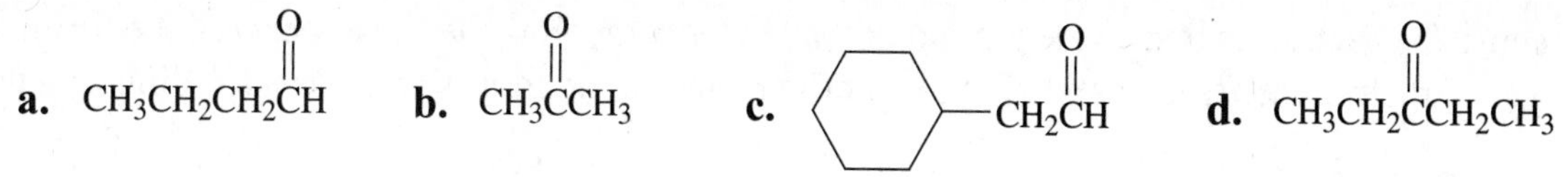

12. aldol addition — aldol condensation

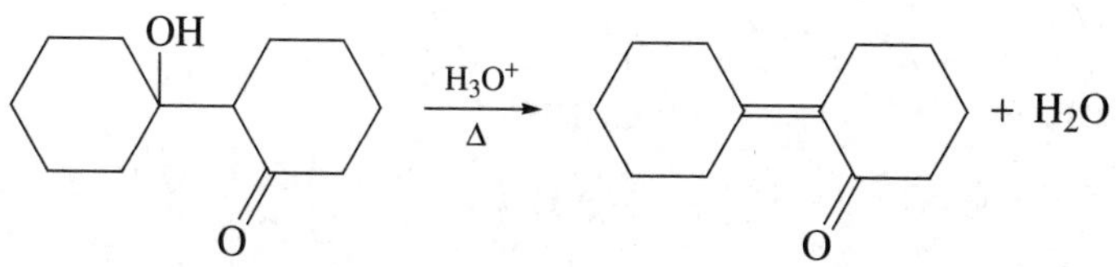

13. **a.** Solved in the text.

b. 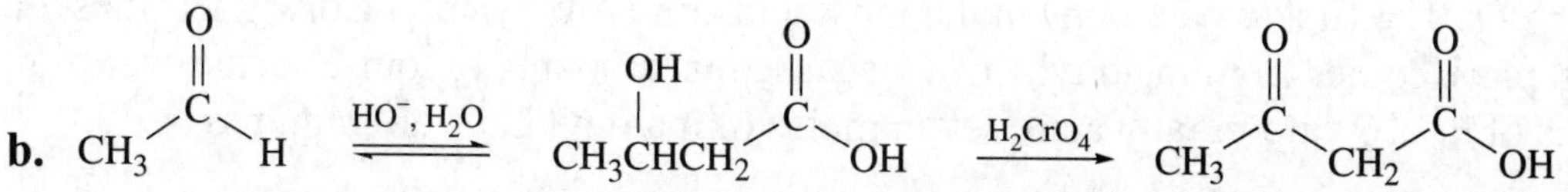

14. 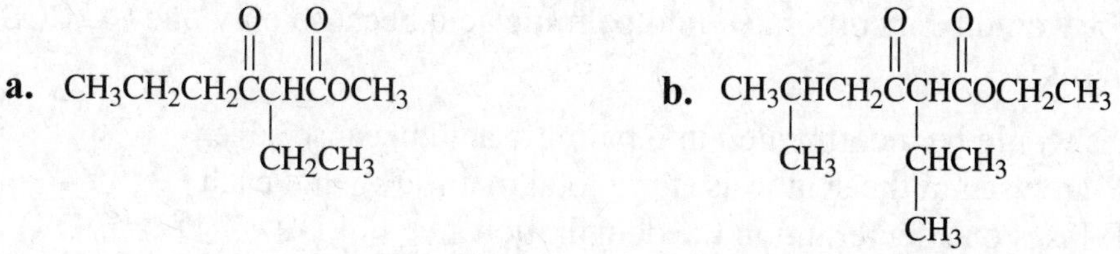

15. "**B**" and "**D**" cannot undergo a Claisen condensation because they do not have an α-hydrogen.

"**A**" cannot undergo a Claisen condensation because it does not have an α-hydrogen on an sp^3 carbon.

16. 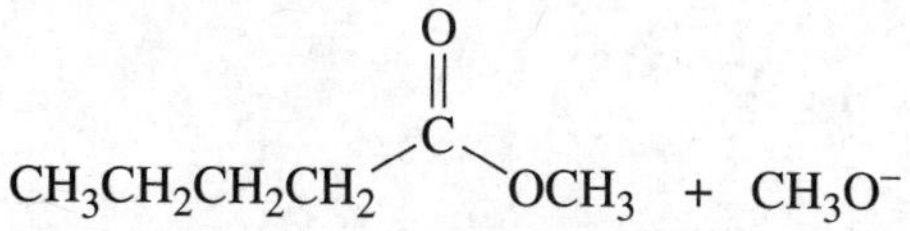

17. **A** and **D** can be decarboxylated.

B cannot be decarboxylated, because it doesn't have a COOH group.

C cannot be decarboxylated because the electrons left behind if it were cannot be delocalized onto an oxygen.

18. **a.** methyl bromide **b.** benzyl bromide **c.** isobutyl bromide

19. Solved in the text.

20. **a.** ethyl bromide **b.** pentyl bromide **c.** benzyl bromide

21. Because the catalyst is hydroxide ion rather than an enzyme, four stereoisomers will be formed since two asymmetric centers are created in the product. One of the four is fructose −1,6-diphosphate.

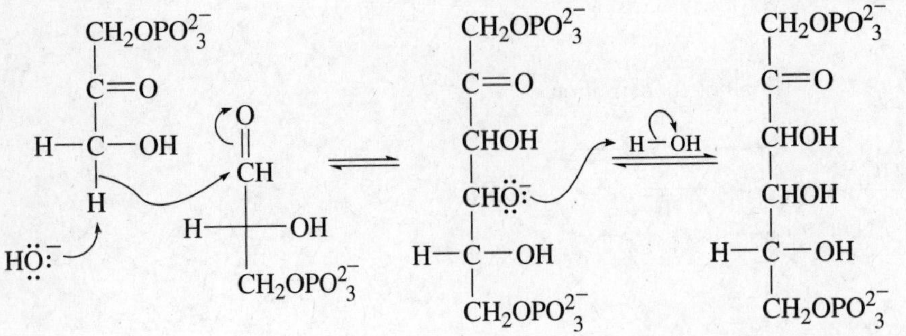

22. **Seven** moles. The first two carbons in the fatty acid come from acetyl CoA. Each subsequent two-unit piece comes from malonyl CoA. Because this amounts to fourteen carbons for the synthesis of the 16-carbon fatty acid, seven moles of malonyl CoA are required.

23. a. **Three** deuteriums would be incorporated into palmitic acid because only one CD_3COSR is used in the synthesis.

b. **Seven** deuteriums would be incorporated into palmitic acid because seven $^-OOCCD_2COSR$ are used in the synthesis (for a total of 14 D's), and each $^-OOCCD_2COSR$ loses one deuterium in the dehydration step (14 D's – 7 D's = 7 D's).

24. a. $CH_3\overset{O}{\overset{\|}{C}}CH_2\overset{O}{\overset{\|}{C}}OCH_2CH_3$ b. 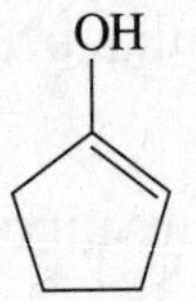 c. $CH_3CH_2CH_2CH_2\overset{O}{\overset{\|}{C}}OH$

25. a. $CH_3CH_2\overset{OH}{\overset{|}{C}}{=}CH\overset{O}{\overset{\|}{C}}CH_2CH_3$ and $CH_3CH{=}\overset{OH}{\overset{|}{C}}CH_2\overset{O}{\overset{\|}{C}}CH_2CH_3$

more stable
because the double bonds are conjugated

b. 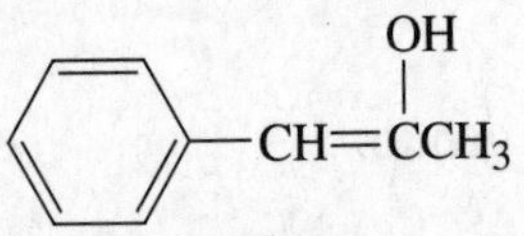 and 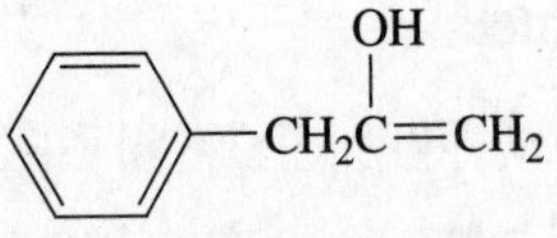

more stable
because the double bond is conjugated with the benzene ring

c.

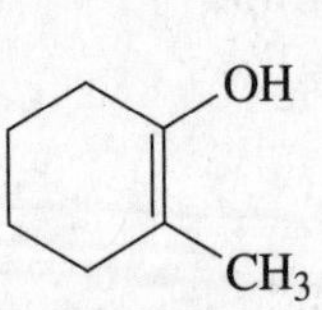 and 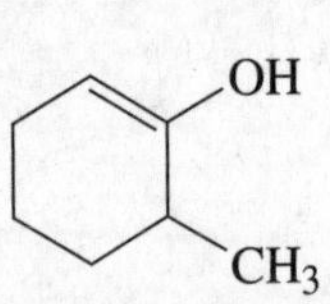

more stable

because the sp^2 carbons are attached to a greater number of alkyl substituents

26. **a.** 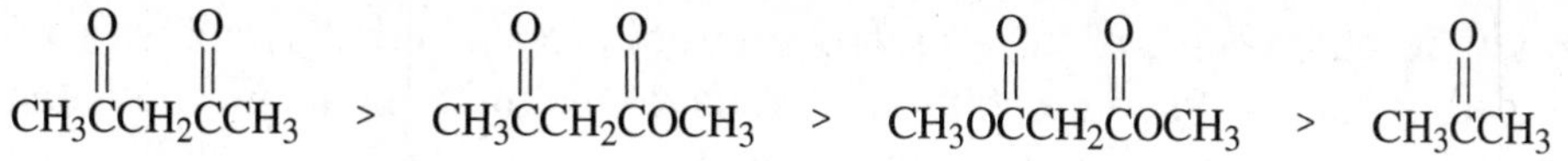

b. 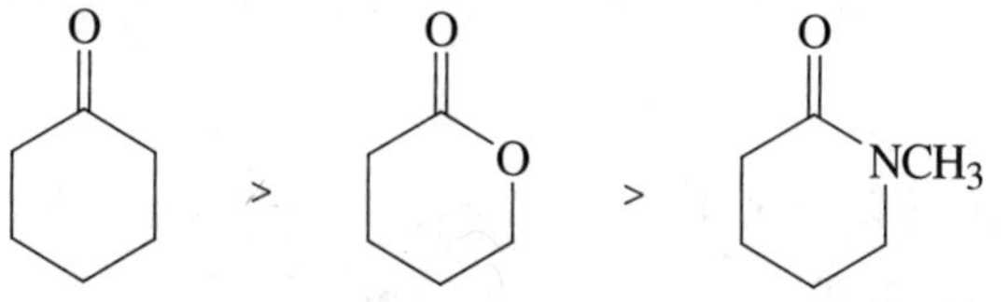

The ketone is the strongest acid because there is no competition for delocalization of the electrons that are left behind when the *α*-hydrogen is removed. The cyclic amide is the weakest acid because nitrogen, being less electronegative, can better accommodate a positive charge and, therefore, is better than oxygen at delocalizing its lone pair onto the carbonyl oxygen. Therefore, nitrogen is better at competing with the electrons left behind when an *α*-hydrogen is removed for delocalization onto the carbonyl oxygen.

c. 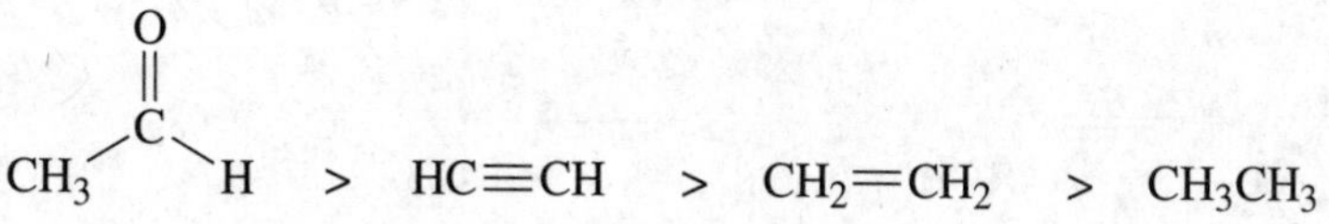

27.

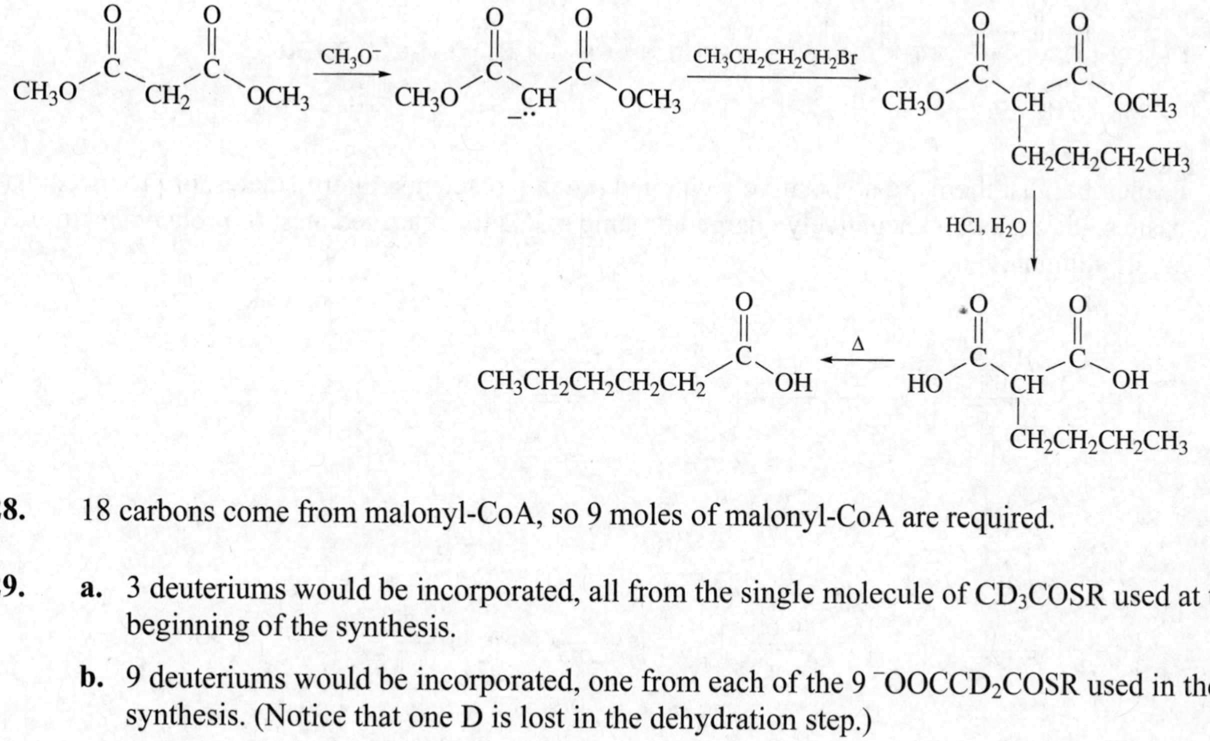

28. 18 carbons come from malonyl-CoA, so 9 moles of malonyl-CoA are required.

29. **a.** 3 deuteriums would be incorporated, all from the single molecule of CD_3COSR used at the beginning of the synthesis.

b. 9 deuteriums would be incorporated, one from each of the 9 $^-OOCCD_2COSR$ used in the synthesis. (Notice that one D is lost in the dehydration step.)

30. The electrons left behind when a proton is removed from propene are delocalized—they are shared by two carbon atoms. In contrast, the electrons left behind when a proton is removed from an alkane are localized—they belong to a single (carbon) atom. The base with delocalized electrons is more stable so it has the stronger conjugate acid; that is, propene is a stronger acid than an alkane.

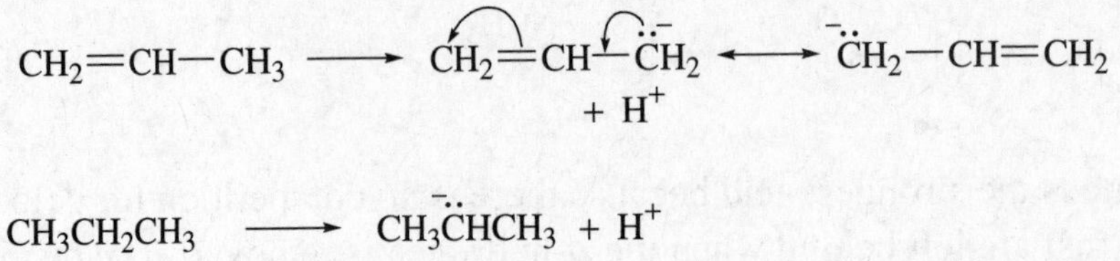

Propene, however, is not as acidic as the carbon acids in Table 13.1, because the electrons left behind when a proton is removed from these carbon acids are delocalized onto an oxygen or a nitrogen, which are both more electronegative than carbon and, therefore, better able to accommodate the electrons.

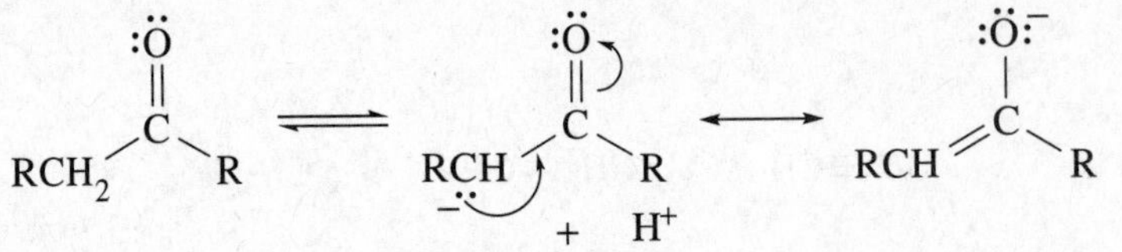

31.

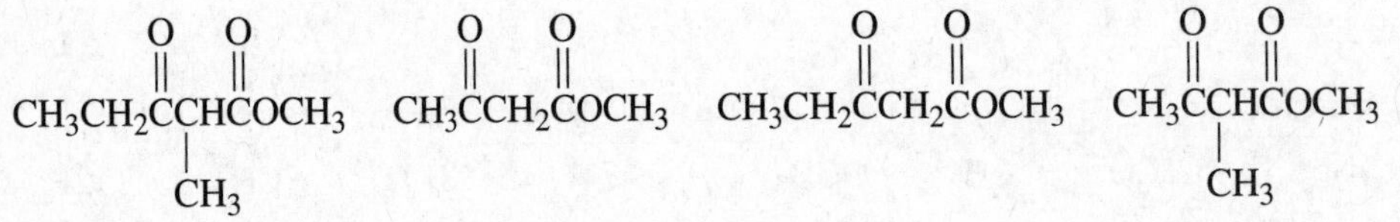

32. Remember that there are no positively charged organic reactants, intermediates, or products in a basic solution, and no negatively charged organic reactants, intermediates, or products in an acidic solution.

a.

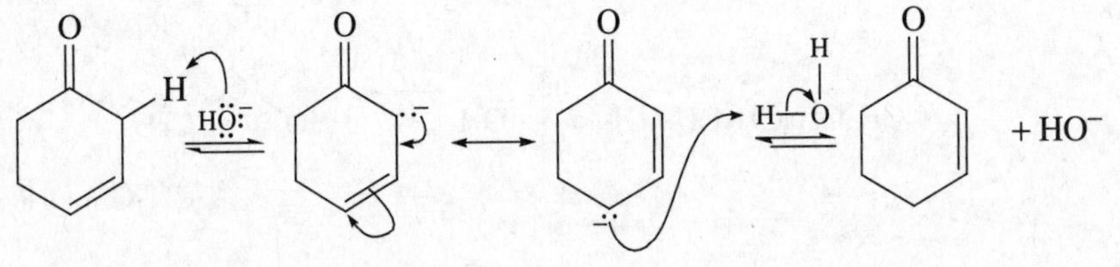

b.

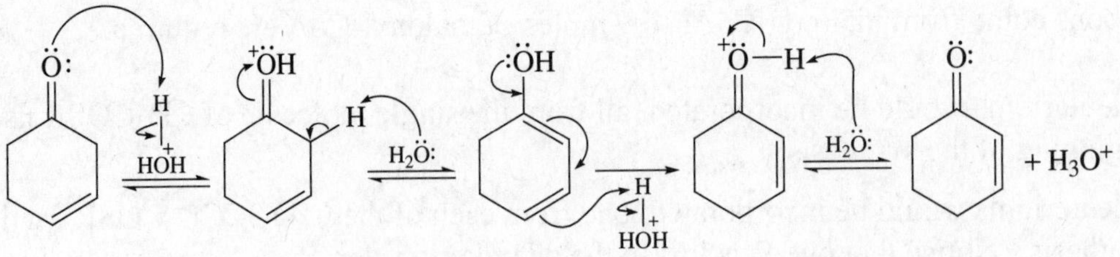

33.

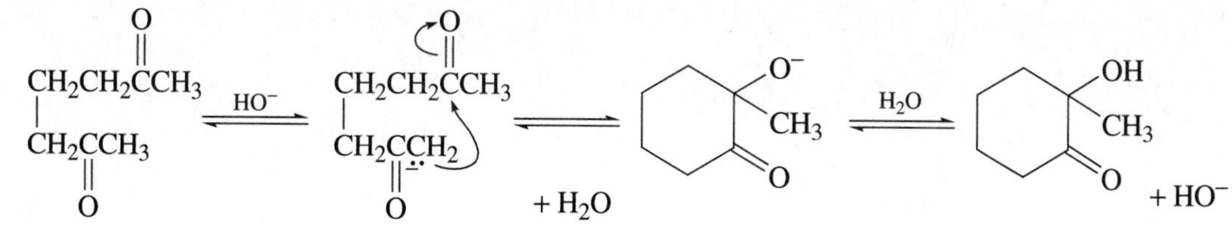

2,6-heptanedione

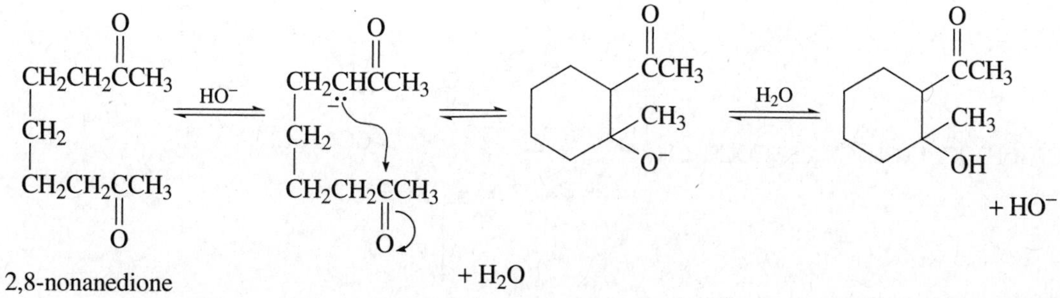

2,8-nonanedione

34. **a.** In a basic solution, the ketone will be in equilibrium with its enol tautomer; when the enol tautomer forms, the asymmetric center is lost. When the enol tautomer reforms the ketone, it can form the *R* and *S* enantiomer equally as easily, so a racemic mixture is obtained.

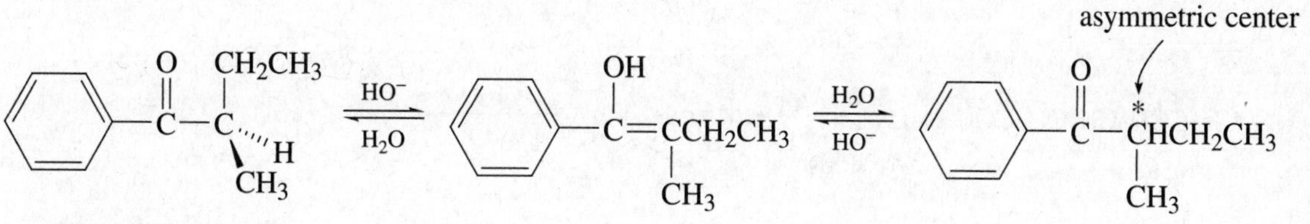
asymmetric center

(*R*)-2-methyl-1-phenyl-1-butanone

b. You need a ketone that has an α-carbon that is an asymmetric center. A racemic mixture will be formed when an α-hydrogen is removed from the asymmetric carbon.

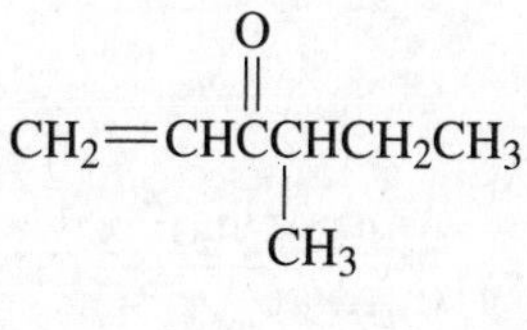

35.

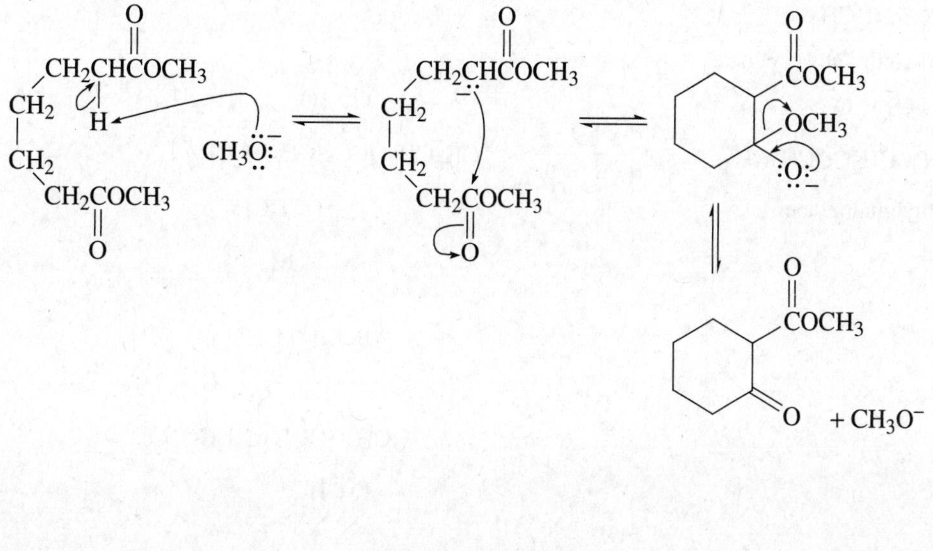

36.

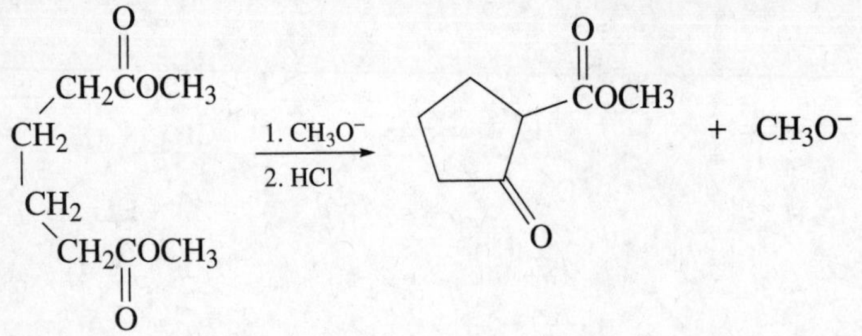
1. CH_3O^-
2. HCl
+ CH_3O^-

37. **a.**

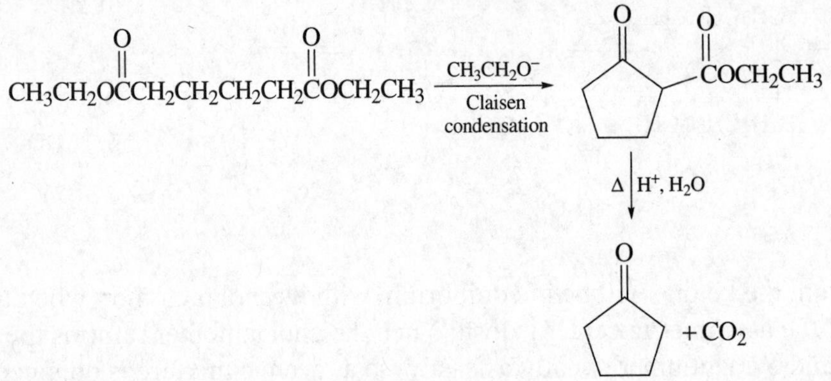
$CH_3CH_2O^-$
Claisen condensation
Δ H^+, H_2O
+ CO_2

b.

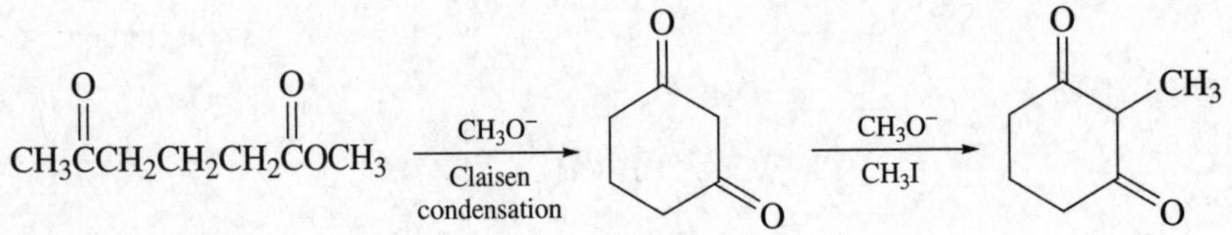
CH_3O^-
Claisen condensation
CH_3O^-
CH_3I

38. **a.**

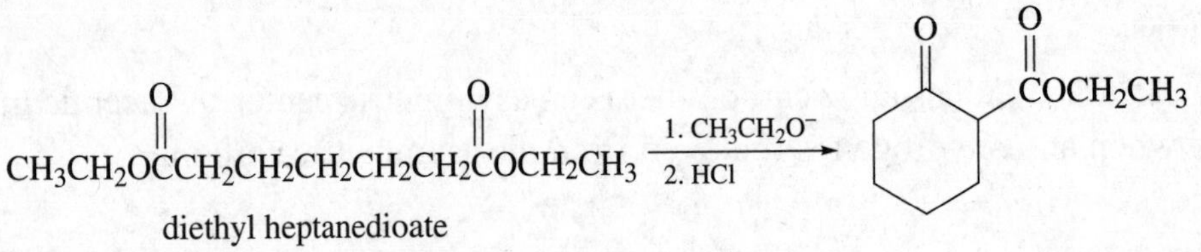
1. $CH_3CH_2O^-$
2. HCl
diethyl heptanedioate

b.

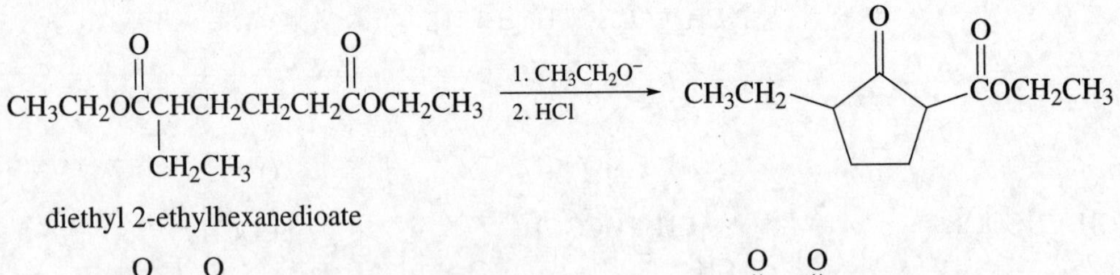
1. $CH_3CH_2O^-$
2. HCl
diethyl 2-ethylhexanedioate

c.

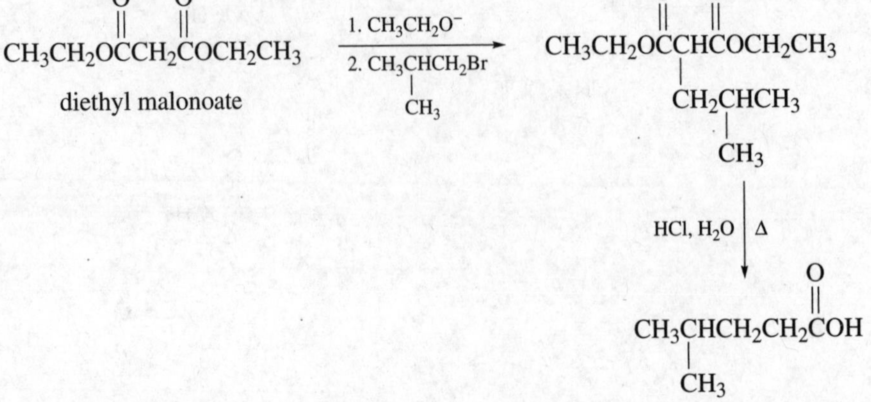
1. $CH_3CH_2O^-$
2. CH_3CHCH_2Br (CH_3)
diethyl malonoate
HCl, H_2O Δ

d. 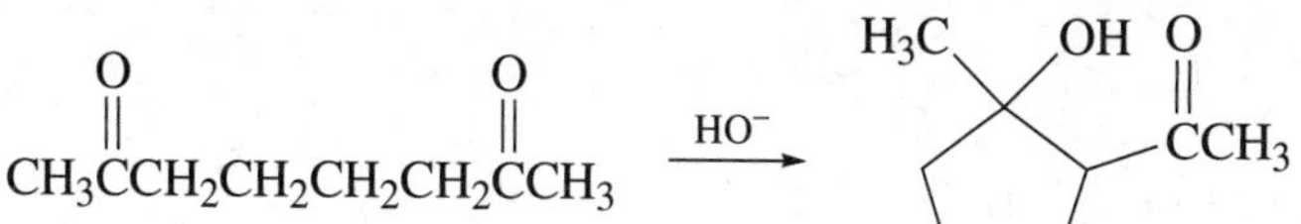

2,7-octanedione

39. Decarboxylation of the *β*-dicarboxylic acid would require a higher temperature because the electrons left behind when CO_2 is removed are not as readily delocalized onto the carbonyl oxygen because a lone pair on the second OH group can also be delocalized onto that oxygen.

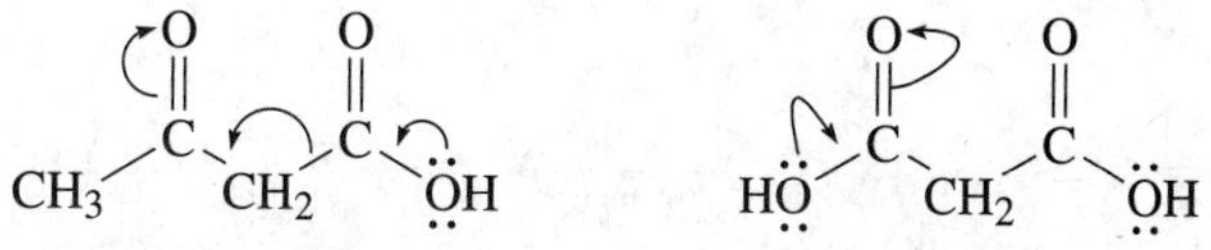

β-keto acid   *β*-dicarboxylic acid

40.

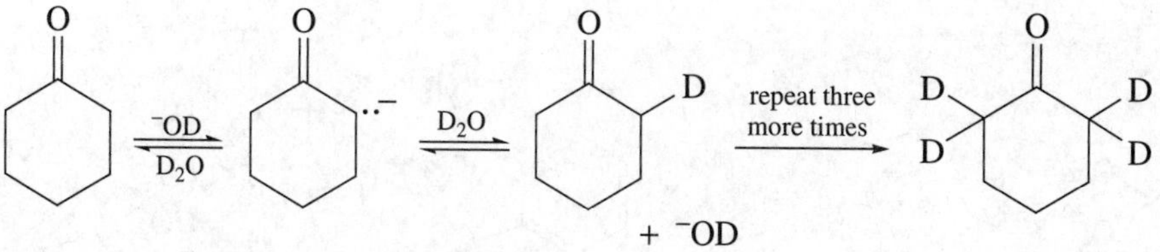

41. The synthesis would require 5-bromo-2-methylpropane, a tertiary alkyl halide. An S_N2 reaction cannot be done on a tertiary alkyl halide, because only elimination and no substitution occurs. (Section 9.11 of the text.)

42.

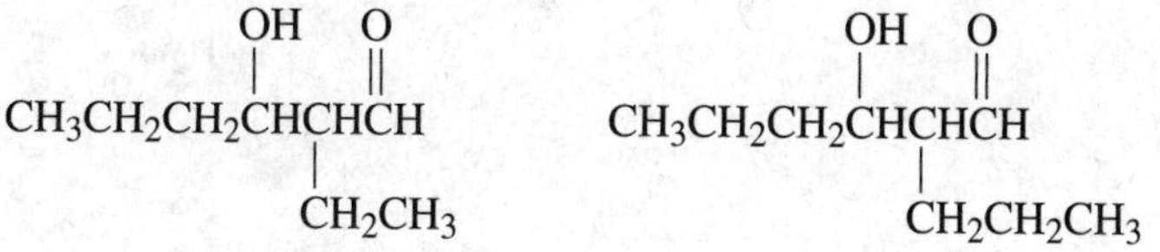

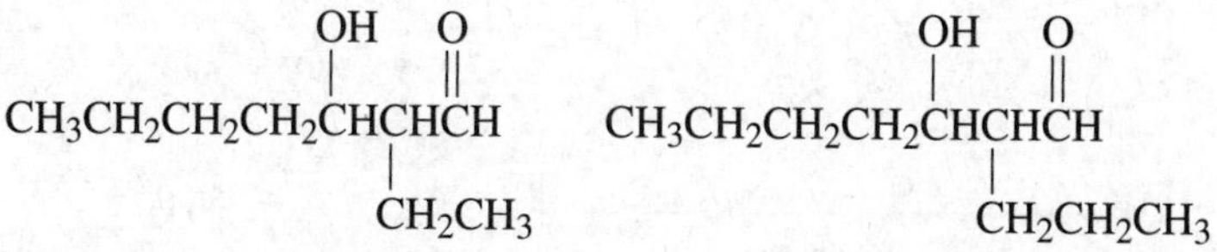

43. **a.**

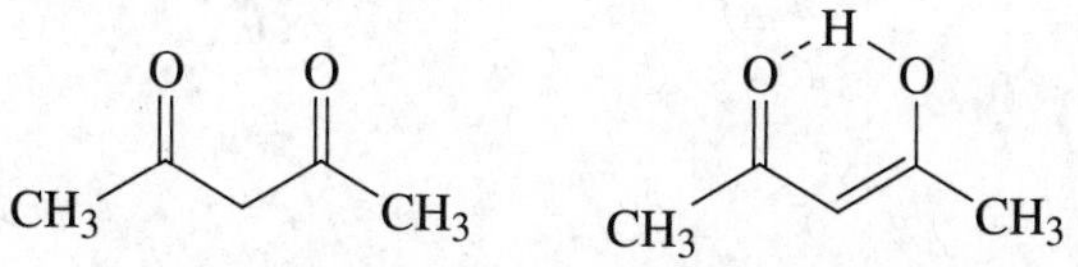

keto tautomer   enol tautomer

2,4-pentanedione

b. The enol tautomer of 2,4-pentanedione is more stable than most enol tautomers because it can form an intramolecular hydrogen bond that stabilizes it.

44.

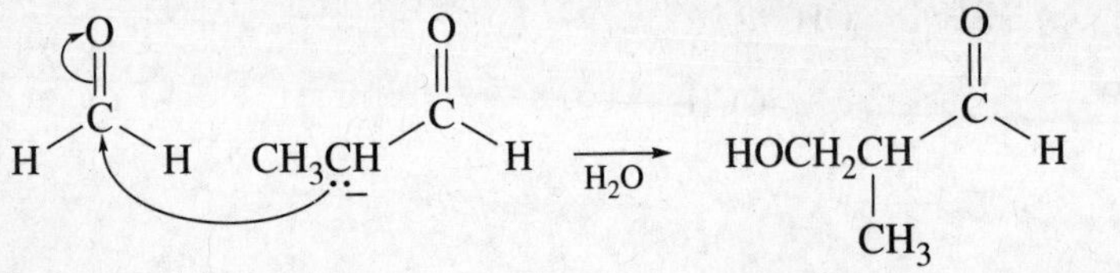

45. The middle carbonyl group is hydrated because it is stabilized by the electron-withdrawing carbonyl groups on either side of it.

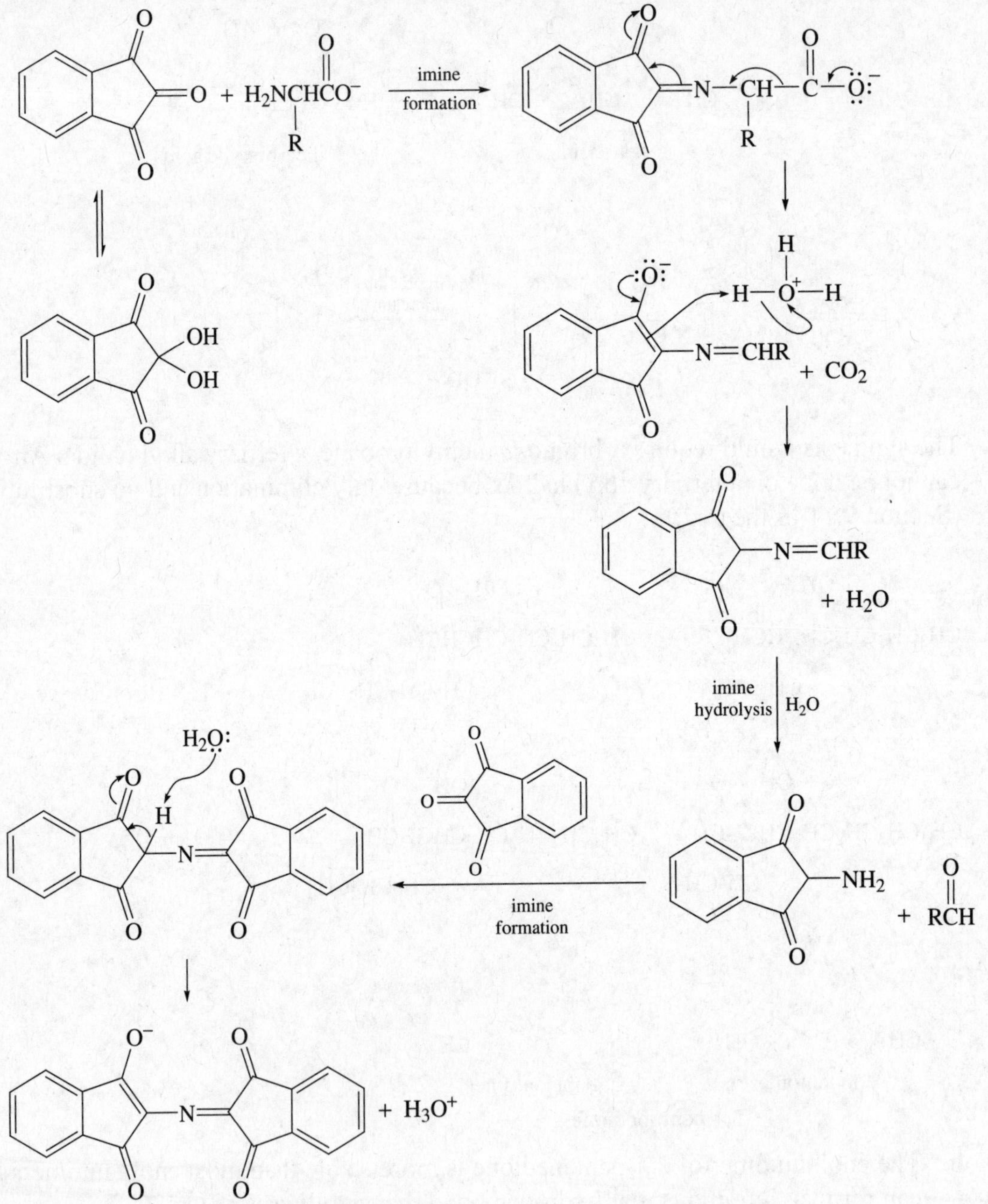

Chapter 13 Practice Test

1. Rank the following compounds in order of decreasing acidity. (Label the most acidic #1.)

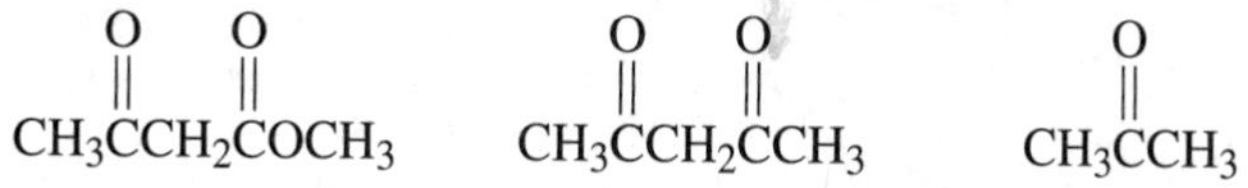

2. Give a structure for each of the following:

    a. the most stable enol tautomer of 2,4-pentanedione

    b. a β-keto ester

3. Give the product of each of the following reactions.

    a. 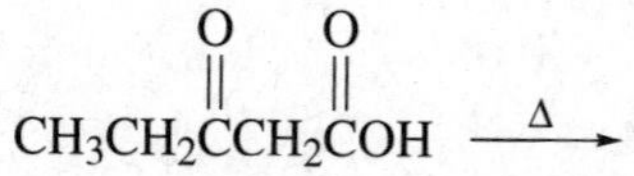
$$\underset{\text{O} \quad\quad \text{O}}{CH_3CH_2\overset{\parallel}{C}CH_2\overset{\parallel}{C}OH} \xrightarrow{\Delta}$$

    b. $2\ CH_3CH_2CH_2\overset{\overset{\displaystyle O}{\parallel}}{C}OCH_3 \xrightarrow[\text{2. HCl}]{\text{1. CH}_3\text{O}^-}$ 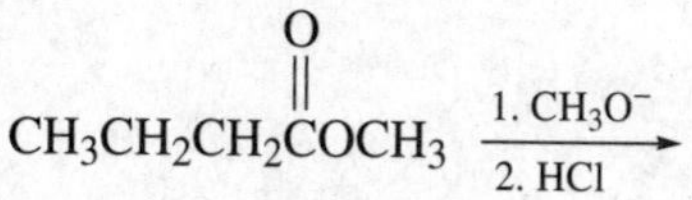

    c. $CH_3CH_2O\overset{\overset{\displaystyle O}{\parallel}}{C}CH_2\overset{\overset{\displaystyle O}{\parallel}}{C}OCH_2CH_3 \xrightarrow[\substack{\text{2. CH}_3\text{CH}_2\text{CH}_2\text{Br} \\ \text{3. HCl, H}_2\text{O, }\Delta}]{\text{1. CH}_3\text{CH}_2\text{O}^-}$ 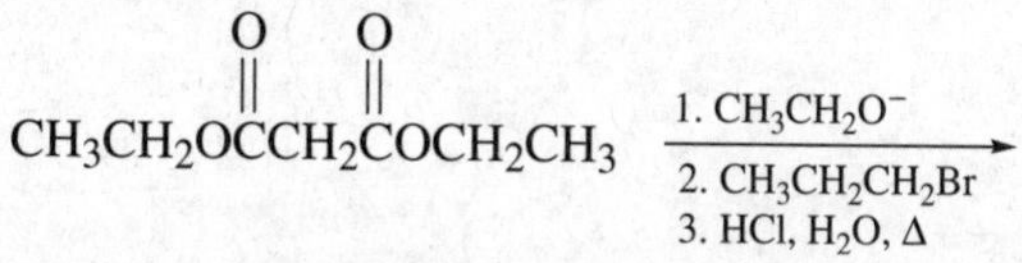

4. Give an example of each of the following:

    a. an aldol addition

    b. an aldol condensation

    c. a Claisen condensation

    d. a malonic ester synthesis

    e. an acetoacetic ester synthesis

5. Give the four β-hydroxyalcohols formed from the following two aldehydes:

$$\underset{\underset{\displaystyle CH_3}{|}}{CH_3CHCH_2CH_2\overset{\overset{\displaystyle O}{\parallel}}{C}H} + CH_3CH_2CH_2CH_2\overset{\overset{\displaystyle O}{\parallel}}{C}H \xrightarrow[\text{H}_2\text{O}]{\text{HO}^-}$$

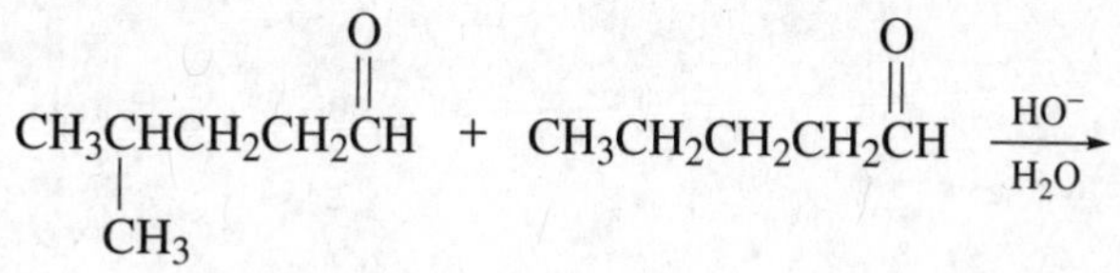

CHAPTER 14
Determining the Structures of Organic Compounds

1. Only positively charged fragments are accelerated through the analyzer tube.

$CH_3CH_2\overset{+}{C}H_2$ $\quad [CH_3CH_2CH_3]^{\bullet +}$ $\quad \overset{+}{C}H_2CH{=}CH_2$

2. Peaks should occur at *m/z* = 57 for loss of an ethyl group (86 − 29 = 57), and at *m/z* = 71 for loss of a methyl group (86 − 15 = 71).

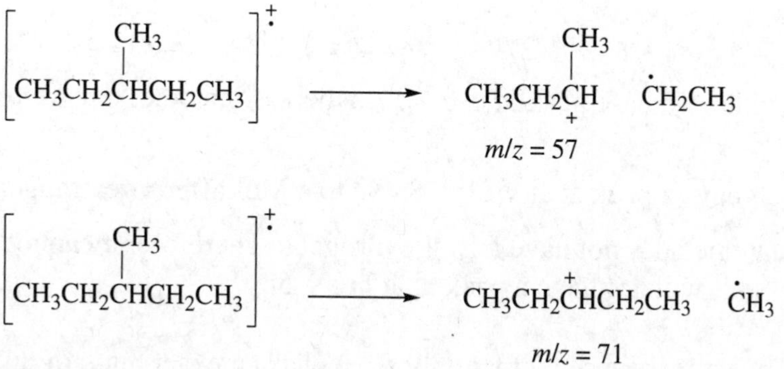

m/z = 57
m/z = 71

A secondary carbocation is forned in both cases. Because an ethyl radical is more stable than a methyl radical, the base peak will most likely be at *m/z* = 57.

3. Solved in the text.

4. The equal heights of the M and M + 2 peaks indicate that the compound contains bromine. Loss of a bromine atom from either the M peak (122-79) or the M + 2 peak (124-81) gives a peak with *m/z* = 43, which indicates a propyl group. Thus, we can conclude that the primary alkyl halide responsible for the spectrum is 1-bromopropane.

5. Because the compound contains chlorine, the M + 2 peak is one third the size of the M peak. Loss of a chlorine atom from either the M + 2 peak (80 − 37) or the M peak (78 − 35) gives a peak with *m/z* = 43.

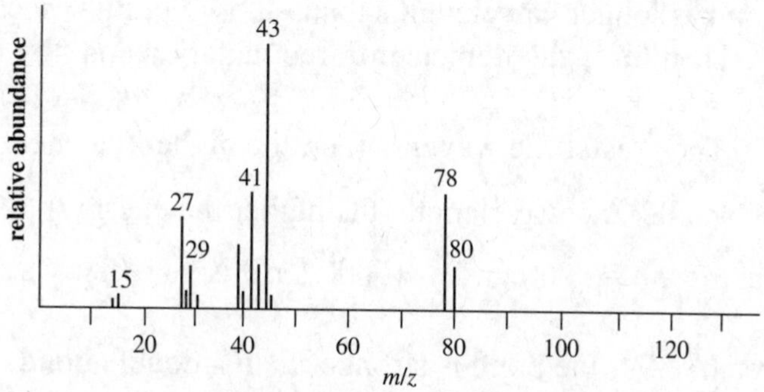

relative abundance
43
41
27
29
15
78
80
20
40
60
80
100
120
m/z

Mass spectrum of 1-chloropropane

6. The molecular ion with *m/z* = 86 indicates that the ketone has five carbons. The fact that Figure 14.5 shows a peak at *m/z* = 43 for loss of a propyl group (86 − 43) indicates that it is the mass spectrum of either 2-pentanone or 3-methyl-2-butanone, since each of these has a propyl group.

The fact that the spectrum has a peak at $m/z = 58$, indicating loss of ethene, indicates that the compound has a γ-hydrogen that enables it to undergo a rearrangement.

Therefore, it must be 2-pentanone, since 3-methyl-2-butanone does not have a γ-hydrogen.

7. All three ketones will have a molecular ion with $m/z = 86$.

3-pentanone 2-pentanone 3-methyl-2-butanone

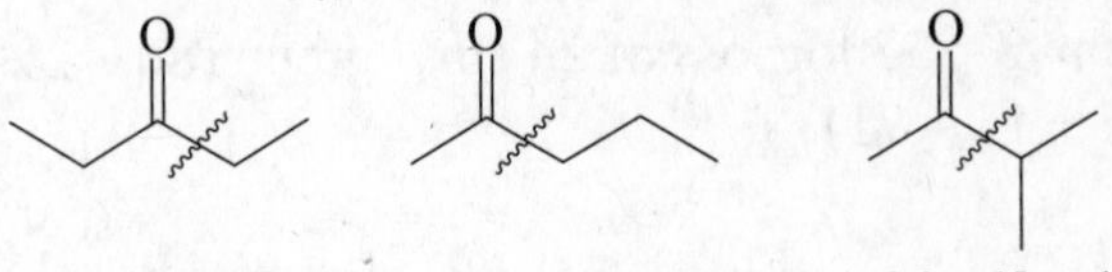

86 – 29 = 57 86 – 43 = 43 86 – 43 = 43

3-Pentanone will have a base peak at $m/z = 57$, whereas the other two ketones will have base peaks at $m/z = 43$.

2-Pentanone will have a peak at at $m/z = 58$ due to a McLafferty rearrangement.

3-Methyl-2-butanone does not have any γ-hydrogens. Therefore, it cannot undergo a McLafferty rearrangement, so it will not have a peak at at $m/z = 58$.

8. The calculated exact masses show that only C_6H_{14} has an exact mass of 86.10955.

C_6H_{14}	6(12.00000) =	72.00000
	14(1.007825) =	14.10955
		86.10955

$C_4H_6O_2$	4(12.00000) =	48.00000
	6(1.007825) =	6.04695
	2(15.9949) =	31.9898
		86.03675

$C_4H_{10}N_2$	4(12.00000) =	48.00000
	10(1.007825) =	10.07825
	2(14.0031) =	28.0062
		86.08445

9. The wavelength is the distance from the top of one wave to the top of the next wave. We see that **"a"** has a longer wavelength than **"b"**.

Infrared radiation has longer wavelengths than ultraviolet light because infrared radiation is lower in energy. Therefore, **"a"** depicts infrared radiation and **"b"** depicts ultraviolet light.

10. **a.** 2000 cm^{-1} (The greater the wavenumber, the higher the energy.)

b. 8 μm (The smaller the wavelength, the higher the energy.)

c. 2 μm, because 2 μm = 5000 cm^{-1} (5000 cm^{-1} is a greater wavenumber than 3000 cm^{-1}.)

11. **a.** C≡C stretch A triple bond is stronger than a double bond, so it takes more energy to stretch a triple bond.

b. C—H stretch It requires more energy to stretch a bond than to bend it.

c. C≡N stretch A triple bond is stronger than a double bond, so it takes more energy to stretch a triple bond.

d. C=O stretch A double bond is stronger than a single bond, so it takes more energy to stretch a double bond.

12. The C—O bond in 1-hexanol occurs at a smaller wavenumber because it is a pure single bond, whereas the C—O bond in pentanoic acid has some double bond character, which makes it harder to stretch. Because it is harder to stretch, its absorption band is at a larger wavenumber.

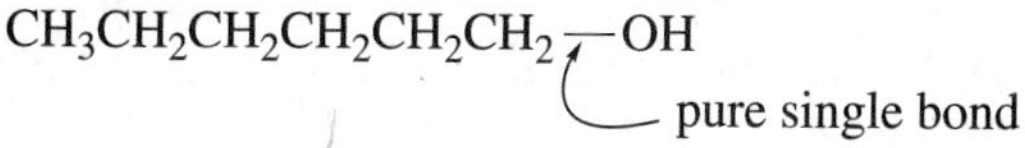
$CH_3CH_2CH_2CH_2CH_2CH_2$—OH
pure single bond

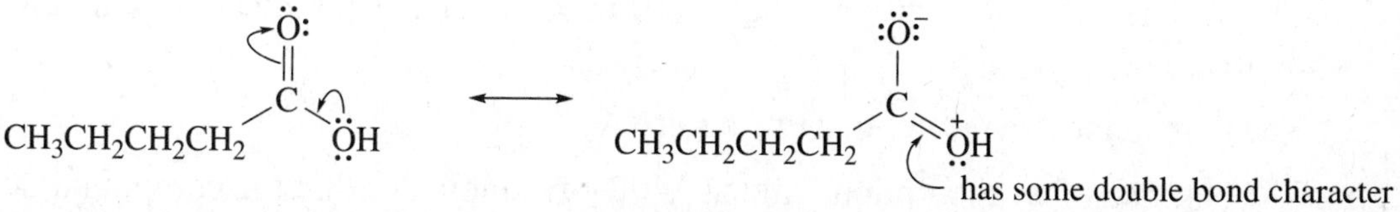
$CH_3CH_2CH_2CH_2$
$CH_3CH_2CH_2CH_2$
has some double bond character

13. **a.** The C—O stretch of phenol because bond has partial double-bond character as a result of electron delocalization. The C—O bond of cyclohexanol is a pure single bond, because all its electrons are localized.

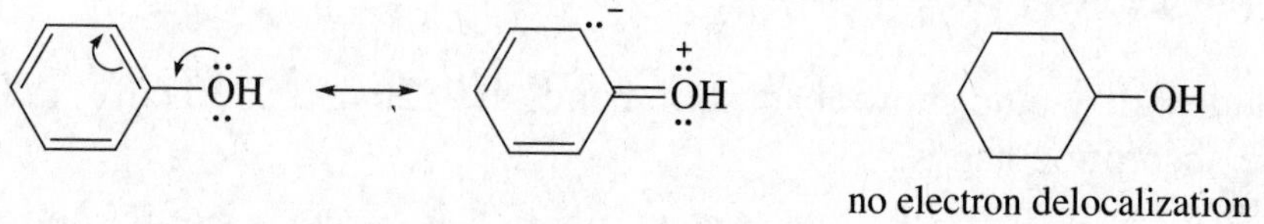
OH
OH
OH
no electron delocalization

b. The C=O stretch of a ketone because the bond has more double-bond character. The double bond character of the C=O bond of the amide is reduced by electron delocalization.

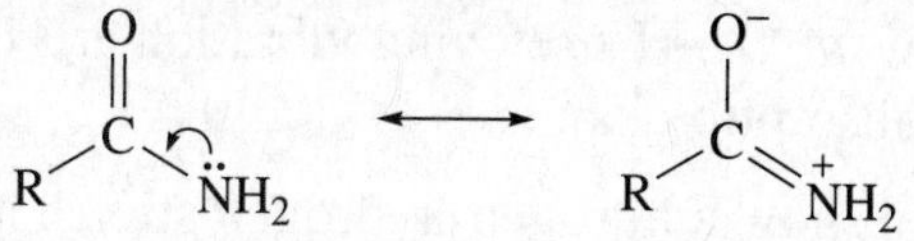

c. The C—O stretch, because a bond stretches at a higher frequency than it bends.

14. A carbonyl (C=O) group bonded to an sp^3 carbon will exhibit an absorption band at a higher frequency because a carbonyl group bonded to an sp^2 carbon will have less double-bond character as a result of electron delocalization.

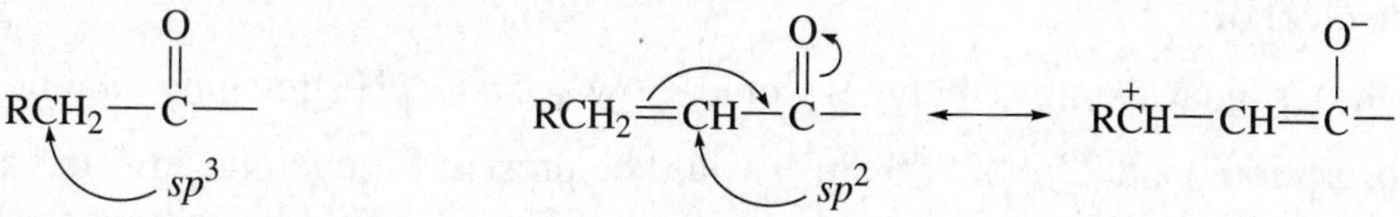
sp3
sp2

15. The stretching vibration of a C=O bond will be more intense because it is more polar than a C=C bond.

16. The O—H group of a carboxylic acid can form both intermolecular and intramolecular hydrogen bonds, whereas an alcohol can form only intermolecular hydrogen bonds.

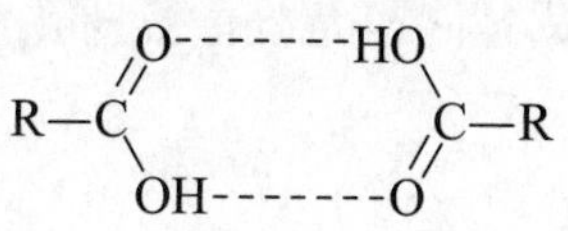

intramolecular hydrogen bonds

Therefore, the extent of hydrogen bonding is greater in a carboxylic acid, and hydrogen bonded OH groups have broader absorption bands.

17. **a.** The absorption band at 1700 cm^{-1} indicates that the compound has a carbonyl group.

The absence of an absorption band at 3300 cm^{-1} indicates that the compound is not a carboxylic acid.

The absence of an absorption band at 2700 cm^{-1} indicates that the compound is not an aldehyde.

The absence of an absorption band at 1100 cm^{-1} indicates that the compound is not an ester or an amide.

The compound, therefore, must be a **ketone**.

b. The absence of an absorption band at 3400 cm^{-1} indicates that the compound does not have an N—H bond.

The absence of an absorption band between 1700 cm^{-1} and 1600 cm^{-1} indicates that the compound is not an amide.

The compound, therefore, must be a **tertiary amine**.

18. **a.** An aldehyde would show absorption bands at 2820 and 2720 cm^{-1}. A ketone would not have these absorption bands.

b. Toluene would show an sp^3 C—H stretch slightly to the right of 3000 cm^{-1}. Benzene would not show an absorption band in this region.

c. Cyclohexene would show a carbon-carbon double-bond stretching vibration at 1680–1600 cm^{-1} and an sp^2 C—H stretching vibration at 3100–3020 cm^{-1}. Cyclohexane would not show these absorption bands.

d. A primary amine would show a nitrogen-hydrogen stretch at 3500–3300 cm^{-1}, and a tertiary amine would not have this absorption band.

19. **a.** A broad absorption band at 3300–2500 cm^{-1} would be present for the carboxylic acid and absent for the ester.

b. An absorption band at 1780–1650 cm^{-1} would be present for the carboxylic acid and absent for the alcohol.

c. Only the terminal alkyne (1-butyne) would show an *sp* C—H stretching vibration at 3300 cm^{-1}.

d. An absorption band 2960–2850 cm^{-1} would be present for cyclohexene and absent for benzene. (Also, an absorption band at 1500–1430 cm^{-1} would be present for benzene and absent for cyclohexene.)

20. **a.** 2 **c.** 4 **e.** 3

b. 1 **d.** 3 **f.** 3

21. **A** would give two signals, **B** would give one signal, and **C** would give three signals.

22. **a.** $\frac{600 \text{ Hz}}{300 \text{ MHz}} = 2.0 \text{ ppm}$

b. The answer would still be 2.0 ppm, because the chemical shift is independent of the operating frequency of the spectrometer.

23. The chemical shift is independent of the operating frequency. Therefore, if the two signals differ by **1.5 ppm** in a 300-MHz spectrometer, they will still differ by **1.5 ppm** in a 100-MHz spectrometer.

24. Magnesium is less electronegative than silicon. (See Table 1.3 on page 8 of the text.) Therefore, the peak for $(CH_3)_2Mg$ would be upfield from the TMS peak.

25. **a.** and **b.**

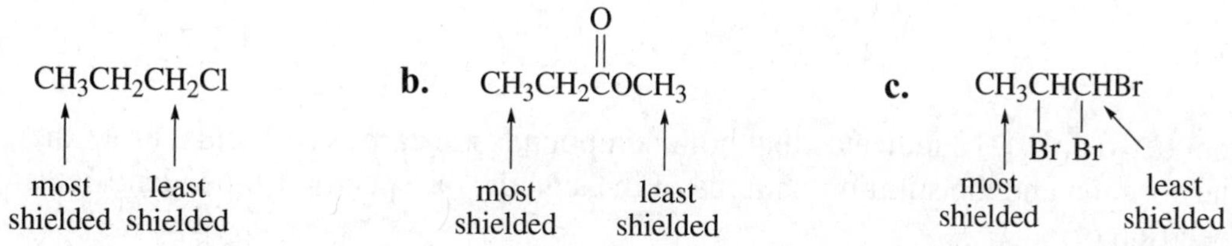

$CH_3CH_2CH_2Cl$
most shielded
least shielded
b. $CH_3CH_2COCH_3$
most shielded
least shielded
c. $CH_3CHCHBr$
Br Br
most shielded
least shielded

26. The signal farthest downfield in both spectra is the signal for the hydrogens bonded to the carbon that is also bonded to the halogen. Because chlorine is more electronegative than iodine, the farthest downfield signal should be farther downfield in the ^1H NMR spectrum for 1-chloropropane than in the ^1H NMR spectrum for 1-iodopropane.

Therefore, the **first spectrum** in Figure 14.23 is the ^1H NMR spectrum for **1-iodopropane**, and the **second spectrum** in the Figure is the ^1H NMR spectrum for **1-chloropropane**.

27.

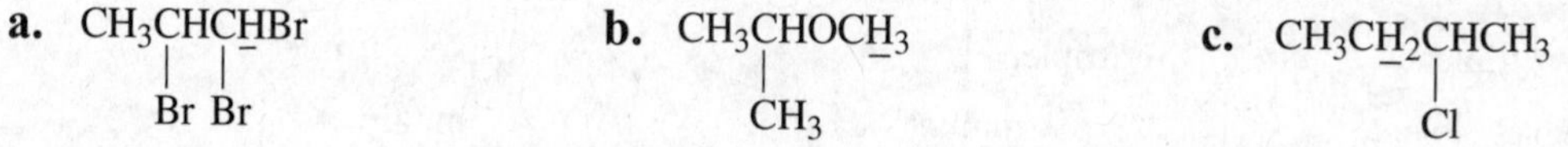

28. **a.**

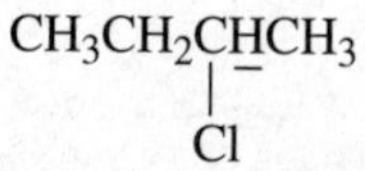

In the same environment, a methine H has a greater chemical shift than a methylene H.

b.

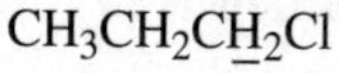

Chlorine is more electronegative than bromine.

29.

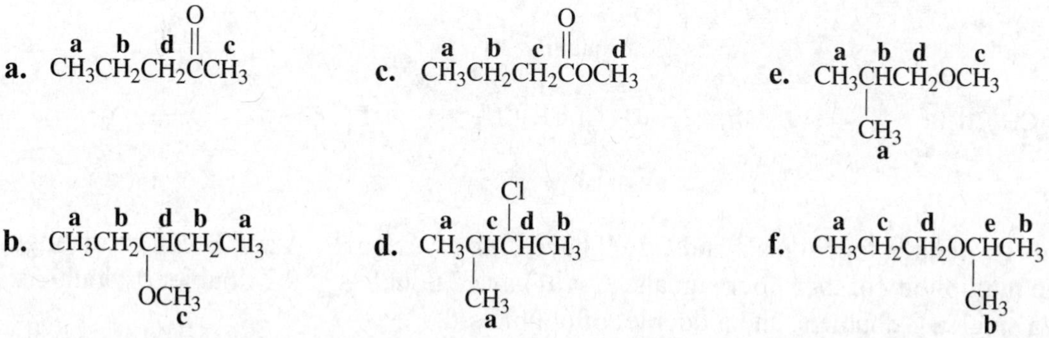

30. Each of the compounds would show two signals, but the ratio of the integrals for the two signals would be different for each of the compounds. The ratio of the integrals for the signals given by the first compound would be 2:9 (or 1:4.5), the ratio of the integrals for the signals given by the second compound would be 2:6 (or 1:3), and the ratio of the integrals for the signals given by the third compound would be 3:6 (or 1:2).

31. The heights of the integrals for the signals in the spectrum shown in Figure 14.25 are about 3.5 and 5.2. The ratio of the integrals, therefore, is 5.2/3.5 = 1.5. This matches the ratio of the integrals calculated for B. (Later we will see that a signal at ~7 ppm is characteristic of a benzene ring.)

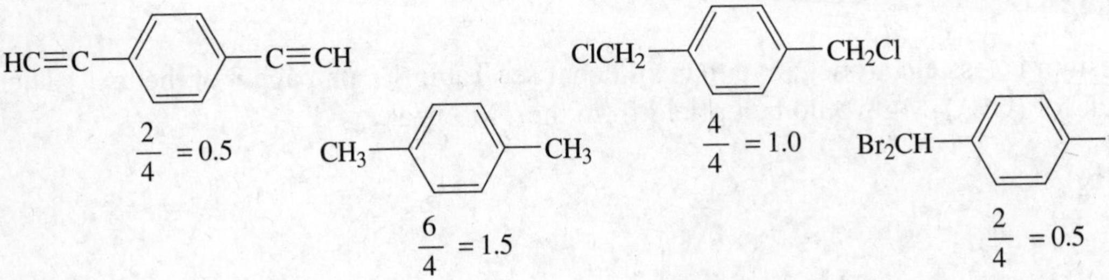

32. The signal between 10–12 indicates that both compounds are carboxylic acids. From the molecular formula and the splitting patterns of the signals, the spectra can be identified as the ^{1}H NMR spectrum of:

33. **a.** 3 signals **b.** 2 signals **c.** 3 signals

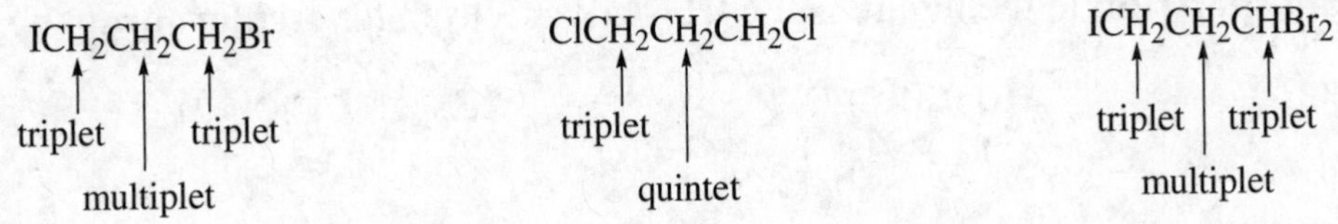

34.

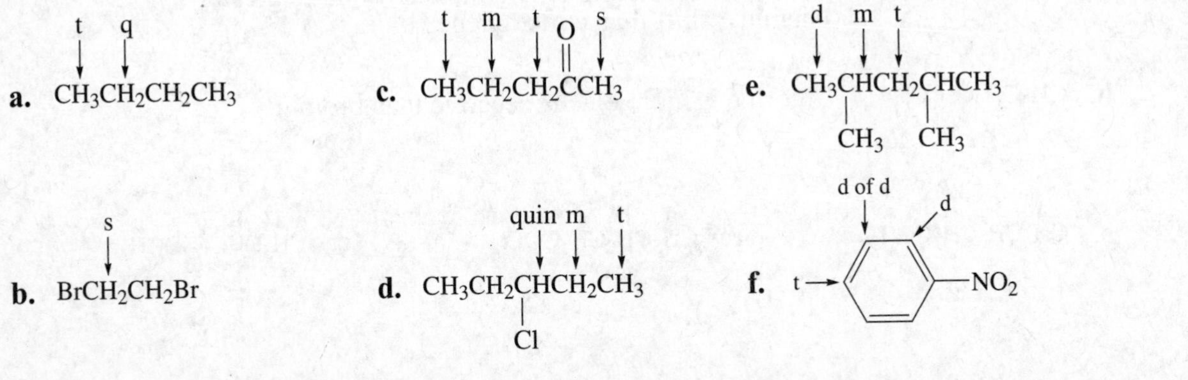

35. A and B will have four signals and C will have only two signals. A and B can be distinguished by the multiplicity of their four signals. A will have 2 doublets and 2 doublet of doublets; B will have a singlet, 2 doublets, and a doublet of doublets.

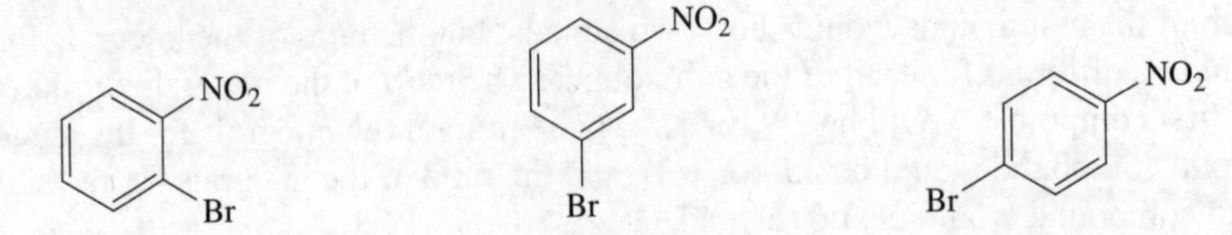

4 signals — 2 doublets and 2 doublet of doublets

4 signals — 1 singlet, 2 doublets and 1 doublet of doublets

2 signals — 2 doublets

36.

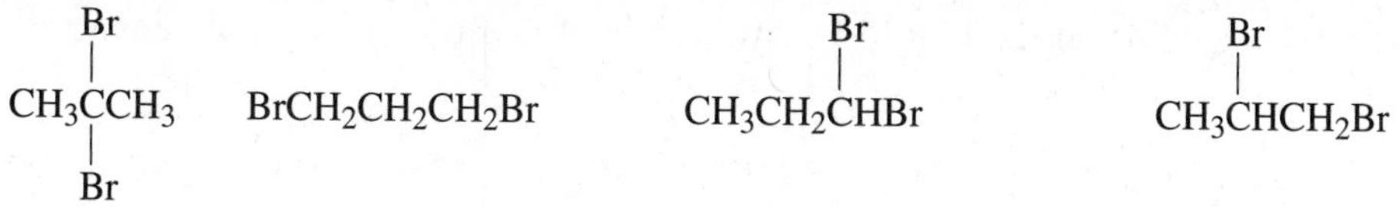

1 signal | **2 signals** | **3 signals** 2 triplets and a multiplet | **3 signals** 2 doublets and a multiplet

37. The IR spectrum indicates a benzene ring (1600 cm^{-1} and ~1460 cm^{-1}) with hydrogens on sp^2 carbons, and no carbonyl group. The absorption bands in the 1250–1000 cm^{-1} region suggests there are two C—O single bonds, one with no double bond character and one with some double bond character.

The two singlets in the ^{1}H NMR spectrum with about the same integration suggest two methyl groups, one of which is adjacent to an oxygen. That the benzene ring protons (6.7–7.1 ppm) consist of two doublets indicates a 1,4-disubstituted benzene ring.

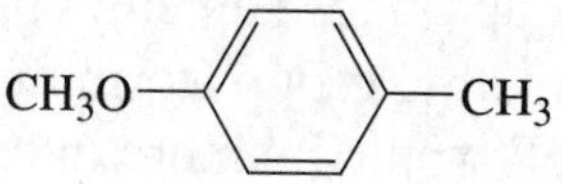

38. **a.** **1.** 3 **3.** 4 **5.** 2

2. 3 **4.** 3 **6.** 3

b. An arrow is drawn to the carbon that gives the signal at the lowest frequency.

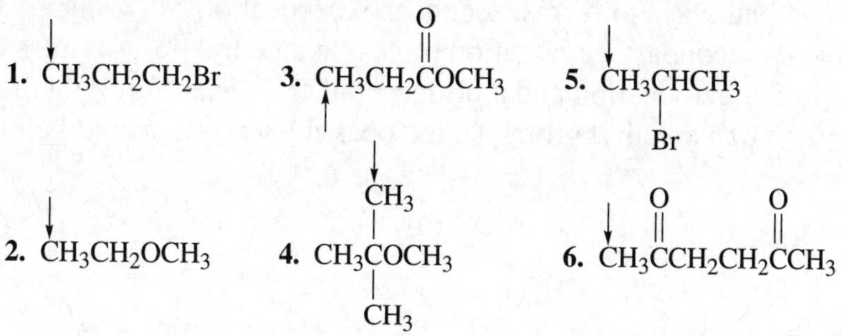

39. Each spectrum is described going from left to right.

1. triplet sextet triplet

2. triplet quartet singlet

3. triplet quartet singlet

40. The signal at 210 is for a carbonyl carbon. There are ten other carbons in the compound and five other signals. That suggests the compound is a ketone with identical five-carbon alkyl groups.

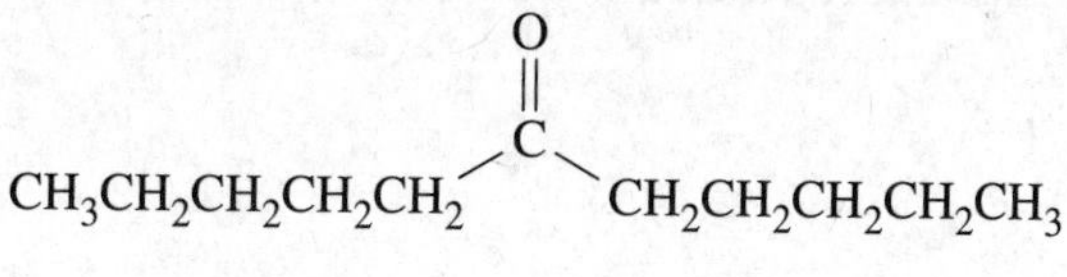

41. **a.** An absorption band at ~2700 cm^{-1} would be present for the aldehyde and absent for the ketone.

b. Absorption bands at 1600 cm^{-1} and 1500–1430 cm^{-1} and at 3100–3020 cm^{-1} would be present for the compound with the benzene ring and absent for the compound with the cyclohexane ring. An absorption band at 2960–2850 cm^{-1} would be present for the compound with the cyclohexane ring and absent for the compound with the benzene ring.

c. Two absorption bands at 3500–3300 cm^{-1} would be present for the amide and absent for the ester.

d. An absorption band at 3650–3200 cm^{-1} would be present for the alcohol and absent for the ether.

42. The molecular ion peak for these compounds is $m/z = 86$; the peak at $m/z = 57$ is due to loss of an ethyl group (86 − 29), and the peak at $m/z = 71$ is due to loss of a methyl group (86 − 15).

a. 3-Methylpentane will be more apt to lose an ethyl group (forming a secondary carbocation and a primary radical) than a methyl group (forming a secondary carbocation and a methyl radical), so the peak at $m/z = 57$ would be more intense than the peak at $m/z = 71$.

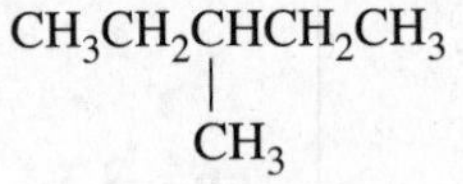

3-methylpentane

b. 2-Methylpentane has two pathways to form a secondary carbocation by losing a methyl group, but it cannot form a secondary carbocation by losing an ethyl group. (Loss of an ethyl group would form a primary carbocation and a primary radical.) Therefore, it will be more apt to lose a methyl group than an ethyl group, so the peak at $m/z = 71$ would be more intense than the peak at $m/z = 57$.

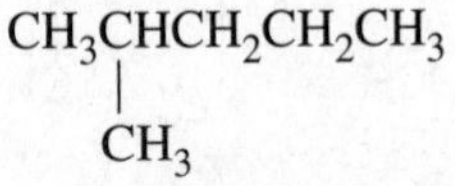

2-methylpentane

43. **a.** The absorption band at ~2100 cm^{-1} indicates a carbon-carbon triple bond, and the absorption band at ~3300 cm^{-1} indicates a hydrogen bonded to an *sp* carbon.

$$CH_3CH_2CH_2CH_2C{\equiv}CH$$

b. The absence of an absorption band at ~2700 cm^{-1} indicates that the compound is not an aldehyde, and the absence of a broad absorption band in the vicinity of 3000 cm^{-1} indicates that the compound is not a carboxylic acid. The ester and the ketone can be distinguished by the absorption band that is present at ~1200 cm^{-1} that indicates the carbon-oxygen single bond of an ester.

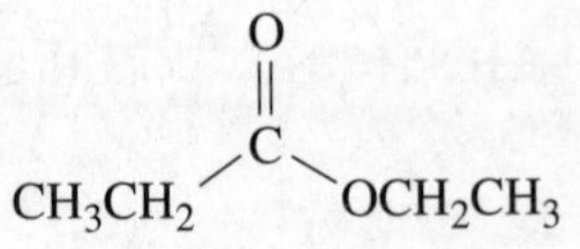

44. Enovid would have its carbonyl stretch at a higher frequency. The carbonyl group in Norlutin has some single-bond character because of the conjugated double bonds. This causes the carbon-oxygen bond to be easier to stretch than the carbon-oxygen bond in Enovid that is not involved in electron delocalization.

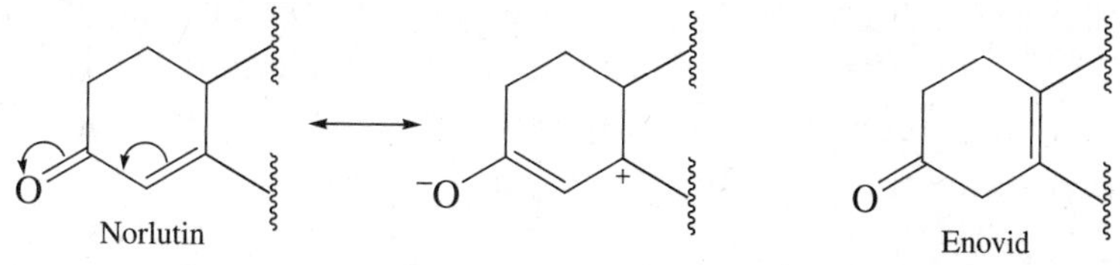
Norlutin
Enovid

45. **a.** An absorption band at ~1250 cm^{-1} would be present for the ester and absent for the ketone.

b. The C=O absorption band will be at a higher wavenumber for the *β,γ*-unsaturated ketone (1720 cm^{-1}) than for the *α,β*-unsaturated ketone (1680 cm^{-1}), since the double bonds in the latter are conjugated.

c. The alkene would have absorption bands at 1680–1600 cm^{-1} and at 3100–3020 cm^{-1} that the alkyne would not have. The alkyne would have an absorption band at 2260–2100 cm^{-1} that the alkene would not have.

d. Only the first compound has a hydrogen on an sp^2 carbon, so only that compound will show an absorption band at ~3050 cm^{-1}.

46. **a.** If the reaction had occurred, the intensity of the absorption bands at ~1700 cm^{-1} (due to the carbonyl group) and at ~2700 cm^{-1} (due to the aldehyde hydrogen) of the reactant would have decreased. If all the aldehyde had reacted, these absorption bands would have disappeared.

b. If all the NH_2NH_2 had been removed, there would be no N—H absorption at ~3400 cm^{-1}.

47. The peak at $m/z = 57$ will be more intense for 2,2-dimethylpropane than for 2-methylbutane or for pentane because the peak at $m/z = 57$ is due to loss of a methyl group: loss of a methyl group from 2,2-dimethylpropane results in the formation of a tertiary carbocation, whereas loss of a methyl group from 2-methylbutane and pentane results in the formation of a secondary and primary carbocation, respectively.

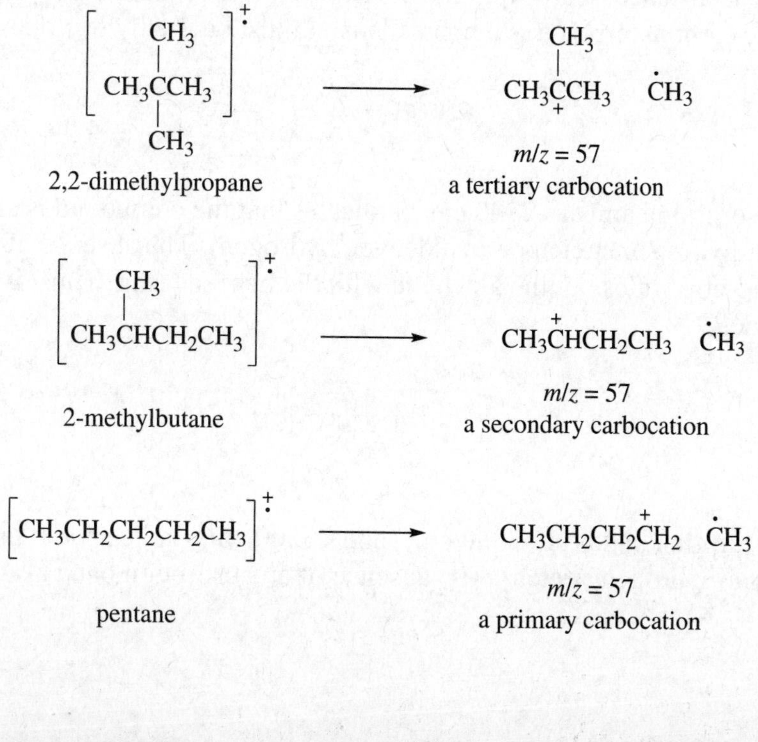
2,2-dimethylpropane
m/z = 57
a tertiary carbocation
2-methylbutane
m/z = 57
a secondary carbocation
pentane
m/z = 57
a primary carbocation

Notice that the mass spectrum of 2-methylbutane can be distinguished from those of the other isomers by the peak at $m/z = 43$. The peak at $m/z = 43$ will be most intense for 2-methylbutane because such a peak is due to loss of an ethyl group, which results in the formation of a secondary carbocation. Pentane gives a less intense peak at $m/z = 43$ because loss of an ethyl group from pentane forms a primary carbocation. 2,2-Dimethylpropane cannot form a peak at $m/z = 43$, because it does not have an ethyl group.

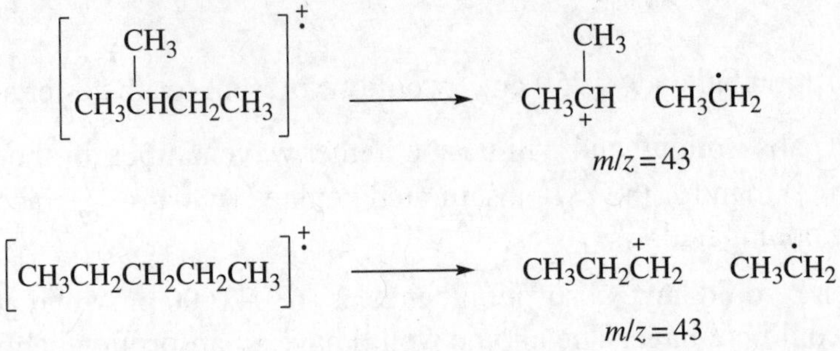
CH3
CH3CHCH2CH3
CH3
CH3CH CH3CH2
m/z = 43
CH3CH2CH2CH2CH3
CH3CH2CH2 CH3CH2
m/z = 43

48. **a.** The broad absorption band at ~3300 cm^{-1} is characteristic of the oxygen-hydrogen stretch of an alcohol, and the absence of absorption bands at ~1600 cm^{-1} and at ~3100 cm^{-1} indicates that it is not the alcohol with a carbon-carbon double bond.

$CH_3CH_2CH_2CH_2OH$

b. The absorption band at ~1685 cm^{-1} indicates a carbon-oxygen double bond. The absence of a strong and broad absorption band at ~3000 cm^{-1} rules out the carboxylic acid, and the absence of an absorption band at ~2700 cm^{-1} rules out the aldehyde. Thus, it must be the ketone.

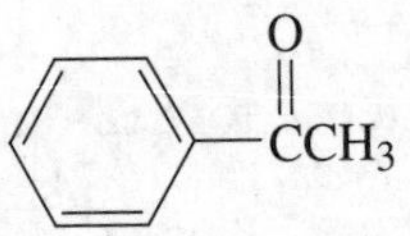
O
CCH3

c. The absorption band at ~1700 cm^{-1} indicates a carbon-oxygen double bond. The absence of an absorption band at ~1600 cm^{-1} rules out the ketones with the benzene or cyclohexene rings. The absence of absorption bands at ~2100 cm^{-1} and ~3300 cm^{-1} rules out the ketone with the carbon-carbon triple bond. Thus, it must be 4-ethylcyclohexanone.

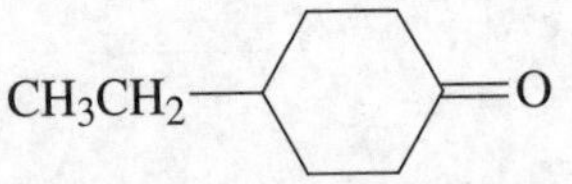
CH3CH2
O

49. **a.** The absorption band at ~2700 cm^{-1} indicates that the compound is an aldehyde (carbon-hydrogen stretch of an aldehyde hydrogen). The absence of an absorption band at ~1600 cm^{-1} rules out the aldehyde with the benzene ring. Thus, it must be the other aldehyde.

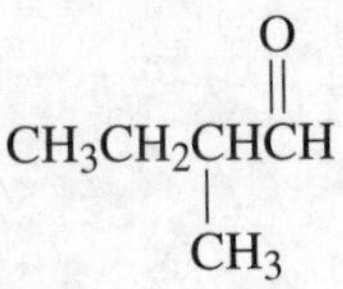
O
CH3CH2CHCH
CH3

b. The absorption bands at ~3350 cm^{-1} and ~3200 cm^{-1} indicate that the compound is an amide (nitrogen-hydrogen stretch). The absence of an absorption band at ~3050 cm^{-1} indicates that the compound does not have hydrogens bonded to sp^2 carbons. Therefore, it is not the amide that has a benzene ring.

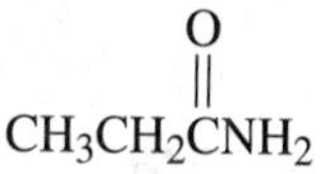

c. The absence of absorption bands at ~1600 cm^{-1} and ~1500 cm^{-1} indicates that the compound does not have a benzene ring. Thus, it must be the ketone. This is confirmed by the absence of an absorption band at ~1380 cm^{-1}, indicating that the compound does not have a methyl group.

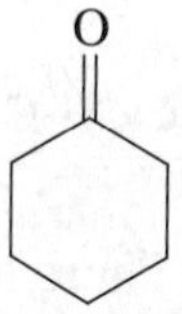

50. The compound would have an M and an M + 4 peak of equal intensity, and there would be an M + 2 peak that has twice the intensity of the M peak.

51. The broad absorption band at ~3300 cm^{-1} indicates that the compound has an OH group. The absence of absorption at ~2950 cm^{-1} indicates the compound does not have any hydrogens bonded to sp^3 carbons. Therefore, the compound is phenol.

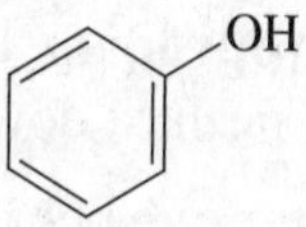

52. $CH_2{=}CHCH_2CH_2CH{=}CH_2$ 1,5-hexadiene; $CH_3CH{=}CHCH{=}CHCH_3$ 2,4-hexadiene

The easiest way to distinguish the two compounds is by the presence or absence of an absorption band at ~1370 cm^{-1} due to the methyl group that 2,4-hexadiene has but that 1,5-hexadiene does not have. In addition, 2,4-hexadiene has conjugated double bonds and, therefore, its double bonds have some single-bond character due to electron delocalization. Consequently, they are easier to stretch than the isolated double bonds of 1,5-hexadiene. Thus, the carbon-carbon double bond stretch of 2,4-hexadiene will be at a lower wavenumber than the carbon-carbon double-bond stretch of 1,5-hexadiene.

53. The frequency of the electromagnetic radiation used in NMR spectroscopy is lower than that used in IR and UV/Vis spectroscopy.

54. **a.** **1.** 5 **2.** 5 **3.** 4 **4.** 2 **b.** **1.** 7 **2.** 7 **3.** 5 **4.** 2

55. **a.** 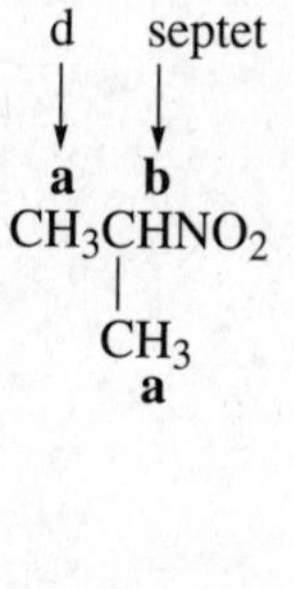

b. 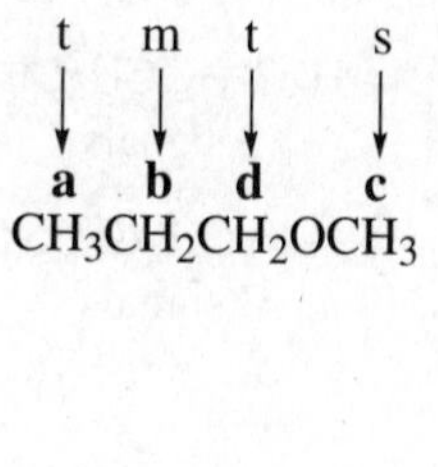

c. 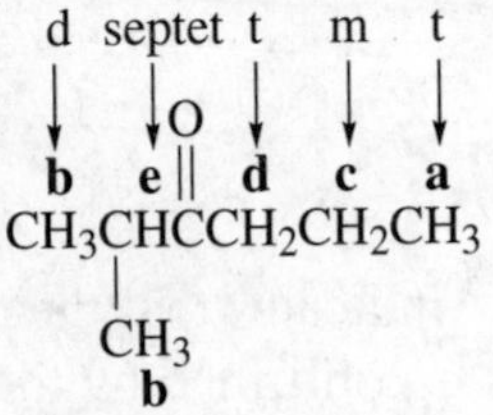

d. 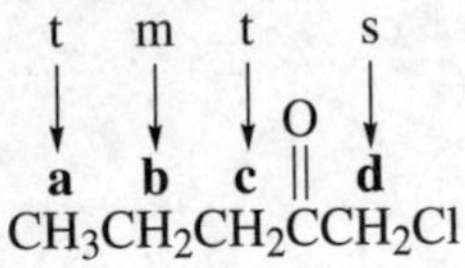

e. 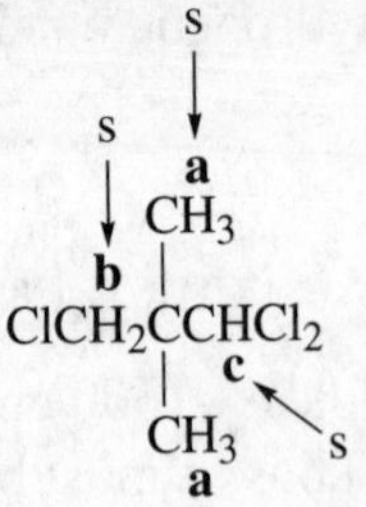

f. 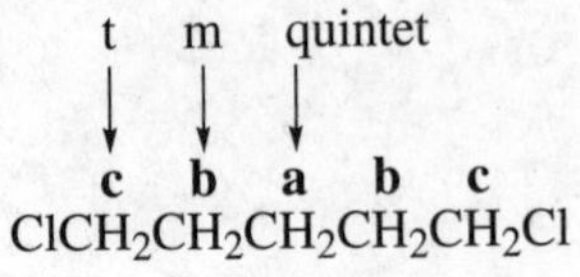

56. **a.** $CH_3OCH_2CH_2OCH_3$

b. There are 2 possibilities.

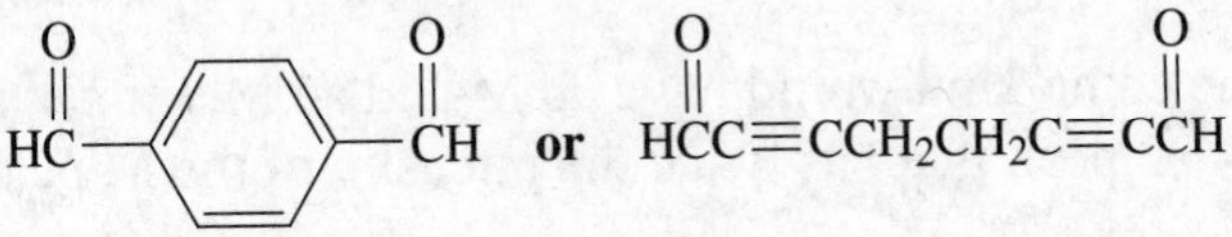

57. **a.** The spectrum must be that of **2-bromopropane** because the NMR spectrum has two signals and the signal farthest upfield is a doublet.

b. The spectrum must be that of **1-nitropropane** because the NMR spectrum has three signals and the signals farthest upfield and farthest downfield are triplets.

c. The spectrum must be that of **2-butanone** because the NMR spectrum has three signals and the signals are a triplet, a singlet, and a quartet.

58.

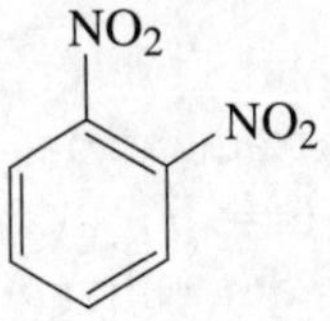 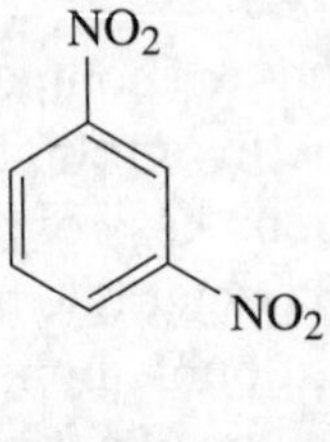 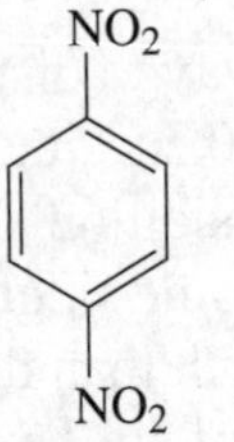

a. ^1H NMR	2 signals	3 signals	1 signal
b. ^1C NMR	3 signals	4 signals	2 signals

59. **a.** 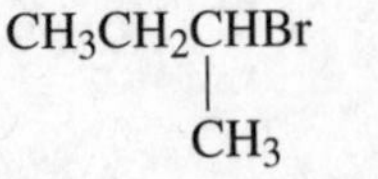

b. $CH_3CH_2CH_2CH_2Br$

c. 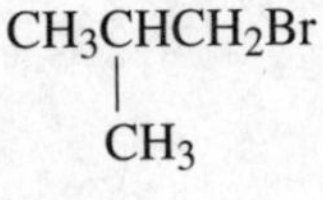

60. **a.** $CH_3CH_2CH_2OCH_3$ (4 signals) and $CH_3CH_2OCH_2CH_3$ (2 signals)

b. $BrCH_2CH_2CH_2Br$ (2 signals) and $BrCH_2CH_2CH_2NO_2$ (3 signals)

c. 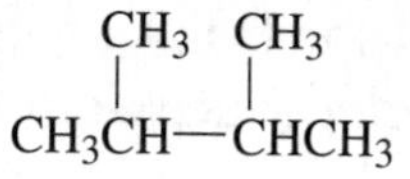 and 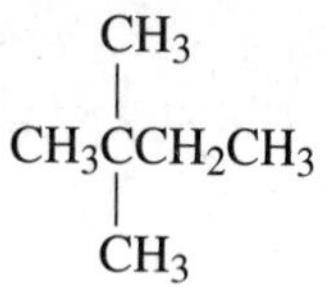

2 signals — 3 signals

d. 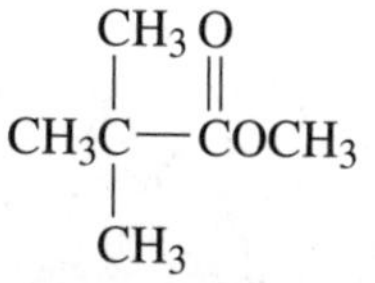 and 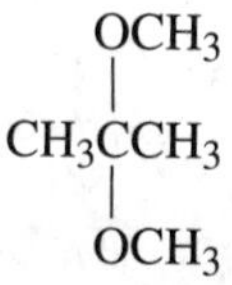

2 signals with integration 3:1 — 2 signals with integration 1:1

61. **a.** 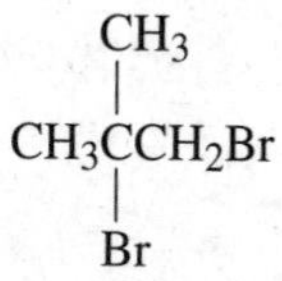 **b.** 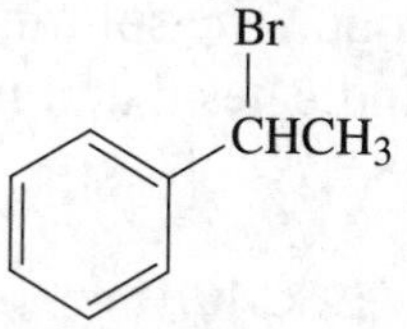 **c.** 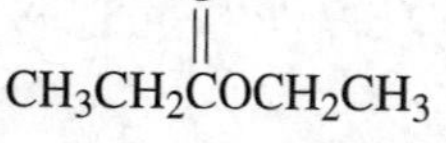

62. 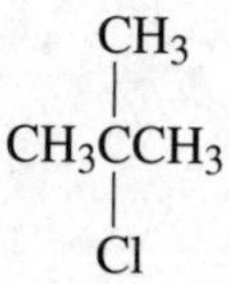 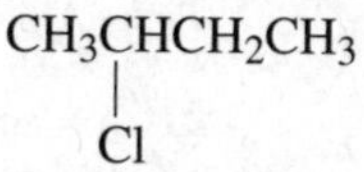

tert-butyl chloride
Compound A

sec-butyl chloride
Compound B

63. **a.** 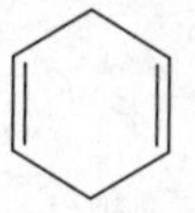 and 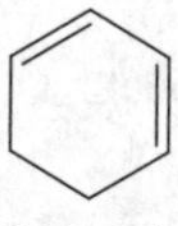

2 signals — 3 signals

b. 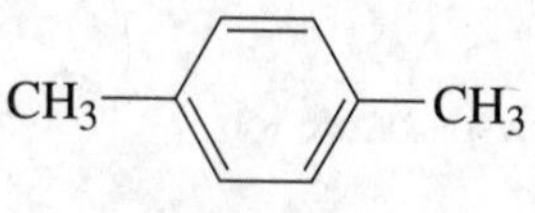 and 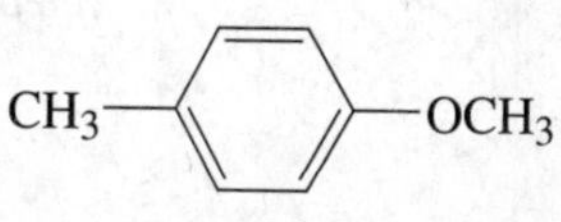

2 signals — 4 signals

64. **a.** 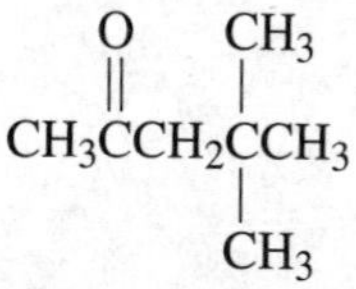 **b.** 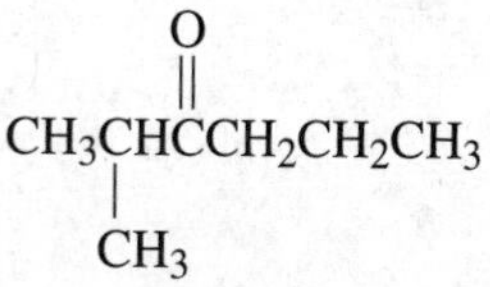 **c.** 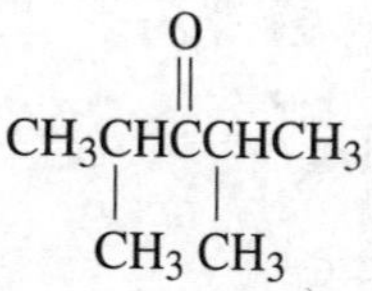

65. Propyl formate (**B**) is easy to distinguish from the other esters because it is the only one that will show four signals.

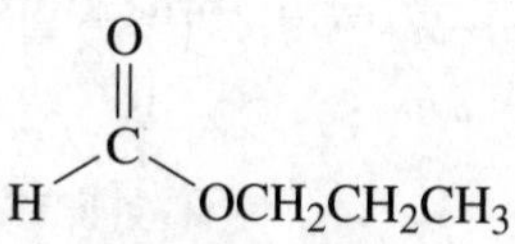

4 signals

The other three esters show three signals. Isopropyl formate (**D**) can be distinguished by its unique splitting pattern: a singlet, a doublet, and a septet.

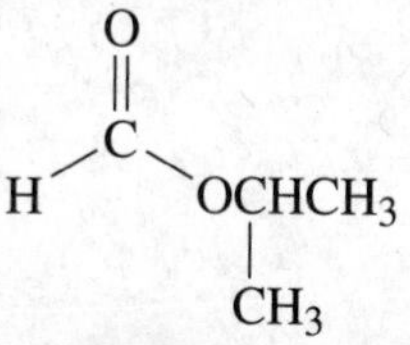

3 signals
singlet, doublet, septet

The splitting patterns of the other two esters are the same: a singlet, a triplet, and a quartet. They can be distinguished because the peak farthest downfield in A is a quartet, whereas the peak farthest downfield in C is a singlet.

$$\underset{\substack{CH_3 \qquad\; OCH_2CH_3}}{C}\!\!\!\overset{\displaystyle O}{\overset{\|}{}}$$

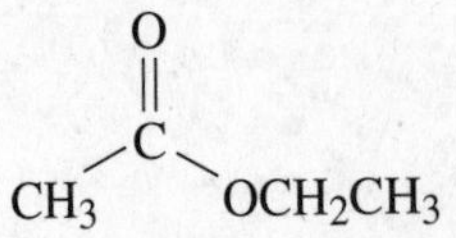

3 signals
singlet, triplet, quartet
The peak farthest downfield
is a quartet.

$$\underset{\substack{CH_3CH_2 \qquad OCH_3}}{C}\!\!\!\overset{\displaystyle O}{\overset{\|}{}}$$

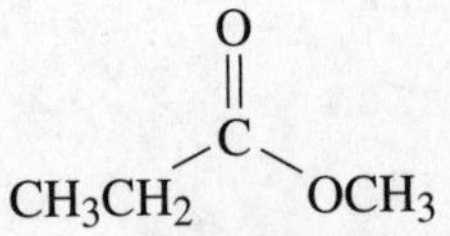

3 signals
singlet, triplet, quartet
The peak farthest downfield
is a singlet.

66. The singlet at 210 indicates a carbonyl group. The splitting of the other two signals indicates an isopropyl group. The molecular formula indicates that it must have two isopropyl groups.

$$\underset{\substack{| \quad | \\ CH_3\ CH_3}}{CH_3CHCCHCH_3}\overset{\displaystyle O}{\overset{\|}{}}$$

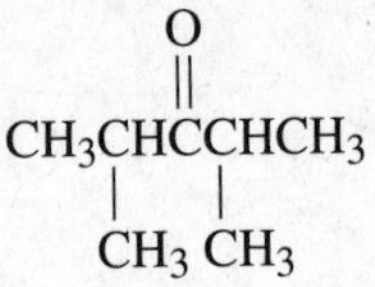

67. It is the ^{1}H NMR spectrum of *tert*-butyl methyl ether.

$$CH_3Br \;+\; CH_3\underset{CH_3}{\overset{CH_3}{\underset{|}{\overset{|}{C}}}}\!\!-O^- \;\longrightarrow\; CH_3\underset{CH_3}{\overset{CH_3}{\underset{|}{\overset{|}{C}}}}\!\!-OCH_3$$

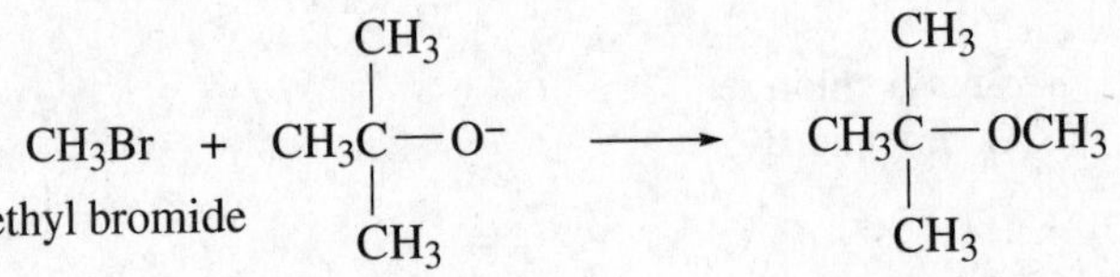

methyl bromide   *tert*-butoxide ion   *tert*-butyl methyl ether

68. 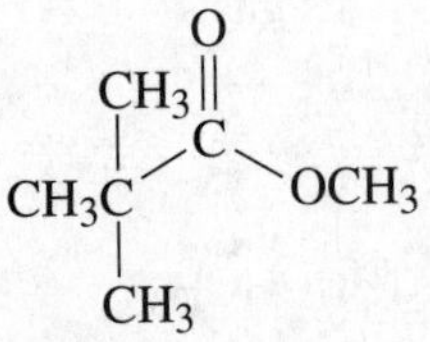
$$CH_3\underset{CH_3}{\overset{CH_3}{\underset{|}{\overset{|}{C}}}}\underset{}{\overset{O}{\overset{\|}{C}}}\!\!-OCH_3$$

69. If addition of HBr to propene follows the rule that says the electrophile adds to the sp^2 carbon that is bonded to the greater number of hydrogens, the product of the reaction will give an NMR spectrum with two signals (a doublet and a septet). If addition of HBr does not follow the rule, the product will give an NMR spectrum with three signals (two triplets and a multiplet).

$$CH_3CH{=}CH_2 \;+\; HBr \;\longrightarrow\; \underset{}{\overset{Br}{\overset{|}{}}}CH_3CHCH_3 \qquad CH_3CH_2CH_2Br$$

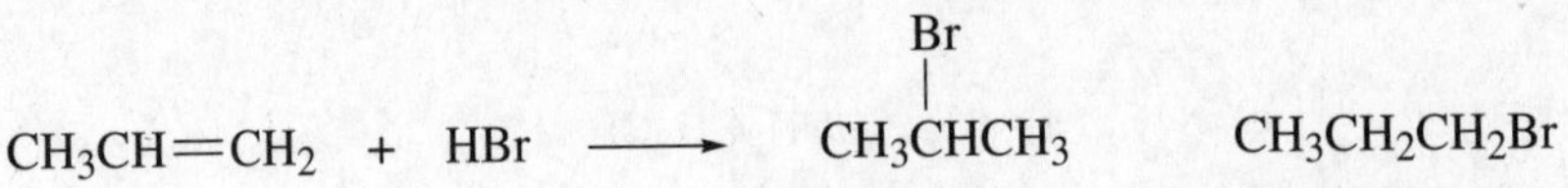

follows the rule   does not follow the rule
2 signals   **3 signals**

Chapter 14 Practice Test

1. Give one IR absorption band that could be used to distinguish each of the following pairs of compounds. Indicate the compound for which the band would be present.

a. $CH_3CH_2CH_2\overset{O}{\overset{\|}{C}}H$ and $CH_3CH_2CH_2\overset{O}{\overset{\|}{C}}CH_3$

b. $CH_3CH_2CH_2CH_2OH$ and $CH_3CH_2CH_2OCH_3$

c. $CH_3CH_2CH_2\overset{O}{\overset{\|}{C}}OCH_3$ and $CH_3CH_2CH_2\overset{O}{\overset{\|}{C}}CH_3$

d. $CH_3CH_2CH{=}CHCH_3$ and $CH_3CH_2C{\equiv}CCH_3$

e. $CH_3CH_2C{\equiv}CH$ and $CH_3CH_2C{\equiv}CCH_3$

2. Indicate whether each of the following is true or false:

a. An O—H bond will show a more intense absorption band than an N—H bond. T F

b. Light of 2 μm is of higher energy than light of 3 μm. T F

c. It takes more energy for a bending vibration than for a stretching vibration. T F

d. The signals on the right of an NMR spectrum are shielded compared to the signals on the left. T F

e. Dimethyl ketone has the same number of signals in its ^{1}H NMR spectrum as in its ^{13}C NMR spectrum. T F

f. In the ^{1}H NMR spectrum of the compound shown below, the lowest frequency signal is a singlet. T F

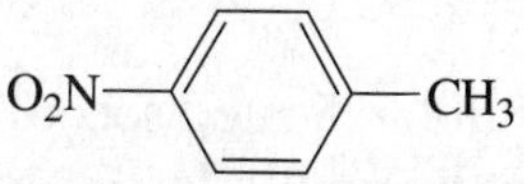
O2N
CH3

g. The M $^+$2 peak of an alkyl bromide is half the height of the M peak. T F

h. The greater the frequency of the signal, the greater its chemical shift in ppm. T F

i. An absorption band at 1150–1050 cm^{-1} would be present for an ether and absent for an alkane. T F

3. How could you distinguish between the IR spectra of the following compounds?

a.

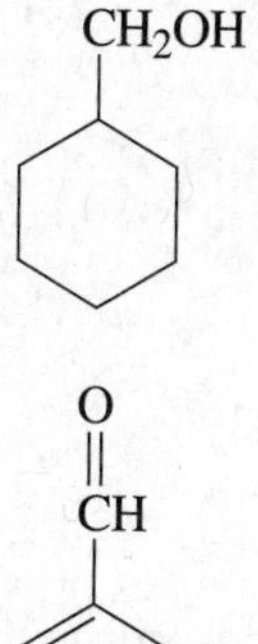
CH2OH

and

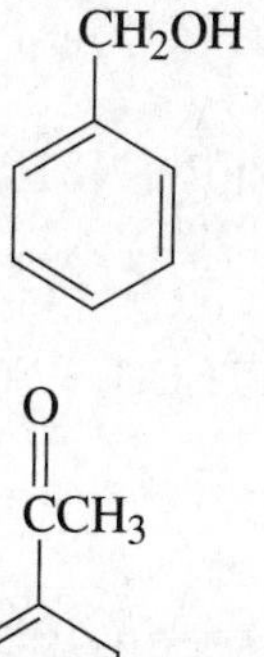
CH2OH

b.

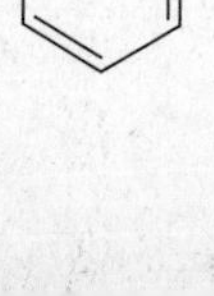
O
CH

and

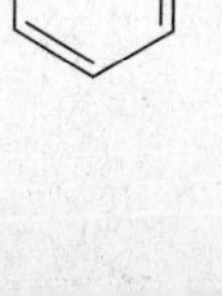
O
CCH3

c. 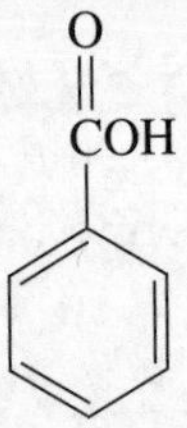 and 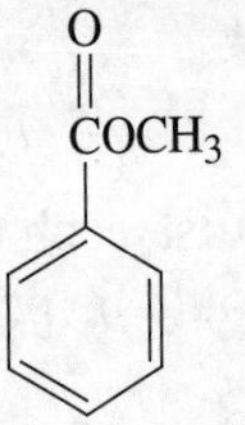

d. 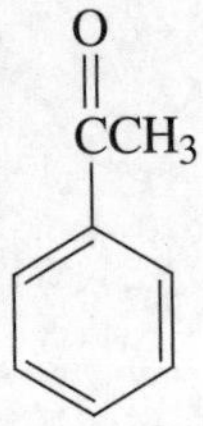 and 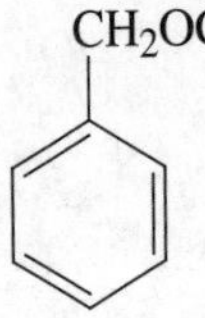

e. 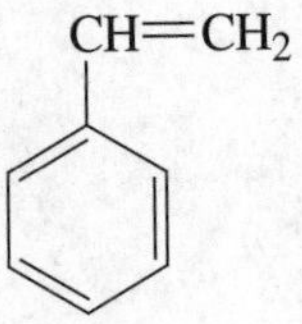 and 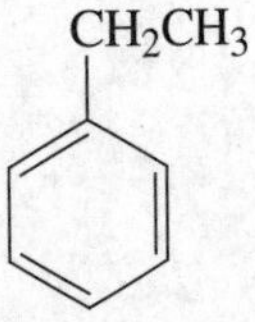

4. How many signals would you expect to see in the ^1H NMR spectrum of each of the following compounds?

$CH_3CH_2CH_2C(=O)CH_3$ $\quad$ $CH_3CH_2CH(Cl)CH_2CH_3$ $\quad$ $BrCH_2CH_2Br$

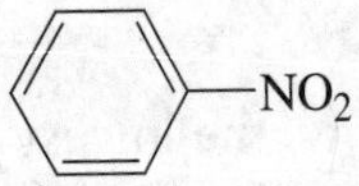 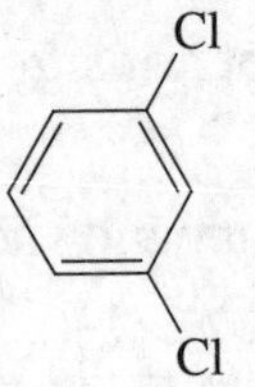 $CH_3CH(CH_3)CH_2CH(CH_3)CH_3$

5. Indicate the multiplicity of each of the indicated sets of protons. (that is, indicate whether it is a singlet, doublet, triplet, quartet, quintet, multiplet, or doublet of doublets.)

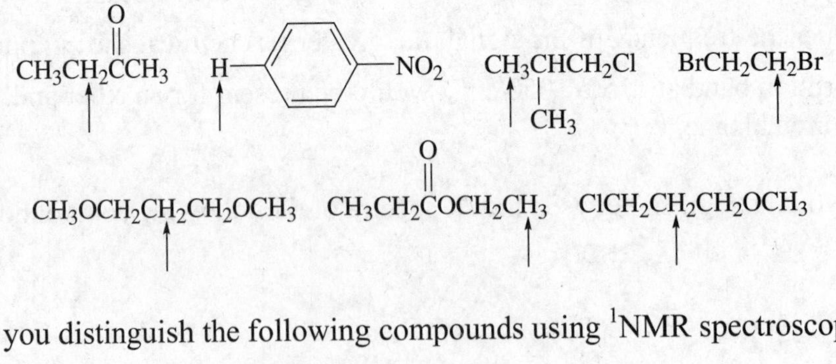

6. How could you distinguish the following compounds using ^{1}NMR spectroscopy?

$CH_3C(=O)OCH_2CH_3$ $\quad$ $CH_3CH_2C(=O)OCH_3$ $\quad$ $HC(=O)OCH_2CH_2CH_3$

7. For each compound:

a. Indicate the number of signals you would expect to see in its ^1H NMR spectrum.

b. Indicate the hydrogen or set of hydrogens that would give the highest frequency (farthest downfield) signal.

c. Indicate the multiplicity of that signal.

d. Indicate the relative integrals going from left to right across the spectrum.

a. $CH_3CH_2CH_2Cl$ **b.** $CH_3CH_2\overset{\overset{\displaystyle O}{\|}}{C}OCH_3$ **c.** $CH_3\underset{\underset{\displaystyle Br}{|}}{C}HCH_3$

8. For each compound in Problem 7:

a. Indicate the number of signals you would expect to see in its proton-decoupled ^{13}C NMR spectrum.

b. Indicate the carbon that would give the highest frequency (farthest downfield) signal.

c. Indicate the multiplicity of that signal in a proton-coupled ^{13}C NMR spectrum.

CHAPTER 15

The Organic Chemistry of Carbohydrates

1. D-Ribose is an aldopentose.

D-Sedoheptulose is a ketoheptose.

D-Mannose is an aldohexose.

2.

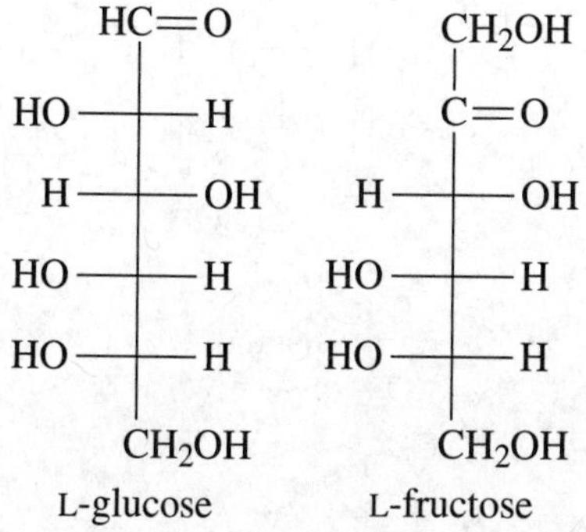
L-glucose
L-fructose

3. **a.** D-ribose **b.** L-talose

4. D-psicose

5. **a.** A ketoheptose has four asymmetric centers ($2^4 = 16$ stereoisomers).

b. An aldoheptose has five asymmetric centers ($2^5 = 32$ stereoisomers).

c. A ketotriose has no asymmetric centers; therefore, it has no stereoisomers.

6. D-tagatose, D-galactose, and D-talose

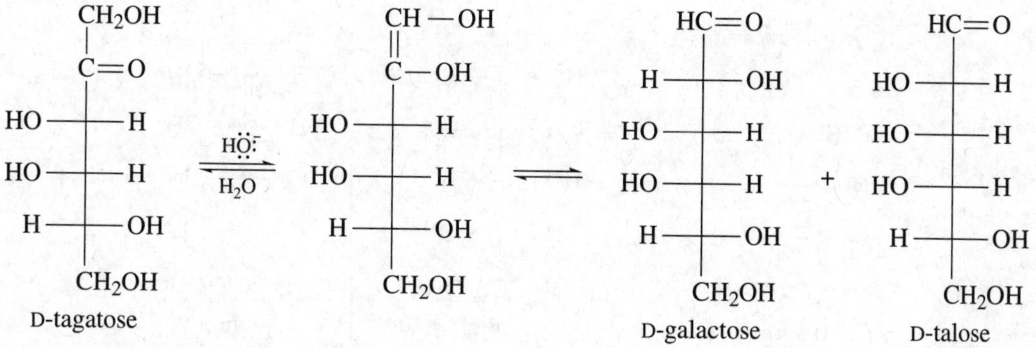
D-tagatose
D-galactose
D-talose

7. Removal of an α-hydrogen creates an enol that can enolize back to the ketone (using the OH at C-2) or can enolize to an aldehyde (using the OH at C-1). The aldehyde has a new asymmetric center (indicated by an *); one of the C-2 epimers is D-glucose, and the other is D-mannose.

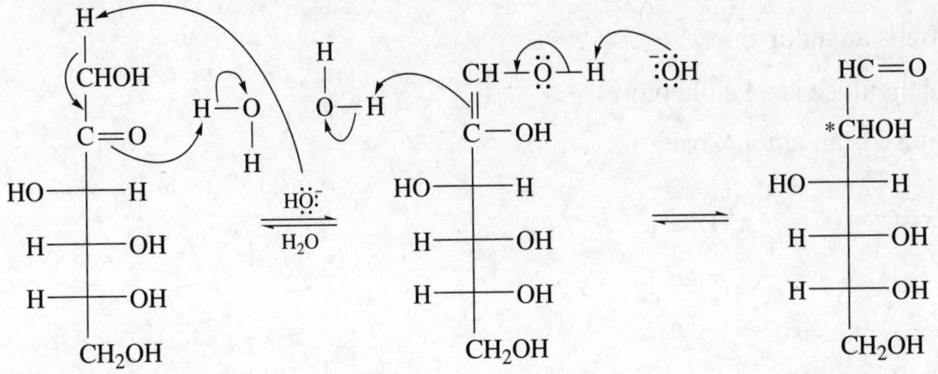

D-fructose D-glucose D-mannose

8. **a.** When D-idose is reduced, D-iditol is formed.

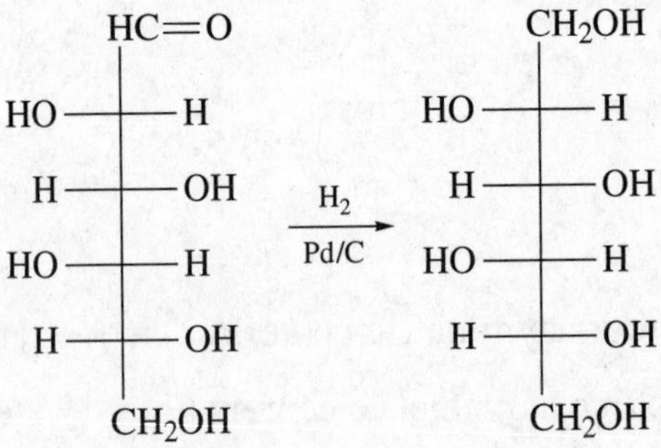

D-idose D-iditol

b. When D-sorbose is reduced, C-2 becomes an asymmetric center, so both D-iditol and the C-2 epimer of D-iditol (D-gulitol) are formed.

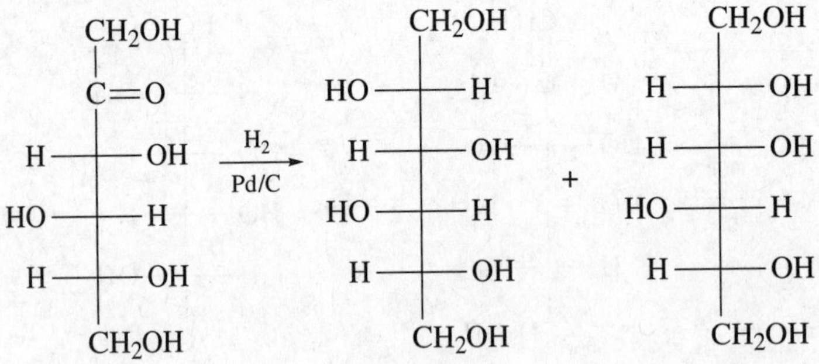

D-sorbose D-iditol D-gulitol

9. L-Galactose is reduced to the same alditol as D-galactose is reduced to.

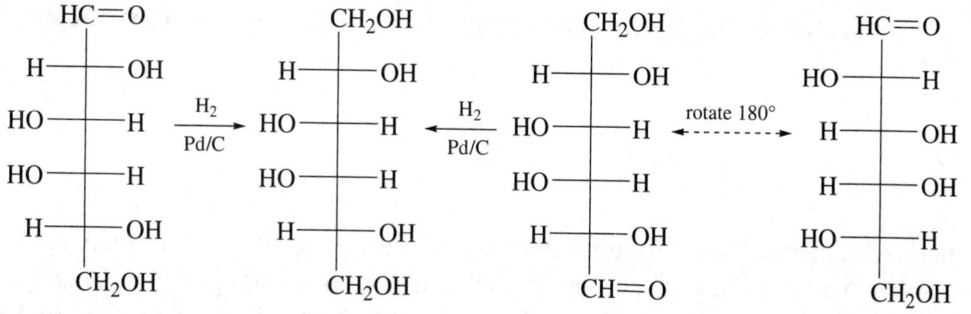
HC=O
H—OH
HO—H
HO—H
H—OH
CH₂OH
H₂
Pd/C
CH₂OH
H—OH
HO—H
HO—H
H—OH
CH₂OH
H₂
Pd/C
CH₂OH
H—OH
HO—H
HO—H
H—OH
CH=O
rotate 180°
HC=O
HO—H
H—OH
H—OH
HO—H
CH₂OH

D-galactose L-galactose

10. The ketohexose (D-tagatose) with the same configuration at C-3, C-4, and C-5 as D-talose will give the same alditol as D-talose. The other alditol is the one with the opposite configuration at C-2.

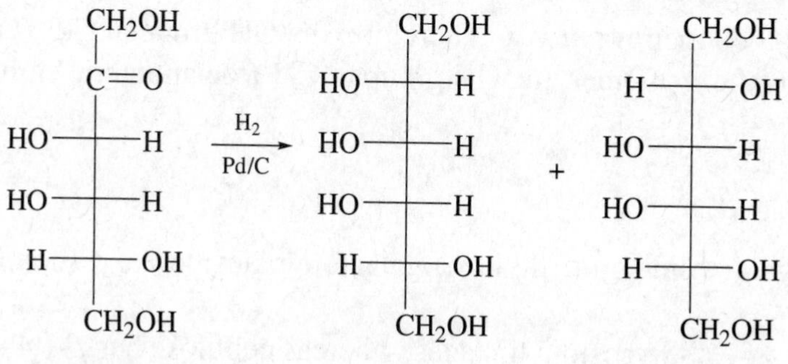
CH₂OH
C=O
HO—H
HO—H
H—OH
CH₂OH
H₂
Pd/C
CH₂OH
HO—H
HO—H
HO—H
H—OH
CH₂OH
+
CH₂OH
H—OH
HO—H
HO—H
H—OH
CH₂OH

D-tagatose D-talitol D-galactitol

11. **a.** L-gulose

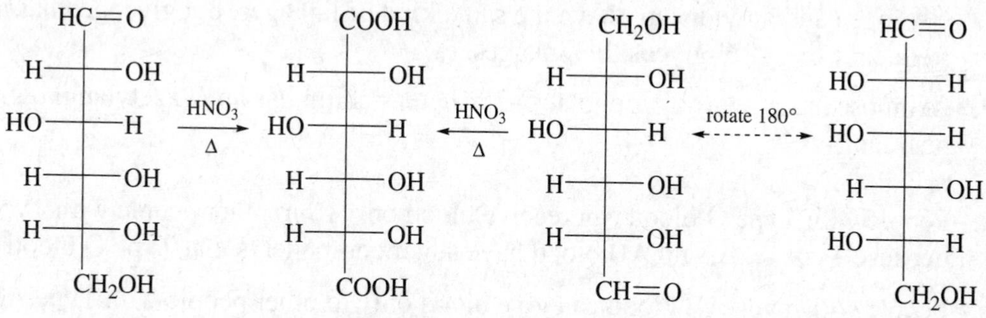
HC=O
H—OH
HO—H
H—OH
H—OH
CH₂OH
HNO₃
Δ
COOH
H—OH
HO—H
H—OH
H—OH
COOH
HNO₃
Δ
CH₂OH
H—OH
HO—H
H—OH
H—OH
CH=O
rotate 180°
HC=O
HO—H
HO—H
H—OH
HO—H
CH₂OH

D-glucose D-glucaric acid L-gulose

b. L-Gularic acid, because when D-glucaric is turned upside down it would be called L-gularic acid.

c. D-allose and L-allose, D-altrose and D-talose, L-altrose and L-talose, D-galactose and L-galactose.

12. **a.** D-gulose and D-idose **b.** L-xylose and L-lyxose

13. **a.** Solved in the text

b. The cyclic hemiacetal has one asymmetric center; therefore, it has two stereoisomers.

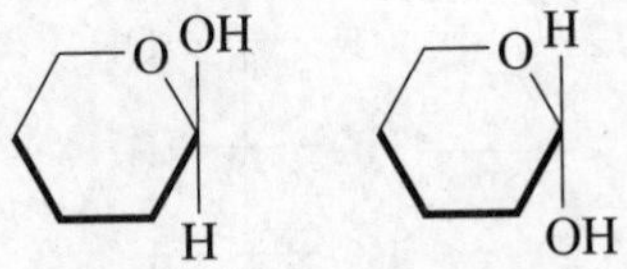

14. First recall that in the chair conformer: an α-anomer has the anomeric carbon in the axial position, whereas a β-anomer has the anomeric carbon in the equatorial position. Then recall that glucose has all its OH groups in equatorial positions. Now this question can be answered easily.

a. Solved in the text.

b. Idose differs in configuration from glucose at C-2, C-3, and C-4. Therefore, the OH groups at C-2, C-3, and C-4 in β-D-idose are in the axial position.

c. Allose is a C-3 epimer of glucose. Therefore, the OH group at C-3 is in the axial position and, since it is the α-anomer, the OH group at C-1 (the anomeric carbon) is also in the axial position.

15. **a.** Solved in the text.

b. methyl α-D-galactoside or methyl α-D-galactopyranoside (nonreducing)

16. **a.** Amylose has α-1,4'-glycosidic linkages, whereas cellulose has β-1,4'-glycosidic linkages.

b. Amylose has α-1,4'-glycosidic linkages, whereas amylopectin has both α-1,4'-glycosidic linkages and α-1,6'-glycosidic linkages.

c. Glycogen and amylopectin have the same kind of linkages, but glycogen has a higher frequency of α-1,6'-glycosidic linkages.

d. Cellulose has a hydroxy group at C-2, whereas chitin has an *N*-acetylamino group at that position.

17. **a.** People with Type O blood can receive blood only from other people with Type O blood, because Type A, B, and AB blood have sugar components that Type O blood does not have.

b. People with Type AB blood can give blood only to other people with Type AB blood, because Type AB blood has sugar components that Type A, B, or O blood does not have.

18. **a.**

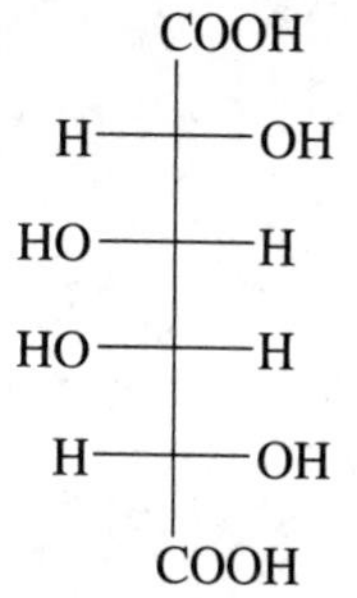

c.

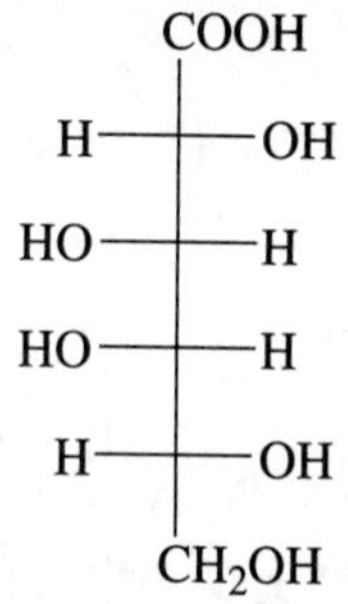

b.

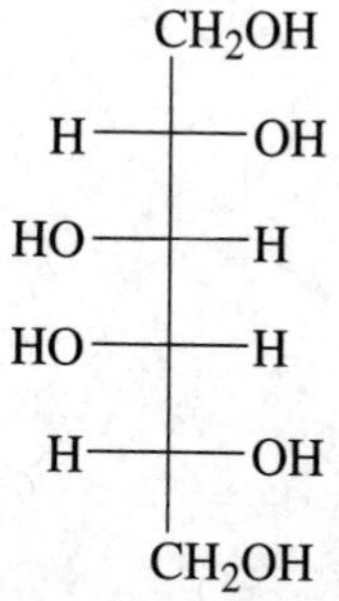

d.

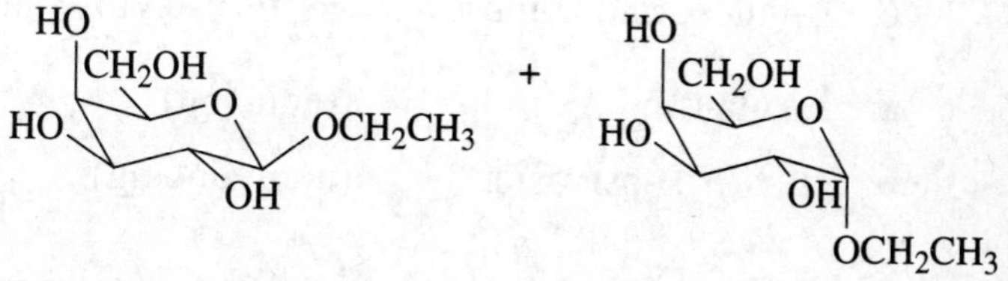

19. L-allose

20. **a.** D-lyxose **b.** D-talose **c.** D-psicose

21. **a.** D-ribose and L-ribose, D-arabinose and L-arabinose, D-xylose and L-xylose, D-lyxose and L-lyxose

b. D-arabinose, L-arabinose, D-lyxose, and L-lyxose

22. D-Altrose is reduced to the same alditol as D-talose.

The easiest way to answer this question is to draw D-talose and its alditol. Then draw the monosaccharide with the same configuration at C-2, C-3, C-4, and C-5 as D-talose, reversing the functional groups at C-1 and C-6. (Put the alcohol group at the top and the aldehyde group at the bottom.) The resulting Fischer projection can be rotated 180° in the plane of the paper, and the monosaccharide can then be identified.

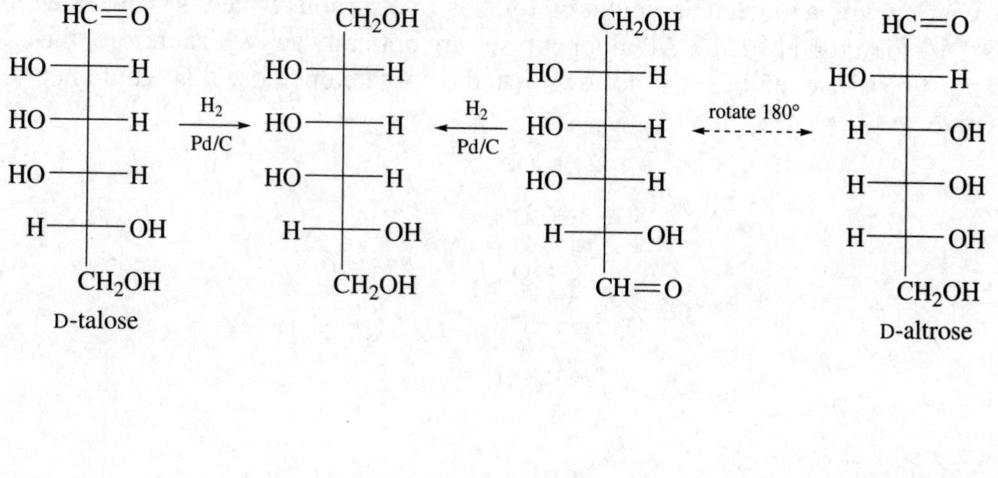
H2
Pd/C
H2
Pd/C
rotate 180°

D-talose D-altrose

23. D-fructose

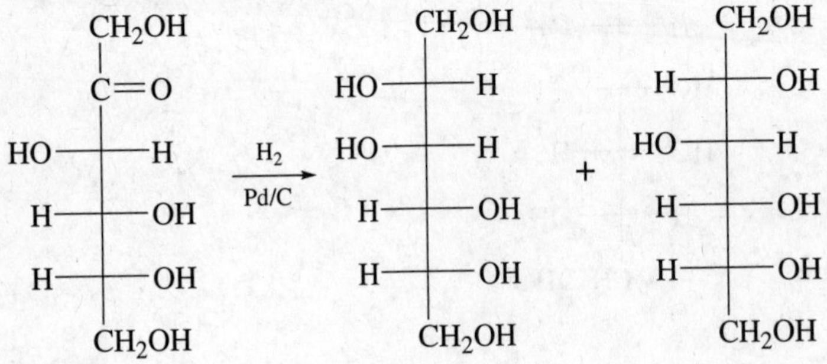

D-fructose D-mannitol D-glucitol

24. **a.** D-gulose and D-idose **b.** L-xylose and L-lyxose

25. **a.** α-D-talose (reducing)

b. ethyl β-D-psicoside (nonreducing)

26. It can form a five-membered ring hemiacetal using the OH group on C-4, or a six-membered ring hemiacetal using the OH group on C-5.

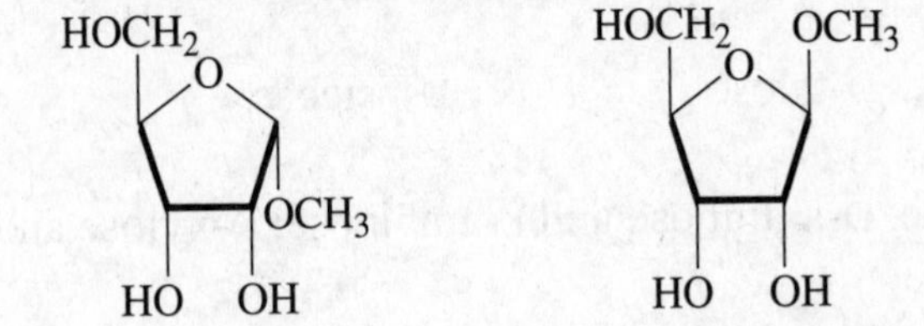

methyl α-D-ribofuranoside methyl β-D-ribofuranoside

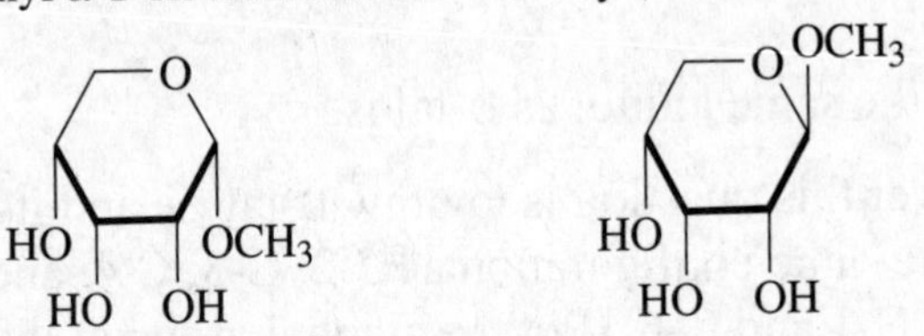

methyl α-D-ribopyranoside methyl β-D-ribopyranoside

27. A monosaccharide with a molecular weight of 150 has five carbons (five C's = 60, five O's = 80, and 10 H's = 10 for a total of 150). All aldopentoses are optically active. Therefore, the compound must be a ketopentose. The following is the only ketopentose that would not be optically active.

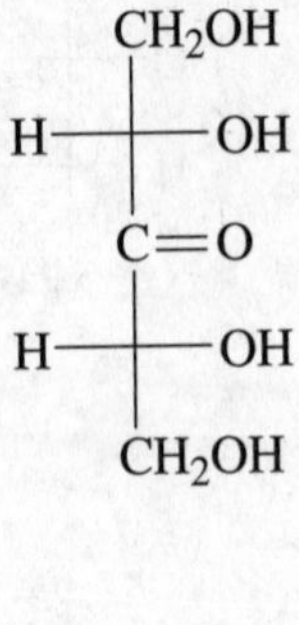

28.

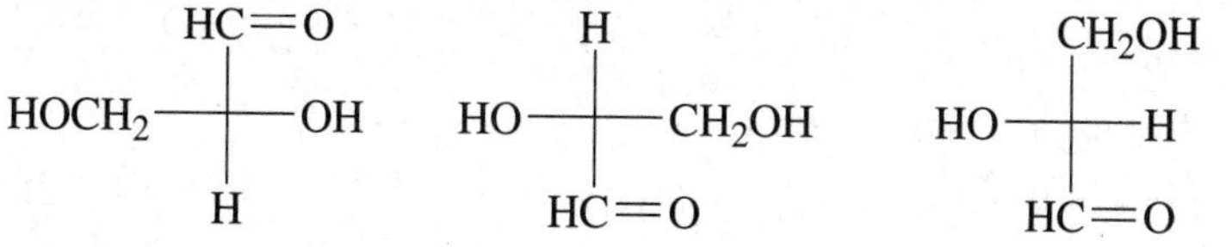

L-glyceraldehyde L-glyceraldehyde D-glyceraldehyde

29. Because the intermediate formed when glucuronic acid loses water has a planar sp^2 carbon, both the α- and β-glucuronides will be formed.

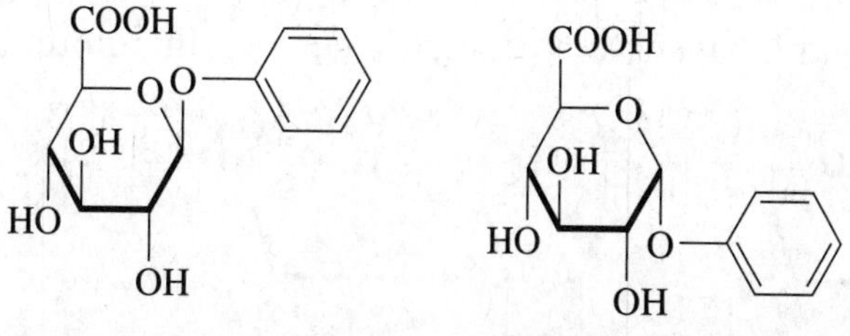

phenyl β-D-glucuronide phenyl α-D-glucuronide

30. She can take a sample of one of the sugars and oxidize it with nitric acid to an aldaric acid or reduce it with sodium borohydride to an alditol. If the product is optically active, the sugar was D-lyxose. If the product is not optically active, the sugar was D-xylose.

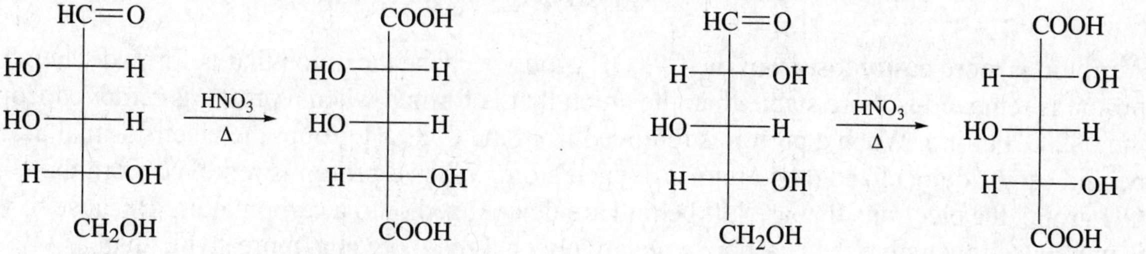

D-lyxose **optically active** D-xylose **optically inactive**

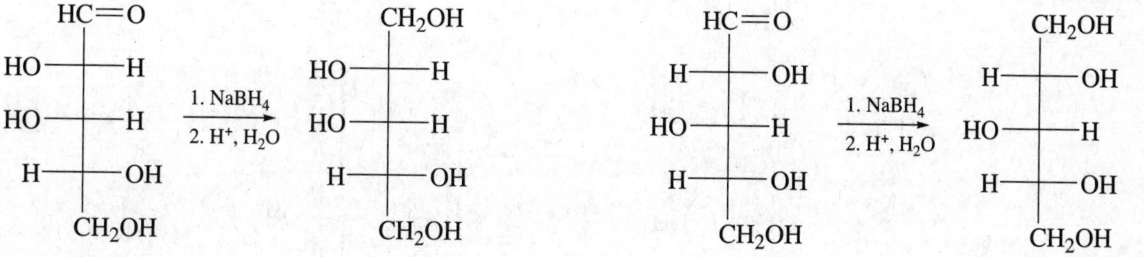

D-lyxose **optically active** D-xylose **optically inactive**

31.

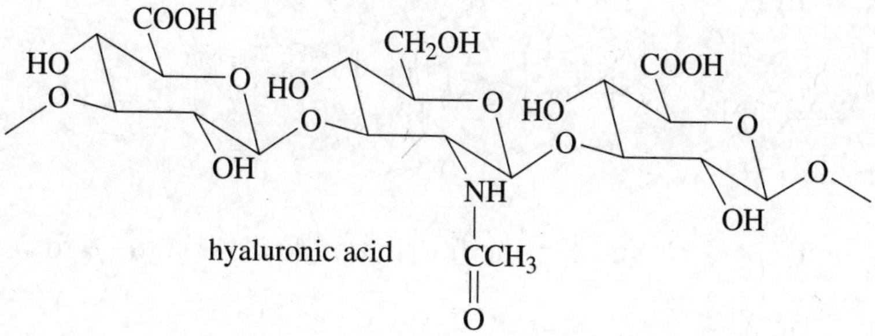

hyaluronic acid

32. **10** aldaric acids

Each of the following pairs forms the same aldaric acid:

D-allose and L-allose	L-altrose and L-talose
D-galactose and L-galactose	D-glucose and L-gulose
D-altrose and D-talose	L-glucose and D-gulose

Thus twelve aldohexoses form six aldaric acids. The other four aldohexoses each form a distinctive aldaric acid, and (6 + 4 = 10).

33. The hemiacetals in **a** and **b** have two asymmetric centers; therefore, each has four stereoisomers.

a. 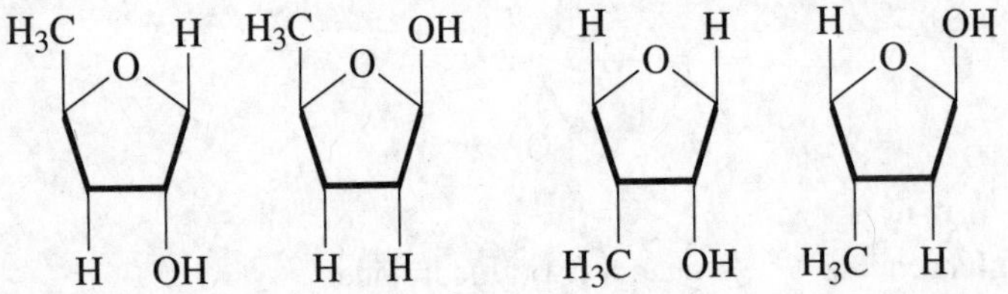

b. 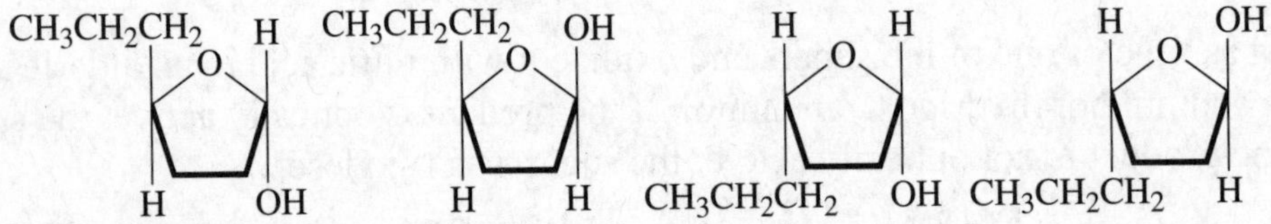

34. A proton is more easily lost from the C-3 OH group because the anion that is formed when the proton is removed is more stable than the anion that is formed when a proton is removed from the C-2 OH group. When a proton is removed from the C-3 OH group, the electrons that are left behind can be delocalized onto another oxygen atom. When a proton is removed from the C-2 OH group, the electrons that are left behind are delocalized onto a carbon atom. Because oxygen is more electronegative than carbon, a negatively charged oxygen is more stable than a negatively charged carbon. Recall that the more stable the base, the more acidic is its conjugate acid.

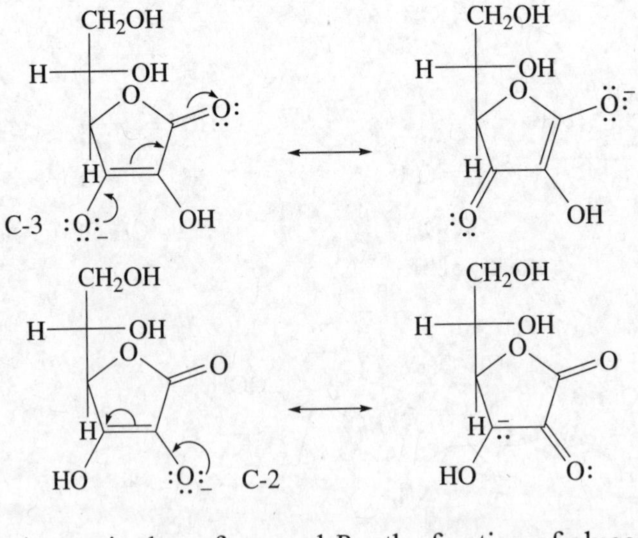

35. Let A = the fraction of glucose in the α-form and B = the fraction of glucose in the β-form.

$$A + B = 1$$
$$B = 1 - A$$

specific rotation of A = 112.2°
specific rotation of B = 18.7°

specific rotation of the equilibrium mixture = 52.7°

specific rotation of the mixture = specific rotation of A × fraction of glucose in the α-form + specific rotation of B × fraction of glucose in the β-form

$52.7 = 112.2\,A + (1 - A)\,18.7$

$52.7 = 112.2\,A + 18.7 - 18.7\,A$

$34.0 = 93.5\,A$

$A = 0.36$

$B = 0.64$

This calculation shows that 36% is in the α-form and 64% is in the β-form.

36. D-Altrose will most likely exist as a furanose because

(1) the furanose is particularly stable, because all the large substituents are trans to each other, which minimizes steric strain and

(2) the pyranose has two of its OH groups in the unstable axial position.

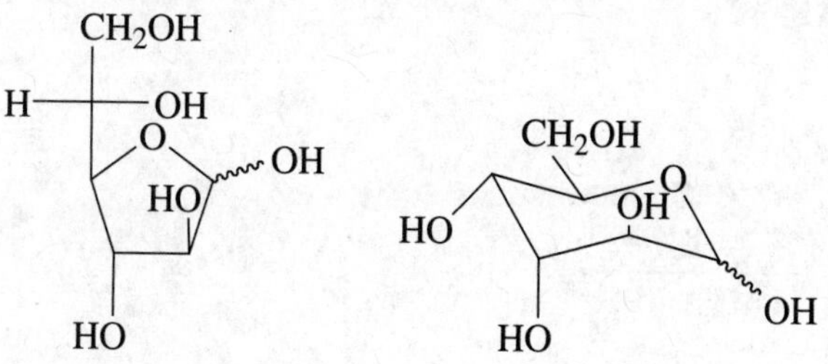
CH2OH
H
OH
O
OH
HO
HO
CH2OH
O
OH
HO
OH
HO

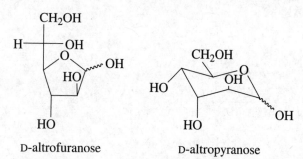

D-altrofuranose D-altropyranose

Chapter 15 Practice Test

1. Give the product(s) of each of the following reactions:

a. 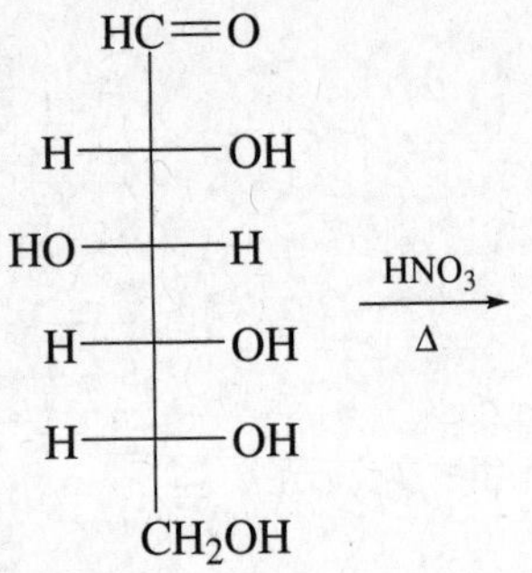

b. 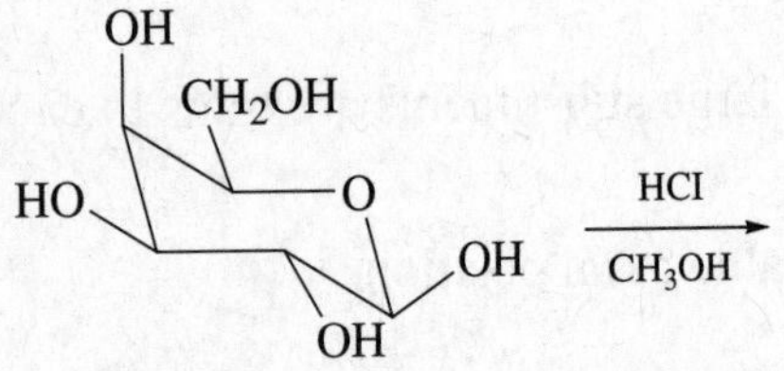

c. 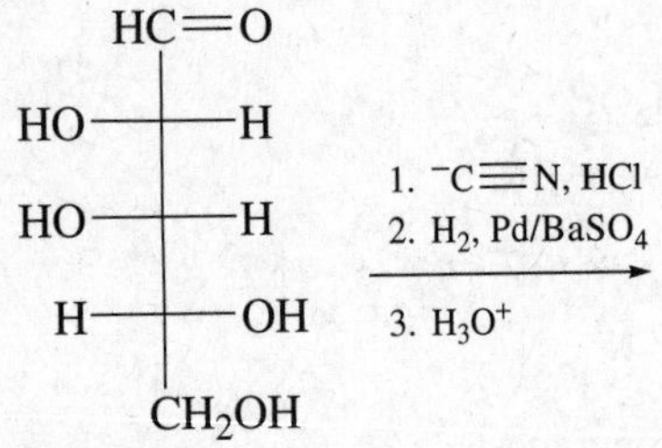

d. 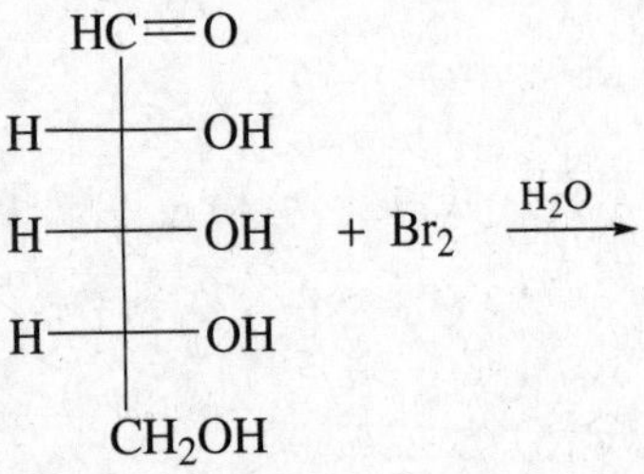

2. Indicate whether the following statements are true or false:

a. Glycogen contains α-1,4'- and β-1,6'-glycosidic linkages.	T	F
b. D-Mannose is a C-1 epimer of D-glucose.	T	F
c. D-Glucose and L-glucose are anomers.	T	F
d. D-Erythrose and D-threose are diastereomers.	T	F

3. Which of the following sugars will form an optically active aldaric acid?

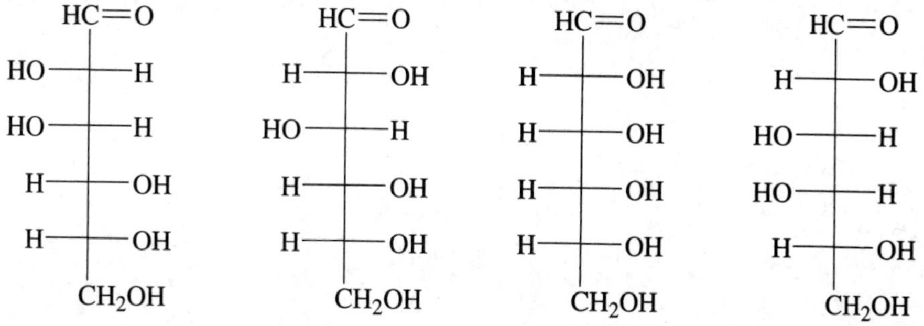
HC=O
HO—H
HO—H
H—OH
H—OH
CH2OH
HC=O
H—OH
HO—H
H—OH
H—OH
CH2OH
HC=O
H—OH
H—OH
H—OH
H—OH
CH2OH
HC=O
H—OH
HO—H
HO—H
H—OH
CH2OH

4. When crystals of D-fructose are dissolved in a basic aqueous solution, two aldohexoses are obtained. Identify the aldohexoses.

5. D-Talose and ______________ are reduced to the same alditol.

6. What is the main structural difference between amylose and cellulose?

7. What aldohexoses are formed from a Kiliani-Fischer synthesis starting with D-xylose?

8. What aldohexose is the C-3 epimer of D-glucose?

9. Draw the most stable chair conformer of β-D-allopyranose, the C-3 epimer of β-D-glucose.

CHAPTER 16
The Organic Chemistry of Amino Acids, Peptides, and Proteins

1. L-Alanine is (*S*)-alanine

2. Solved in the text.

3. isoleucine
naturally occurring L-isoleucine is (2*S*,3*S*)-isoleucine

4. The electron-withdrawing $^+NH_3$ substituent on the α-carbon increases the acidity of the carboxyl group.

5. An amino acid is insoluble in diethyl ether (a relatively nonpolar solvent) because an amino acid exists as a zwitterion (it has both a + and a – charge) at neutral pH.

6. **a.** Solved in the text **b.** $H_2NC(=O)CH_2CH_2CH(^+NH_3)C(=O)O^-$ **c.** $NH_2C(=^+NH_2)NHCH_2CH_2CH_2CH(^+NH_3)C(=O)O^-$

7. **a.** $HOC(=O)CH_2CH_2CH(^+NH_3)C(=O)OH$ **c.** $^-OC(=O)CH_2CH_2CH(^+NH_3)C(=O)O^-$

b. $HOC(=O)CH_2CH_2CH(^+NH_3)C(=O)O^-$ **d.** $^-OC(=O)CH_2CH_2CH(NH_2)C(=O)O^-$

8. **a.** asparagine $pI = \dfrac{2.02 + 8.84}{2} = \dfrac{10.86}{2} = 5.43$

b. arginine $pI = \dfrac{9.04 + 12.48}{2} = \dfrac{21.52}{2} = 10.76$

c. serine $pI = \dfrac{2.21 + 9.15}{2} = \dfrac{11.36}{2} = 5.68$

d. aspartate (aspartic acid) $pI = \dfrac{2.09 + 3.86}{2} = \dfrac{5.95}{2} = 2.98$

9. **a.** Asp (pI = 2.98) **b.** Arg (pI = 10.76)

10. From the products of the reaction of an amino acid with ninhydrin in Section 16.5, we see that the amino acid loses both its amino group and its carboxyl group, leaving the side chain (R) of the amino acid attached to an aldehyde CHO group. Therefore, valine forms 2-methylpropanal when treated with ninhydrin.

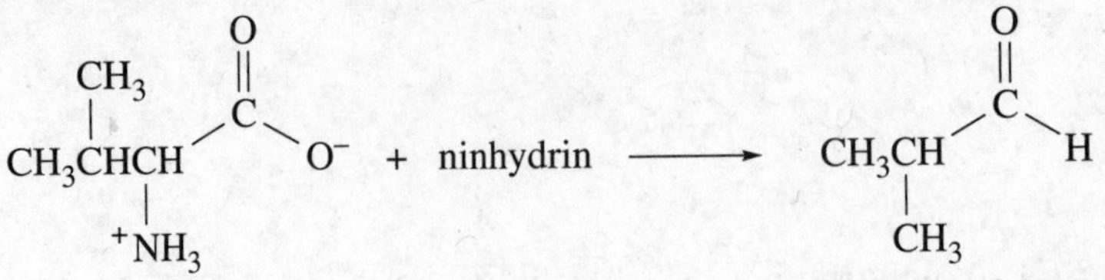

valine 2-methylpropanal

11. **a.** Glutamate (an amino acid with a negatively charged side chain) will be repelled by the negatively charged sulfonic acid groups.

b. Leucine will be retained longer by the relatively nonpolar benzene rings because it is more nonpolar than alanine.

12.

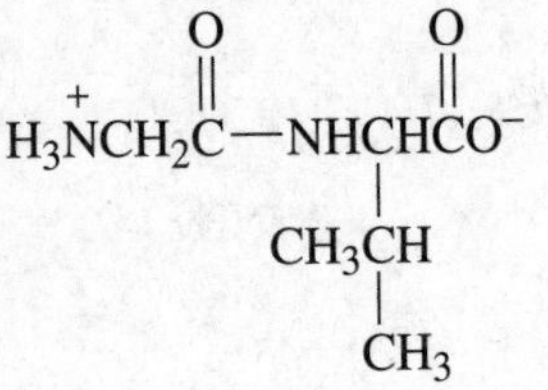

Gly-Val

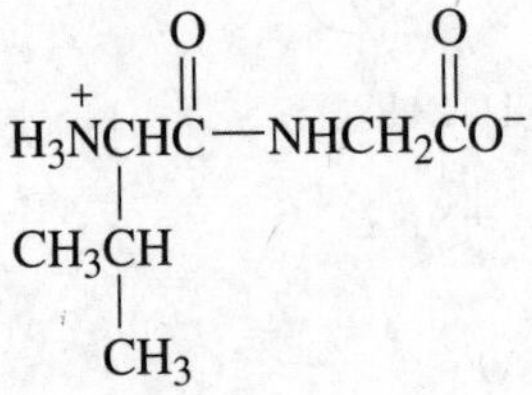

Val-Gly

13.

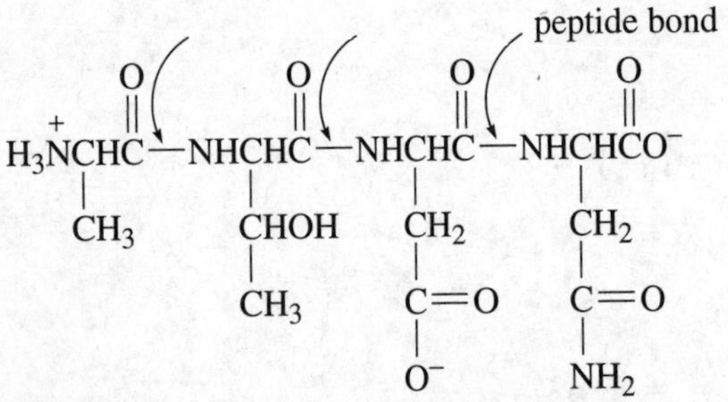
peptide bond

14. **A-G-M** **A-M-G** **M-G-A** **M-A-G** **G-A-M** **G-M-A**

15. The bonds on either side of the *α*-carbon can freely rotate. In other words, the bond between the *α*-carbon and the carbonyl carbon and the bond between the *α*-carbon and the nitrogen (the bonds indicated by arrows) can freely rotate.

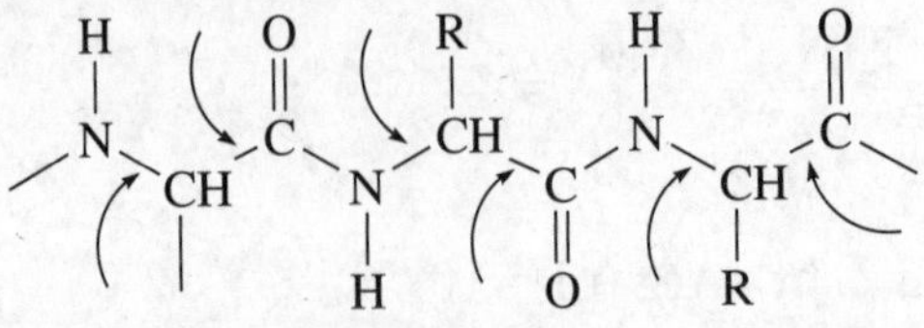

16. **a.** glutamate, cysteine, and glycine

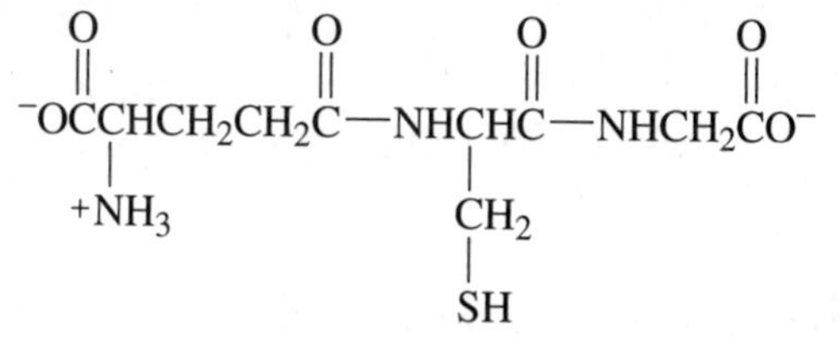

glutathione

b. In forming the amide linkage, the amino group of cysteine reacts with the γ-carboxyl group rather than with the α-carboxyl group of glutamic acid.

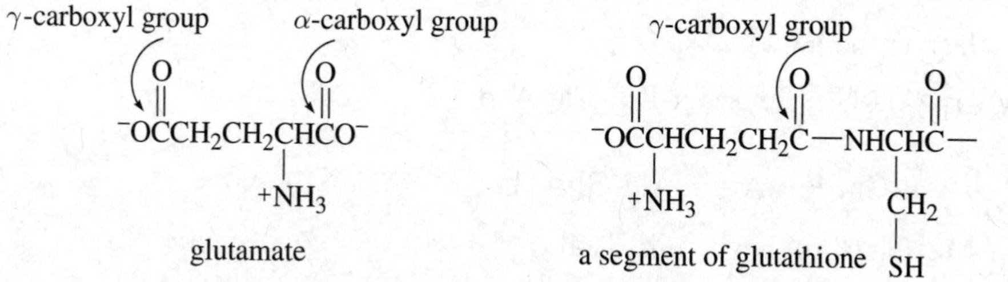
γ-carboxyl group
α-carboxyl group
γ-carboxyl group
glutamate
a segment of glutathione

17. **Leu-Val** and **Val-Val** will be formed because the amino group of leucine is not reactive (so leucine could not be the C-terminal amino acid) but the amino group of valine would react equally easily with the carboxyl group of leucine and the carboxyl group of valine.

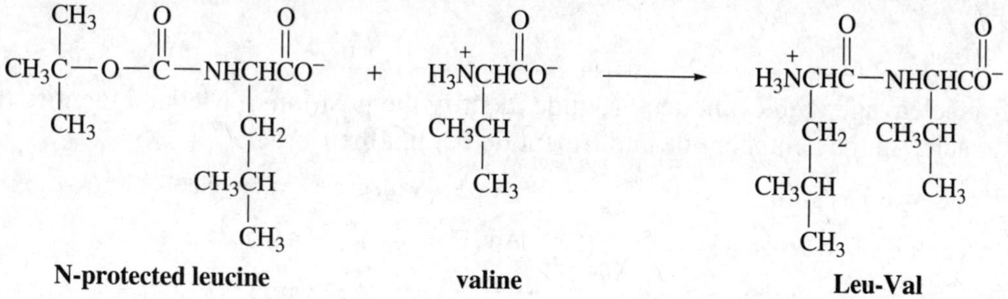
N-protected leucine
valine
Leu-Val

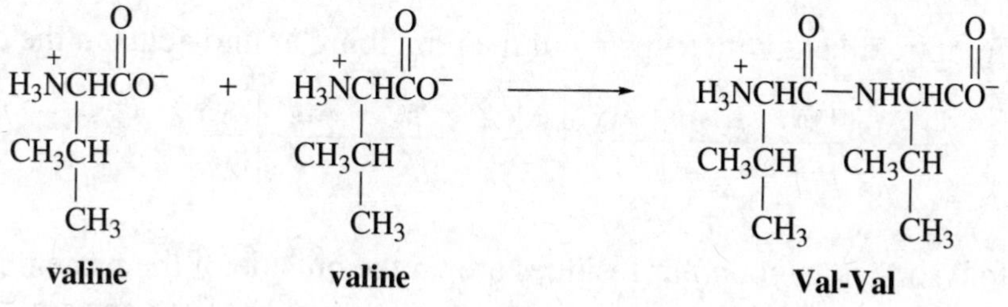
valine
valine
Val-Val

18. 5.8%

2	**3**	**4**	**5**	**6**	**7**	**8**	**9**
70%	49%	34%	24%	17%	12%	8.2%	5.8%
	.70 × .70	.70 × .49	.70 × .34	.70 × .24	.70 × .17	.70 × .12	.70 × .082

19. Treatment with Edman's reagent would release two PTH-amino acids in approximately equal amounts.

20. Knowing that the N-terminal amino acid is Gly, look for a peptide fragment that contains Gly.

"f" tells you that the 2nd amino acid is Arg.

"e" tells you the next two are either Ala-Trp or Trp-Ala. However, **"d"** tells you that Glu is next to Ala, so 3 and 4 must be Trp-Ala and the 5th amino acid is Glu.

"g" tells you the 6th amino acid is Leu.

"h" tells you the next two are Met-Pro or Pro-Met. However, **"c"** tells you that Pro is next to Val, so 7 and 8 must be Met-Pro and the 9th amino acid is Val.

"b" tells you the last amino acid is Asp,

Gly-Arg-Trp-Ala-Glu-Leu-Met-Pro-Val-Asp

21. **a.** His-Lys Leu-Val-Glu-Pro-Arg Ala-Gly-Ala

b. Leu-Gly-Ser-Met-Phe-Pro-Tyr Gly-Val

22. Solved in the text.

23. The data from treatment with Edman's reagent and carboxypeptidase A identify the first and last amino acids.

Leu ___ ___ ___ ___ ___ ___ Ser

The data from cleavage with cyanogen bromide identify the position of Met and identify the other amino acids in the pentapeptide and tripeptide but not their order.

cleavage with cyanogen bromide

Arg, Lys, Tyr | Arg, Phe

Leu ___ ___ ___ Met | ___ ___ Ser

The data from treatment with trypsin put the remaining amino acids in the correct position.

Leu Tyr Lys | Arg | Met Phe Arg | Ser

24. It would fold so that its nonpolar residues are on the outside of the protein in contact with the nonpolar membrane and its polar residues are on the inside of the protein.

25. **a.** A cigar-shaped protein has the greatest surface area to volume ratio, so it has the highest percentage of polar amino acids.

b. A subunit of a hexamer would have the smallest percentage of polar amino acids because part of the surface of the subunit can be on the inside of the hexamer and, therefore, have nonpolar amino acids on its surface.

26. **a.** 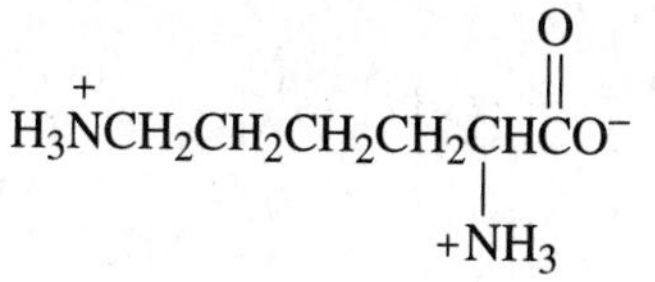

c. 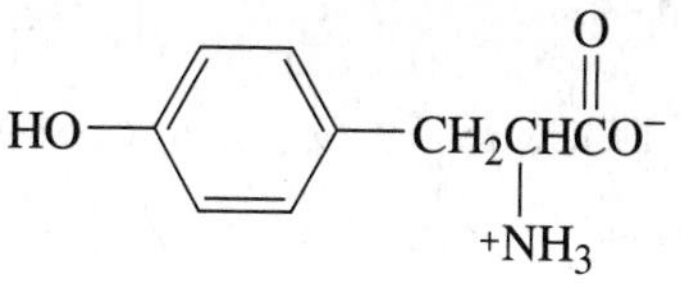

b. 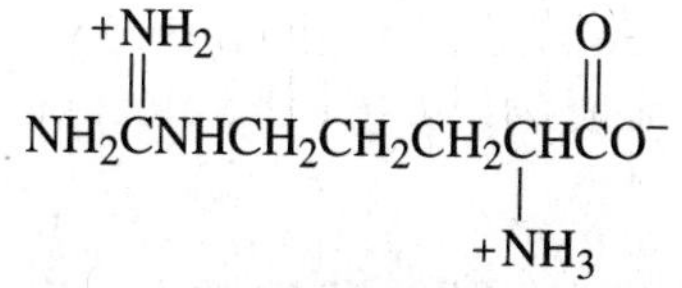

27. The pK_a of the OH group can be ignored because it is ~15, so there will be no ionization until the pH is ~ 13. Thus, it does not have to be considered in calculating the pI.

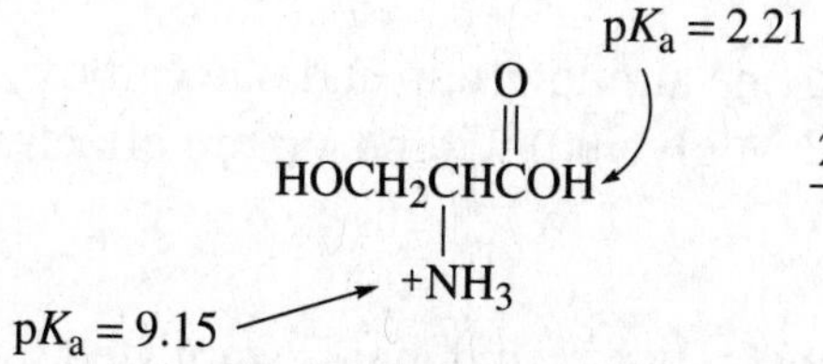

$$\frac{2.21 + 9.15}{2} = \frac{11.36}{2} = 5.68$$

28. **a.** Val-Arg-Gly-Met-Arg-Ala Ser

b. Ser-Phe-Lys-Met Pro-Ser-Ala-Asp

c. Arg Ser-Pro-Lys Lys Ser-Glu-Gly

29. **a.** L-aspartate is (*S*)-aspartate

b. The α-carbons of all the L-amino acids except cysteine have the *S*-configuration. Similarly, the α-carbon of all the D-amino acids except cysteine has the *R*-configuration.

(In all the amino acids except cysteine, the amino group has the highest priority and the carboxyl group has the second-highest priority. In cysteine, the thiomethyl group has a higher priority than the carboxyl group because sulfur has a greater atomic number than oxygen.)

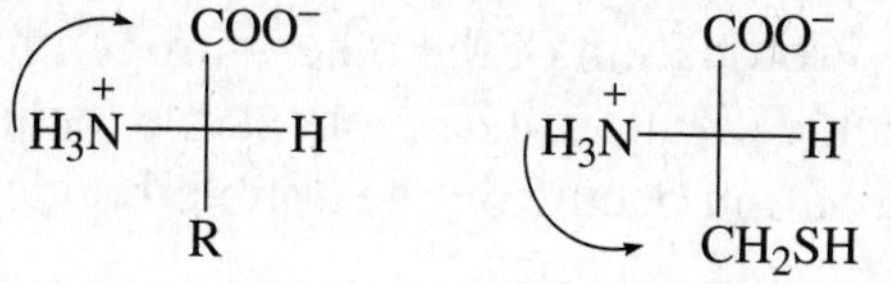

cysteine

30. 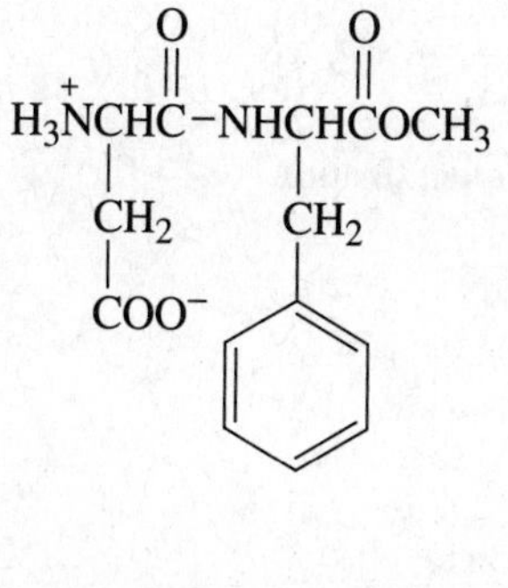

31. 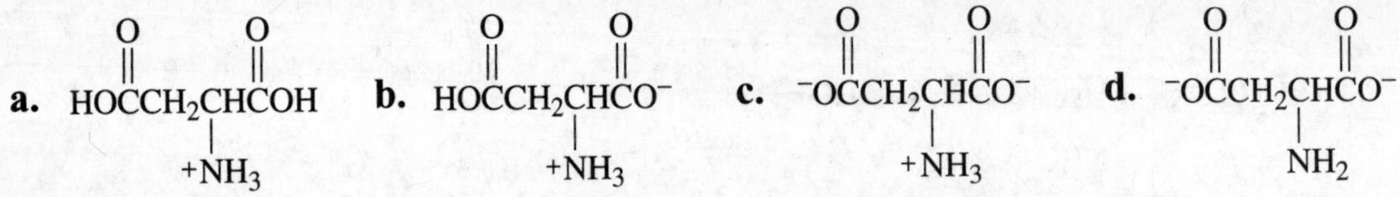
a. HOCCH2CHCOH
+NH3
b. HOCCH2CHCO−
+NH3
c. −OCCH2CHCO−
+NH3
d. −OCCH2CHCO−
NH2

32. **a.** The carboxyl group of the aspartate side chain is a stronger acid than the carboxyl group of the glutamate side chain because the aspartate side chain is closer to the electron-withdrawing protonated amino group.

b. The protonated lysine side chain is a stronger acid than the protonated arginine side chain. The protonated arginine side chain has less of a tendency to lose a proton because the positive charge on its side chain is delocalized over three nitrogen atoms; in other words, the positive charge is shared by three nitrogen atoms.

33. Each compound has two groups that can act as a buffer, one amino group and one carboxyl group. Thus, the compound in higher concentration (0.2 M glycine) will be a more effective buffer.

34. **a.** Protonation of the doubled-bonded nitrogen forms a conjugate acid that is stabilized by electron delocalization. Protonation of either of the other nitrogen atoms forms a conjugate acid that is not stabilized by electron delocalization. The more stable conjugate acid is the one that is more readily formed, so protonation occurs on the double-bonded nitrogen.

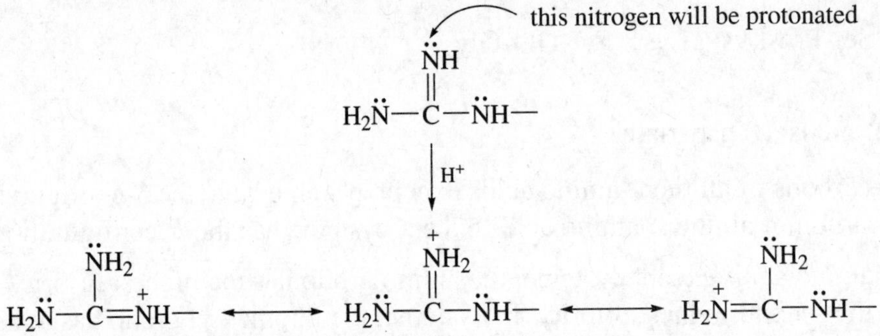
this nitrogen will be protonated
H+

b. Protonation of the doubled-bonded nitrogen forms a conjugate acid that is stabilized by electron delocalization. Protonation of the other nitrogen forms a conjugate acid that is not stabilized by electron delocalization. The more stable conjugate acid is the one that is more readily formed, so protonation occurs on the double-bonded nitrogen.

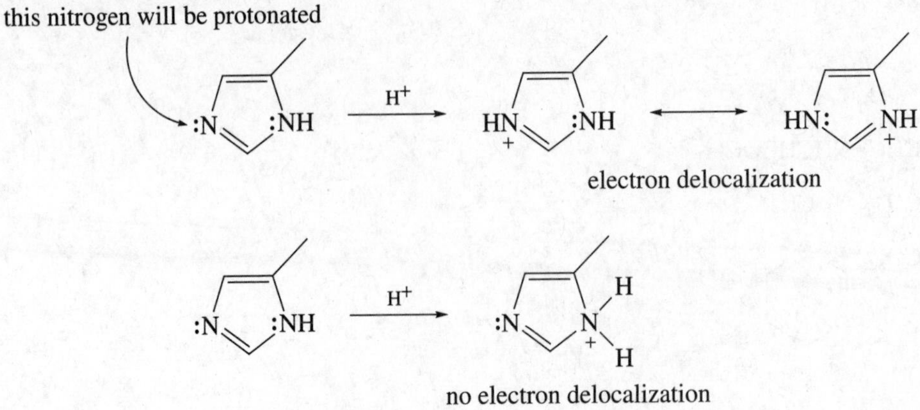
this nitrogen will be protonated
H+
electron delocalization
H+
no electron delocalization

35. **a.** 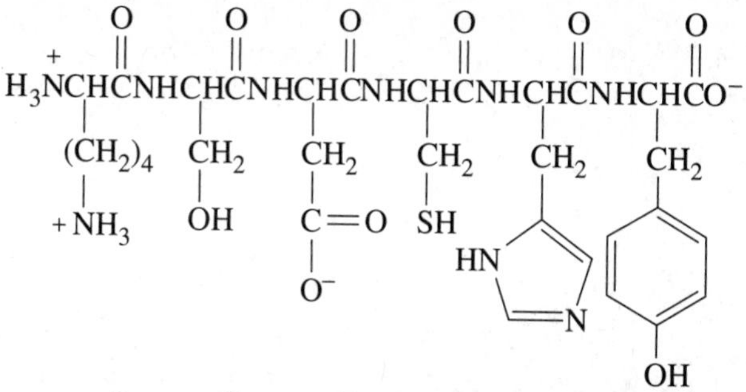

b. 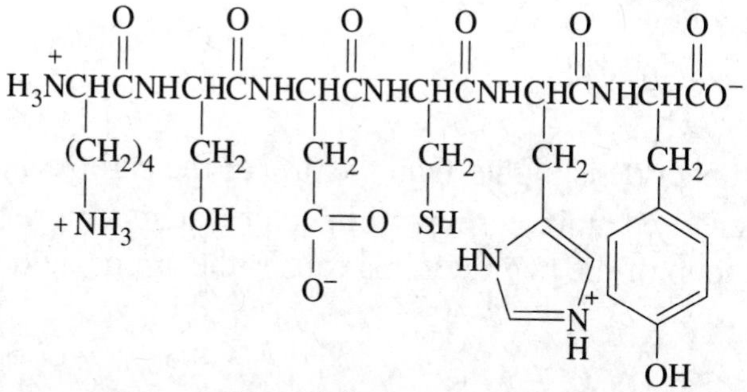

c. 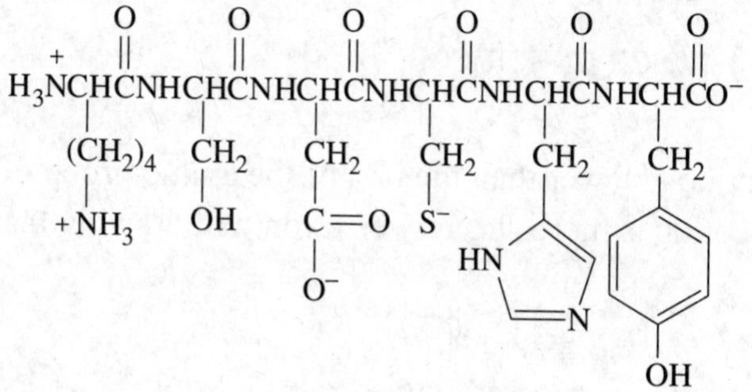

36. **a.** Asp, because its side chain has the lowest pK_a

b. Met, because at pH = 6.20 Met is farther away from the pH at which it has no net charge (pI of Met = 5.75, pI of Gly = 5.97).

37. Cation-exchange chromatography releases amino acids in order of their pI values. The amino acid with the lowest pI is released first because at a given pH it will be the amino acid with the highest concentration of negative charge, and negatively charged molecules are not bound by the negatively charged resin. The relatively nonpolar resin will release polar amino acids before nonpolar amino acids.

a. Asp (pI = 2.98) is more negative than Ser (pI = 5.68).

b. Gly is more polar than Ala.

c. Val is more polar than Leu.

d. Tyr is more polar than Phe.

38. Ser-Glu-Leu-Trp-Lys-Ser-Val-Glu-His-Gly-Ala-Met

From the experiment with carboxypeptidase, we know the C-terminal amino acid is Met.

"l" tells us the amino acid adjacent to Met is Ala

"e" tells us the next amino acid is Gly

"b" tells us the next amino acid is His

"g" tells us the next amino acid is Glu

"j" tells us the next amino acid is Val

"c" tells us the next amino acid is Ser

"i" tells us the next amino acid is Lys

"a" tells us the next amino acid is Trp

"h" tells us the next amino acid is Leu

"k" tells us the next amino acid is Glu

"d" tells us the next (N-terminal) amino acid is Ser

39. The pK_a of the carboxylic acid of the dipeptide is higher than the pK_a of the carboxylic acid of the amino acid because the positively charged amino group of the amino acid is more strongly electron withdrawing than the amide group of the peptide. This causes the amino acid to be a stronger acid and have a lower pK_a.

$$H_3\overset{+}{N}CH_2\overset{\overset{O}{\|}}{C}OH \qquad H_3\overset{+}{N}CH_2\overset{\overset{O}{\|}}{C}NHCH_2\overset{\overset{O}{\|}}{C}OH \leftarrow \text{higher}$$

The pK_a of the amino group of the peptide is lower than the pK_a of the amino group of the amino acid because the amide group of the peptide is more strongly electron withdrawing than the carboxylate group of the amino acid.

$$H_3\overset{+}{N}CH_2\overset{\overset{O}{\|}}{C}O^- \qquad \text{lower} \rightarrow H_3\overset{+}{N}CH_2\overset{\overset{O}{\|}}{C}NHCH_2\overset{\overset{O}{\|}}{C}O^-$$

40. **a.** Trypsin cleaves on the C-side of arginine and lysine residues unless Pro is at the cleavage site (to the right of Arg or Lys). Therefore, 7 fragments will be formed by the cleavage.

b. The N-terminal end of each fragment will be positively charged because of the $^+NH_3$ group, the C-terminal end will be negatively charged because of the COO^- group; arginine and lysine residues will be positively charged; aspartate and glutamate residues will be negatively charged; the overall change on each fragment is shown below. Anions bind most strongly to an anion-exchange column. Therefore, fragment F (the most positively charged fragment) will come off the column first, and fragment C (the most negatively charged fragment) will come of the column last.

A	$\underset{+}{G}$ly-Ser-$\underset{-}{A}$sp-Ala-Leu-Pro-Gly-Ile-Thr-Ser-$\underset{+-}{A}$rg	overall charge = 0
B	$\underset{+-}{A}$sp-Val-Ser-$\underset{+-}{L}$ys	overall charge = 0
C	$\underset{+}{V}$al-$\underset{-}{G}$lu-Tyr-Phe-$\underset{-}{G}$lu-Ala-Gly-$\underset{+-}{A}$rg	overall charge = −1
D	$\underset{+}{S}$er-$\underset{-}{G}$lu-Phe-$\underset{+-}{L}$ys	overall charge = 0

```
E  Glu-Pro-Arg                                      overall charge = 0
   + −     + =

F  Leu-Tyr-Met-Lys                                  overall charge = +1
   +           + −

G  Val-Glu-Gly-Arg-Pro-Val-Ser-Ala-Gly-Leu-Trp      overall charge = 0
   +   −       +                           −
```

41. First mark off where the chains would have been cleaved by chymotrypsin (C-side of Phe, Trp, Tyr).

Val-Met-Tyr|Ala-Cys-Ser-Phe|Ala-Glu-Ser

Ser-Cys-Phe|Lys-Cys-Trp|Lys-Tyr|Cys-Phe|Arg-Cys-Ser

Then, from the pieces obtained from cleavage of the polypeptide by chymotrypsin before the disulfide bridges were broken, you can determine where the disulfide bridges are.

```
Val-Met-Tyr-Ala-Cys-Ser-Phe-Ala-Glu-Ser
                 |
                 S
                 |
                 S
                 |
Ser-Cys-Phe-Lys-Cys-Trp-Lys-Tyr-Cys-Phe-Arg-Cys-Ser
    |______________S—S______________________|
```

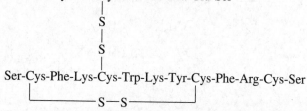

42. **a.** Acid-catalyzed hydrolysis indicates the peptide contains 12 amino acids.

___ ___ ___ ___ ___ ___ ___ ___ ___ ___ ___ ___

b. Treatment with Edman's reagent indicates that Val is the N-terminal amino acid.

Val ___ ___ ___ ___ ___ ___ ___ ___ ___ ___ ___

c. Treatment with carboxypeptidase A indicates that Ala is the C-terminal amino acid.

Val ___ ___ ___ ___ ___ ___ ___ ___ ___ ___ Ala

d. Treatment with cyanogen bromide indicates that Met is the 5th amino acid with Arg, Gly, Ser in an unknown order in positions 2, 3, and 4.

Val ___ ___ ___ Met ___ ___ ___ ___ ___ ___ Ala
(Arg, Gly, Ser)

e. Treatment with trypsin indicates that Arg is the 3rd amino acid, and Ser is 2nd, Gly is 4th, Tyr is 6th, and Lys is 7th. Since Lys is in the terminal fragment means that cleavage did not occur at Lys, so Pro must be at lysine's cleavage site.

Val Ser Arg Gly Met Tyr Lys ___ ___ ___ ___ Ala
(Lys-Pro, Phe, Ser)

f. Treatment with chymotrypsin indicates that Phe is the 10th amino acid and Ser is the 11th.

Val Ser Arg Gly Met Tyr Lys Lys Pro Phe Ser Ala

43.

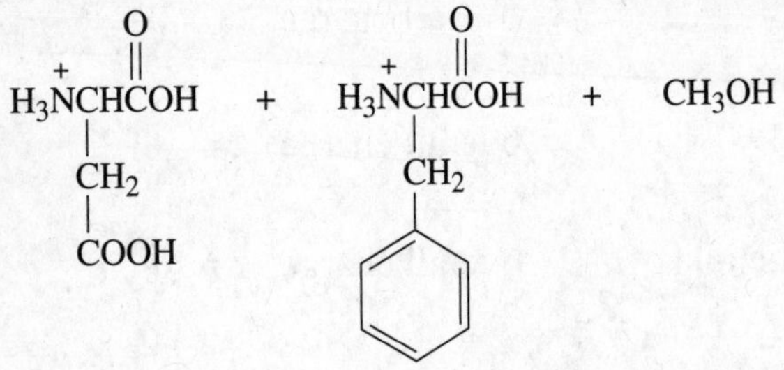
H3NCHCOH
CH2
COOH
+
H3NCHCOH
CH2
+
CH3OH

44. The student is correct. At the pI, the total of the positive charges on the tripeptide's amino groups must be 1 to balance the one negative charge of the carboxyl group. When the pH of the solution is equal to the pK_a of a lysine residue, the three lysine groups each have one-half a positive charge for a total of one and one-half positive charges. Thus, the solution must be more basic than this in order to have just one positive charge.

45.

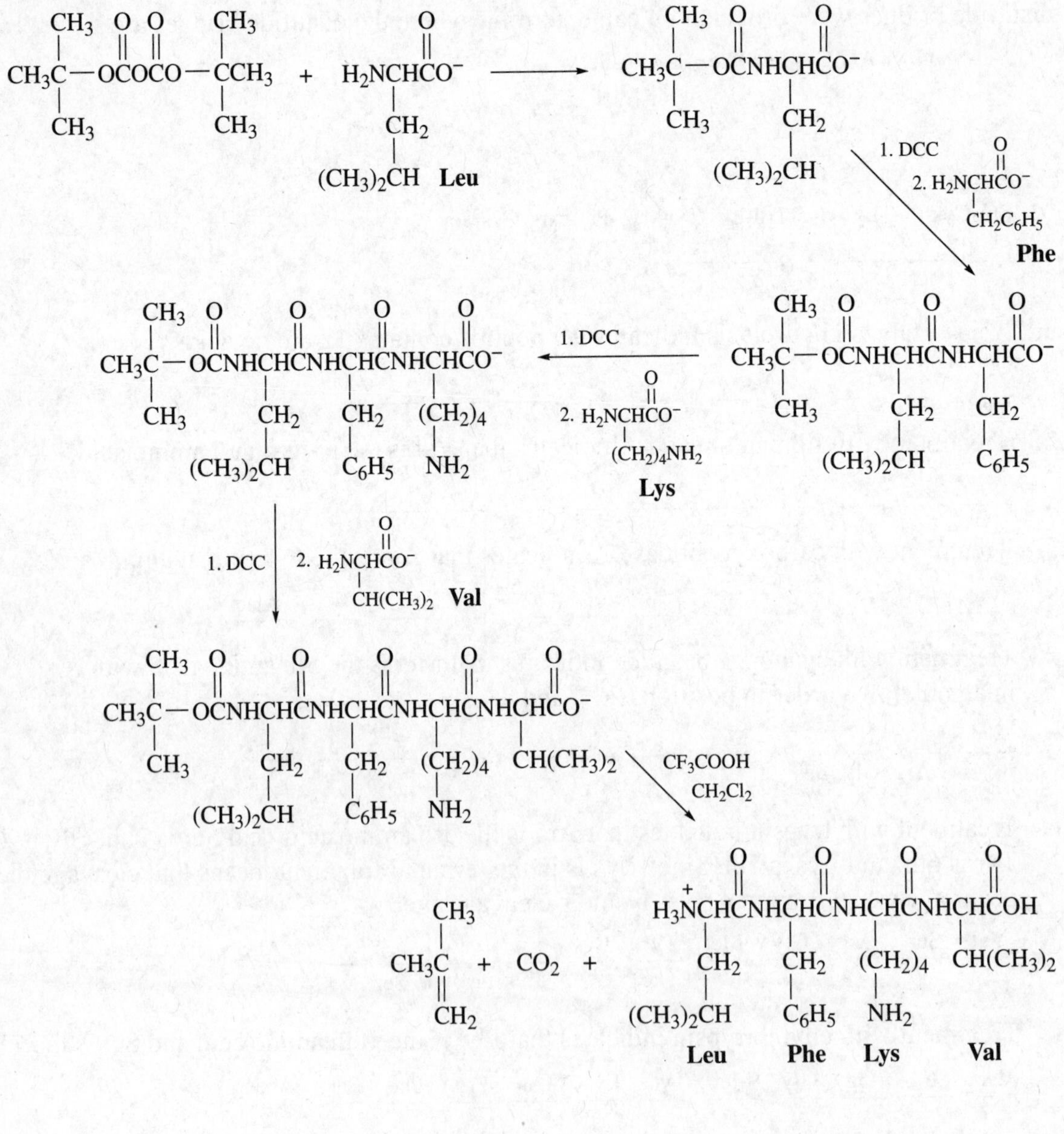
Leu
1. DCC
2. H2NCHCO−
CH2C6H5
Phe
1. DCC
2. H2NCHCO−
(CH2)4NH2
Lys
1. DCC
2. H2NCHCO−
CH(CH3)2
Val
CF3COOH
CH2Cl2
CO2
Leu
Phe
Lys
Val

46. **a.**

2	**3**	**4**	**5**	**6**	**7**	**8**	**9**
70%	**49%**	**34%**	**24%**	**17%**	**12%**	**8.2%**	**5.8%**

b.

2	**3**	**4**	**5**	**6**	**7**	**8**	**9**	**10**	**11**	**12**	**13**	**14**	**15**
80%	**64%**	**51%**	**41%**	**33%**	**26%**	**21%**	**17%**	**13%**	**11%**	**8.6%**	**6.9%**	**5.5%**	**4.4%**

47. A proline residue cannot fit into a helix because the bond between the proline nitrogen and the α-carbon cannot rotate since it is in a ring. Not being able to rotate about this bond makes proline unable to fit into a helix.

48. In each case, the two adjacent amino acids each have a negative charge. Two adjacent side chains will be close to one another if they are in a helix, so they will repel each other if they have the same charge.

49. You would expect both serine and cysteine to have a lower pK_a than alanine since a hydroxymethyl and a thiomethyl group are more electron withdrawing than a methyl group.

Because oxygen is more electronegative than sulfur, one would probably expect serine to have a lower pK_a than cysteine. The fact that cysteine has the lower pK_a can be explained by stabilization of serine's carboxyl proton by hydrogen bonding to the β-OH group of serine, which decreases its tendency to be removed by base.

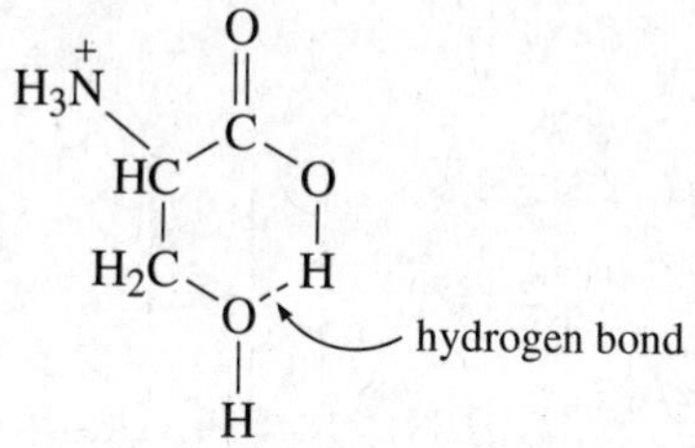
H3N
HC
C
O
O
H2C
H
O
H
hydrogen bond

50. **a.** The C-terminal end of the protein contains 3 nonpolar amino acids and 4 polar amino acids.

Each of the seven amino acids has two atoms that can form hydrogen bonds for a total of 14 atoms that can form hydrogen bonds. Gln and Asp each have two additional groups, and each serine one additional group that can form hydrogen bonds. Thus, the terminal end of the protein has 3 hydrophobic groups and can form 20 hydrogen bonds. So there are 20 hydrogen bonds formed between protein groups and water that must be broken before the protein groups can hydrogen bond to each other. So 20 hydrogen bonds are broken and 20 hydrogen bonds are formed (10 hydrogen bonds from 20 atoms involved in intramolecular hydrogen bonds, and 10 hydrogen bonds from 20 liberated water molecules forming hydrogen bonds with each other).

Thus, ΔG° comes only from removing the three hydrophobic groups from water $(3 \times -4) =$ **-12 kcal/mol.**

b. If two of the polar groups do not form intramolecular bonds in the interior of the protein, 20 hydrogen bonds will be broken, but only 19 will be formed, so $\Delta G° = -12 + 3 = $ **−9 kcal/mol**.

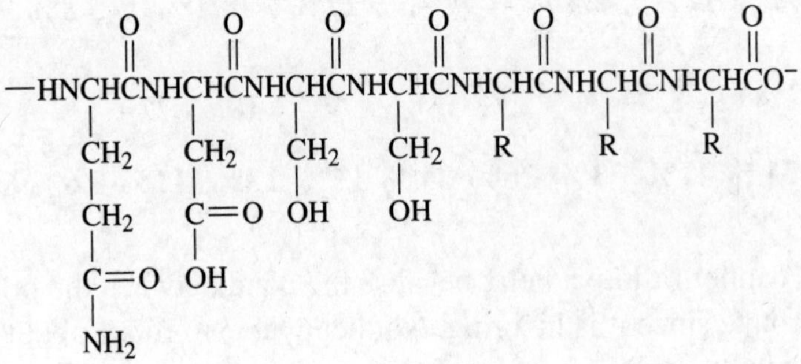

Chapter 16 Practice Test

1. Give the structure of the following amino acids at pH = 7:

   a. glutamic acid **b.** lysine **c.** isoleucine **d.** arginine **e.** asparagine

2. Draw the form of histidine that predominates at:

   a. pH = 1 **b.** pH = 4 **c.** pH = 8 **d.** pH = 11

3. Answer the following:

   a. Alanine has a pI = 6.02 and serine has a pI = 5.68. Which would have the highest concentration of positive charge at pH = 5.50?

   b. Which amino acid is the only one that does not have an asymmetric center?

   c. Which are the two most nonpolar amino acids?

   d. Which amino acid has the lowest pI?

4. Why does the carboxyl group of alanine have a lower pK_a than the carboxyl group of propanoic acid?

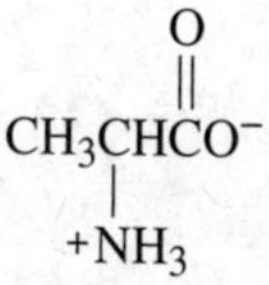

pK_a = 2.2

alanine

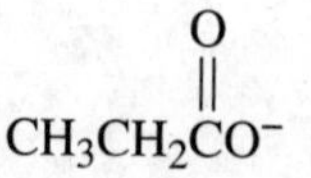

pK_a = 4.7

propanoic acid

5. Indicate whether each of the following is true or false:

   a. A cigar-shaped protein has a greater percentage of polar residues than a spherical protein. T F

   b. Naturally occurring amino acids have the L-configuration. T F

   c. There is free rotation about a peptide bond. T F

6. Draw the compound obtained from mild oxidation of cysteine.

7. Define the following:

   a. the primary structure of a protein

   b. the tertiary structure of a protein

   c. the quaternary structure of a protein

8. Calculate the pI of each of the following amino acids:

   a. phenylalanine (pK_a's = 2.16, 9.18) **b.** arginine (pK_a's = 2.17, 9.04, 12.48)

9. From the following information, determine the primary sequence of the decapeptide:

   a. Acid hydrolysis gives: Ala, 2 Arg, Gly, His, Ile, Lys, Met, Phe, Ser.

   b. Reaction with Edman's reagent liberates Ala.

c. Reaction with carboxypeptidase A liberates Ile.

d. Reaction with cyanogen bromide (cleaves on the C-side of Met):

1. Gly, 2 Arg, Ala, Met, Ser

2. Lys, Phe, Ile, His

e. Reaction with trypsin (cleaves on the C-side of Arg and Lys):

1. Arg, Gly

2. Ile

3. Phe, Lys, Met, His

4. Arg, Ser, Ala

f. Reaction with thermolysin (cleaves on the N-side of Leu, Ile, Phe, Trp, Tyr):

1. Lys, Phe

2. 2 Arg, Ser, His, Gly, Ala, Met

3. Ile

CHAPTER 17
How Enzymes Catalyze Reactions • The Organic Chemistry of the Vitamins

1. Solved in the text.

2. **2** and **3** are bases, so they can help remove a proton.

3. Only a primary amine can form an imine, so only **2** can form an imine with the substrate.

Notice that **1** cannot form an imine because the lone pair on the NH_2 group is delocalized onto the oxygen atom, so this NH_2 group is not a nucleophile. It is also not a base, that is why it cannot remove a proton in Problem 2.

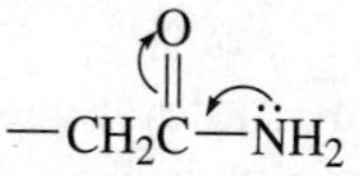

4. The secondary alcohol is oxidized to a ketone.

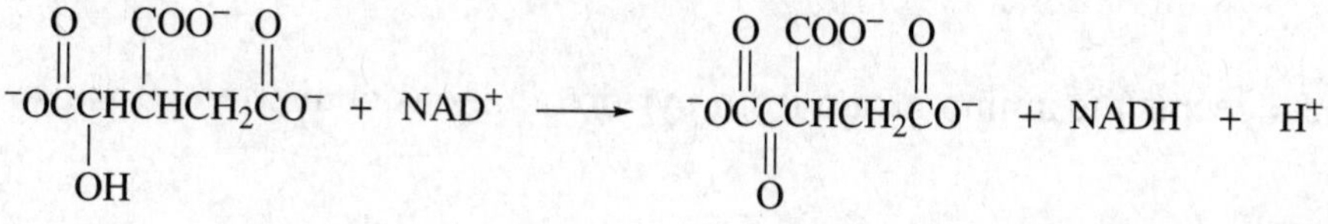

5. The ketone is reduced to a secondary alcohol. Remember that it is easier to reduce a ketone than a carboxylic acid.

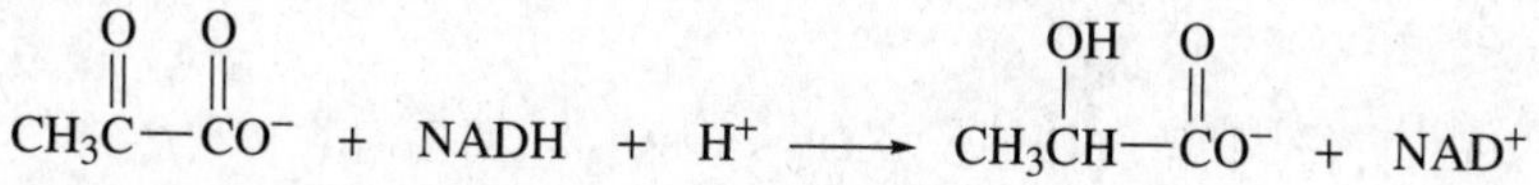

6. **a.** FAD has seven conjugated double bonds. (The conjugated double bonds are indicated *.)

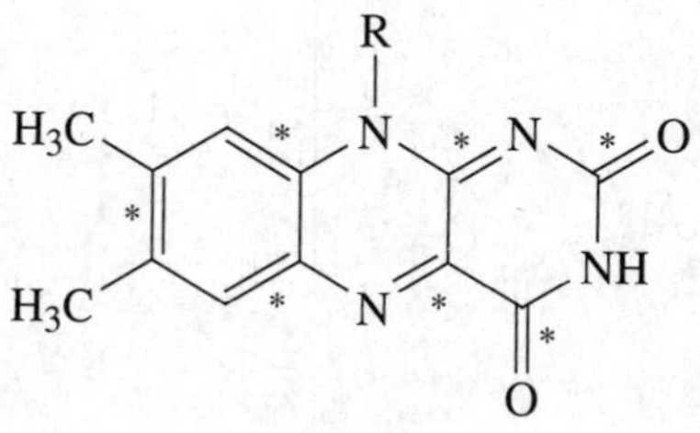

b. $FADH_2$ has three conjugated double bonds. It also has two conjugated double bonds that are isolated from the other three.

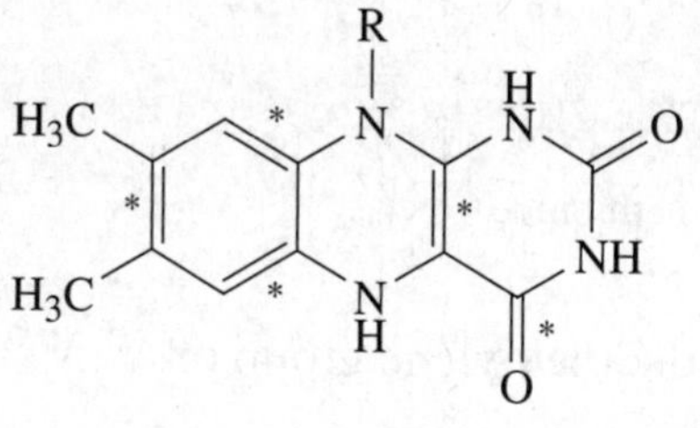

7. FAD oxidizes the two thiol groups to a disulfide.

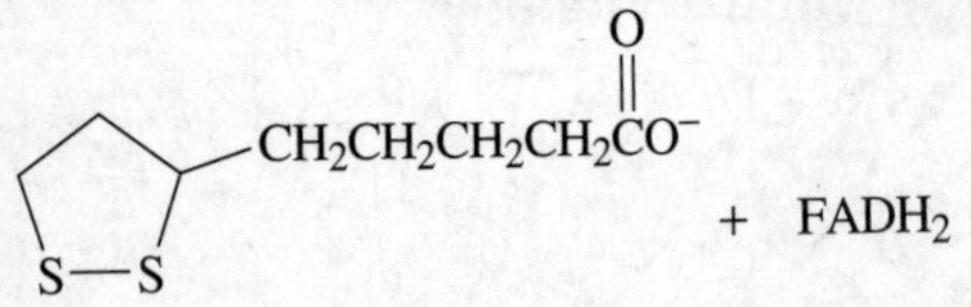

8. 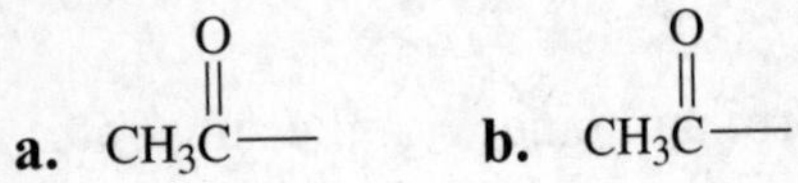

9. 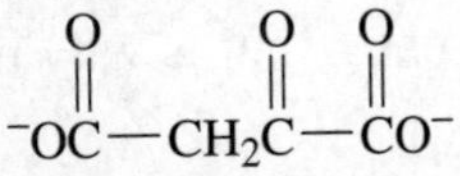

10. 9, because one mole of arachidic acid is made from one mole of acetyl-CoA and nine moles of malonyl-CoA

11. Solved in the text.

12. The α-keto group that accepts the amino group from pyridoxamine is converted into an amino group.

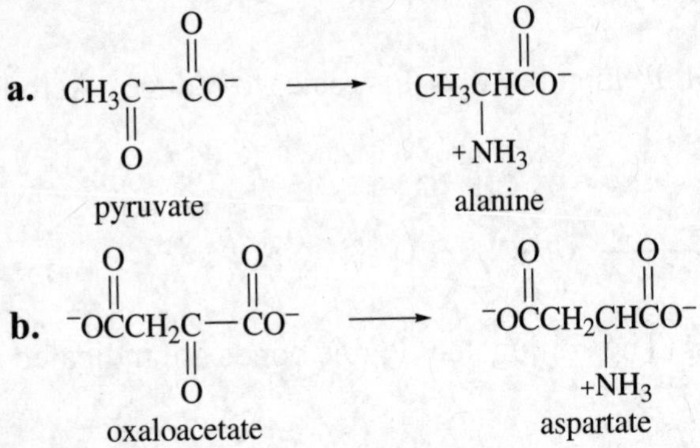
a.
pyruvate
alanine
b.
oxaloacetate
aspartate

13. 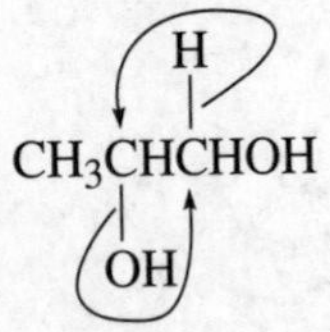

14. It is called tetrahydrofolate because it is formed by adding four hydrogens to folate (folic acid).

15. 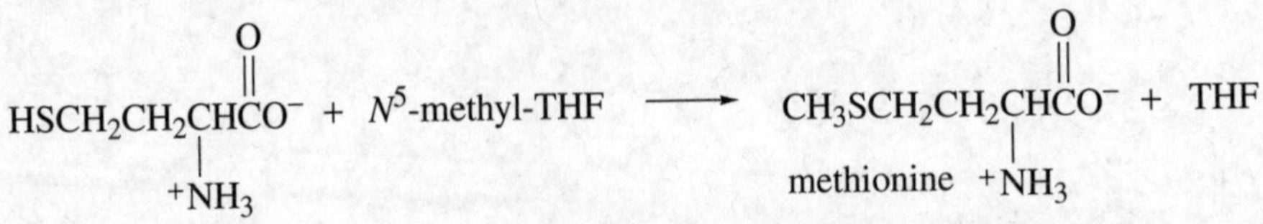
methionine

16. The methyl group in thymidine comes from the methylene group of N^5,N^{10}-methylene-THF.

17. **a.** niacin **b.** riboflavin (vitamin B_2) **c.** pyridoxine (vitamin B_6) **d.** folate

18. NAD^+ and FAD

19. thiamine pyrophosphate and pyridoxal phosphate

20. $-CH_3$ methyl $-CH_2-$ methylene $-\overset{\displaystyle O}{\overset{\|}{C}}H$ formyl

21.

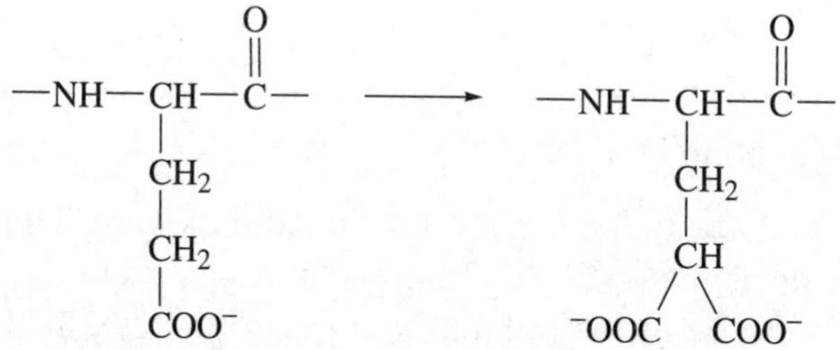

22. biotin and vitamin KH_2

23. **a.** biotin **b.** coenzyme B_{12}

24. $RCH{=}CH\overset{\displaystyle O}{\overset{\|}{C}}SR + FADH_2 \longrightarrow RCH_2CH_2\overset{\displaystyle O}{\overset{\|}{C}}SR + FAD$

25.

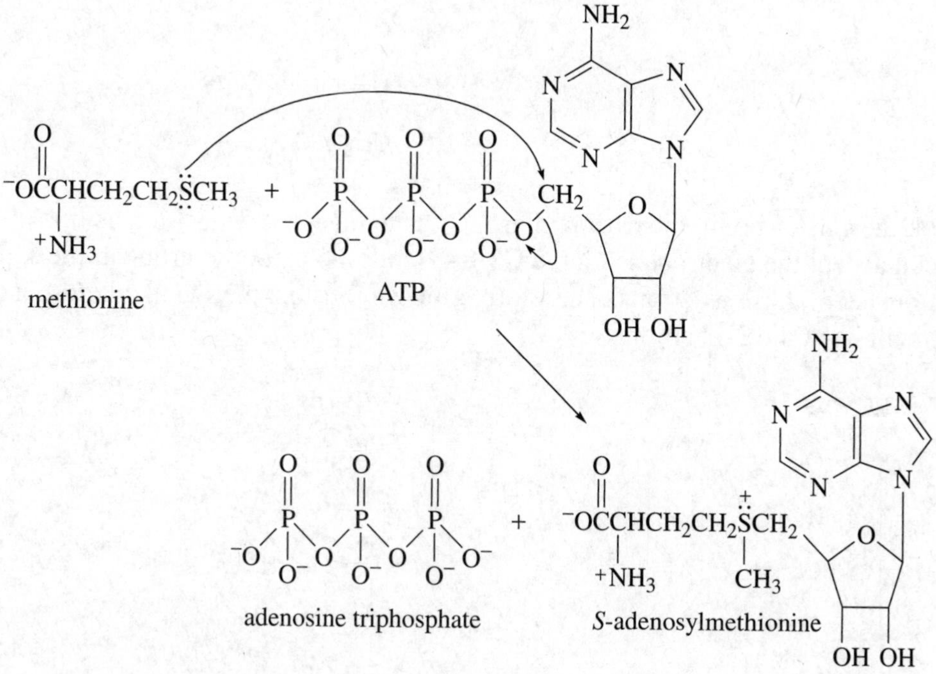
methionine
ATP
adenosine triphosphate
S-adenosylmethionine

26. **a.** pyridoxal phosphate **b.** biotin

27. The compound on the right can more easily lose CO_2 because the electrons left behind when CO_2 is eliminated are delocalized onto the positively charged nitrogen of the pyridine ring.

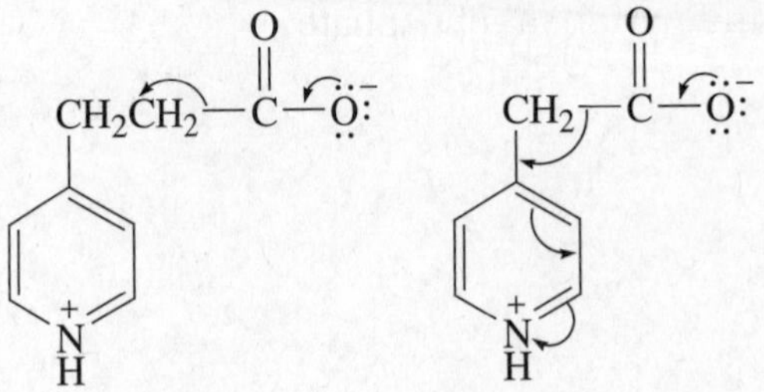

28. The positively charged nitrogen atom of the imine serves as an electron sink to accept the electrons that are left behind when the C3-C4 bond breaks.

In the absence of the imine, the electrons would be delocalized onto a neutral oxygen. The neutral oxygen is not as electron-withdrawing as the positively charged nitrogen. In other words, the electron sink that is present as a result of imine formation makes it easier to break the C—C bond.

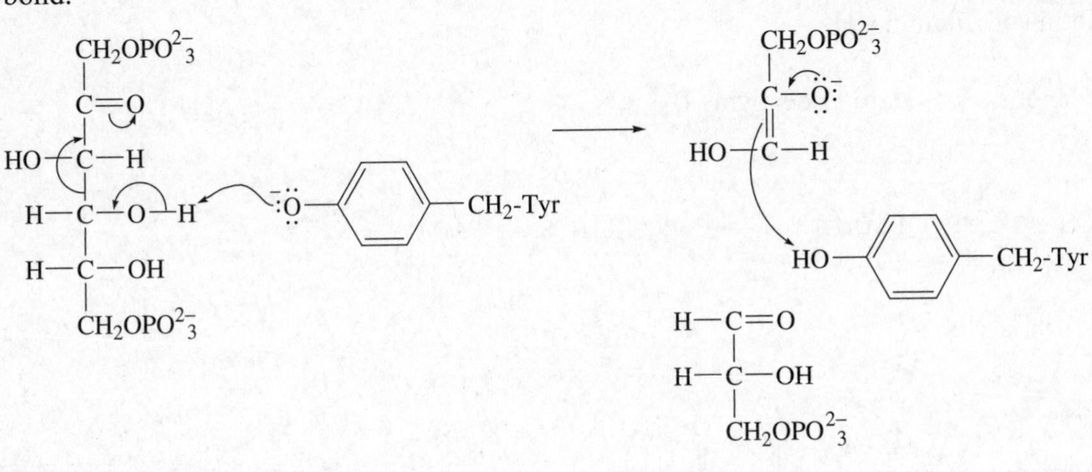

29. In order to break the C3-C4 bond, the carbonyl group has to be at the 2-position as it is in fructose, so it can accept the electrons from the C3-C4 bond; the carbonyl group at the 1-position in glucose cannot accept those electrons. Therefore, glucose must isomerize to a ketose, so the carbonyl group will be at the 2-position.

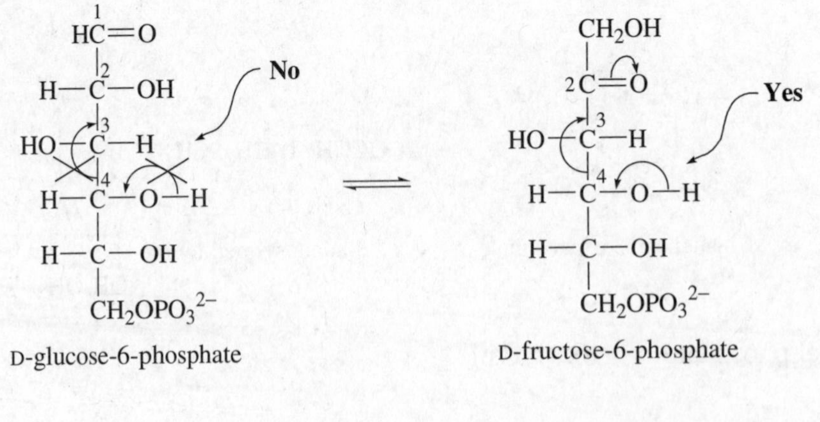
No
Yes
D-glucose-6-phosphate
D-fructose-6-phosphate

Chapter 17 Practice Test

1. What two coenzymes put carboxyl groups on their substrates?

2. What two coenzymes allow electrons to be delocalized?

3. What are the three one-carbon groups that tetrahydrofolate coenzymes put on their substrates?

4. Indicate whether each the following statements is true or false:

a. An acid catalyst donates a proton to the substrate and a base catalyst removes a proton from the substrate. T F

b. $FADH_2$ is an oxidizing agent. T F

c. Thiamine pyrophosphate is vitamin B_6. T F

d. Vitamin KH_2 is the coenzyme that puts a carboxyl group on the β-carbon of glutamate residues. T F

e. The reactant of an enzyme-catalyzed reaction is called a substrate. T F

f. NADH is a reducing agent. T F

CHAPTER 18
The Organic Chemistry of Metabolic Pathways

1. Solved in the text.

2.

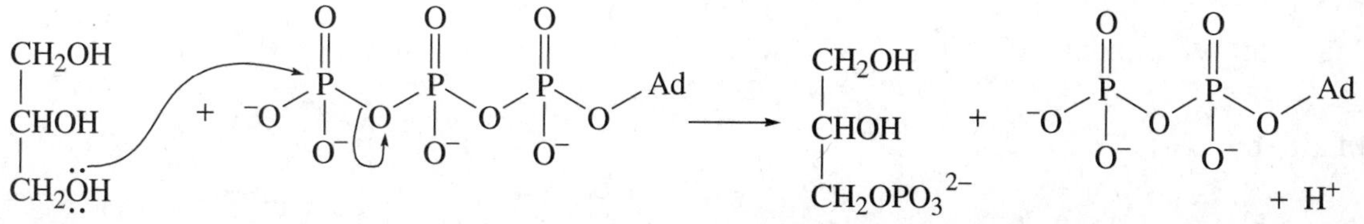

glycerol glycerol-3-phosphate

3.

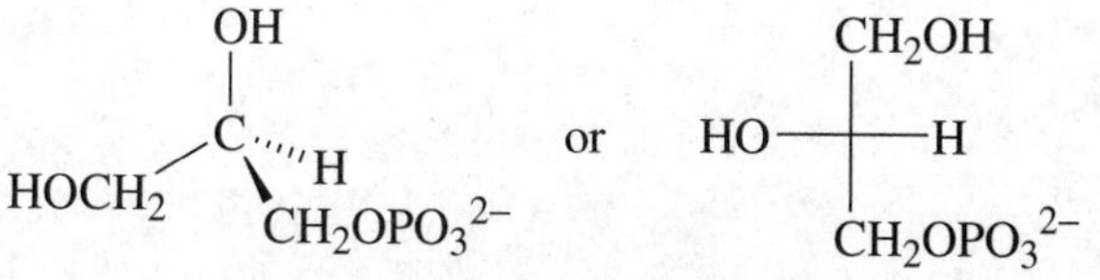

(*R*)-glycerol-3-phosphate

4. The resonance contributor on the right shows that the β-carbon of the α,β-unsaturated carbonyl compound has a partial positive charge and the α-carbon has no charge. The nucleophilic OH group, therefore, is attracted to the positively charged β-carbon.

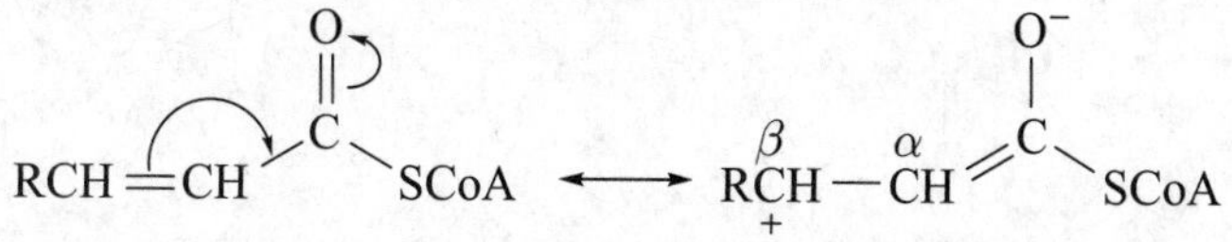

5. eight

6. seven: one mole of NADH will be obtained from each of the following steps

$$16 \xrightarrow{1} 14 \xrightarrow{2} 12 \xrightarrow{3} 10 \xrightarrow{4} 8 \xrightarrow{5} 6 \xrightarrow{6} 4 \xrightarrow{7} 2$$

(each step + 2)

7.

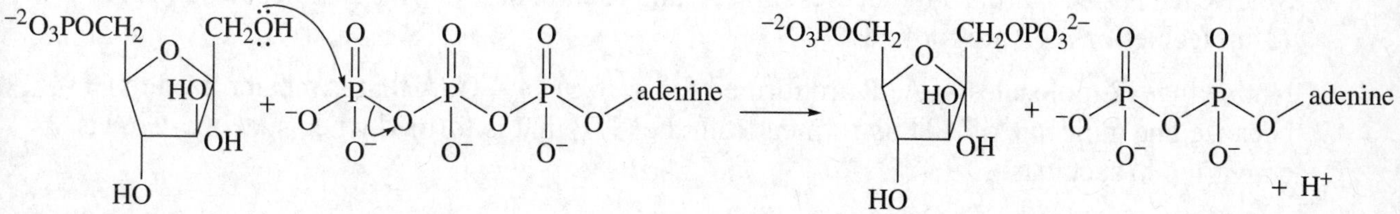

fructose-6-phosphate ATP fructose-1,6-diphosphate ADP

8. **a.** conversion of glucose to glucose-6-phosphate
conversion of fructose-6-phosphate to fructose-1,6-diphosphate

b. conversion of 1,3-diphosphoglycerate to 3-phosphoglycerate
conversion of 2-phosphoenolpyruvate to pyruvate

9. two; each mole of D-glucose is converted, in several steps, to two moles of glyceraldehyde-3-phosphate, and each mole of glyceraldehyde-3-phosphate requires one mole of NAD^+ for it to be eventually converted to a mole of pyruvate.

10. a ketone

11. thiamine pyrophosphate

12. an aldehyde

13.

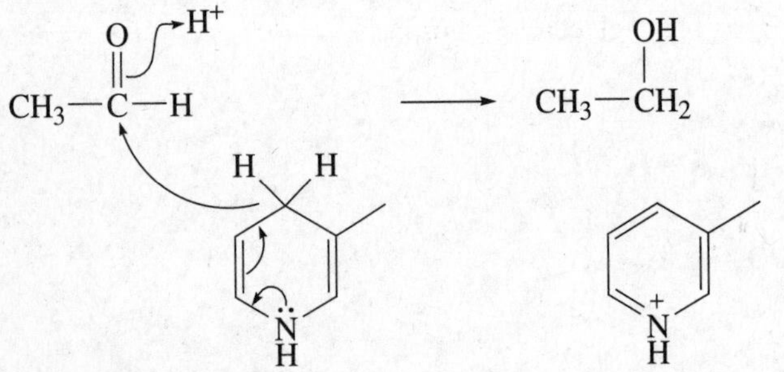
H+
O
CH3—C—H
H H
N
H
OH
CH3—CH2
N
H

14. pyridoxal phosphate

15.

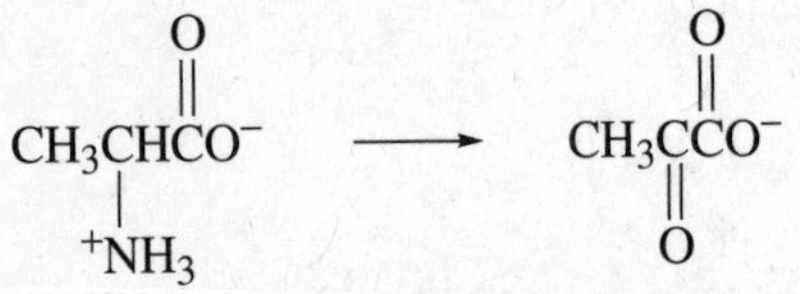
O
CH3CHCO−
+NH3
O
CH3CCO−
O

alanine pyruvate

16. a secondary alcohol

17. two

18. citrate and isocitrate

19. When one molecule of acetyl-CoA is converted to two molecules of CO_2 and one molecule of CoASH, 3 molecules of NADH and one molecule of $FADH_2$ are formed.

Since each NADH forms 3 molecules of ATP and each $FADH_2$ forms 2 molecules of ATP, 11 molecules of ATP are formed.

Notice that 12 molecules of ATP are formed for every acetyl-CoA that is metabolized to CO_2, because one molecule of ATP is formed from the GTP that is formed when succinyl-CoA is converted to succinate.

20. **a.** a catabolic pathway **b.** a catabolic pathway

21.

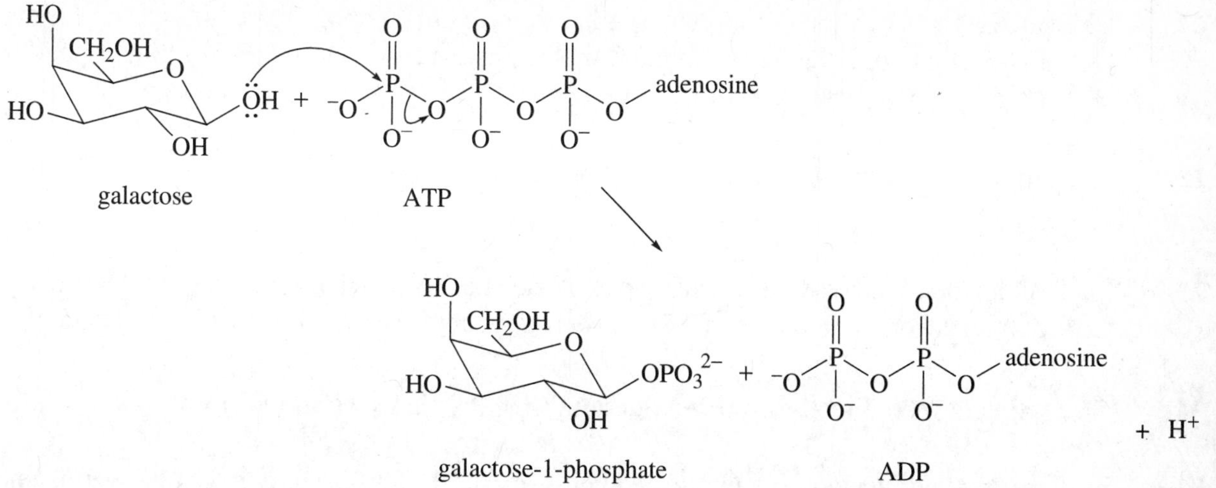
galactose
ATP
adenosine
galactose-1-phosphate
ADP
adenosine
+ H^+

22. Pyruvate is a good leaving group because the electrons left behind when pyruvate leaves are delocalized.

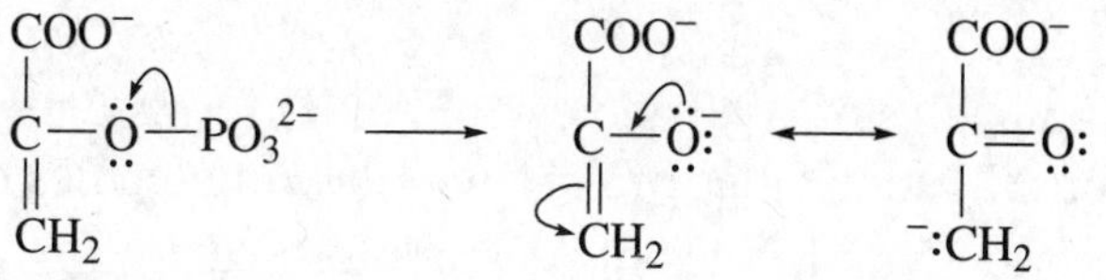

23. The hydrogen on the α-carbon.

24. The conversion of citrate to isocitrate.
The conversion of fumarate to (*S*)-malate

25. The label will be on the phosphate group that is attached to the enzyme that catalyzes the isomerization of 3-phosphoglycerate to 2-phosphoglycerate.

26. The mechanism in Section 17.3 shows that aldolase cleaves glucose between C-3 and C-4.

The bottom three carbons (C-4, C-5, and C-6) become pyruvate. Therefore, C-4 is the carboxyl group of pyruvate.

The top three carbons (C-1, C-2, and C-3) become dihydroxyacetone phosphate (the phosphate group is on C-1 of what was glucose). When dihydroxyacetone phosphate is converted to glyceraldehyde-3-phosphate, the phosphate group is on C-3 of glyceraldehyde-3-phosphate, which means that the aldehyde carbon of glyceraldehyde-3-phosphate is C-3 of glucose.

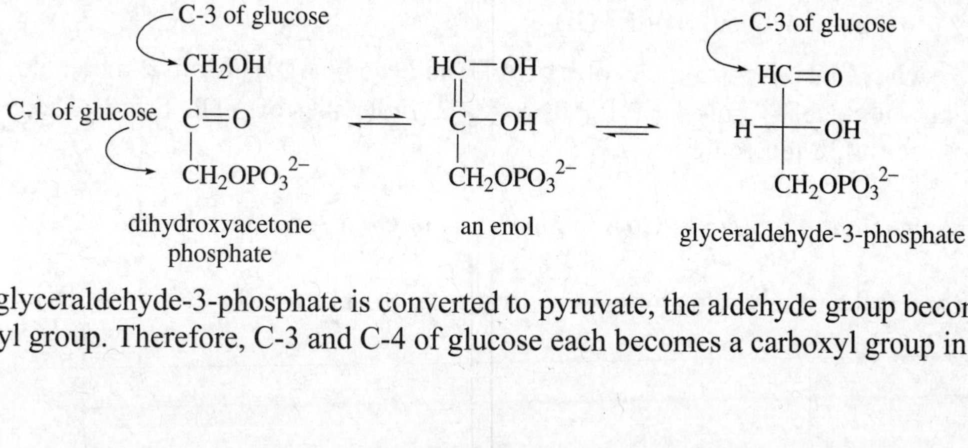
C-3 of glucose
C-1 of glucose
C-3 of glucose
dihydroxyacetone phosphate
an enol
glyceraldehyde-3-phosphate

When glyceraldehyde-3-phosphate is converted to pyruvate, the aldehyde group becomes a carboxyl group. Therefore, C-3 and C-4 of glucose each becomes a carboxyl group in pyruvate.

27. Pyruvate loses its carboxyl group when it is converted to ethanol. Since the carboxyl group is C-3 or C-4 of glucose, half of the ethanol molecules contain C-1 and C-2 of glucose and the other half contain C-5 and C-6 of glucose.

28. The β-oxidation of one molecule of a 16-carbon fatty acyl-CoA will form 8 molecules of acetyl-CoA.

29. Each molecule of acetyl-CoA forms two molecules of CO_2. Therefore, the 8 molecules of acetyl-CoA obtained from a molecule of a 16-carbon fatty acyl-CoA will form 16 molecules of CO_2.

30. No ATP is formed directly from β-oxidation. (See Problems 31 and 32.)

31. Each molecule of acetyl-CoA that is cleaved from the 16-carbon fatty acyl-CoA forms one molecule of $FADH_2$ and one molecule of NADH. Since a 16-carbon fatty acyl-CoA undergoes 7 cleavages, 7 molecules of $FADH_2$ and 7 molecules of NADH are formed from the 16-carbon fatty acyl-CoA.

32. Since each NADH forms 3 molecules of ATP and each $FADH_2$ forms 2 molecules of ATP, the 7 molecules of NADH form 21 molecules of ATP and the 7 molecules of $FADH_2$ form 14 molecules of ATP. Therefore, 35 molecules of ATP are formed.

33. We have seen that each molecule acetyl-CoA that enters the citric acid cycle forms 12 molecules of ATP (Section 18.7 and Problem 19). A molecule of a 16-carbon fatty acid will form 8 molecules of acetyl-CoA. These will form 96 molecules of ATP. When these are added to the 35 molecules of ATP formed from the NAD and $FADH_2$ generated in β-oxidation (Problem 32), we see that 131 molecules of ATP are formed.

34. Each molecule of glucose, while being converted to 2 molecules of pyruvate, forms 2 molecules of ATP and 2 molecules of NADH.

The 2 molecules of pyruvate form 2 molecules of NADH while being converted to 2 molecules of acetyl-CoA.

Each molecule of acetyl-CoA that enters the citric acid cycle forms 3 molecules of NADH, one molecule of $FADH_2$, and one molecule of ATP (see Problem 19). Therefore the 2 molecules of acetyl-CoA obtained from glucose form 6 molecules of NADH, 2 molecules of $FADH_2$, and 2 molecules of ATP.

Therefore, each molecule of glucose forms 4 molecules of ATP, 10 molecules of NADH $(2 + 2 + 6)$, and 2 molecules of $FADH_2$.

Since each NADH forms 3 molecules of ATP and each $FADH_2$ forms 2 molecules of ATP, one molecule of glucose forms $4 + (10 \times 3) + (2 \times 2)$ molecules of ATP. That is, each molecule of glucose forms 38 molecules of ATP.

35. The conversion of propionyl-CoA to methylmalonyl-CoA requires biotin.

36. The conversion of methylmalonyl-CoA to succinyl-CoA requires coenzyme B_{12}.

Chapter 18 Practice Test

1. Which of the following are not citric acid cycle intermediates: fumarate, acetate, citrate?

2. Which provide energy to the cell, anabolic reactions or catabolic reactions?

3. Draw the structure of the compound obtained when the following amino acid undergoes transamination.

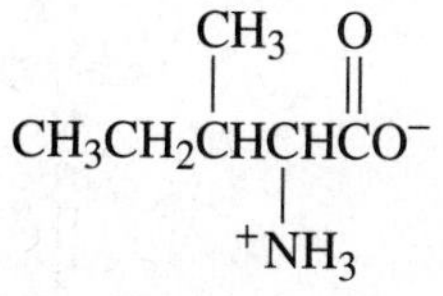
CH3 O
CH3CH2CHCHCO−
+NH3

4. What compounds are formed when proteins undergo the first stage of catabolism?

5. What compounds are formed when a fatty acid undergoes the second stage of catabolism?

6. Indicate whether each of the following statements is true or false:

   a. Each molecule of $FADH_2$ forms 3 molecules of ATP in the fourth stage of catabolism. T F

   b. $FADH_2$ is oxidized to FAD. T F

   c. NAD^+ is oxidized to NADH. T F

   d. Acetyl-CoA is a citric acid cycle intermediate. T F

CHAPTER 19
The Organic Chemistry of Lipids

1. **a.** Stearic acid has the higher melting point because it has two more methylene groups (giving it a greater surface area) than palmitic acid.

   b. Palmitic acid has the higher melting point because it does not have any carbon-carbon double bonds, whereas palmitoleic acid has a cis double bond that prevents the molecules from packing closely together.

   c. Oleic acid has the higher melting point because it has one double bond, while linoleic acid has two double bonds, which give greater interference to close packing of the molecules.

2. Glyceryl tripalmitate has a higher melting point because the carboxylic acid components are saturated and can, therefore, pack more closely together than the unsaturated carboxylic acid components of glyceryl tripalmitoleate.

3. To be optically inactive, the fat must have a plane of symmetry. Therefore, identical fatty acids must be at C-1 and C-3. Therefore, stearic acid must be at C-1 and C-3.

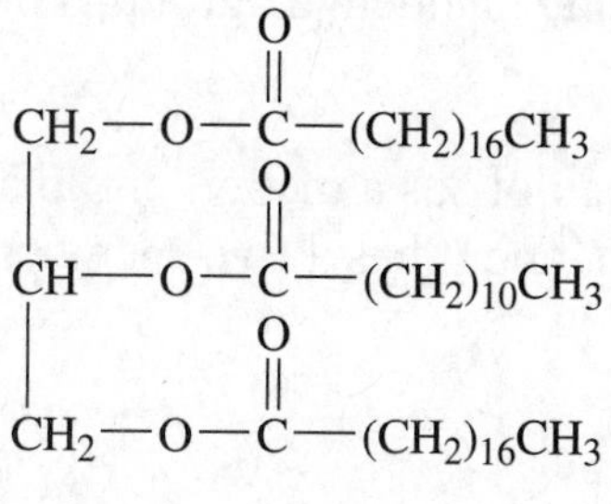

$CH_2-O-C(=O)-(CH_2)_{16}CH_3$
$CH-O-C(=O)-(CH_2)_{10}CH_3$
$CH_2-O-C(=O)-(CH_2)_{16}CH_3$

4. To be optically active, the fat must not have a plane of symmetry. Therefore, identical fatty acids must be at C-1 and C-2 (or at C-2 and C3). Therefore, stearic acid must be at C-1 and C-2 (or at C-2 and C3). It does not make any difference if lauric acid is at C-1 or at C-3 because both compounds would be the same.

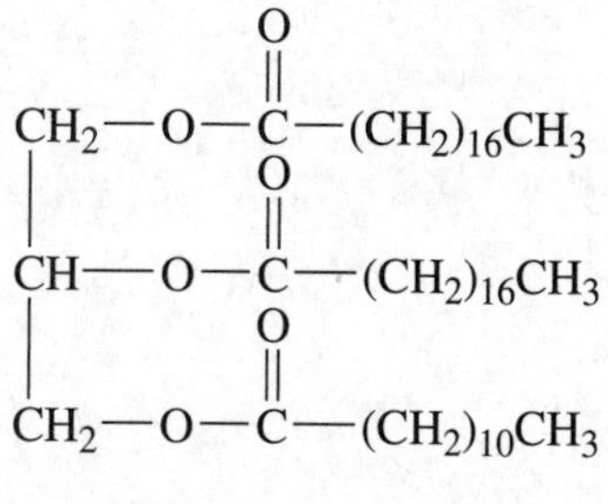

$CH_2-O-C(=O)-(CH_2)_{16}CH_3$
$CH-O-C(=O)-(CH_2)_{16}CH_3$
$CH_2-O-C(=O)-(CH_2)_{10}CH_3$

5.

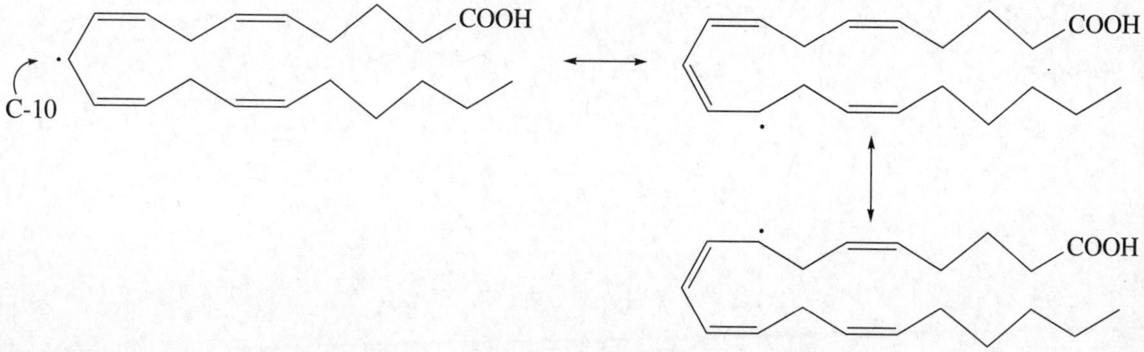

COOH
C-10
COOH
COOH

6. Because the interior of a membrane is nonpolar and the surface of a membrane is polar, integral proteins will have a higher percentage of nonpolar amino acids.

7. The bacteria could synthesize phosphoacylglycerols with more saturated fatty acids because these triacylglycerols would pack more tightly in the lipid bilayer and, therefore, would have higher melting points and be less fluid.

8. Membranes must be kept in a semifluid state in order to allow transport across them. Cells closer to the hoof of an animal are going to be in a colder average environment than cells closer to the body. Therefore, the cells closer to the hoof have a higher degree of unsaturation to give them a lower melting point so the membranes will not solidify at the colder temperature.

9. **a.** Solved in the text. **b.** a monoterpene

10. **a.** Solved in the text. **b.**

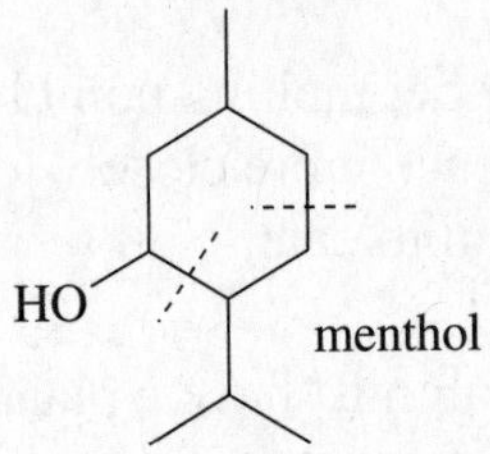
HO
menthol

11. two ketone groups, a double bond, an aldehyde group, a primary alcohol, and a secondary alcohol

12. Compared to testosterone, stanozolol has a methyl group attached to the five-membered ring, it does not have the double bond, and it has a five-membered ring with two nitrogens in place of the ketone group.

13. **a.**

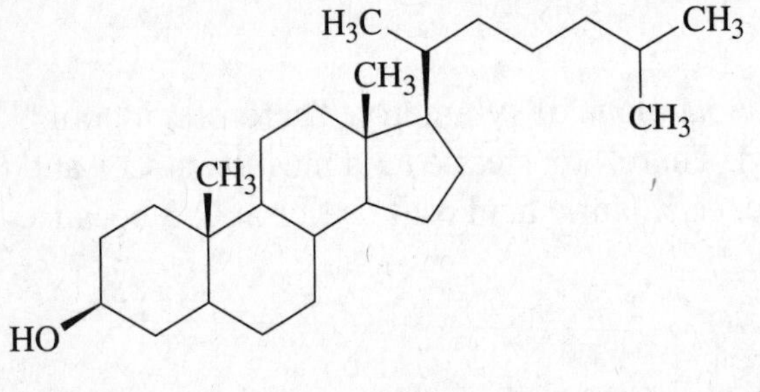
H3C
CH3
CH3
CH3
CH3
HO

b.

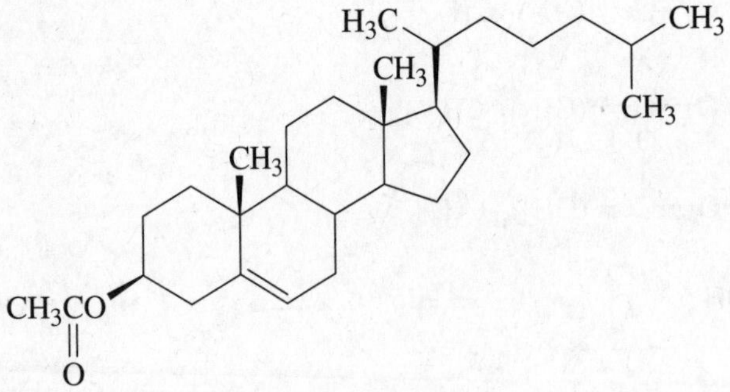
H3C
CH3
CH3
CH3
CH3
CH3CO
O

c.

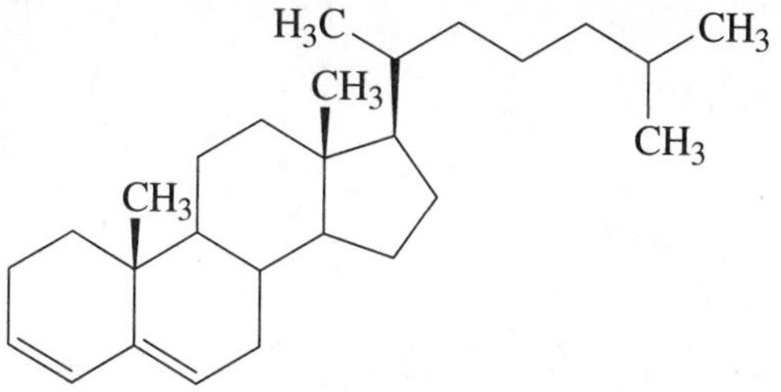

d.

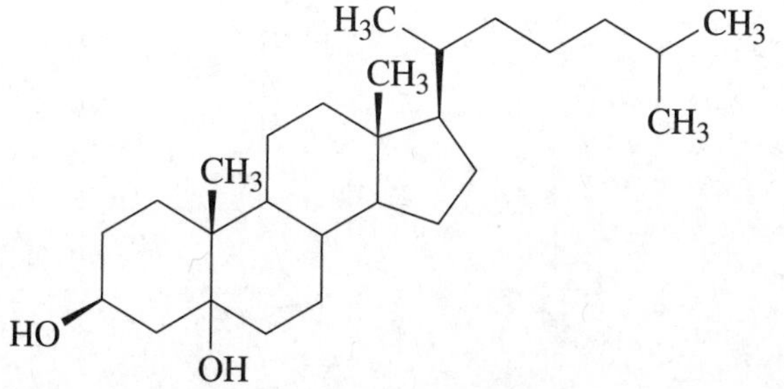

e.

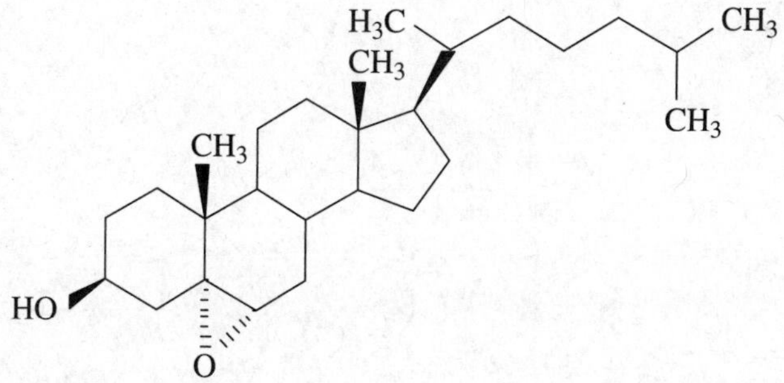

14. All triacylglycerols do not have the same number of asymmetric centers. If the carboxylic acid components at C-1 and C-3 of glycerol are not identical, the triacylglycerol has one asymmetric center (C-2). If the carboxylic acid components at C-1 and C-3 of glycerol are identical, the triacylglycerol has no asymmetric centers.

15.

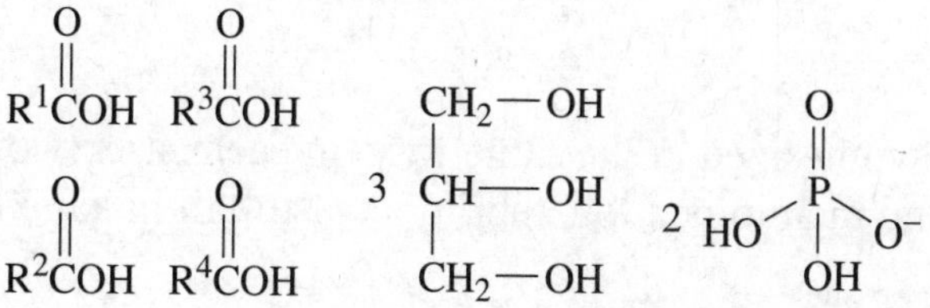

16.

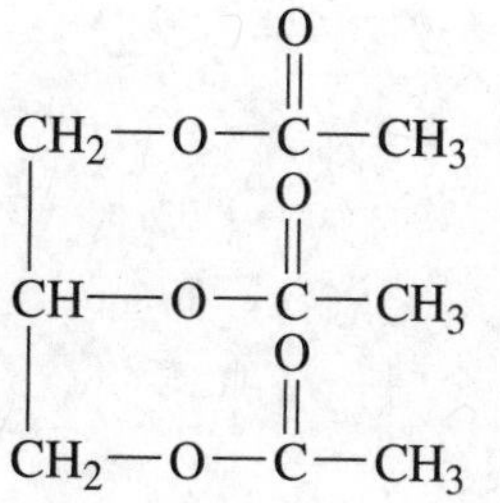

The structure at left has a molecular formula = $C_9H_{14}O_6$ and a molecular weight = 218.

Subtracting 218 from the total molecular weight gives the molecular weight of the methylene (CH_2) groups in the triacylgcerol.

$$722 - 218 = 504$$

Dividing 504 by the molecular weight of a methylene group (14) will give the number of methylene groups.

$$\frac{504}{14} = 36$$

Since there are 36 methylene groups, each fatty acid in the triacylglycerol has 12 methylene groups.

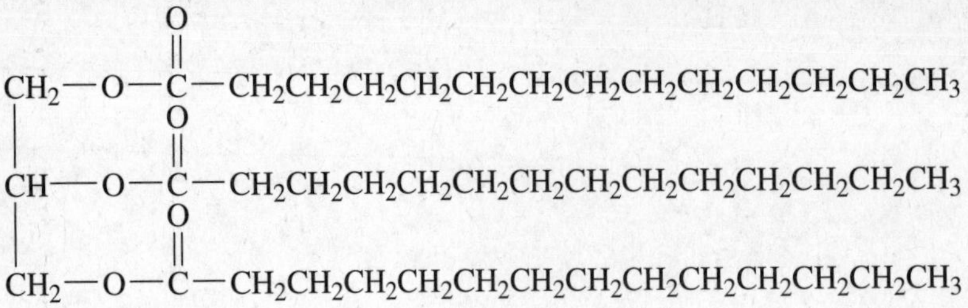

nutmeg

17.

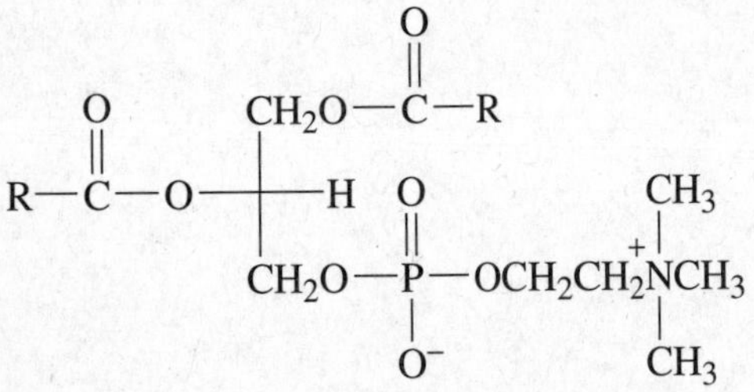

18.

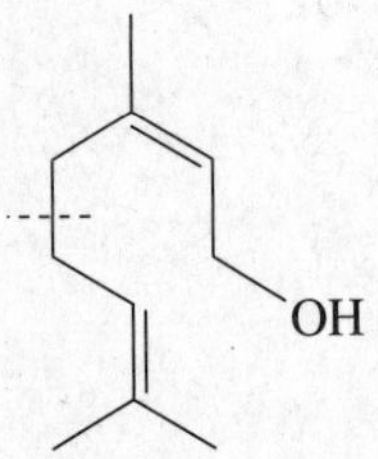

19. The fact that the tail-to-tail linkage occurs in the exact center of the molecule indicates that the two halves are synthesized (in a head-to-tail fashion) and then joined together in a tail-to-tail linkage.

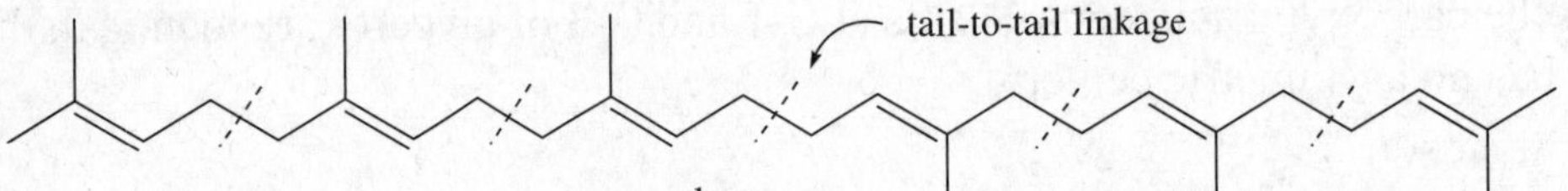
tail-to-tail linkage

squalene

20. Squalene, lycopene, and *β*-carotene are all synthesized in the same way. In each case, two halves are synthesized (in a head-to-tail fashion) and then joined together in a tail-to-tail linkage.

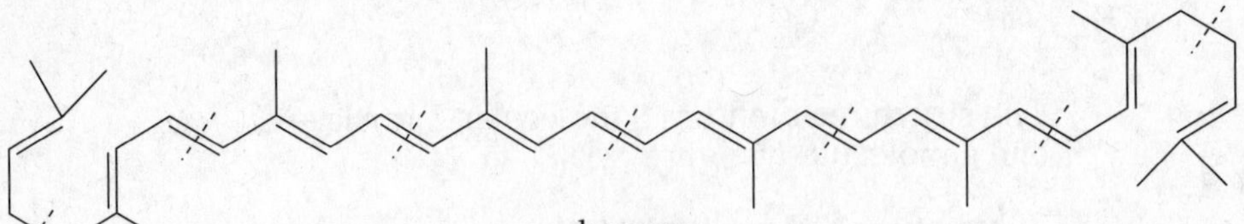

lycopene

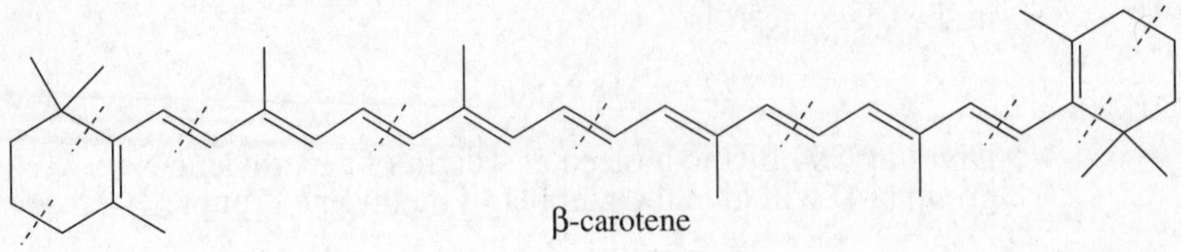

β-carotene

21.

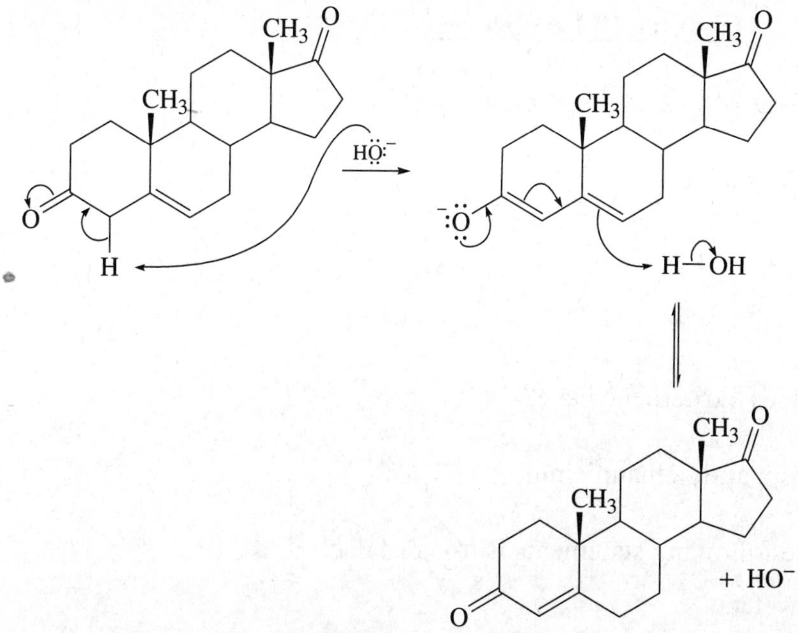

CH3
O
CH3
HO:⁻
H
:O:
H—OH
+ HO⁻

22. **a.** There are three triacylglycerols in which one of the fatty acid components is lauric acid and two are myristic acid. Myristic acid can be at C-1 and C-3 of glycerol, in which case the triacylglycerol does not have any asymmetric centers. If myristic acid is at C-1 and C-2 of glycerol, C-2 is an asymmetric center, and consequently, two enantiomers are possible for the compound.

b. There are six triacylglycerols in which one of the fatty acid components is lauric acid, one is myristic acid, and one is palmitic acid. The three possible arrangements are shown below (with the fatty acid components abbreviated as L, M, and P). Since each has an asymmetric center (indicated by a star), each can exist as a pair of enantiomers for a total of six triacylglycerols.

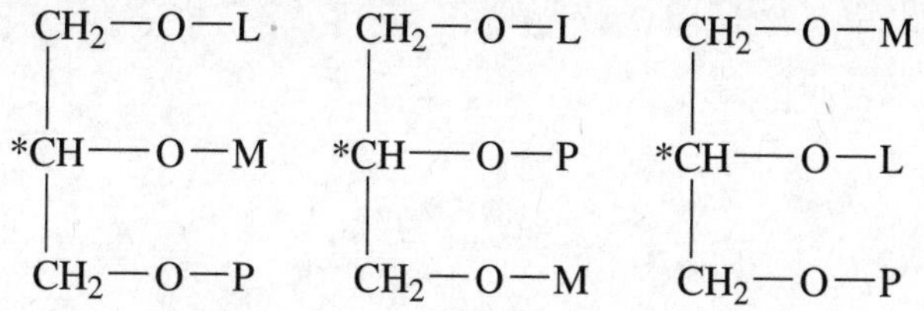

CH2—O—L
*CH—O—M
CH2—O—P
CH2—O—L
*CH—O—P
CH2—O—M
CH2—O—M
*CH—O—L
CH2—O—P

23.

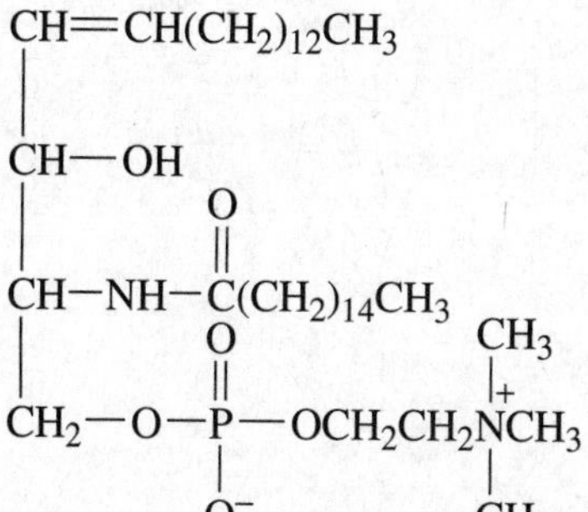

CH=CH(CH2)12CH3
CH—OH
CH—NH—C(CH2)14CH3
CH2—O—P—OCH2CH2NCH3
CH3
CH3
O⁻

24.

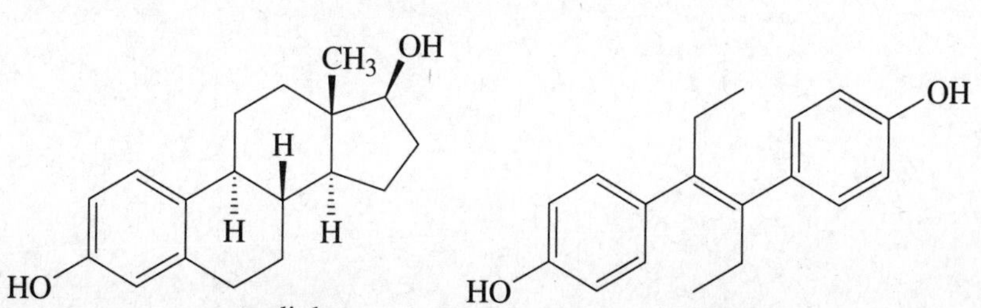

CH3
OH
H
H
H
HO
OH
HO

estradiol DES

Chapter 19 Practice Test

1. Explain why the melting points of fats are higher than those of oils.

2. Mark off the isoprene units in squalene.

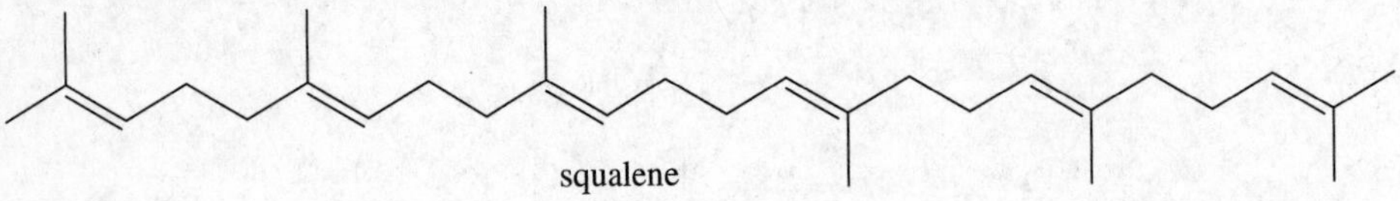

squalene

3. How many isoprene units does a triterpene have?

4. Draw the structure of a phosphatidylethanolamine.

5. Indicate whether each of the following statements is true or false:

a. Lipids are insoluble in water.	T	F
b. Cholesterol is the precursor of all other steroids.	T	F
c. Saturated fatty acids have higher melting points than unsaturated fatty acids.	T	F
d. Vitamin K is a water-soluble vitamin.	T	F
e. Fats have a higher percentage of saturated fatty acids than do oils.	T	F

CHAPTER 20
The Chemistry of the Nucleic Acids

1. a.

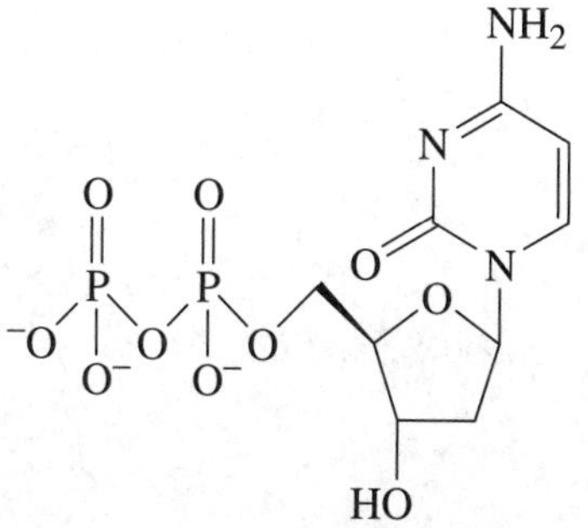

dCDP

b.

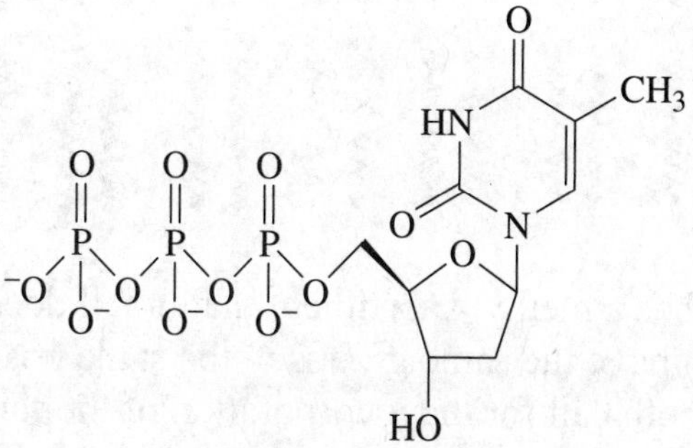

dTTP

c.

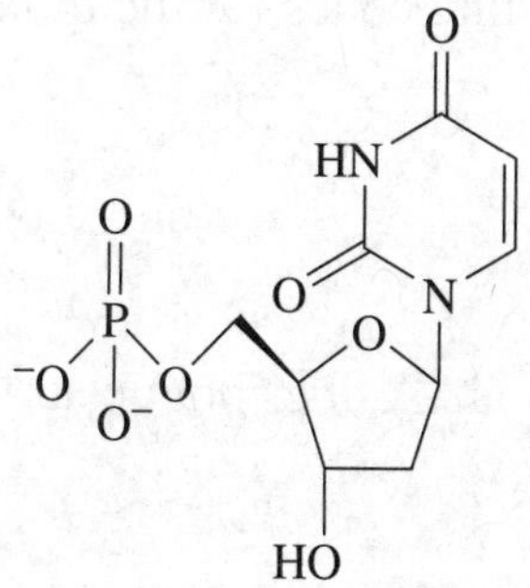

dUMP

d.

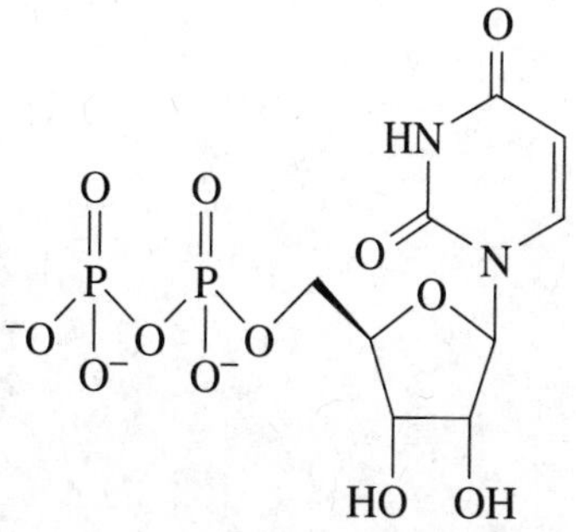

UDP

e.

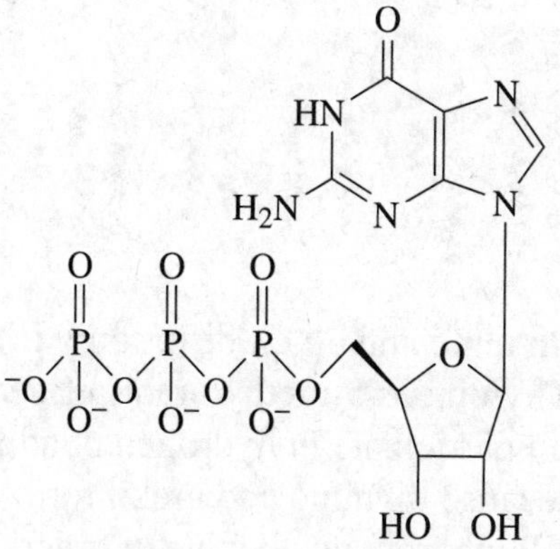

guanosine 5'-triphosphate
GTP

f.

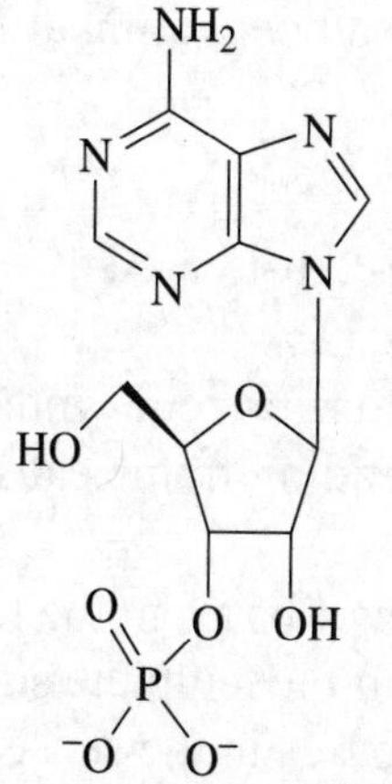

adenosine 3'-monophosphate

2. **a.** 3′—C—C—T—G—T—T—A—G—A—C—G—5′
   b. guanine

3.

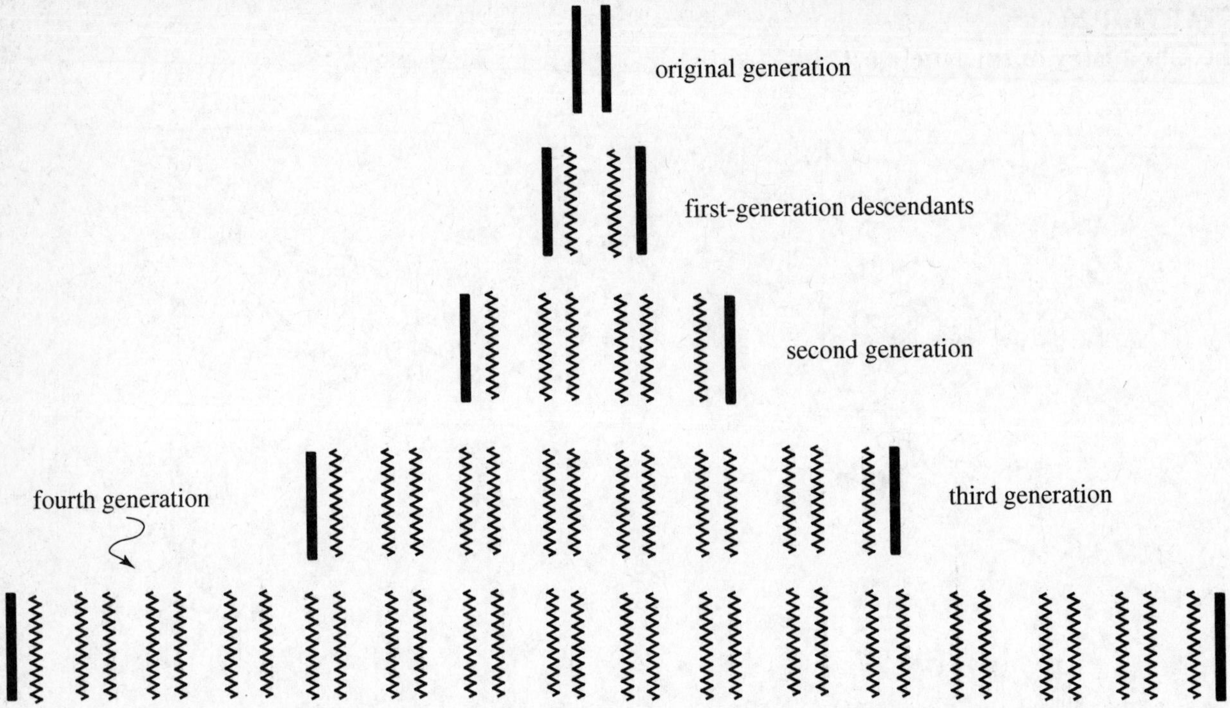
original generation
first-generation descendants
second generation
third generation
fourth generation

4. Thymine and uracil differ only in that thymine has a methyl substituent that uracil does not have. (Thymine is 5-methyluracil.) Because they both have the same groups in the same positions that can participate in hydrogen bonding, they will both call for the incorporation of the same purine. Because thymine and uracil form one hydrogen bond with guanine and two with adenine, they will incorporate adenine in order to maximize hydrogen bonding.

5. Because methionine is known to be the first base incorporated into the oligopeptide, the mRNA sequence is read beginning at AUG, since that is the only codon that codes for methionine.

Met-Asp-Pro-Val-Ile-Lys-His

6. Met-Asp-Pro-Leu-Leu-Asn

7. It does not cause protein synthesis to stop, because the sequence UAA does not occur within a triplet. The reading frame causes the triplets to be AU**U** and **AA**A.

8. The sequence of bases in the template strand of DNA specify the sequence of bases in mRNA, so the bases in the template strand and the bases in mRNA are complementary. Therefore, the sequence of bases in the sense strand of DNA are identical to the sequence of bases in mRNA, except wherever there is a U in mRNA, there is a T in the sense strand of DNA.

5′— G-C-A-T-G-G-A-C-C-C-C-G-T-T-A-T-T-A-A-A-C-A-C—3′

9. **a** is the only sequence that has a chance of being recognized by a restriction endonuclease because it is the only one that has the same sequence of bases in the 5' → 3' direction that the complementary strand has in the 5' → 3' direction.

ACGCGT
TGCGCA

10. All the fragments will end in "G".

^{32}P—TCCGAGGTCACTAGG

^{32}P—TCCGAGGTCACTAG

^{32}P—TCCGAGG

^{32}P—TCCGAG

^{32}P—TCCG

11. **a.** guanosine 3'-monophosphate **c.** 2'-deoxyadenosine 5'-monophosphate

b. cytidine 5'-diphosphate **d.** 2'-deoxythymidine

12. Lys-Val-Gly-Tyr-Pro-Gly-Met-Val-Val

13. 3′—TTT—CAA—CCG—ATG—GGG—CCT—TAC—CAC—CAG—5′

14. 5′—AAA—GTT—GGC—TAC—CCC—GGA—ATG—GTG—GTC—3′

15. **a.** isoleucine **c.** valine

b. aspartate **d.** valine

16. The third base in each codon has some variability.

mRNA 5'-GG(UCA or G)UC(UCA or G)CG(UCA or G)GU(UCA or G)CA(U or C)GA(A or G)-3'
or AG(U or C) AG(A or G)

DNA
template 3'-CC(AGT or C)AG(AGT or C)GC(AGT or C)CA(AGT or C)GT(A or G)CT(T or C)-5'
or TC(A or G) TC(T or C)

sense 5'-GG(TCA or G)TC(TCA or G)CG(TCA or G)GT(TCA or G)CA(T or C)GA(A or G)-3'
or AG(T or C) AG(A or G)

Notice that Ser and Arg are two of three amino acids that can be specified by six different codons.

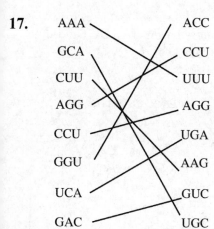

17.

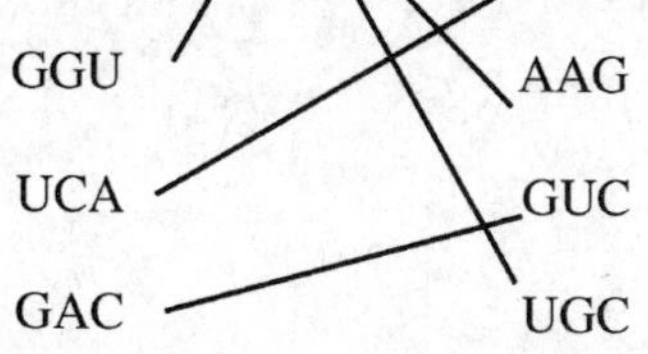
AAA
GCA
CUU
AGG
CCU
GGU
UCA
GAC
ACC
CCU
UUU
AGG
UGA
AAG
GUC
UGC

18.

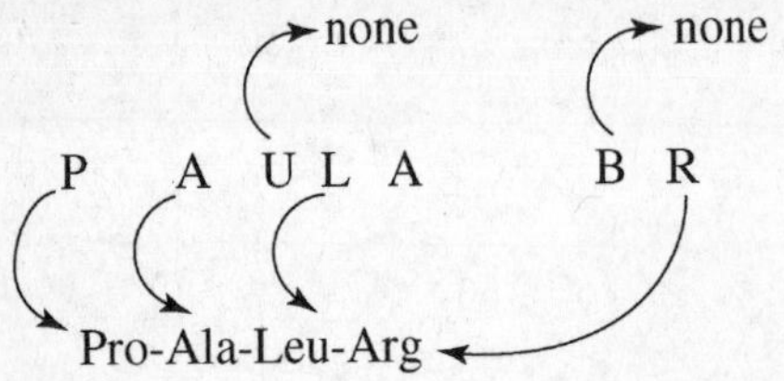
none
none
P A U L A
B R
Pro-Ala-Leu-Arg

mRNA CC(UCA or G)GC(UCA or G)CU(UCA or G)CG(UCA or G)
or UU(A or G) AG(A or G)

DNA (sense strand) CC(TCA or G)GC(TCA or G)CT(TCA or G)CG(TCA or G)
or TT(A or G) AG(A or G)

Note that because mRNA is complementary to the template strand of DNA, which is complementary to the sense strand, the sense strand of DNA and mRNA have the same sequence of bases (except DNA has a T where RNA has a U).

Also note that Leu and Arg are specified by six codons.

19. **a.** CC and GG **c.** CA and TG

CA and TG are formed in equal amounts, since A pairs with T and C pairs with G. (Remember that the dinucleotides are written in the 5' → 3' direction. That is why f. is not a correct answer; it would require A to pair with A and T to pair with T.)

correct: 5' C A 3' / 3' G T 5' not correct: 5' T A 3' / 3' T A 5'

20. The normal and mutant peptides would have the following base sequence in their mRNA.

normal: CA(AG) UA(UC) GG(UCAG) AC(UCAG) CG(UCAG) UA(UC) GU(UCAG)

mutant: CA(AG) UC(UCAG) GA(AG) CC(UCGA) GG(UCGA) AC(UCAG)

a. The middle nucleotide (A) in the second triplet was deleted. This means that an A was deleted in the sense strand of DNA or a T was deleted in the template strand of DNA.

b. The mRNA for the mutant peptide has an unused 3'-terminal two-letter code, U(UCGA). The last amino acid in the octapeptide of the normal fragment is leucine, so its last triplet is UU(AG) or CU(UCAG).

This means that the triplet for the last amino acid in the mutant is U(UCGA)(UC) and that the last amino acid in the mutant is one of the following: Phe, Ser, Tyr, or Cys.

21.

	Met	Asp	Pro	Val	Ile	Lys	His
<u>codons</u>	AUG	GAU	CCU	GUU	AUU	AAA	CAU
		GAC	CCC	GUC	AUC	AAG	CAC
			CCA	GUA	AUA		
			CCG	GUG			

<u>anticodons</u>

Note that the anticodons are stated in the 5’→ 3’ direction. For example, the anticodon of AUG is stated as CAU (not UAC).

codon 5′ A U G 3′

anticodon 3′ U A C 5′

CAU	AUC	AGG	AAC	AAU	UUU	AUG
	GUC	GGG	GAC	GAU	CUU	GUG
		UGG	UAC	UAU		
		CGG	CAC			

22. The NH_2 groups could serve either as hydrogen bond acceptors, using the nonbonding electrons on nitrogen, or as hydrogen bond donors, using the hydrogen bonded to the nitrogen.

The **A**, **D**, and **A/D** designations show that the maximum number of hydrogen bonds that can form are two between thymine and adenine and three between cytosine and guanine. Notice that uracil and thymine have the same designations.

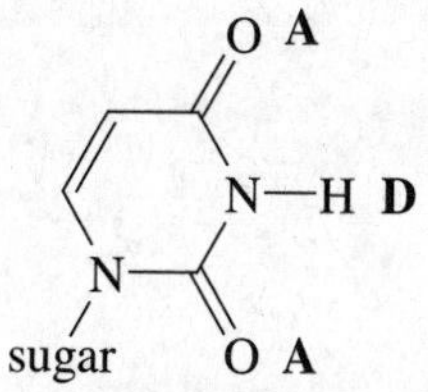
O A
N—H D
N
sugar
O A

uracil

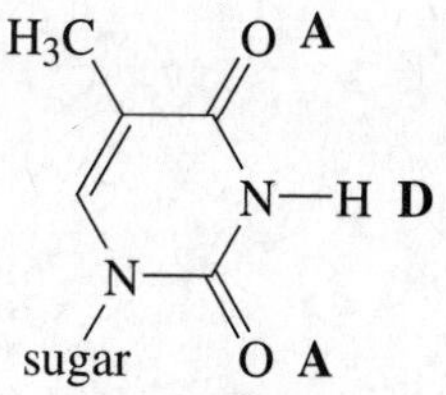
H3C
O A
N—H D
N
sugar
O A

thymine

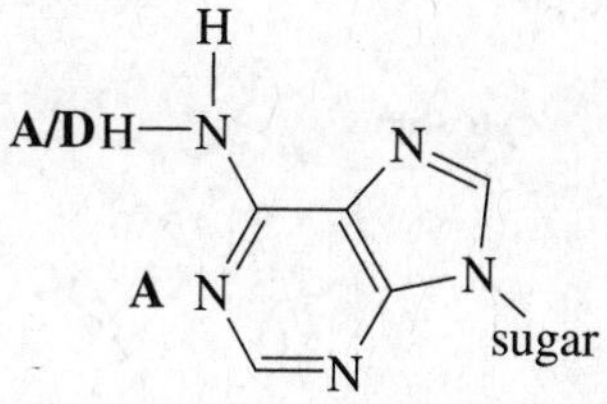
H
A/D H—N
N
A N
N
sugar
N

adenine

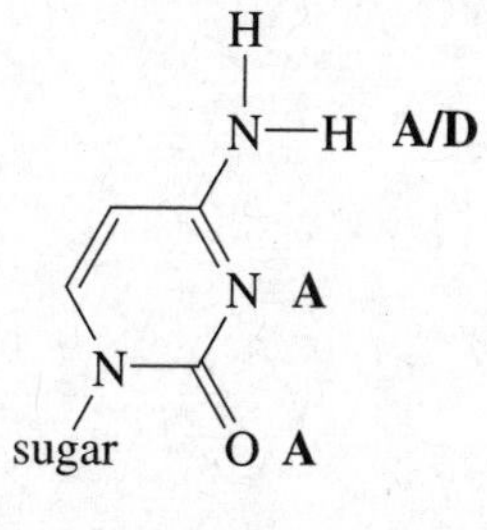
H
N—H A/D
N A
N
sugar
O A

cytosine

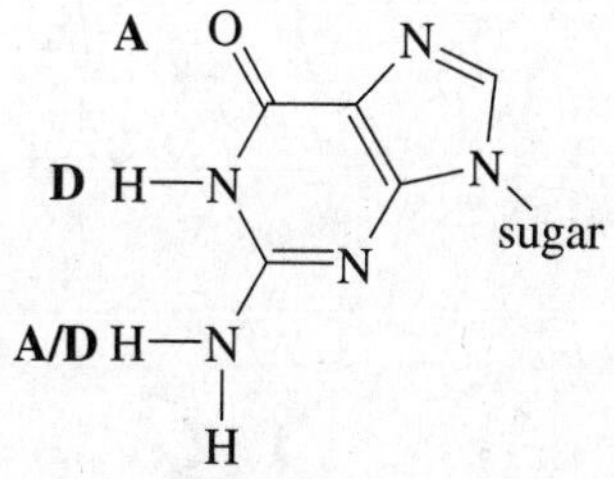
A O
N
D H—N
N
sugar
N
A/D H—N
H

guanine

23. If the bases existed in the enol form, the OH groups and NH_2 groups could act either as hydrogen bond acceptors or as hydrogen bond donors.

The maximum number of hydrogen bonds that could form between the bases if they were in the enol form is one between thymine and adenine and two between cytosine and guanine.

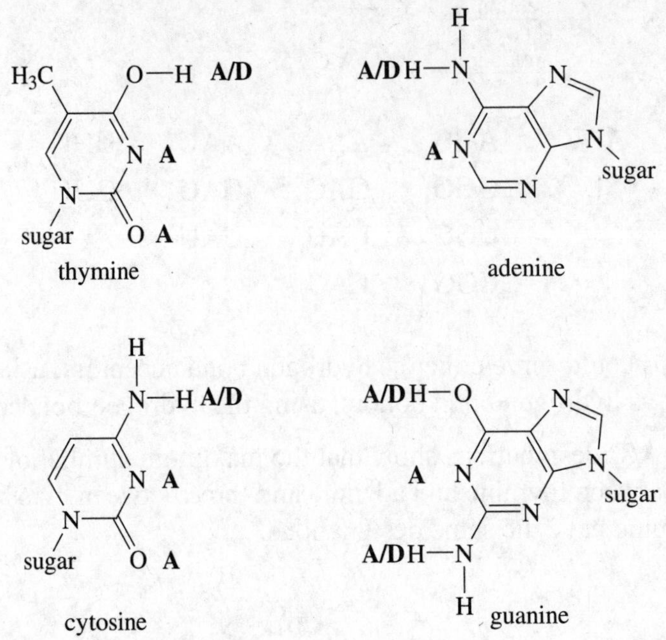

H₃C
O—H A/D
N A
N
sugar
O A
thymine
H
A/D H—N
N
N
A N
N
sugar
adenine
H
N—H A/D
N A
N
sugar
O A
cytosine
A/D H—O
N
N
A N
N
sugar
A/D H—N
H
guanine

24.

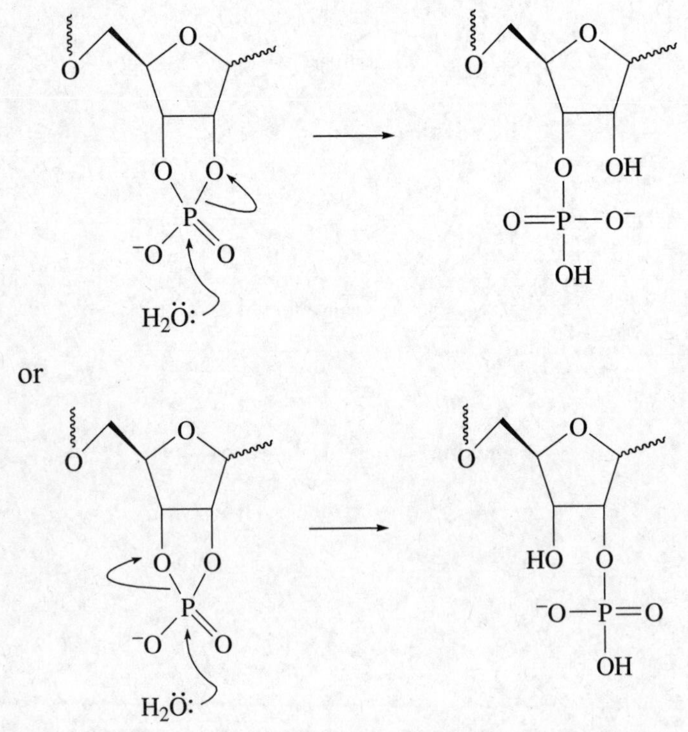

O
O
O O
P
⁻O O
H₂Ö:
O
O
O OH
O=P—O⁻
OH
or
O
O
O O
P
⁻O O
H₂Ö:
O
O
HO O
⁻O—P=O
OH

25.

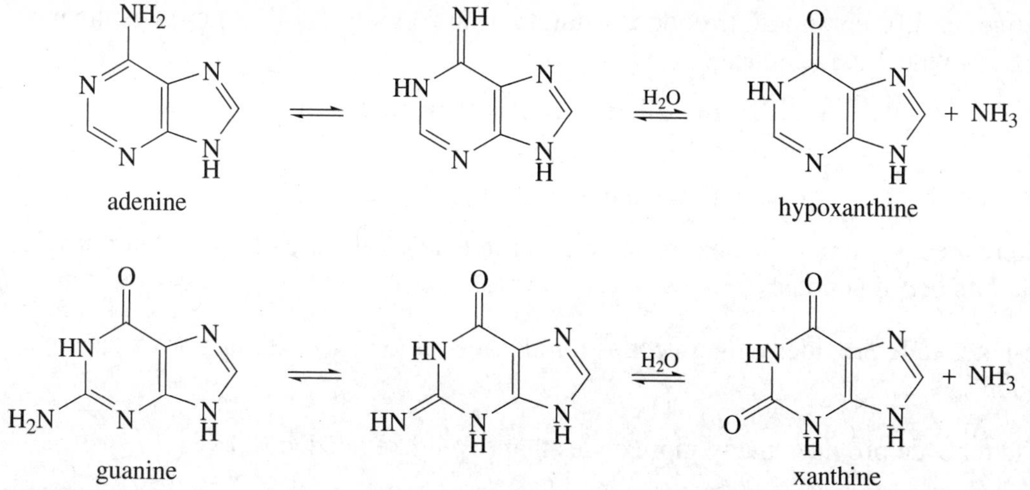

adenine
hypoxanthine
H_2O
$+ NH_3$
guanine
xanthine
H_2O
$+ NH_3$

26. Thymine does not have an amino substituent on the ring, which means that it cannot form an imine. Deamination involves hydrolyzing an imine linkage to a carbonyl group and ammonia.

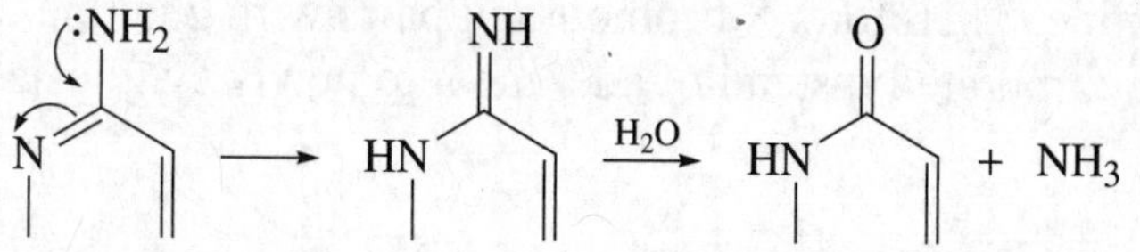

H_2O
$+ NH_3$

27. The ring is protonated at its most basic position. In the case of a purine, this is the 7-position. In the next step, the bond between the heterocyclic base and the sugar breaks, with the anomeric carbocation being stabilized by the ring oxygen's nonbonding electrons.

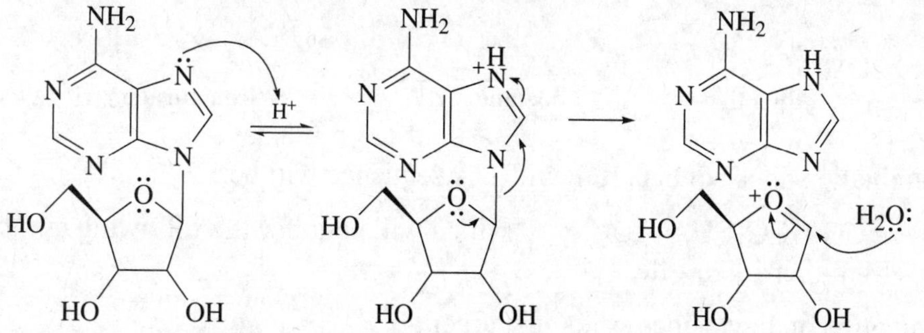

H^+
H_2O:

The mechanism is exactly the same for pyrimidines except that the initial protonation takes place at the 1-position.

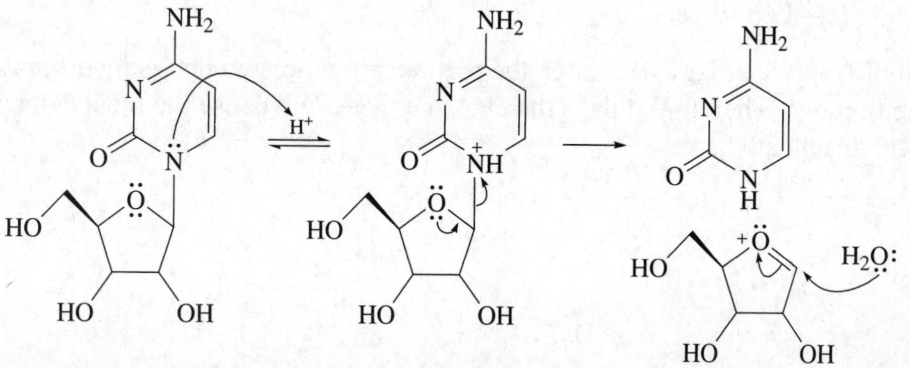

H^+
H_2O:

28. The number of different possible codons using four nucleotides is $(4)^n$ where n is the number of letters (nucleotides) in the code.

for a two-letter code: $(4)^2 = 16$
for a three-letter code: $(4)^3 = 64$
for a four-letter code: $(4)^4 = 256$

Since there are 20 amino acids that must be specified, a two-letter code would not provide an adequate number of codons.

A three-letter code provides enough codes for all the amino acids and also provides the necessary stop codons.

A four-letter code provides many more codes than would be needed.

29. 5-Bromouracil is incorporated into DNA in place of thymine because of their similar size. Thymine pairs with adenine via two hydrogen bonds. 5-Bromouracil exists primarily in the enol form. The enol can form only one hydrogen bond with adenine, but it can form three hydrogen bonds with guanine. Therefore, 5-bromouracil pairs with guanine. Because 5-bromouracil causes guanine to be incorporated instead of adenine into newly synthesized DNA strands, it causes mutations.

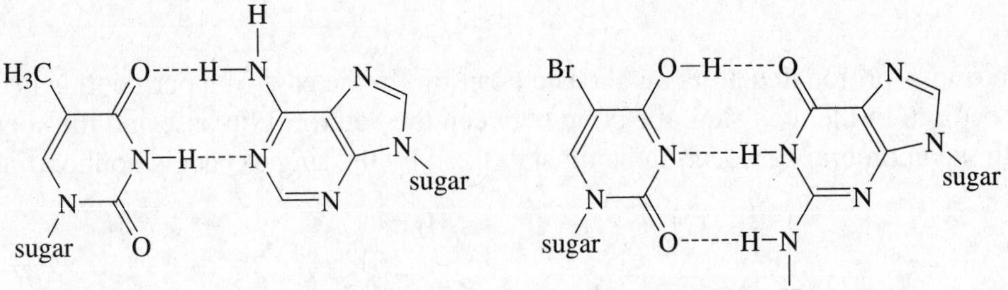
H
H₃C
O----H—N
N
N—H----N
N
sugar
N
N
sugar
O
Br
O—H----O
N
N------H—N
N
sugar
N
N
sugar
O------H—N
H

thymine **adenine** **5-bromouracil** **guanine**

30. If deamination does not occur, the mRNA sequence will be:

AUG-UCG-CUA-AUC which will code for the following tetrapeptide
Met - Ser - Leu - Ile

Deamination of a cytosine results in a uracil.
If the cytosines are deaminated, the mRNA sequence will be:

AUG-UUG-UUA-AUU which will code for the following tetrapeptide
Met - Leu - Leu - Ile

The only cytosine that will change the particular amino acid that is incorporated into the peptide is the first one. Therefore, this is the cytosine that could cause the most damage to an organism if it were deaminated.

31. A ribosome, which is a particle on which the biosynthesis of proteins takes place, contains a binding site for the growing peptide chain and a binding site for the next amino acid to be incorporated into the chain.

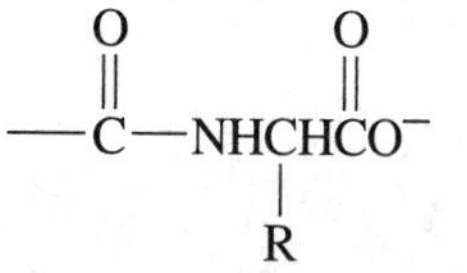

peptide binding site

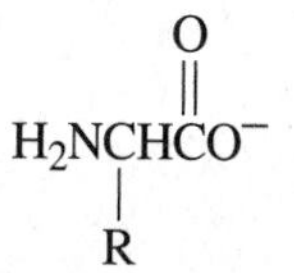

amino acid binding site

All peptide bonds are formed by the reaction of an amino acid with a peptide, except formation of the first peptide bond, which has to be formed by the reaction of two amino acids. Therefore, in the synthesis of the first peptide bond, one of the amino acids has to be a peptide that will fit into the peptide binding site.

The formyl group of *N*-formylmethionine will provide the peptide group that will be recognized by the peptide binding site, and the second amino acid will be bound in the amino acid binding site.

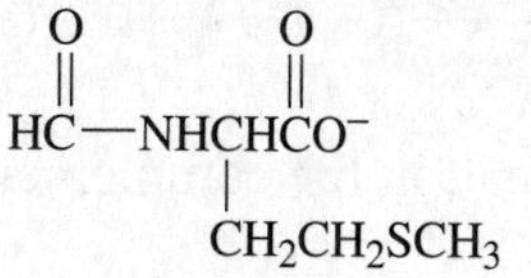

N-formylmethionine

32. It requires energy to break the hydrogen bonds that hold the two chains together, so an enormous amount of energy would be required to unravel the chain completely. However, as the new nucleotides that are incorporated into the growing chain form hydrogen bonds with the parent chain, energy is released, and this energy can be used to unwind the next part of the double helix.

Chapter 20 Practice Test

1. Is the following compound dTMP, UMP, dUMP, or dUTP?

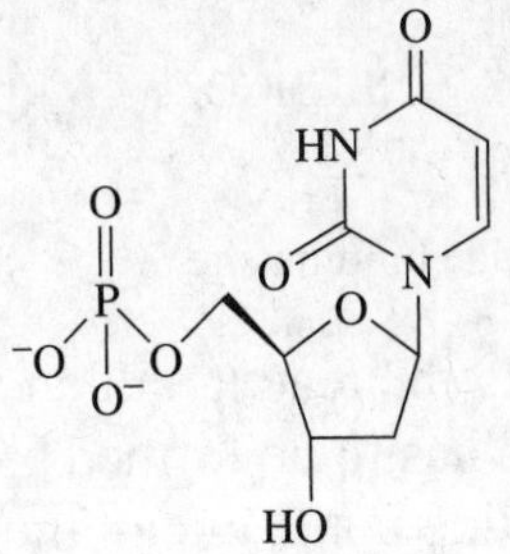

2. If one of the strands of DNA has the following sequence of bases running in the 5'→ 3' direction, what is the sequence of bases in the complementary strand?

5′—A—C—T—T—G—C—A—T—3′

3. What base is closest to the 5'-end in the complementary strand?

4. Indicate whether each of the following statements is true or false:

a. Guanine and cytosine are purines.	T	F
b. The 3'-OH groups allows RNA to be easily cleaved.	T	F
c. The number of A's in double-stranded DNA is equal to the number of T's.	T	F
d. rRNA carries the amino acid that will be incorporated into a protein.	T	F
e. The template strand of DNA is the one transcribed to form mRNA.	T	F
f. The 5'-end of DNA has a free OH group.	T	F
g. The synthesis of proteins from an RNA blueprint is called transcription.	T	F
h. A nucleotide consists of a base and a sugar.	T	F

CHAPTER 21

The Organic Chemistry of Drugs: Discovery and Design

1. **a.** ethyl *para*-aminobenzoate **b.** 2'-deoxy-5-iodouridine

2. There are many possibilities. The following are just a few of them.

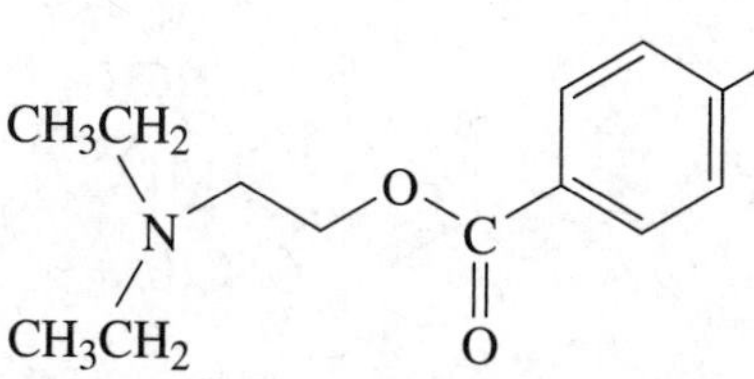

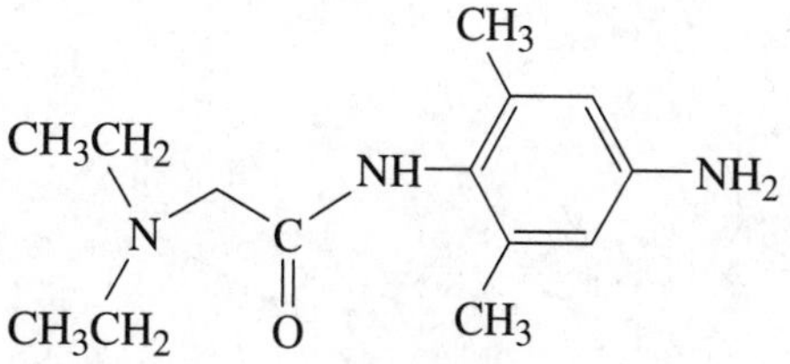

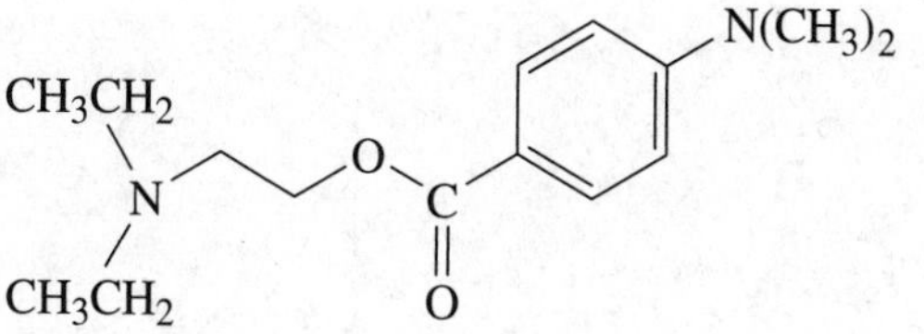

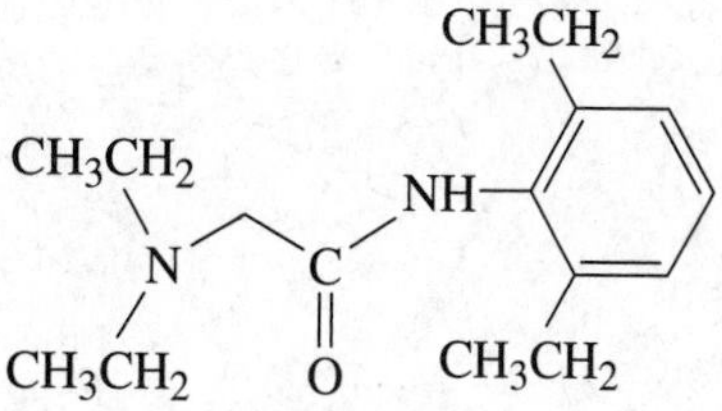

3. The compound on the left has a substituent that can donate electrons by resonance, whereas the compound on the right has a substituent that can withdraw electrons by resonance in the same position. Because the compounds known to be effective tranquilizers have an electron-withdrawing substituent in that position (Cl or NO_2), the compound on the right is more likely to show activity as a tranquilizer.

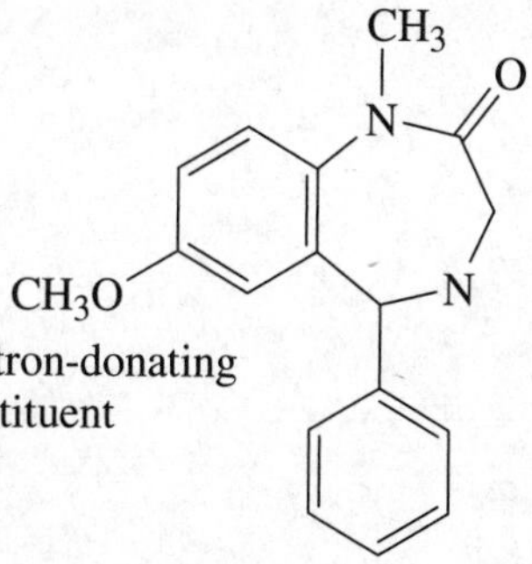

electron-donating substituent

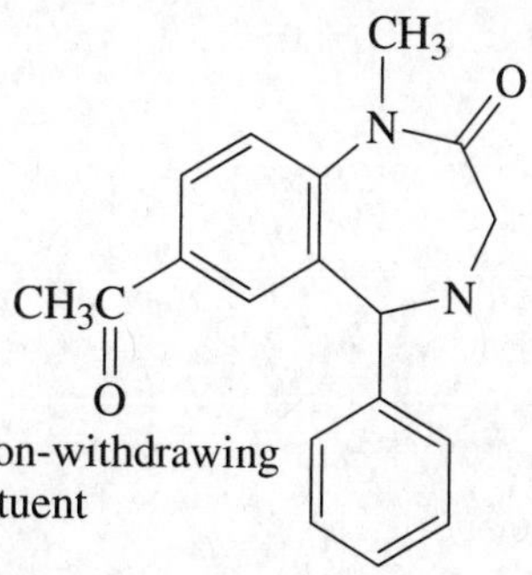

electron-withdrawing substituent

4. Anesthetics have been found to have similar distribution coefficients, which means that they have similar polarities. Since diethyl ether is a known anesthetic, ethyl methyl ether with a polarity similar to that of diethyl ether is more apt to be a general anesthetic than propanol, which is considerably more polar.

5. Tetrahydrocannabinol is a safer drug because it has a higher therapeutic index.

$$\text{therapeutic index} = \frac{\text{lethal dose}}{\text{therapeutic dose}}$$

$$\text{tetrahydrocannabinol} = \frac{2.0\text{ g/kg}}{20\text{ mg/kg}} = \frac{2000\text{ mg/kg}}{20\text{ mg/kg}} = 100$$

$$\text{sodium pentothal} = \frac{100\text{ mg/kg}}{30\text{ mg/kg}} = 3.3$$

6. Aclovir differs from guanosine in that it has a hydroxy-ether substituent rather than a ribose substituent.

Cytosar differs from cytidine in that it has the 2'-OH group in the *β*-position.

Viramid has an unusual heterocyclic base attached to ribose.

Herplex differs from 2'-deoxyuridine in that it has an iodo substituent in the 5-position.

7.

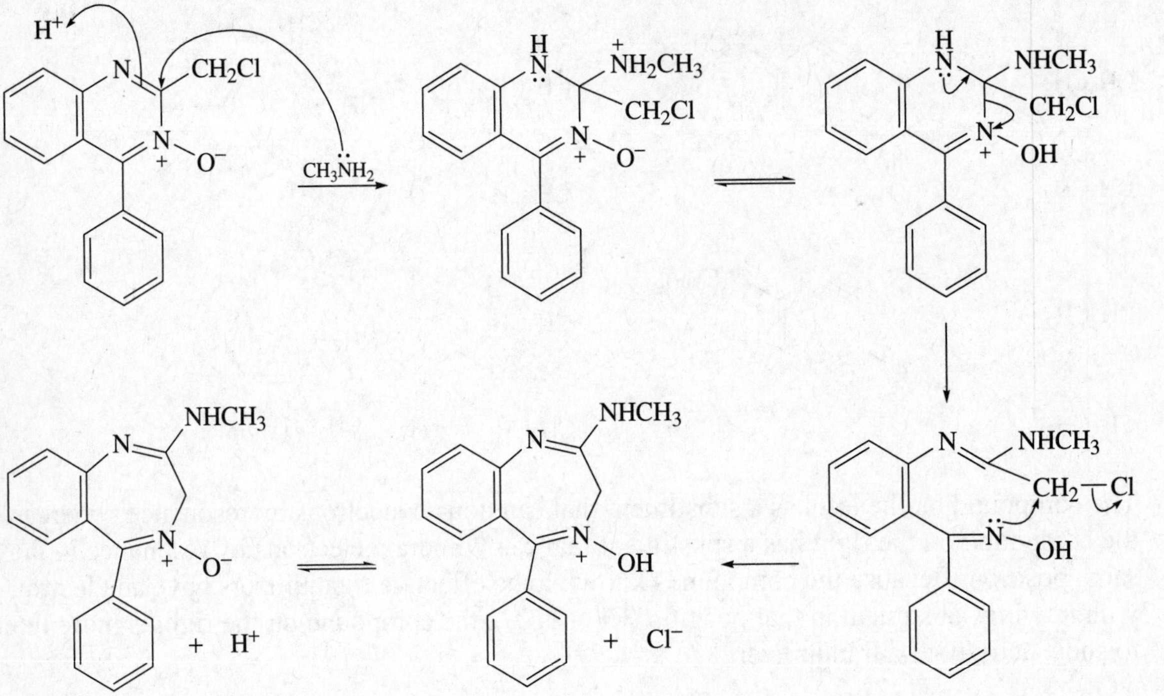

H+
N
CH2Cl
N
O−
CH3NH2
H
N
NH2CH3
CH2Cl
N
O−
H
N
NHCH3
CH2Cl
N
OH
NHCH3
N
N
O−
+ H+
NHCH3
N
N
OH
+ Cl−
N
NHCH3
CH2
Cl
N
OH

Answers to Practice Tests

Chapter 1

1. $\overset{+}{C}H_3$ $\quad$ $\overset{-}{C}H_3$ $\quad$ $\overset{\bullet}{C}H_3$

   sp^2 $\quad$ sp^3 $\quad$ sp^2

2.
```
     ..
    :O:
     ..
H:O:C:O:H
  ..   ..
```

3. $^{+}NH_4$

4. **a.** C—O $\qquad$ **b.** C—F

5. $CH_3CH_2CH_2CH{=}CH_2$ or $CH_3CH_2CH{=}CHCH_3$ or $CH_3\underset{\displaystyle CH_3}{\underset{|}{C}}HCH{=}CH_2$

6. **a.** $1s^2, 2s^2, 2p_x^{\,1}, 2p_y^{\,1}$ $\qquad$ **b.** $1s^2, 2s^1, 2p_x^{\,1}, 2p_y^{\,1}, 2p_z^{\,1}$ $\qquad$ **c.** $1s^2, 2sp^{3\,4}$

7. **a.** A pi bond is stronger than a sigma bond. F
   b. A triple bond is shorter than a double bond. T
   c. The oxygen-hydrogen bonds in water are formed by the overlap of an sp^2 orbital of oxygen with the s orbital of hydrogen. F
   d. A double bond is stronger than a single bond. T
   e. A tetrahedral carbon has bond angles of 107.5°. F

8. O=C=O (C: sp) $\qquad$ $H\overset{O}{\overset{\|}{C}}OH$ (C: sp^2) $\qquad$ HC≡N (C: sp) $\qquad$ CH_3OCH_3 (C: sp^3) $\qquad$ $CH_3CH{=}CH_2$ (C: sp^3)

Chapter 2

1. **a.** HBr $\qquad$ **b.** H_2O $\qquad$ **c.** NH_3 $\qquad$ **d.** $^{-}NH_2$ $\qquad$ **e.** CH_3O^-

2. CH_3COO^- $\qquad$ CH_3CH_2OH $\qquad$ CH_3OH $\qquad$ $CH_3CH_2\overset{+}{N}H_3$

3. **a.** $^{-}NH_2$ $\qquad$ **b.** HO^- $\qquad$ **c.** H_3O^+

4. **a.** $CH_3\overset{H}{\underset{+}{O}}H + NH_3$ $\qquad$ **b.** reactants

5. 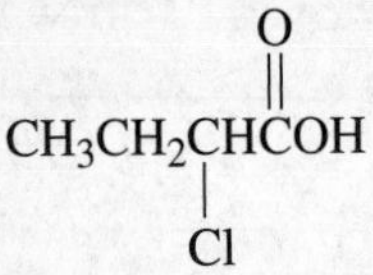

6. a. HO^- is a stronger base than $^-NH_2$. F

b. A Lewis acid is a compound that accepts a share in an electron pair. T

c. $ClCH_2COOH$ is a stronger acid than CH_3COOH T

d. $ClCH_2COOH$ is a stronger acid than $BrCH_2COOH$ T

Chapter 3

1. 3-methyloctane

2. a. 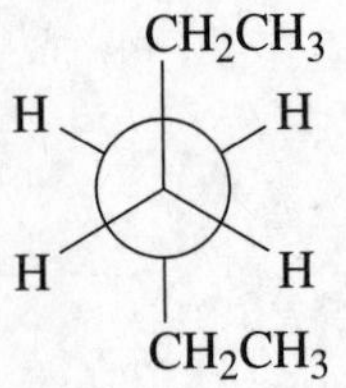

b. 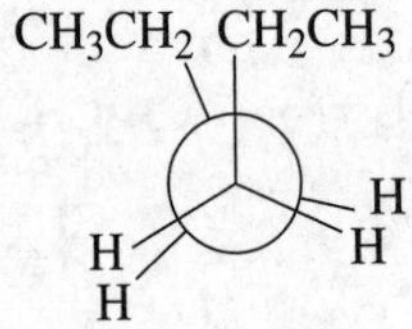

3. a. *sec*-butyl chloride, 2-chlorobutane

b. isopentyl fluoride, 1-fluoro-3-methylbutane

c. cyclopentyl bromide, bromocyclopentane

4. a. $CH_3CH_2CH_2CH_2CH_2Br$ **1** $CH_3CH_2CH_2Br$ **3** $CH_3CH_2CH_2CH_2Br$ **2**

b. $CH_3CH_2CH_2CH_2CH_3$ **3** $CH_3CH_2CH_2CH_2OH$ **1** $CH_3CH_2CH_2CH_2Cl$ **2**

c. 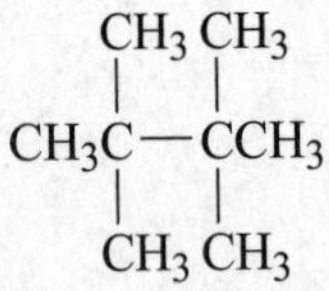 **3** $CH_3CH_2CH_2CH_2CH_2CH_2CH_2CH_3$ **1** 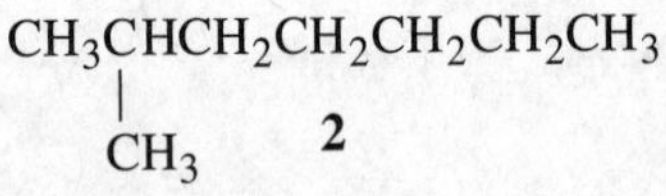 **2**

5. a. 5-bromo-2-methylheptane

b. 3-methylheptane

c. 1-bromo-3-methylcyclopentane

d. 1,4-dichloro-5-methylheptane

6. 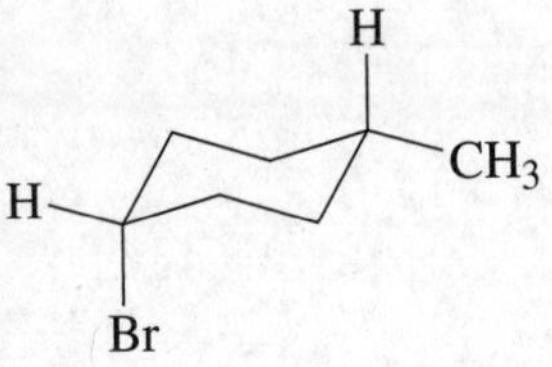

7. 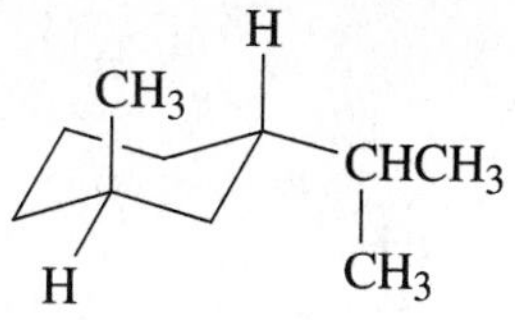

8. **a.** butyl alcohol **b.** butyl alcohol **c.** hexane **d.** propylamine **e.** ethyl alcohol

9. **a.** 2-bromobutane
sec-butyl bromide
b. isopentyl methyl ether
1-methoxy-3-methylbutane
c. 2-methylbutane
isopentane

10. **a.** a staggered conformation
b. the chair conformer of methylcyclohexane with the methyl group in the equatorial position
c. cyclohexane

11. **a.** 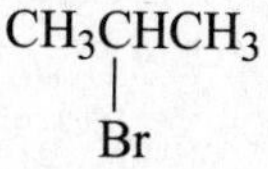
b. $CH_3CH_2NHCH_3$
c. 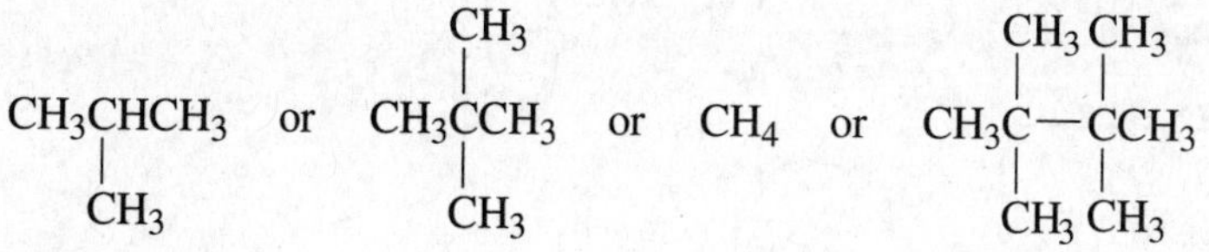
d. 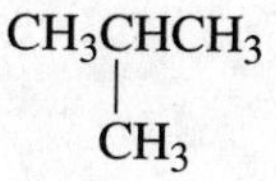
e. 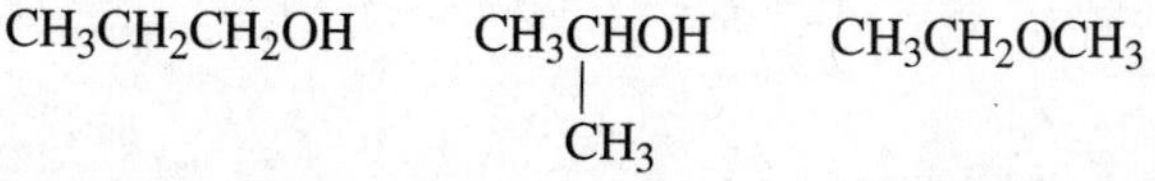

Chapter 4

1. **a.** 4-methyl-1-hexene **b.** 4-bromocyclopentene **c.** 7-methyl-3-nonene **d.** 4-chloro-3-methylcyclohexene

2. 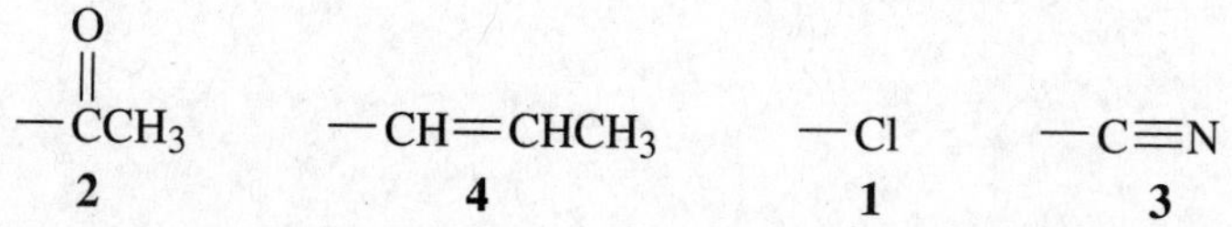

3. **a.** 2-pentene **b.** 3-methyl-1-hexene **c.** 3-methyl-2-pentene **d.** 1-methylcyclohexene

4. **a.** 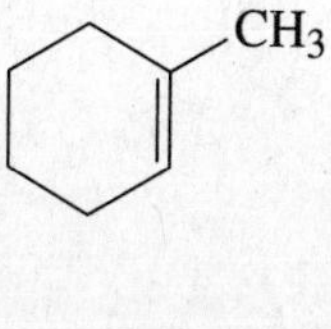 **b.** 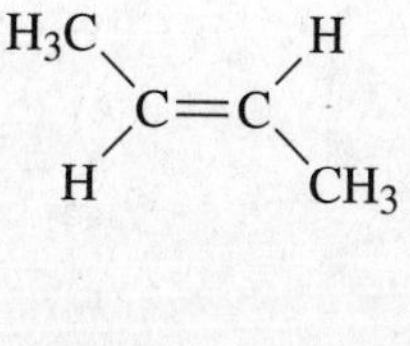

5. b and c

6. 5

7. 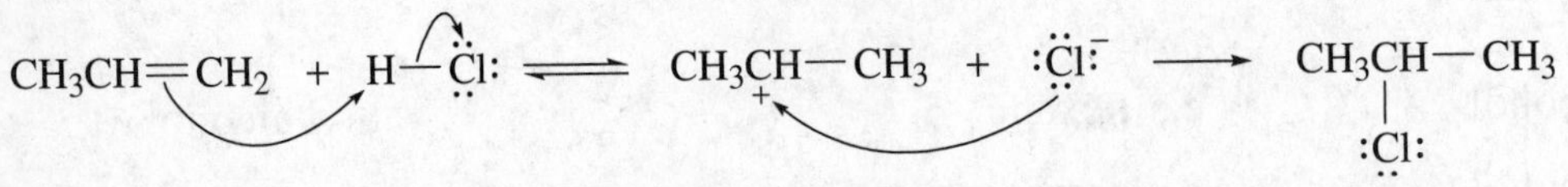

8.
a. Increasing the energy of activation, increases the rate of the reaction. F
b. An exergonic reaction is one with a $-\Delta G°$. T
c. An alkene is an electrophile. F
d. The higher the energy of activation, the more slowly the reaction will take place. T
e. Another name for *trans*-2-butene is *Z*-2-butene. F
f. The more stable the compound, the greater its concentration at equilibrium. T
g. 2,3-Dimethyl-2-pentene is more stable than 3,4-dimethyl-2-pentene. T

9. **a.** *Z* **b.** *Z*

10. **a.** $CH_2{=}CHCH_2OH$

b. 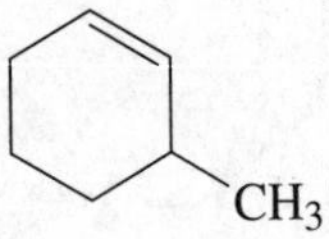

c. 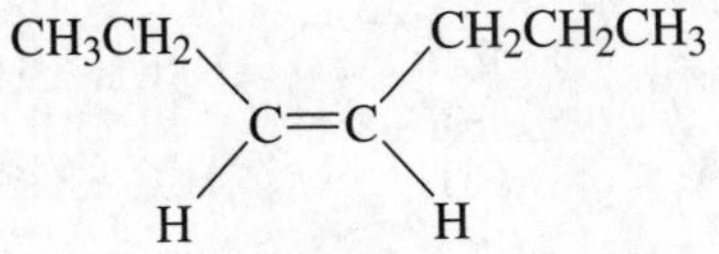

d. $CH_2{=}CHBr$

11. 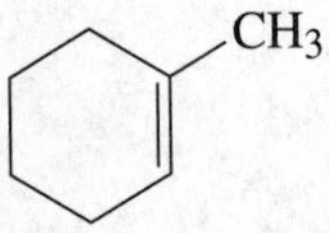

Chapter 5

1. **a.** 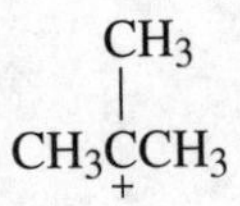

b. $CH_3\overset{+}{C}HCH_3$

2. $CH_3CH_2CH_2CH{=}CH_2$

3. **a.** 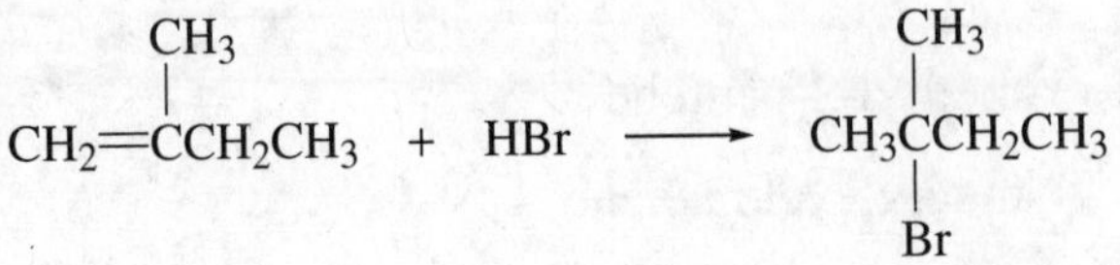

b. 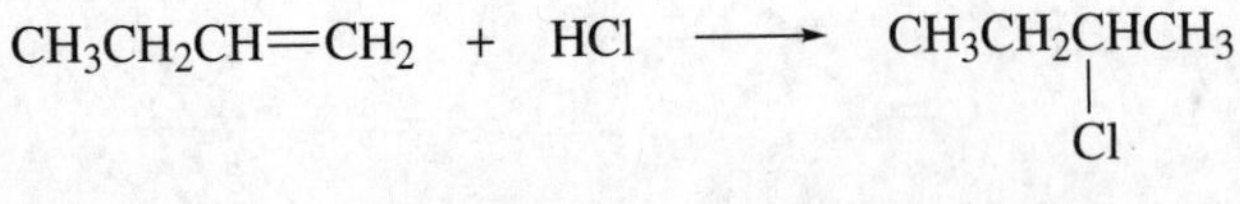

4. 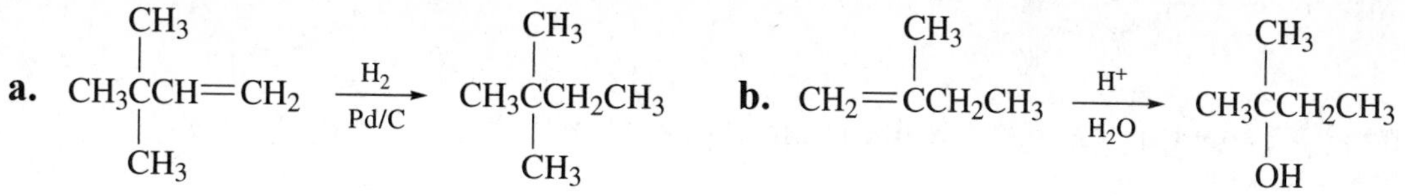

5. 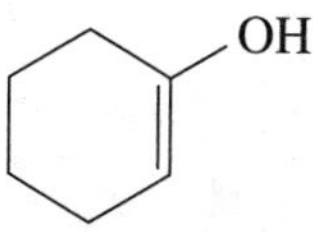

6. **a.** 1-Butyne is more acidic than 1-butene. T

b. An sp^2 carbon is more electronegative than an sp^3 carbon. T

c. Water is a stronger acid than ammonia. T

d. The reaction of 1-butene with HCl will form 1-chlorobutane as the major product. F

7. H_2/Lindlar catalyst

8. 1-bromo-5-methyl-3-hexyne

9. $CH_3CH_2CH_2C{\equiv}CH$

10.

NH_3	$CH_3C{\equiv}CH$	CH_3CH_3	H_2O	$CH_3CH{=}CH_2$
3	**2**	**5**	**1**	**4**

11. 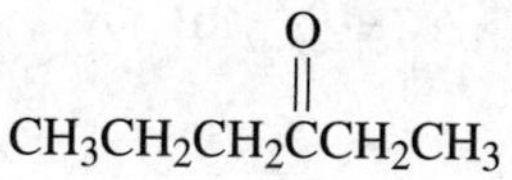 and 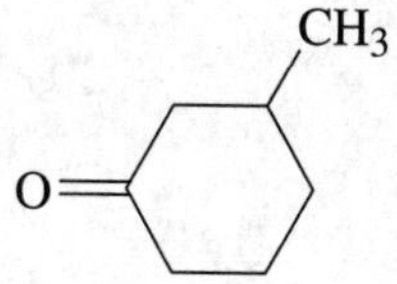

12. **a.** 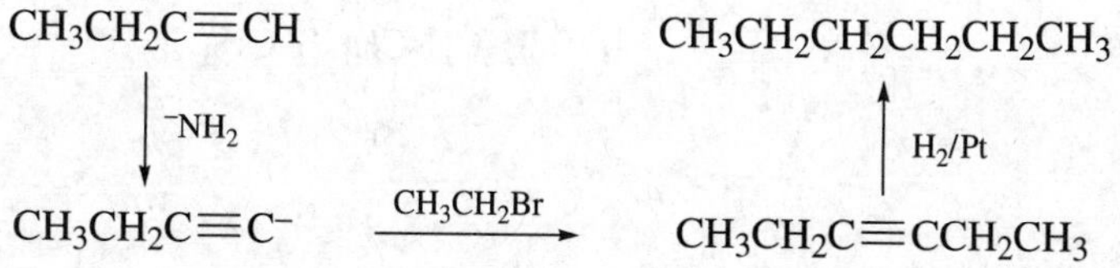

b. 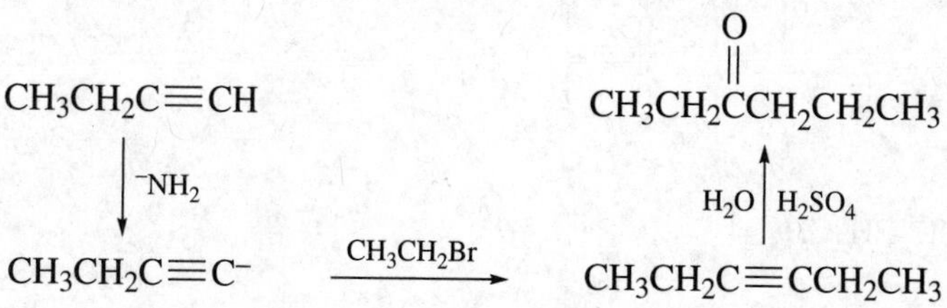

Chapter 6

1. **a.** a pair of enantiomers **b.** a pair of enantiomers

2. –3.0°

3. D

4. 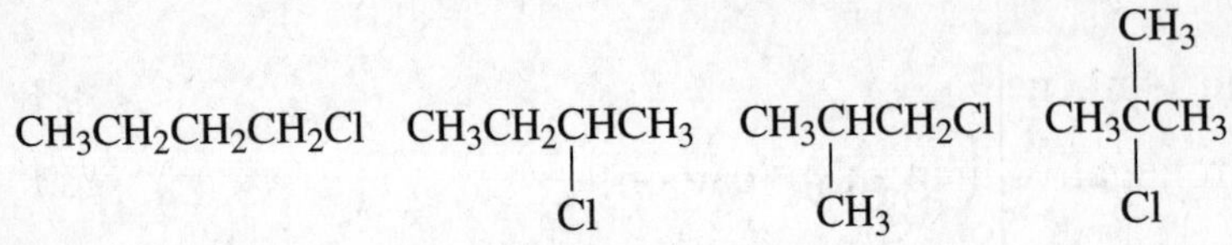

5. **a.** 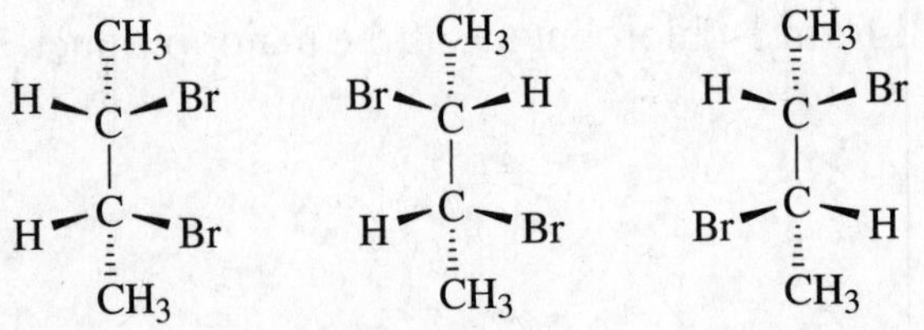

   b. 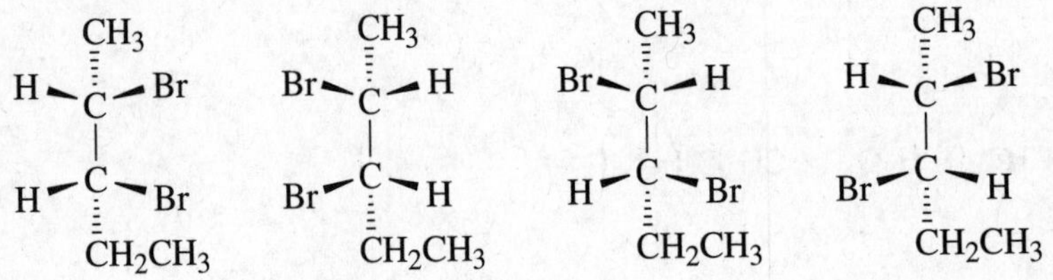

6. **a.** 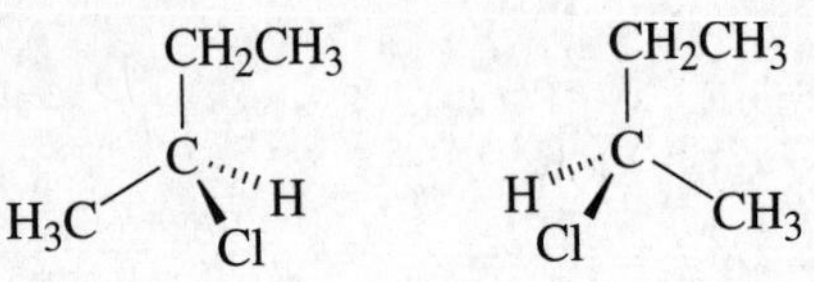

   b. 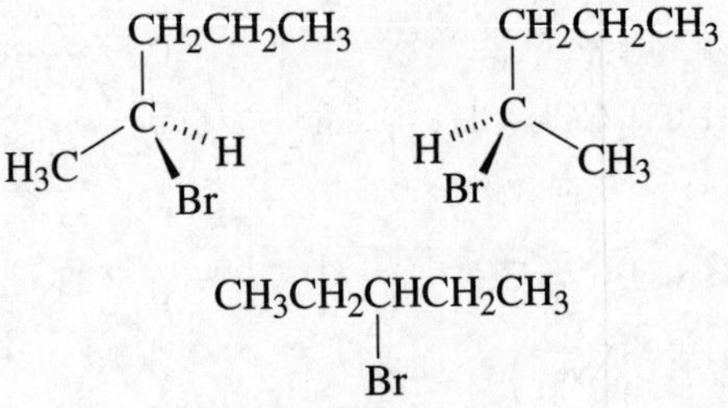

7. **a.** 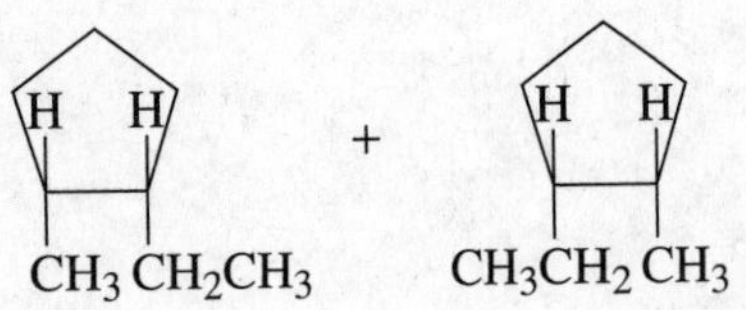

   b. 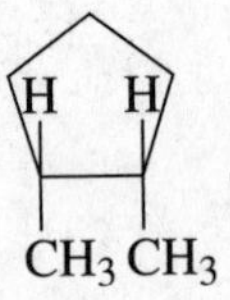

8. 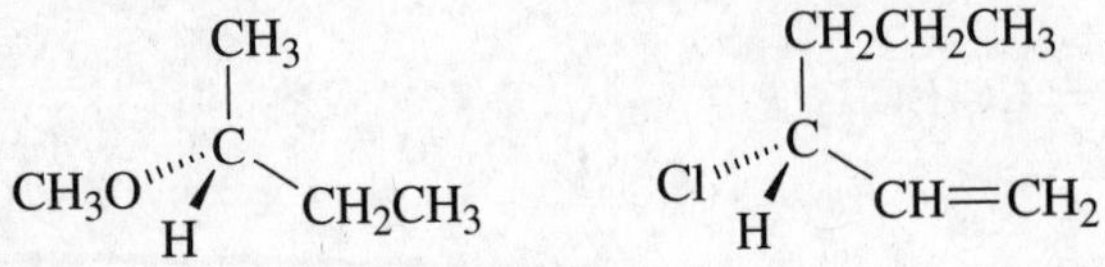

9. 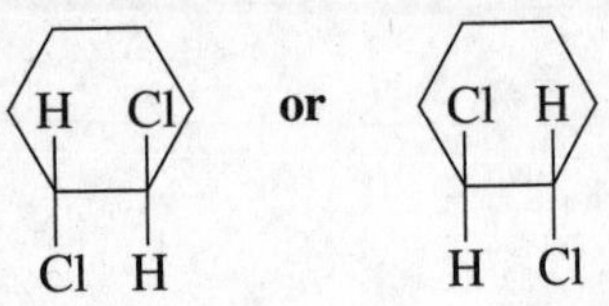

10. a. Diastereomers have the same melting points. F

b. Meso compounds do not rotate polarized light. T

c. 2,3-Dichloropentane has a stereoisomer that is a meso compound. F

d. All compounds with the *R* configuration are dextrorotatory. F

e. A compound with three asymmetric centers can have a maximum of nine stereoisomers. F

Chapter 7

1. a. 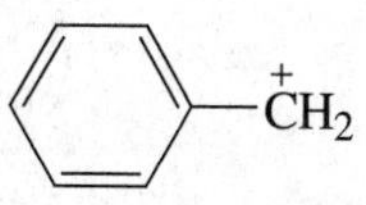

c. $CH_3\bar{C}H\overset{O}{\overset{\|}{C}}CH_3$

e. 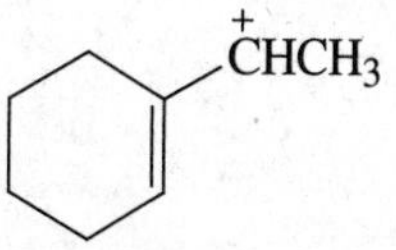

b. $CH_3\bar{C}HC{\equiv}CH$

d. $CH_2{=}CH\overset{+}{C}H_2$

2. a. $CH_3CH{=}CH{-}\ddot{\underset{..}{O}}CH_3 \longleftrightarrow CH_3\bar{C}H{-}CH{=}\overset{+}{O}CH_3$

b. $CH_3CH{=}CH{-}CH{=}CH{-}\overset{+}{C}H_2 \longleftrightarrow CH_3CH{=}CH{-}\overset{+}{C}H{-}CH{=}CH_2$

$\updownarrow$

$CH_3\overset{+}{C}H_2{-}CH{=}CH{-}CH{=}CH_2$

c. ${}^{-}CH_2{-}CH{=}CH{-}\overset{O}{\overset{\|}{C}}H \longleftrightarrow CH_2{=}CH{-}\bar{C}H{-}\overset{O}{\overset{\|}{C}}H \longleftrightarrow CH_2{=}CH{-}CH{=}\overset{O^-}{\overset{|}{C}}H$

3. $CH_3\underset{+}{\overset{CH_3}{\overset{|}{C}}}CH_2CH{=}CH_2$ $CH_2{=}CHCH_2CH{=}CH_2$ $CH_3CH_2NHCH_2CH{=}CHCH_3$

4. c. $CH_3\overset{O}{\overset{\|}{C}}OH$ and $CH_3\overset{O^-}{\overset{|}{C}}{=}\overset{+}{O}H$

5. a. 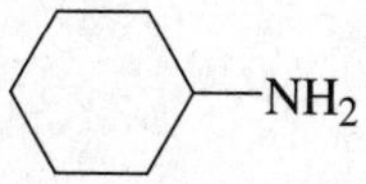

b. 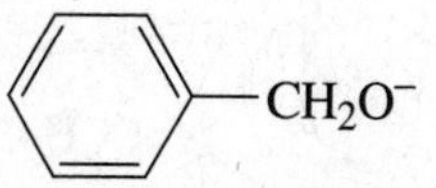

6. a. 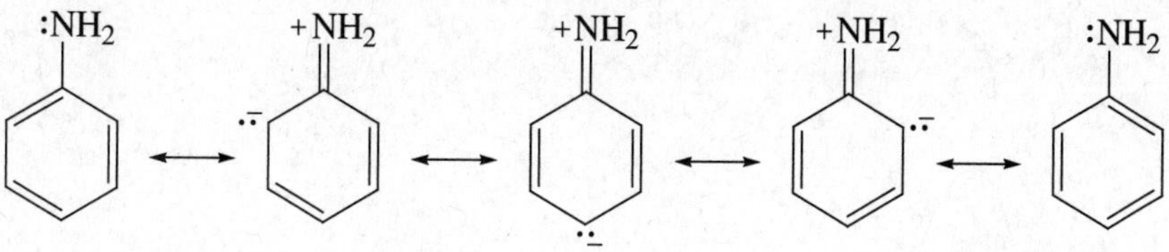

b. 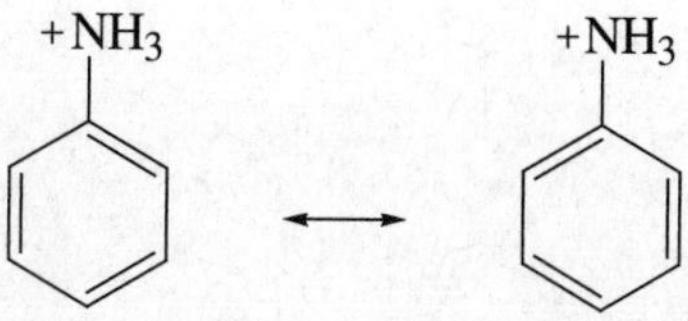

c. 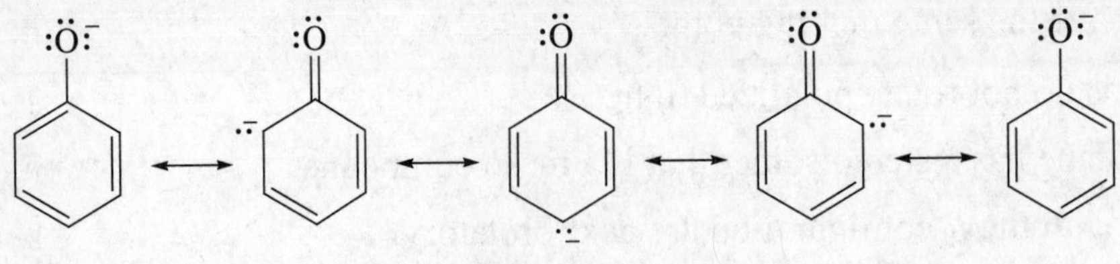

7. **a.**  **b.** $CH_3CH{=}CH_2$ **c.** 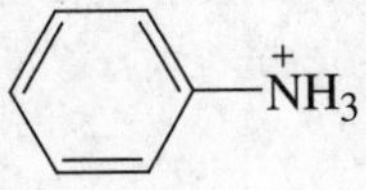

8. **a.** 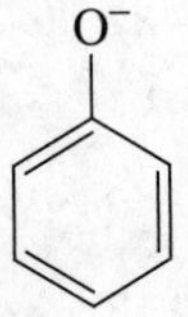 **b.** 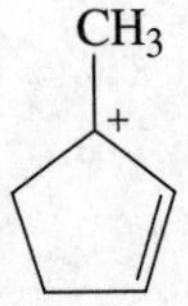

9. 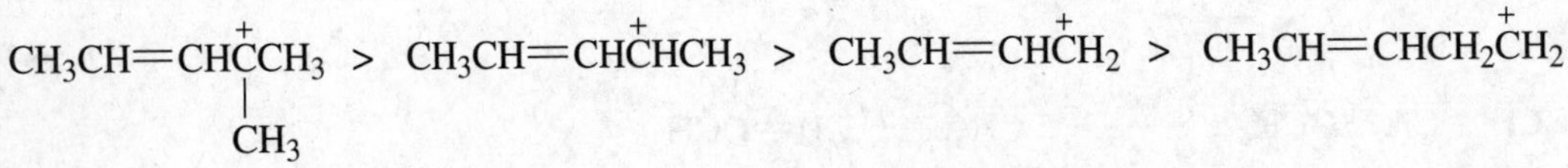

10. **a.** 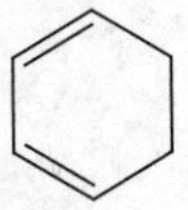 **b.** 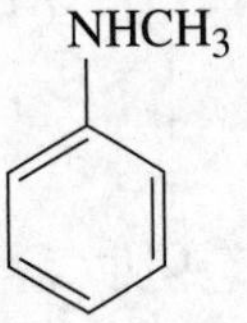 **c.** 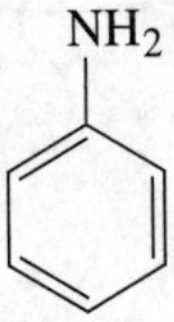 **d.** 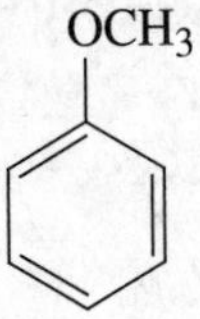

11. 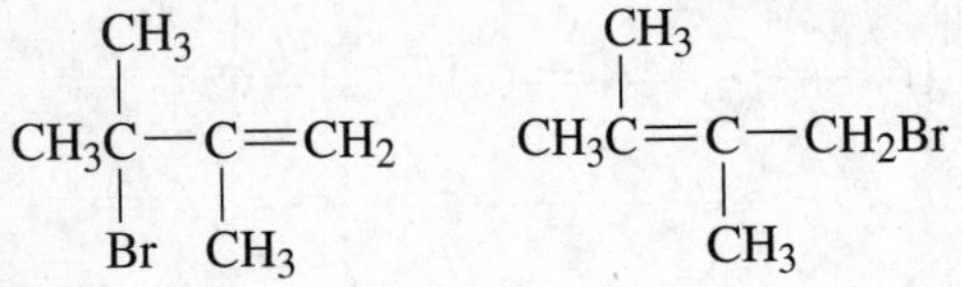

Chapter 8

1. 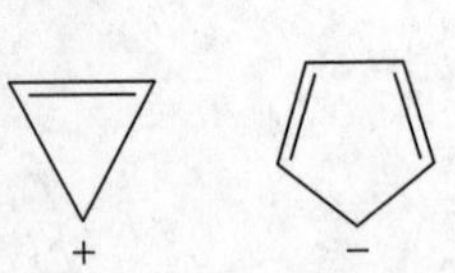

2. 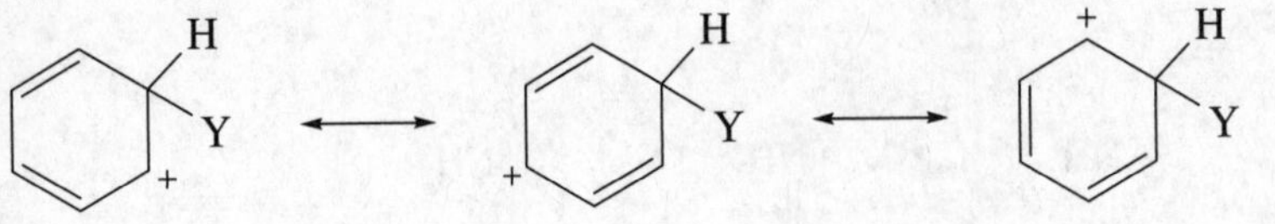

3. **a.** 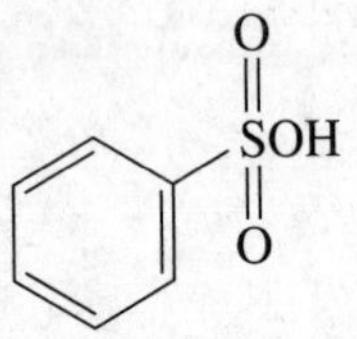 **b.** 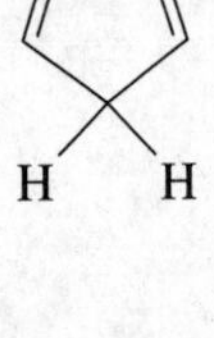

4. **a.** $CH_3\underset{\displaystyle Br}{\underset{|}{C}}HCH_2CH_3$ **b.** $CH_3CH{=}CHCH_3$ and $CH_2{=}CHCH_2CH_3$

5. **a.** 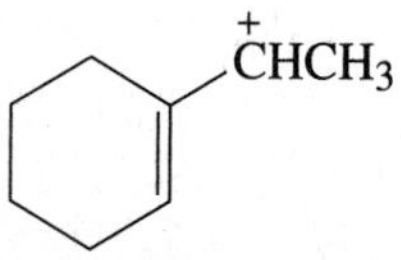 **b.** $CH_3\overset{+}{C}HCH{=}CH_2$ **c.** $CH_3\overset{-}{C}H\overset{\displaystyle O}{\overset{\|}{C}}CH_3$

6. 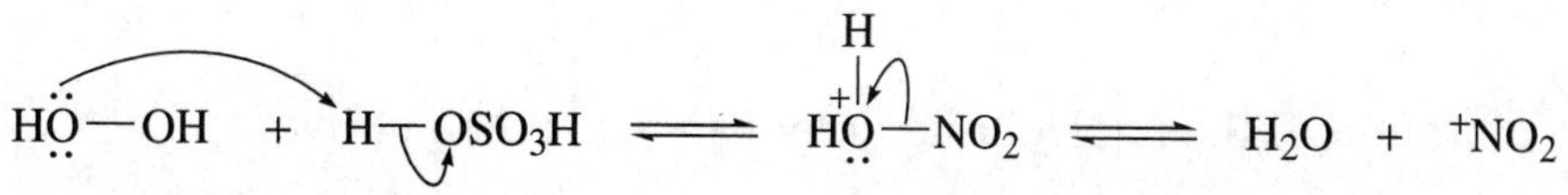

7. **a.** *meta*-nitrotoluene
3-nitrotoluene

b. 1,2,4-tribromobenzene

c. *ortho*-ethylbenzoic acid
2-ethylbenzoic acid

d. *para*-chlorophenol
4-chlorophenol

8. 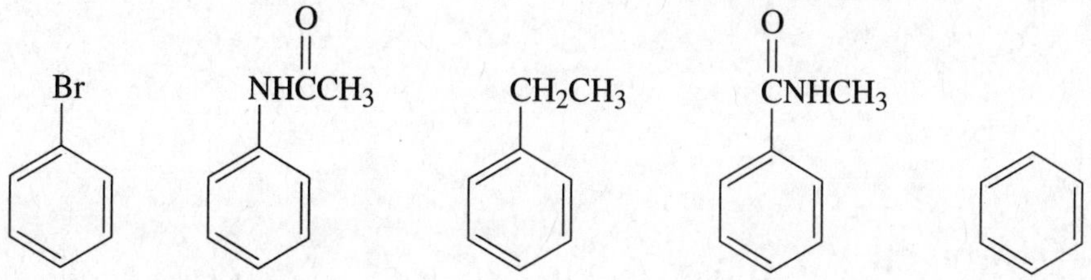

4 1 2 5 3

9. **a.** 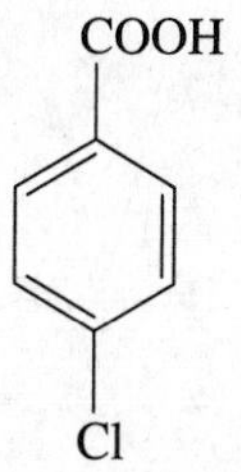 **b.** 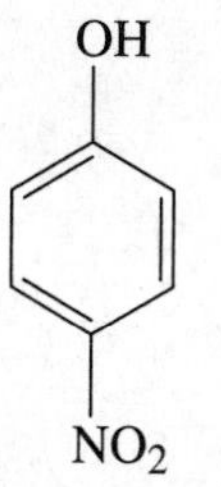 **c.** 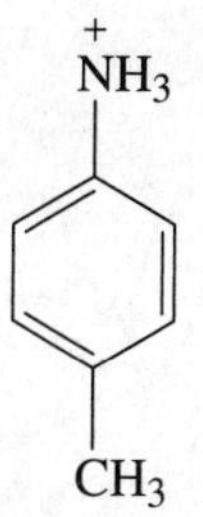 **d.** 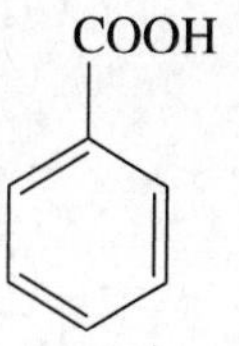

10. *para*-bromoethylbenzene

11. **a.** 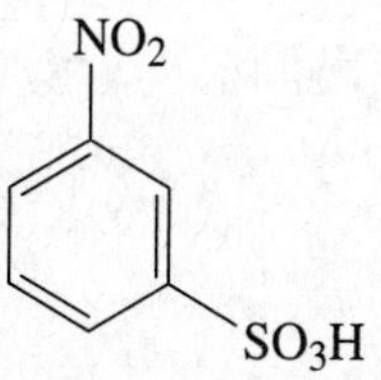 **b.** 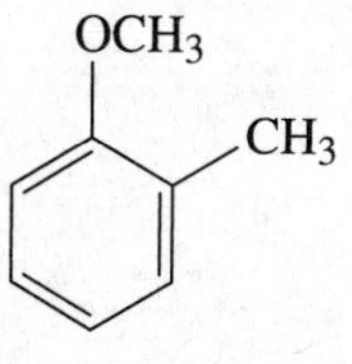 + 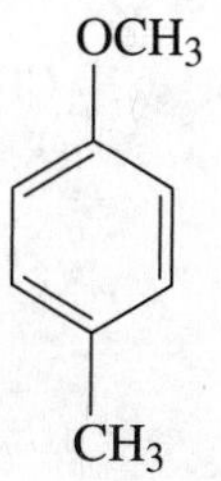 **c.** 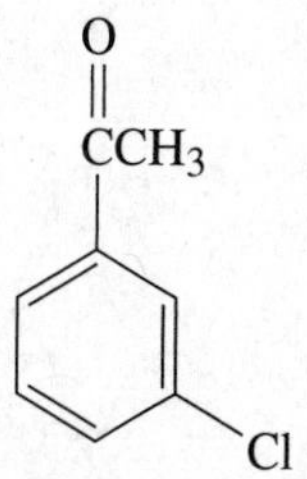

12. **a.** Benzoic acid is more reactive than benzene towards electrophilic substitution. F

b. *para*-Chlorobenzoic acid is more acidic than *para*-methoxybenzoic acid. T

c. A $CH{=}CH_2$ group is a meta director. F

d. *para*-Nitroaniline is more basic than *para*-chloroaniline. F

Chapter 9

1. $\overset{\displaystyle CH_3}{\overset{|}{CH_3CH_2CH_2CH_2CHBr}}$

2. $\overset{\displaystyle CH_3}{\overset{|}{CH_3CH_2CHBr}}$

3. a. Increasing the concentration of the nucleophile favors an S_N1 reaction over an S_N2 reaction. F

   b. Ethyl iodide is more reactive than ethyl chloride in an S_N2 reaction. T

   c. An S_N2 reaction is a two-step reaction. F

4. a. $CH_3CH_2CH_2Cl + HO^-$ b. $CH_3CH_2CH_2I + HO^-$ c. $CH_3CH_2CH_2Br + HO^-$

5. $\overset{\displaystyle CH_3}{\overset{|}{CH_3CHBr}}$

6. a. 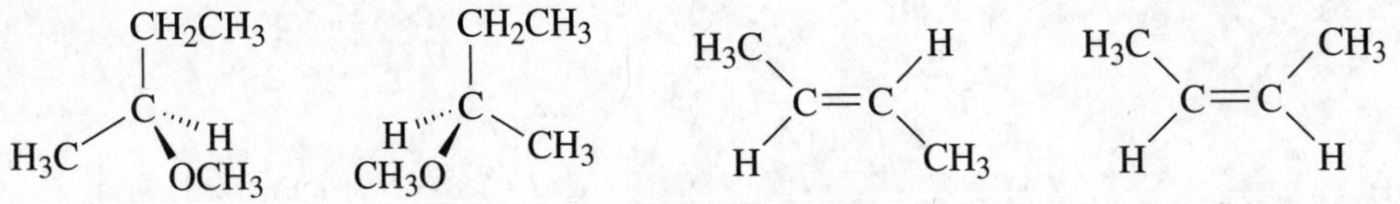

major minor

7. a. $CH_3CH_2\underset{\displaystyle CH_3}{\underset{|}{\overset{\displaystyle CH_3}{\overset{|}{C}}}}O^- + CH_3CH_2CH_2Br$

   b. 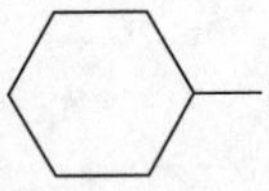 $-O^- + CH_3Br$

8. a. $CH_3\underset{\displaystyle Cl}{\underset{|}{CH}}CH_3 + HO^-$ b. $CH_3CH_2CH_2I + HO^-$ c. $CH_3CH_2CH_2Br + HO^-$

9. a. 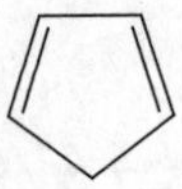 b. $CH_3CH{=}\overset{\displaystyle CH_3}{\overset{|}{C}}CH_3$

Chapter 10

1.   HBr

2.   **a.** 1-ethoxy-2-methylbutane   **c.** 6-methyl-3-heptanol

 b. 1-ethoxy-2-ethylpentane   **d.** 2-cyclohexyl-1-ethanol

3.   2,2-diethyloxirane or 1,2-epoxy-2-ethylbutane

4.   **a.** $\underset{\underset{OCH_3}{|}}{\overset{\overset{CH_2CH_3}{|}}{HOCH_2CCH_2CH_3}}$ **b.** $\underset{\underset{OH}{|}}{\overset{\overset{CH_2CH_3}{|}}{CH_3OCH_2CCH_2CH_3}}$ 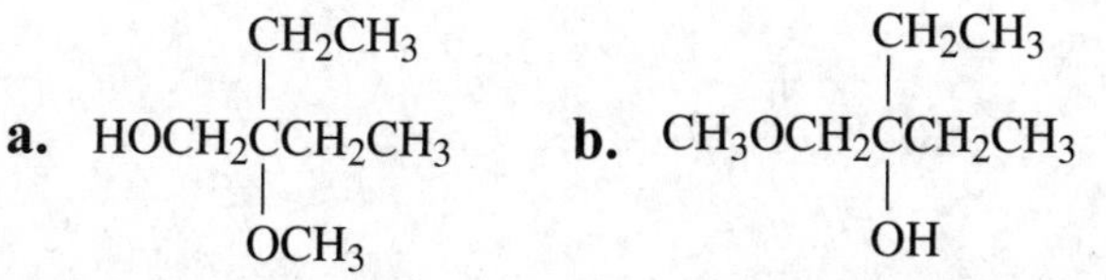

5.   **a.** $\underset{\underset{CH_3}{|}}{\overset{\overset{CH_3}{|}}{CH_3CH_2C=CCH_3}}$ **b.** $\underset{\underset{CH_3}{|}}{\overset{\overset{CH_3}{|}}{CH_3CH_2CH_2C=CCH_3}}$ **c.** $CH_3CH_2CH=CHCH_3$ **d.** 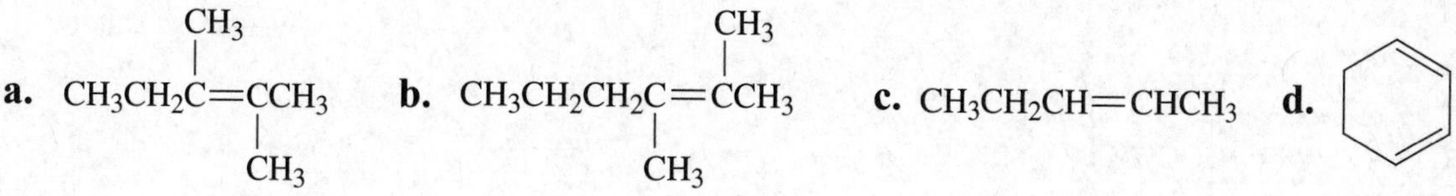

6.   **a.** Tertiary alcohols are easier to dehydrate than secondary alcohols.   T

 b. 1-Methylcyclohexanol reacts more rapidly than 2-methylcyclohexanol with HBr.   T

 c. 1-Butanol forms a ketone when it is oxidized by chromic acid   F

7.   **a.** $\underset{\underset{CH_3}{|}}{\overset{\overset{CH_3}{|}}{CH_3CH_2C-I}}$ + CH_3OH **b.** —OH + ICH_2— 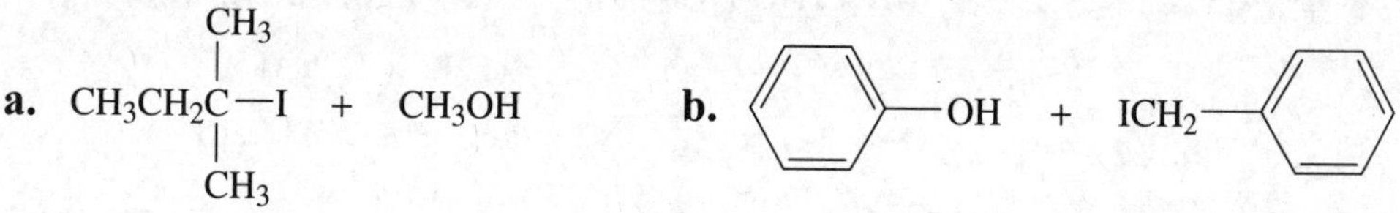

Chapter 11

1.   **a.** $CH_3\overset{\overset{O}{\|}}{C}OCH_3$ **b.** $CH_3\overset{\overset{O}{\|}}{C}O$—   **c.** $CH_3\overset{\overset{O}{\|}}{C}O$——$NO_2$ **d.** $CH_3\overset{\overset{O}{\|}}{C}Cl$ 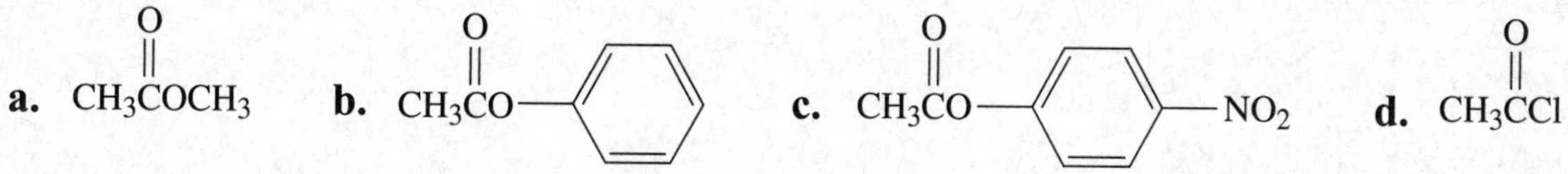

2.   **a.** *N*-ethylpentanamide   **b.** 3-methylpentanoic acid   **c.** methyl 4-phenylbutanoate

3.   **a.** $CH_3\overset{\overset{O}{\|}}{C}Cl$ + $2\ CH_3NH_2$ $\longrightarrow$ $CH_3\overset{\overset{O}{\|}}{C}NHCH_3$ + $CH_3\overset{+}{N}H_3\ Cl^-$

 Any reaction in which a reactant is cleaved as a result of reaction with an amine.

 b. $CH_3\overset{\overset{O}{\|}}{C}Cl$ + H_2O $\longrightarrow$ $CH_3\overset{\overset{O}{\|}}{C}OH$ + · HCl

 Any reaction in which a reactant is cleaved as a result of reaction with water.

c. $CH_3\overset{O}{\overset{\|}{C}}OCH_3 + CH_3CH_2OH \underset{}{\overset{H^+}{\rightleftharpoons}} CH_3\overset{O}{\overset{\|}{C}}OCH_2CH_3 + CH_3OH$

4. a. $CH_3-\overset{O}{\overset{\|}{C}}-OH$ b. $CH_3-\overset{O}{\overset{\|}{C}}-OCH_3$ c. $CH_3-\overset{O}{\overset{\|}{C}}-NH_2$ d. $CH_3-\overset{O}{\overset{\|}{C}}-OH$

5. a. $CH_3CH_2\overset{O}{\overset{\|}{C}}OH + \overset{+}{N}H_4$

b. $CH_3CH_2\overset{O}{\overset{\|}{C}}OH + CH_3CH_2CH_2OH$

c. $CH_3CH_2\overset{O}{\overset{\|}{C}}NHCH_2CH_3 + CH_3CH_2\overset{+}{N}H_3\ Cl^-$

d. 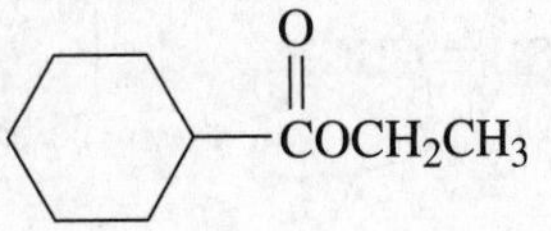 + CH_3OH

e. $CH_3CH_2CH_2\overset{O}{\overset{\|}{C}}OCH_2CH_2CH_3$

f. $CH_3CH_2\overset{O}{\overset{\|}{C}}OH + CH_3\overset{+}{N}H_3$

g. $CH_3CH_2\overset{O}{\overset{\|}{C}}OH$ + 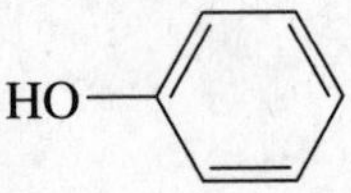

Chapter 12

1. a. 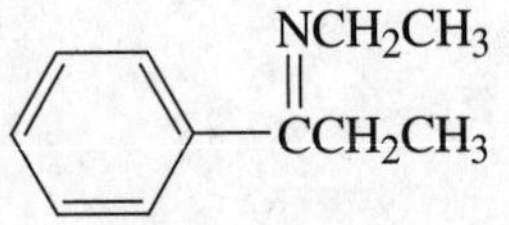 + H_2O

b. 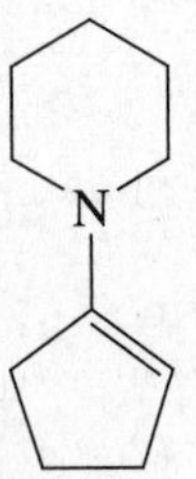 + H_2O

c. $CH_3CH_2CH_2CH_2OH$

d. 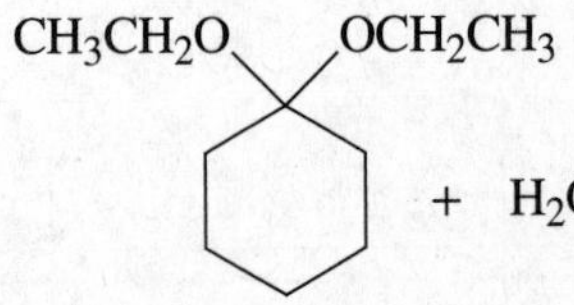 + H_2O

e. $CH_3CH_2\overset{OH}{\overset{|}{\underset{CH_3}{\underset{|}{C}}}}CH_2CH_3$

f. 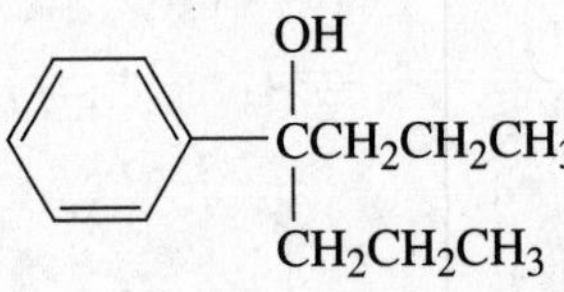

2. $CH_3CH_2\overset{OH}{\overset{|}{\underset{CH_3}{\underset{|}{C}}}}CH_2CH_2CH_3$

3. 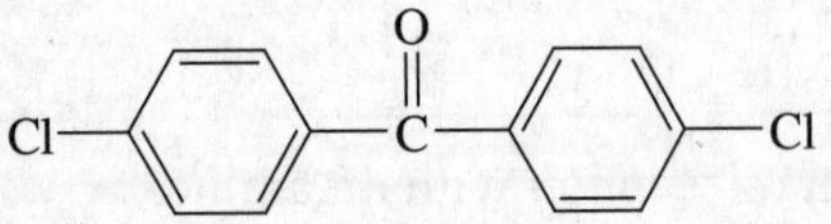

4. a. 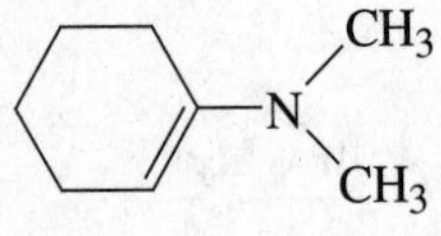 b. $CH_3CH_2\overset{OCH_3}{\overset{|}{\underset{OCH_3}{\underset{|}{CH}}}}$ c. 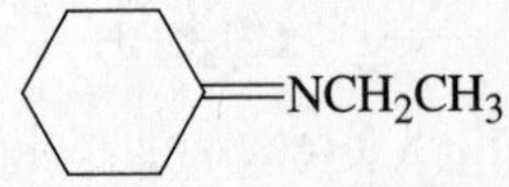 d. $CH_3CH_2\overset{OH}{\overset{|}{\underset{OCH_3}{\underset{|}{CH}}}}$

5. **a.** butanal **b.** 3-pentanone

6.

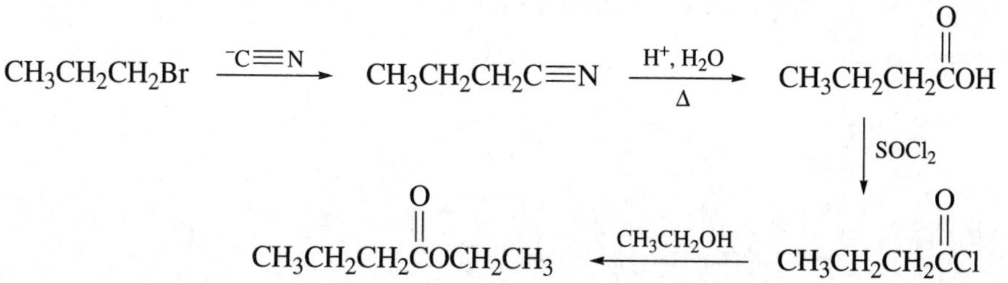
CH3CH2CH2Br
−C≡N
CH3CH2CH2C≡N
H+, H2O
Δ
CH3CH2CH2COH
SOCl2
CH3CH2CH2CCl
CH3CH2OH
CH3CH2CH2COCH2CH3

Chapter 13

1.

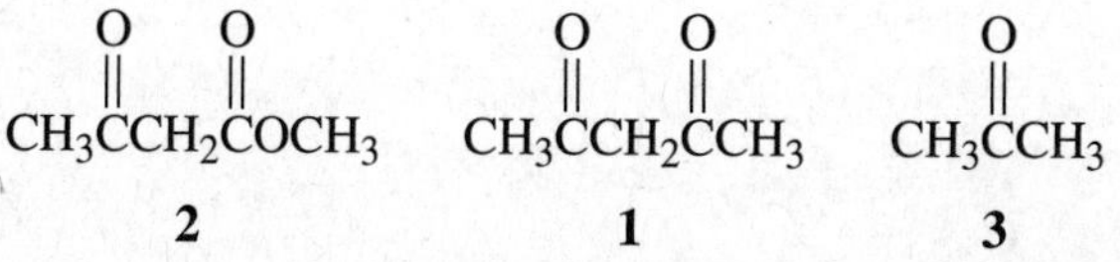
CH3CCH2COCH3
2
CH3CCH2CCH3
1
CH3CCH3
3

2. **a.**

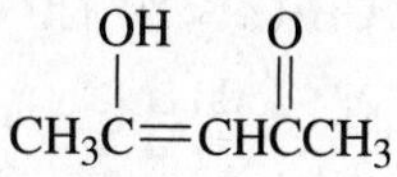
CH3C=CHCCH3

b.

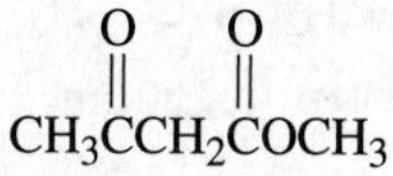
CH3CCH2COCH3

3. **a.**

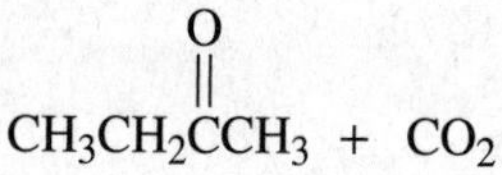
CH3CH2CCH3 + CO2

b.

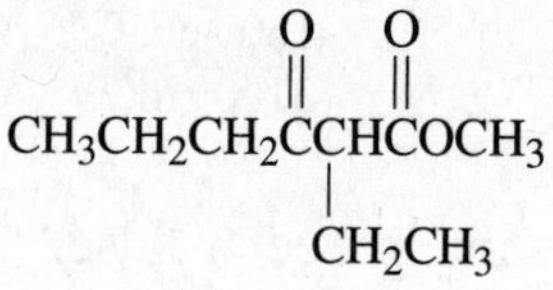
CH3CH2CH2CCHCOCH3
CH2CH3

c.

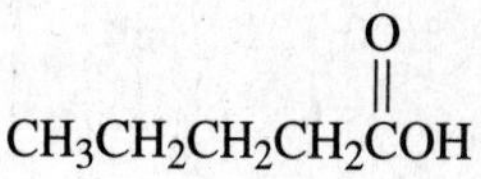
CH3CH2CH2CH2COH

4. **a.**

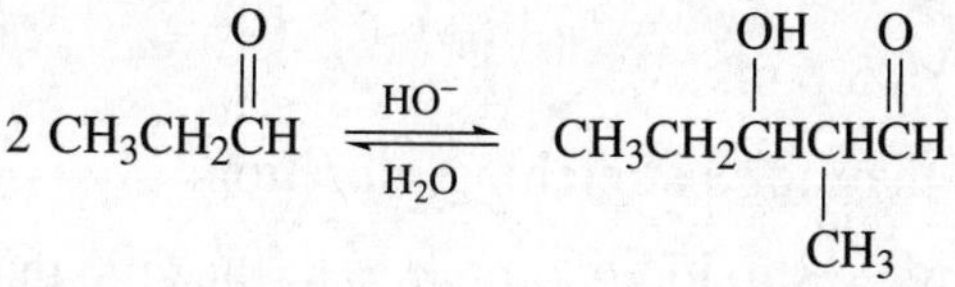
2 CH3CH2CH
HO−
H2O
CH3CH2CHCHCH
CH3

b.

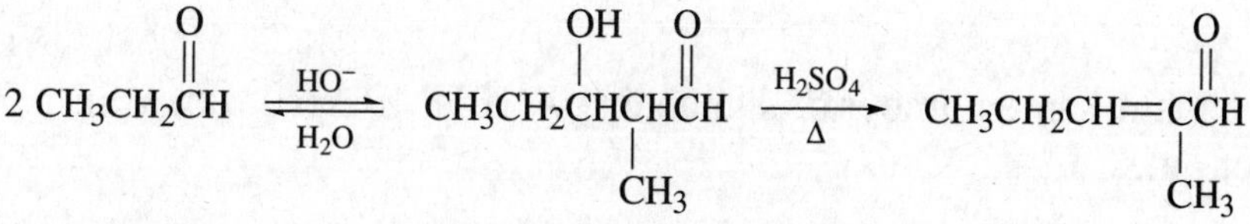
2 CH3CH2CH
HO−
H2O
CH3CH2CHCHCH
CH3
H2SO4
Δ
CH3CH2CH=CCH
CH3

c.

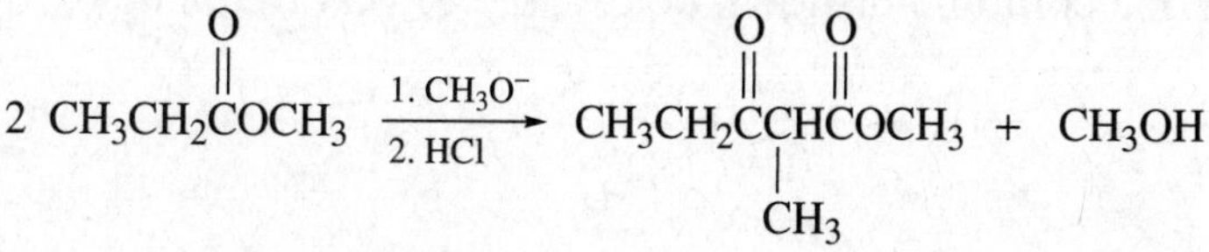
2 CH3CH2COCH3
1. CH3O−
2. HCl
CH3CH2CCHCOCH3 + CH3OH
CH3

d.

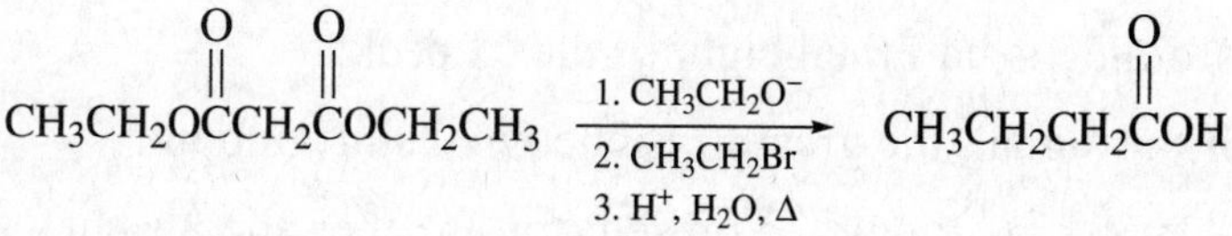
CH3CH2OCCH2COCH2CH3
1. CH3CH2O−
2. CH3CH2Br
3. H+, H2O, Δ
CH3CH2CH2COH

e.

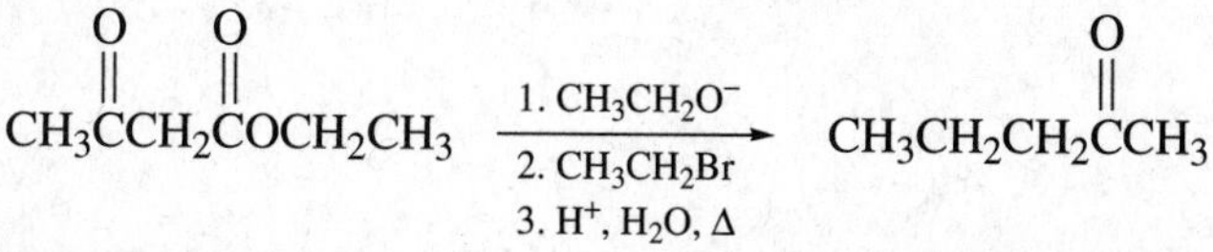
CH3CCH2COCH2CH3
1. CH3CH2O−
2. CH3CH2Br
3. H+, H2O, Δ
CH3CH2CH2CCH3

5.

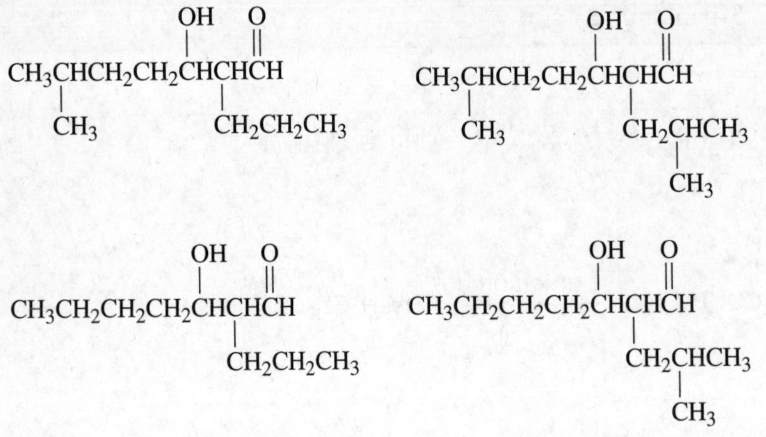

Chapter 14

1. a. $CH_3CH_2CH_2\overset{O}{\overset{\|}{C}}H$ ~2700 cm^{-1}

b. $CH_3CH_2CH_2CH_2OH$ ~3600 – 3200 cm^{-1}

c. $CH_3CH_2CH_2\overset{O}{\overset{\|}{C}}OCH_3$ ~1050 or ~1250 cm^{-1}

d. $CH_3CH_2CH{=}CHCH_3$ ~1600 cm^{-1} ~3100 cm^{-1}

e. $CH_3CH_2C{\equiv}CH$ ~3300 cm^{-1} $CH_3CH_2C{\equiv}CCH_3$ ~2100 cm^{-1}

2. Indicate whether each of the following is true or false.

a. An O—H bond will show a more intense absorption band than an N—H bond. T

b. Light of 2 μm is of higher energy than light of 3 μm. T

c. It takes more energy for a bending vibration than for a stretching vibration. F

d. The signals on the right of an NMR spectrum are shielded compared to the signals on the left. T

e. Dimethyl ketone has the same number of signals in its ^{1}H NMR spectrum as in its ^{13}C NMR spectrum. F

f. In the ^{1}H NMR spectrum of the compound shown below, the lowest frequency signal is a singlet. T

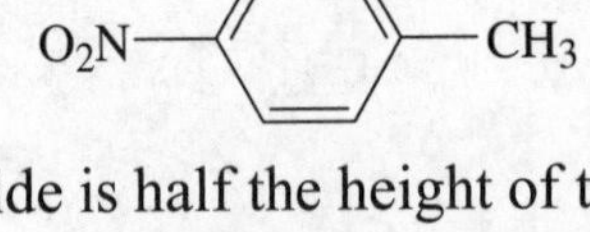
O2N
CH3

g. The M+2 peak of an alkyl bromide is half the height of the M peak. F

h. The greater the frequency of the signal, the greater its chemical shift in ppm. T

i. An absorption band at 1150-1050 cm^{-1} would be present for an ether and absent for an alkane. T

3. **a.**

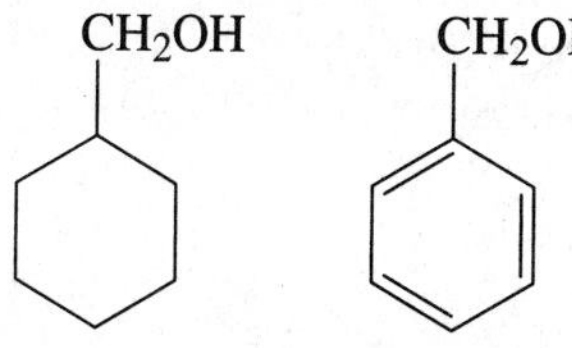

~3100 cm⁻¹ ~1600 and 1500 cm⁻¹

d.

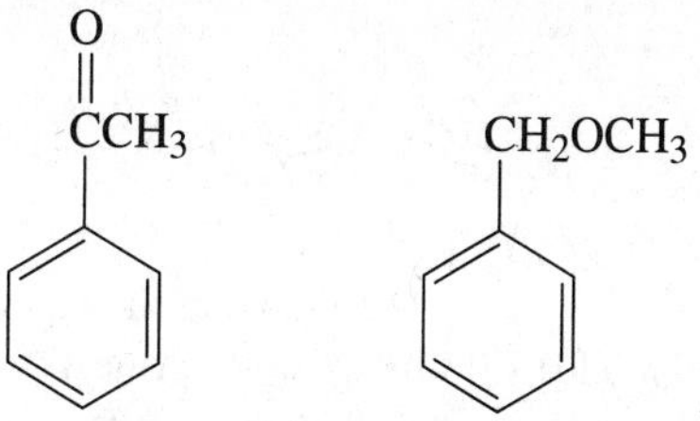

~1700 cm⁻¹ ~1050 cm⁻¹

b.

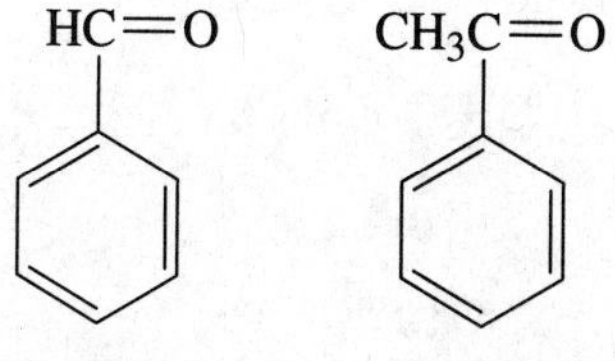

~2700 cm⁻¹ ~2900 cm⁻¹

e.

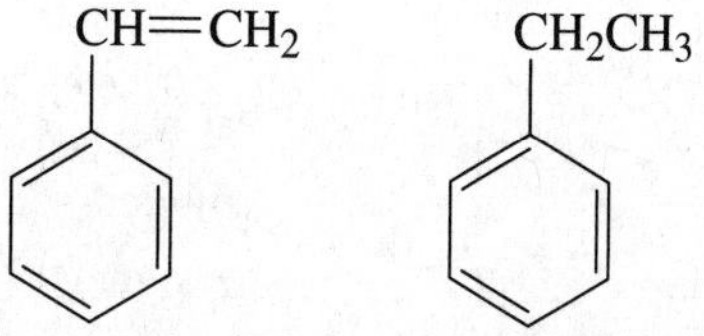

~3100 cm⁻¹ ~2900 cm⁻¹

c.

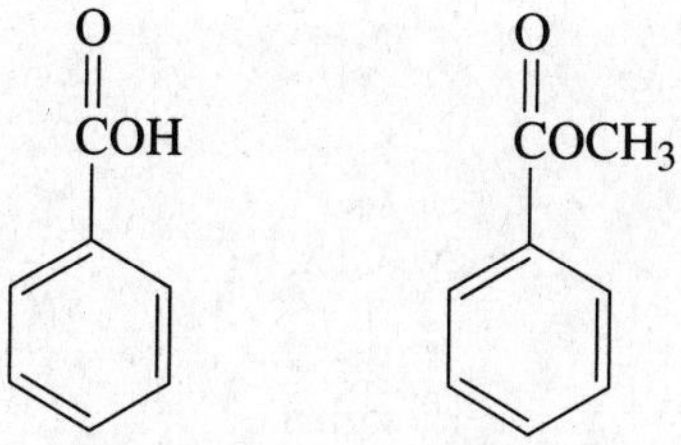

~3300-2500 cm⁻¹ ~2900 cm⁻¹ ~1050 cm⁻¹

4.

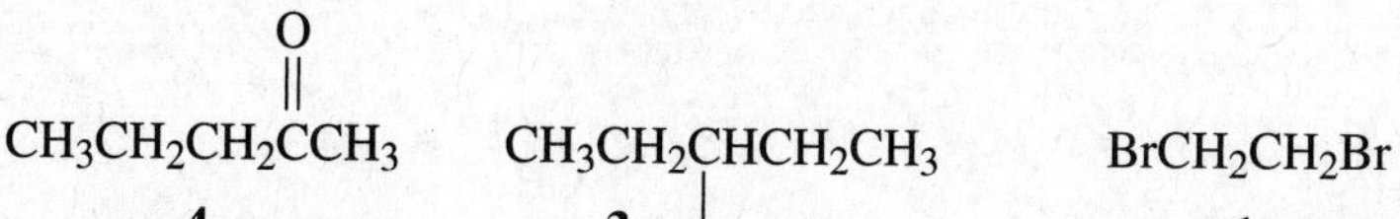

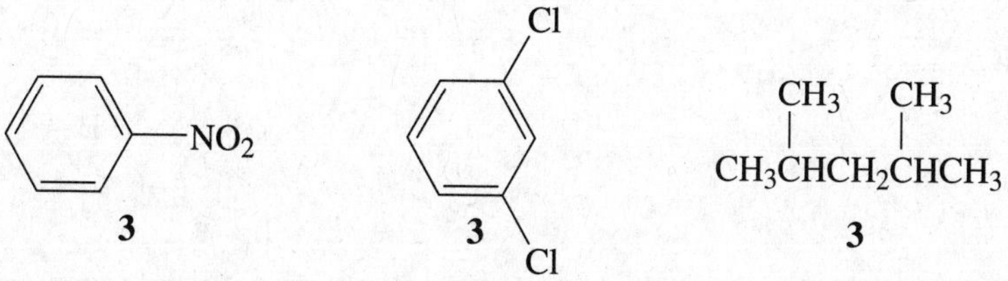
4
3
1
3
3
3

5.

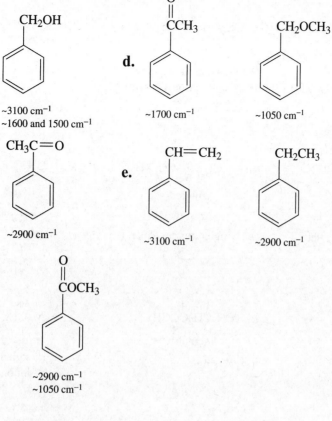

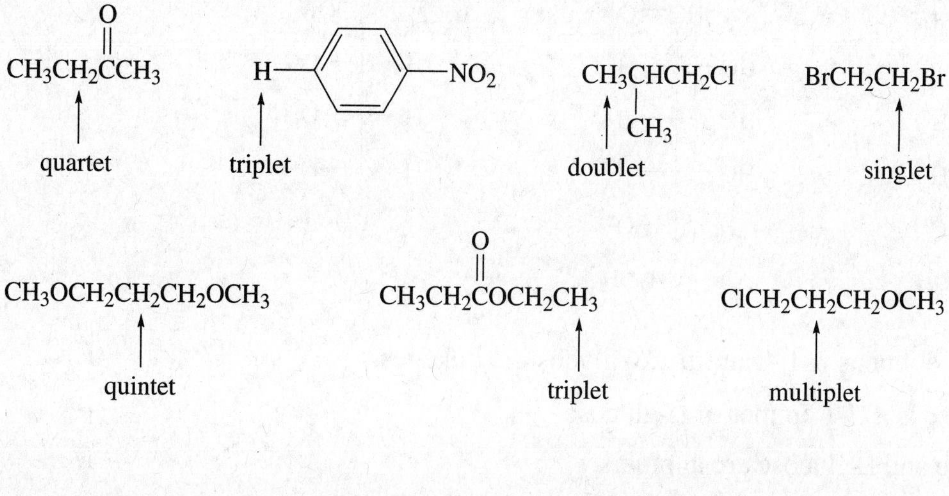
quartet
triplet
doublet
singlet
quintet
triplet
multiplet

6. $CH_3COCH_2CH_3$ (C=O on C-2)
3 signals
The signal at the highest frequency (farthest downfield) is a quartet.

$CH_3CH_2COCH_3$ (C=O on C-3)
3 signals
The signal at the highest frequency (farthest downfield) is a singlet.

$HCOCH_2CH_2CH_3$ (C=O on C-1)
4 signals

7. a. $CH_3CH_2CH_2Cl$
triplet
3 signals
triplet multiplet triplet
2 : 2 : 3

b. $CH_3CH_2COCH_3$
singlet
3 signals
singlet quartet triplet
3 : 2 : 3

c. CH_3CHCH_3 (Br on C-2)
septet
2 signals
septet doublet
1 : 6

8. a. $CH_3CH_2CH_2Cl$
3 signals
triplet

b. $CH_3CH_2COCH_3$
4 signals
singlet

c. CH_3CHCH_3 (Br on C-2)
2 signals
doublet

Chapter 15

1. a.
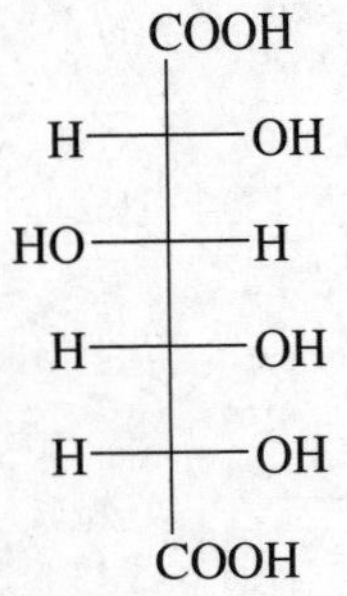

COOH
H
OH
HO
H
H
OH
H
OH
COOH

b.
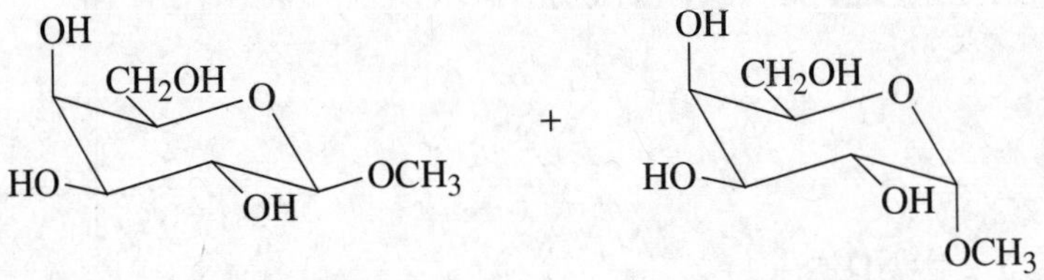

OH
CH2OH
O
HO
OH
OCH3
+
OH
CH2OH
O
HO
OH
OCH3

c.
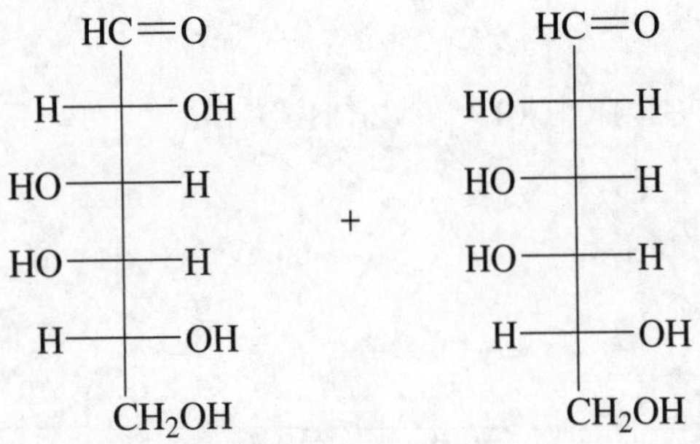

HC=O
H
OH
HO
H
HO
H
H
OH
CH2OH
+
HC=O
HO
H
HO
H
HO
H
H
OH
CH2OH

d.
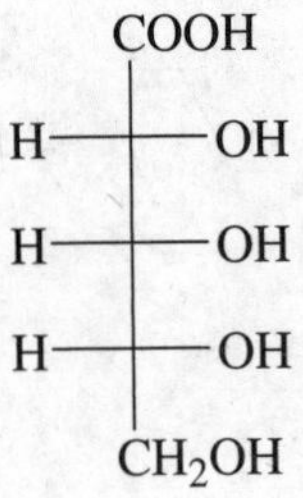

COOH
H
OH
H
OH
H
OH
CH2OH

2. a. Glycogen contains α-1,4' and β-1,6'-glycosidic linkages. F

b. D-Mannose is a C-1 epimer of D-glucose. F

c. D-Glucose and L-glucose are anomers. F

d. D-Erythrose and D-threose are diastereomers. T

3.

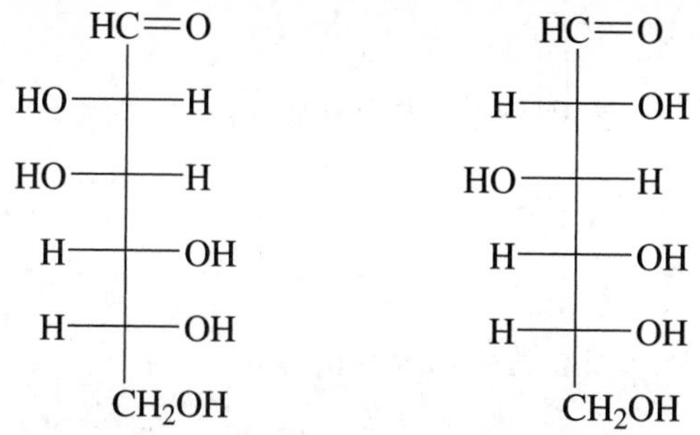

4. D-mannose and D-glucose

5. D-altrose

6. Amylose has α-1,4'-glycosidic linkages, while cellulose has β-1,4'-glycosidic linkages.

7. D-gulose and D-idose

8. D-allose

9.

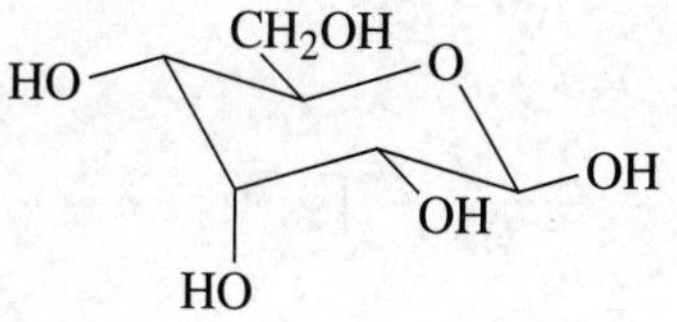

Chapter 16

1.

a.

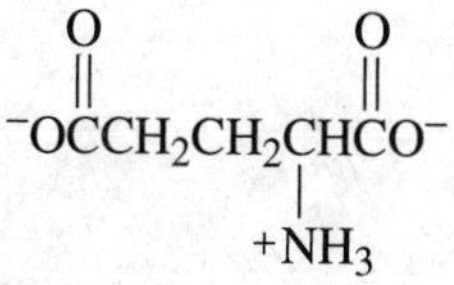

c.

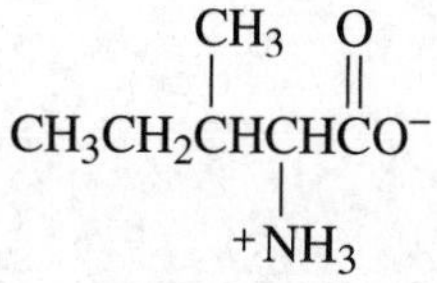

e.

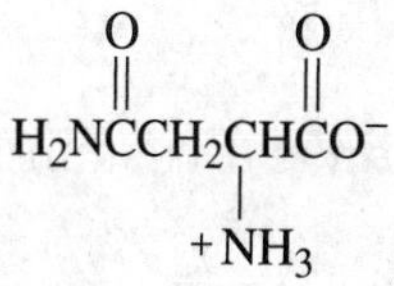

b.

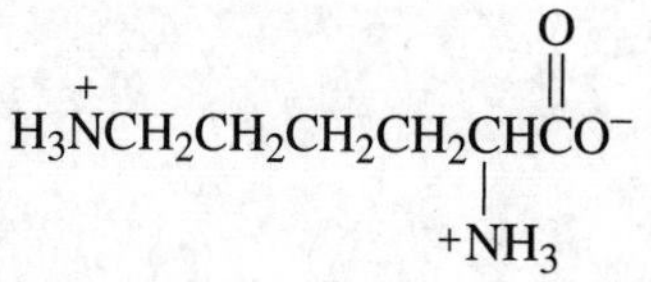

d.

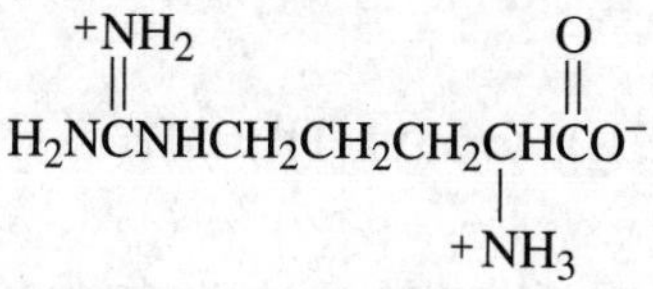

2.

a.

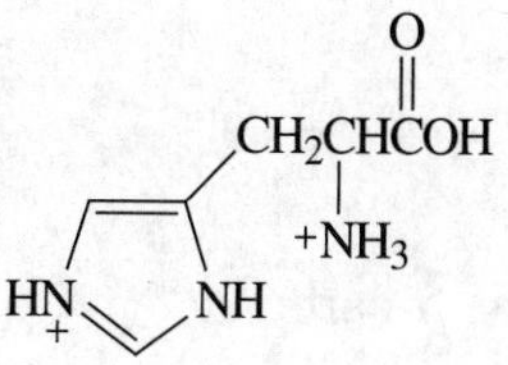

c.

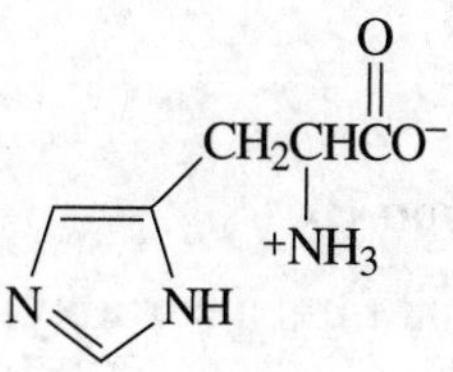

b.

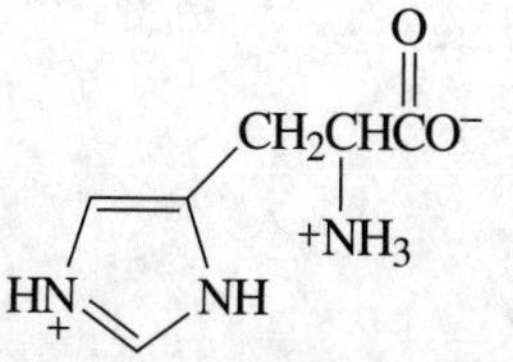

d.

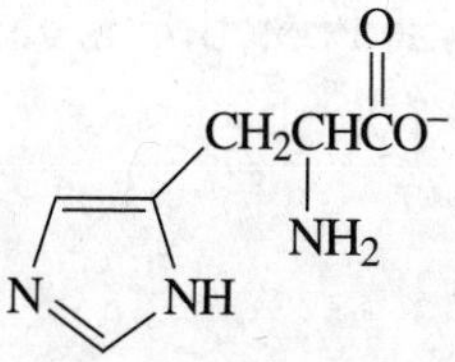

3. **a.** Alanine, because it is farther away from its pI.
   b. glycine
   c. leucine and isoleucine
   d. aspartate (aspartic acid)

4. The electron-withdrawing protonated amino group causes the carboxyl group of alanine to have a lower pK_a.

5. **a.** A cigar shaped protein has a greater percentage of polar residues than a spherical protein. T
   b. Naturally occurring amino acids have the L-configuration. T
   c. There is free rotation about a peptide bond. F

6. 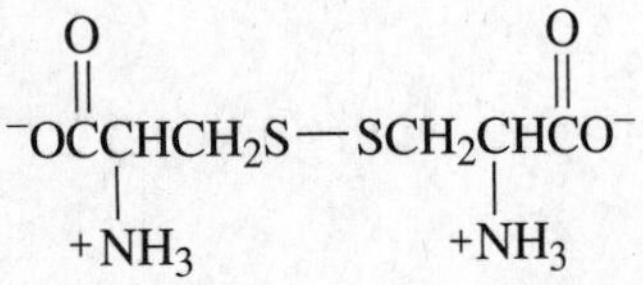

7. **a.** The sequence of the amino acids in the protein chain.
   b. The three-dimensional arrangement of all the atoms in the protein.
   c. A description of the way the subunits of a protein are arranged in space.

8. **a.** $\frac{2.16 + 9.18}{2} = \frac{11.34}{2} = 5.67$
   b. $\frac{9.04 + 12.48}{2} = \frac{21.52}{2} = 10.76$

9. Ala Ser Arg Gly Arg Met His Phe Lys Ile

Chapter 17

1. biotin and vitamin KH_2

2. thiamine pyrophosphate and pyridoxal phosphate

3. methyl (CH_3), methylene (CH_2), formyl (HC=O)

4. **a.** An acid catalyst donates a proton to the substrate and a base catalyst removes a proton from the substrate. T
   b. $FADH_2$ is an oxidizing agent. F
   c. Thiamine pyrophosphate is vitamin B_6. F
   d. Vitamin KH2 is the coenzyme that puts a carboxyl group on the β-carbon of glutamate residues. F
   e. The reactant of an enzyme-catalyzed reaction is called a substrate. T
   f. NADH is a reducing agent. T

Chapter 18

1. acetate

2. catabolic reactions

3.   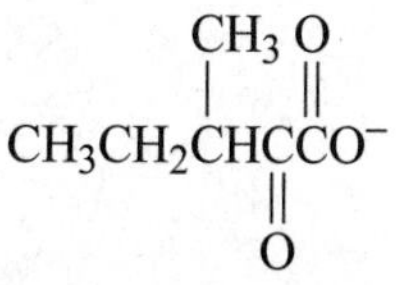
$$\underset{\displaystyle \overset{\displaystyle \|}{O}}{CH_3CH_2\overset{\displaystyle \overset{CH_3}{|}}{C}H\overset{\displaystyle \overset{O}{\|}}{C}CO^-}$$

4. amino acids

5. acetyl-CoA + CO_2

6. a. Each molecule of $FADH_2$ forms 3 molecules of ATP in the
       fourth stage of catabolism.   F
    b. $FADH_2$ is oxidized to FAD.   T
    c. NAD^+ is oxidized to NADH.   F
    d. Acetyl-CoA is a citric acid cycle intermediate.   F

Chapter 19

1. Because fats are composed primarily of saturated fatty acids, the fat molecules can pack closely
    together, which gives them higher melting points.

2.   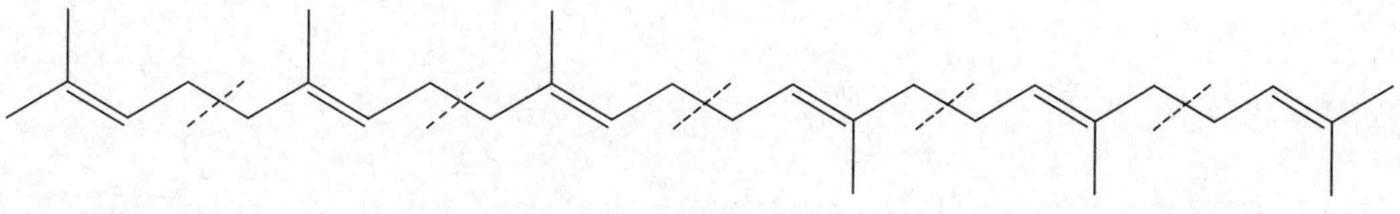

squalene

3. 6

4.   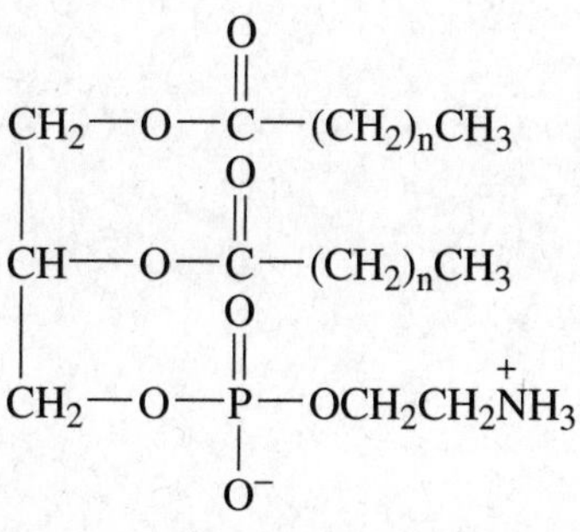
$$
\begin{array}{l}
CH_2-O-\overset{\displaystyle \overset{O}{\|}}{C}-(CH_2)_nCH_3 \\
CH-O-\overset{\displaystyle \overset{O}{\|}}{C}-(CH_2)_nCH_3 \\
CH_2-O-\overset{\displaystyle \underset{\displaystyle \underset{O^-}{|}}{\overset{O}{\|}}}{P}-OCH_2CH_2\overset{+}{N}H_3
\end{array}
$$

5. a. Lipids are insoluble in water.   T
    b. Cholesterol is the precursor of all other steroids.   T

c.	Saturated fatty acids have higher melting points than unsaturated fatty acids.	T
d.	Vitamin K is a water-soluble vitamin.	F
e.	Fats have a higher percentage of saturated fatty acids than do oils.	T

Chapter 20

1. dUMP

2. 5'—A—T—G—C—A—A—G—T—3'

3. A

4.

a.	Guanine and cytosine are purines.	F
b.	The 3'-OH group allows RNA to be easily cleaved.	F
c.	The number of A's in DNA is equal to the number of T's.	T
d.	rRNA carries the amino acid that will be incorporated into a protein.	F
e.	The template strand of DNA is the one transcribed to form mRNA.	T
f.	The 5'-end of DNA has a free OH group.	F
g.	The synthesis of proteins from an RNA blueprint is called transcription.	F
h.	A nucleotide consists of a base and a sugar.	F